Low Intensity Breeding of Native Forest Trees in Argentina

Mario J. Pastorino • Paula Marchelli
Editors

Low Intensity Breeding of Native Forest Trees in Argentina

Genetic Basis for their Domestication and Conservation

 Springer

Editors
Mario J. Pastorino (iD)
Instituto de Investigaciones Forestales y
Agropecuarias Bariloche (IFAB)
INTA-CONICET
National Agricultural Technology Institute -
National Scientific and Technical Research
Council
Bariloche, Rio Negro, Argentina

Paula Marchelli (iD)
Instituto de Investigaciones Forestales y
Agropecuarias Bariloche (IFAB)
INTA-CONICET
National Agricultural Technology Institute -
National Scientific and Technical Research
Council
Bariloche, Rio Negro, Argentina

ISBN 978-3-030-56464-3   ISBN 978-3-030-56462-9   (eBook)
https://doi.org/10.1007/978-3-030-56462-9

This Springer imprint is published by the registered company Springer Nature Switzerland AG
The registered company address is: Gewerbestrasse 11, 6330 Cham, Switzerland

Ámame peteribí
Ama siempre
De esta manera ninguna chica te podría amar.
Piénsalo bien, amigo

Los cuerpos son árboles.
Las mentes son árboles insertos en la tierra
bebiendo el mar, respirando el cielo
y heredando las flores

Love me peteribí
Love always
No girl could love you this way.
Think it well, my friend

The bodies are trees.
Minds are trees inserted in the earth
drinking the sea, breathing the sky
and inheriting the flowers

Carlos Cutaia, Black Amaya, Luis Alberto Spinetta
Pescado Rabioso - Album *Pescado 2*, 1973

"Peteribí" is the common name of *Cordia trichotoma* (Vell.) Arráb. ex Steud. used locally in the Argentine Province of Misiones, a magnificent tree of the Alto Paraná Rainforest and the Yungas Rainforest.

v

Contents

Chapter 1
Native Forests Claim for Breeding in Argentina: General Concepts and Their State

Mario J. Pastorino

1.1 Beginnings of Genetic Studies of Forest Tree Species Native to Argentina

The first genetic studies on forest trees in Argentina were carried out with exotic fast-growing species and in relation to their productive use. These antecedents, conducted mainly by the National Institute of Agricultural Technology (Instituto Nacional de Tecnología Agropecuaria, INTA), date back to the 1950s and were based on field trials of poplars, willows, pines, and eucalypts, which were eventually developed as formal genetic improvement programs (Marcó et al. 2016).

It was not until the 1980s that genetic studies of native forest species began. The initial work was carried out at the University of Buenos Aires, where the first doctoral thesis on population genetics of a native forest species (Saidman 1985) was defended. Since the 1990s, new scientific groups commenced to develop with similar lines of research, in several INTA groups and in other national universities such as Comahue, Misiones, and Córdoba, expanding the objects of study to a variety of genera and species.

In the beginning, the lack of knowledge about the genetic resources of the forest species from Argentina promoted studies of genetic characterization of their natural populations, by means of genetic markers (initially isoenzymes and then RAPD, AFLP, RFLP, and more recently SSR). These first steps advanced toward the study of demo-stochastic evolutionary processes (i.e., drift and gene flow) using neutral genetic markers as tools, and a little later, toward the study of adaptation and phenotypic plasticity through the analysis of variation in quantitative traits in common garden trials. Currently, this type of approach continues to expand the range of species involved and questions addressed, and furthermore, the use of new tools for the

M. J. Pastorino (✉)
Instituto de Investigaciones Forestales y Agropecuarias Bariloche (IFAB) INTA-CONICET, Bariloche, Argentina
e-mail: pastorino.mario@inta.gob.ar

© Springer Nature Switzerland AG 2021
M. J. Pastorino, P. Marchelli (eds.), *Low Intensity Breeding of Native Forest Trees in Argentina*, https://doi.org/10.1007/978-3-030-56462-9_1

generation of genomic resources, the identification of candidate genes, and the analysis of full transcriptomes have been added to study selection processes also with molecular markers (see Chap. 17).

Based on the knowledge gained about the genetic resources of native forest trees but focusing on their use, INTA (with the collaboration of research groups from other institutions) formally initiated domestication programs for the most relevant species in 2006, which finally led to the development of low-intensity breeding programs. Given its national projection, INTA implements these programs throughout the country, covering all forest ecosystems in Argentina. For the selection of species, in addition to practical aspects such as the region of concern of each research group, the ecological and/or productive value (current or potential) of the species considered was weighted.

1.2   Main Forest Ecoregions of Argentina

Following Cabrera and Willink (1973) and subsequent works, seven forest regions in Argentina with physiognomic and floristic identity are schematically recognized (Fig. 1.1). Below we present the basic characteristics of each of them.

1.2.1   Alto Paraná Rainforest

Known in Argentina as "Selva Misionera", it is a subtropical rainforest that corresponds to the southern and interior extension of the Atlantic Forest (i.e., the tropical rainforest developed on the coast of Brazil). This formation covers the southern extreme of Brazil (west of the Serra do Mar), the northeast corner of Argentina, and eastern Paraguay. In Argentina, this sector is crossed by the Sierra de Misiones, which forms the watershed between the Paraná and Uruguay rivers, with a maximum altitude of 843 m asl toward the border with Brazil. Precipitation exceeds 1500 mm annually, with a dry season during winter. The average temperatures range from 16 °C to 22 °C, with the occurrence of frost (sometimes it even snows at the highest points).

Two forest districts are recognized within this ecoregion in Argentina:

1. *Jungles*, formed by trees 20 to 30 m high with vines and epiphytes and with strata of smaller trees and a dense undergrowth composed of bamboo and tree ferns. There are also some species of palm trees. The botanical families Leguminosae, Lauraceae, Myrtaceae, and Meliaceae are preponderant, but no species predominate, since more than 40 tree species are counted per ha. However, *Nectandra lanceolata*, *Balfourodendron riedelianum*, *Cedrela fissilis*, and *Aspidosperma polyneuron* stand out.

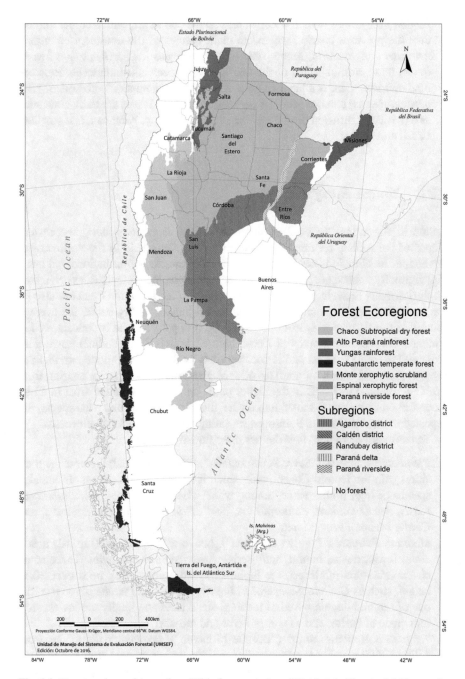

Fig. 1.1 Forest regions of Argentina. (With the permission of Unidad de Manejo del Sistema de Evaluación Forestal, MAyDS)

2. *Paraná pine forests*, which occupy areas above 600 m asl, with a colder climate than the previous district, where the main species of the dominant stratum is *Araucaria angustifolia*, accompanied by *Podocarpus lambertii, Drimys brasiliensis*, and several species of the families Myrtaceae and Lauraceae. Between both districts, there is a large ecotone where the emblematic "yerba mate" (*Ilex paraguariensis*) can be found, i.e., a small tree whose leaves are daily consumed as a particular infusion throughout Argentina, Uruguay, Paraguay, and southern Brazil and Chile.

1.2.2 Yungas Rainforest

Called in Argentina "Selva Tucumano-Oranense," it extends along the eastern slopes of the Andes, forming a narrow strip from Venezuela to the northwest of Argentina. Its climate is cooler than that of adjoining forest formations and has a high humidity, caused not only by the abundant orographic precipitations originated on the humidity brought by the winds coming from the Atlantic Ocean but also by the permanent fogs that characterize this cloud jungle. In Argentina, the average rainfall in this formation is from 1000 to 3000 mm per year and is concentrated in summer, between the months of December and March. Its altitudinal range goes from 400 m to 3000 m asl and is accompanied by the aforementioned gradient of precipitation and also by a gradient of average annual temperatures ranging from 26 °C at the base to 14 °C in the higher parts, with the presence of frost in all its extension and occasional snowfall in the heights. Some genera and even species are repeated in the Alto Paraná Rainforest revealing possible vicariance processes.

The altitudinal gradient thus defines three floristic districts:

1. *Piedmont Rainforest* ("Selva Pedemontana"), up to 500 m asl; it marks the transition with the Humid Chaco, with which it shares some elements. It contains species of high timber value, among which *Cedrela balansae, Amburana cearensis, Anadenanthera colubrina* var. *cebil, Handroanthus impetiginosus*, and *Cordia trichotoma* stand out.
2. *Montane Rainforest* ("Selva Montana"), between 500 and 1800 m asl; it is a cloud forest, dense, humid, and shady, with vines and epiphytes. Three strata characterize this rainforest, the highest formed by emerging trees over 30 m height, such as *Cedrela angustifolia, Juglans nigra*, and *Phoebe porphyria*; the one of the middle composed of medium-sized trees and finally the lowest stratum made of bushes and a layer of herbs and mosses.
3. *Temperate Montane Forest* ("Bosque Montano"), occupying the slopes from 1500 to 2500 m asl, it is composed of conifers and deciduous species and has a much lower biodiversity. There, pure stands of *Podocarpus parlatorei* can be found and, at the highest sites, pure stands of *Alnus acuminata*, finally appearing the small tree *Polylepis australis*, which is the woody species that reaches the transition with the meadows of the alpine tundra.

1.2.3   Chaco Subtropical Dry Forest

Better known simply as the "Chaco," it is the main forest ecoregion in Argentina, where it has its greatest development, extending to Bolivia, Paraguay, and a small portion of Brazil. It occupies a plain with warm continental climate, and average annual temperatures from 20 °C to 23 °C, but usually exceeding 45 °C in summer. Annual rainfall decreases east to west from 1200 mm to 500 mm, with a marked dry season (winter) in the west, thus allowing to recognize a wet Chaco and a dry Chaco. A deciduous xerophilous forest predominates, with a stratum of grasses and numerous cacti and terrestrial bromeliads. Among the main forest species are *Schinopsis balansae*, *Aspidosperma quebracho-blanco*, *Caesalpinia paraguariensis*, *Gonopterodendron sarmientoi*, and numerous species of the genus *Prosopis*, of which *P. alba* is the one with the highest forest value. There are also palm groves of *Copernicia alba* or *Trithrinax campestris*.

1.2.4   Sub-Antarctic Temperate Forest

It develops on both sides of the Andes Mountains, from 34° 50′ S to the south of the Tierra del Fuego archipelago. The Argentine portion is locally called the Patagonian-Andean Forest and extends in a narrow strip on the slopes of the Andes from 36° 50′ S to Ushuaia, between the border with Chile and the Patagonian steppe. The climate is cold temperate, with annual rainfall of 2500 to 700 mm, a dry summer season in the northern portion and a more homogeneous distribution of the precipitations in the south. Precipitation depends primarily on the humidity of the Pacific Ocean and forms an abrupt gradient over distances of less than 50 km.

It is a temperate mountain forest, composed of monotypic or mixed stands of two or three dominant species, with bamboo undergrowth in the northern part and some accompanying shrubs. The genus *Nothofagus* predominates with six species (Fig. 1.2), being *N. pumilio* and *N. antarctica* the most widely distributed. There are also some conifers of great regional importance, such as *Austrocedrus chilensis*, *Araucaria araucana*, and *Fitzroya cupressoides*.

1.2.5   Espinal Xerophytic Forest

It is a xerophilous forest, of exclusively Argentine distribution, similar to Chaco but lower and poorer in species, also with the presence of palm groves. The climate is warm and humid in its northern part but mild and dry in its southern portion: the average rainfall varies from 1100 mm to 350 mm in the year and the average temperatures from 20 °C to 15 °C. Regarding the specific composition of its forest elements, *Prosopis* is its emblematic genus, with several species that define three

Fig. 1.2 Autumnal
specimen of *Nothofagus
obliqua*, one of the species
with the highest breeding
potential from the
Patagonian-Andean Forest.
(Photo: Mario Pastorino)

floristic districts according to their presence and predominance: (1) *District of
Ñandubay* (*Prosopis affinis*), in its northeastern portion (provinces of Corrientes
and Entre Ríos), accompanied by *Prosopis nigra* and palm groves of *Butia yatay*;
(2) *District of Algarrobo* (*P. nigra* and *P. alba*) at the center of this formation; and
(3) *District of Caldén* (*Prosopis caldenia*), in the southern third of this formation.
Other tree species widely present are *Acacia caven*, *Geoffroea decorticans*, and
Celtis spinosa.

1.2.6   Monte Xerophytic Scrubland

It is a scrubland of 1 to 2 m height, which extends over a large arid region of
Argentina, with sandy plains, plateaus, and low mountain slopes. It has a dry climate, somewhat warmer in the northern portion, with average annual temperatures
ranging from 15.5 °C to 13 °C and average rainfall of 80 to 250 mm in the year. It
forms broad ecotones with the dry subtropical forest and the Espinal forest, dominated by several shrubs, such as *Larrea* sp., *Bulnesia retama*, and *Cercidium prae-
cox*. The genus *Prosopis* is present with species such as *P. flexuosa*, *P. chilensis*, and

P. alpataco, which form edaphic communities in which somewhat larger trees can be found.

1.2.7   Paraná Riverside Wetland Forest

This formation is a set of wetland macrosystems of fluvial origin, which occupies the flood plains of the Paraná River and the delta that forms at its mouth on the estuary of the Río de la Plata. It includes forests with floristic elements of the dry subtropical forest and the Alto Paraná Rainforest, dominated by *Salix humboldtiana* and *Tessaria integrifolia* among the outstanding trees and large areas of grasslands and reeds. These formations are commonly included in the previous ecoregions according to the predominance of species.

Of these seven ecoregions identified in Argentina, the first four are those of greatest forest importance, and we will use them to structure this book.

1.3   Native Forest Surface in Argentina

The first reliable estimate of the area covered by native forests in the country was made by the National Agricultural Census of 1937, which reported 37,535,306 ha. Fifty years later, the National Forestry Institute (IFoNa) estimated a decrease of more than two million hectares, reporting an area of 35,180,000 ha. In 1998 Argentina made its first national inventory of native forests, carried out by the Forest Directorate of the Ministry of Environment and Sustainable Development (SAyDS 2005). In this inventory, an area of 31,443,873 ha of native forest was reported, consisting of two inventory classes: forestlands, with 30,309,524 ha, and rural forests, with 1,134,349 ha. For the purposes of this inventory, based on FAO criteria, forestlands were defined as areas with trees at least 7 m high at maturity and at a density such that their canopies cover at least 20% of the surface (the minimum inventory area was 10 ha). Rural forests are defined as remnants of these forests immersed in agricultural landscapes, in fragments smaller than 1000 ha. Because of their use-pressure situation, these fragments are considered with a high probability of disappearing in the short term due to the advance of the agricultural frontier (Montenegro et al. 2005).

However, the inventory defines a third category of interest for this book: "other forest lands," which corresponds mainly to scrublands with various levels of degradation. Notwithstanding, this inventory class include, to a significant extent, groups of trees at very low density (less than 20% coverage of canopies), riparian forests, palm groves, and forest formations of a height of less than 7 m. The latter are usually made up of tree species with a bushy habit due to strong environmental stresses, such as drought, salinity, waterlogging, or high altitude, an example being the Krummholz formation in the treeline. It is necessary to mention this category since it totals 64,975,518 ha throughout the country, of which 66% corresponds to the Monte (all this phytogeographic region corresponds to this category), 20% to the Chaco, and 9% to the Espinal.

Currently, the Second National Inventory of Native Forest is underway, again in charge of the Ministry of Environment and Sustainable Development of Argentina, this time with the establishment of permanent plots in order to create a continuous system of forest inventories. The inventory categories have also been revised, lowering the minimum height of the trees from 7 m to 3 m for the classification of forestland and the minimum inventory area from 10 ha to 0.5 ha (Peri et al. 2019). Until results of this second inventory are obtained, it is those of the first inventory that provide the most reliable information on the natural forest cover of Argentina.

However, deforestation in Argentina has not stopped from 1998 to the present, maintaining significant fluctuations as throughout the entire history of colonization of the country's territory, with long-term trends and conjunctural ups and downs. In the 25-year period between 1990 and 2014, 7,226,000 ha of native forests were lost (MAyDS 2017a), at an average rate of 289,040 ha per year (Fig. 1.3). It should be noted that these forests correspond to the three inventory categories mentioned in the 1998 inventory. However, an estimate performed for the Second Biennial Update Report (BUR 2017) of Argentina to the United Nations Framework Convention on Climate Change (UNFCCC) for the period 2011–2014 showed that the vast majority (77%) of the 877,680 ha deforested in that period corresponds to forestlands, that is, high forests with a canopy cover of not less than 20%. The most current data have been published in the Third BUR (2019) of Argentina, reporting a deforestation surface of 155,851 ha for 2016.

Deforestation is the most obvious symptom of forest loss. However, a degradation process much more complex to identify and inventory usually precedes it. The predominant causes of deforestation and forest degradation in Argentina are mentioned in the following list (MAyDS 2017a):

- The shift of the agricultural frontier toward forest areas, driven by a production model of high profitability and uncertain sustainability. In this cause, the expansion of soybean cultivation plays a preponderant role, in many cases advancing over forest lands with extensive livestock that are deforested for cultivation, in

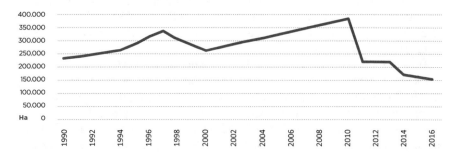

Fig. 1.3 Evolution of the deforested area in Argentina (in ha) between the years 1990 and 2016. (From MAyDS 2017a; BUR 2019)

turn displacing local livestock to more marginal sites due to its inaccessibility, lower water availability, and/or denser forest cover.

- The relocation of livestock production from the Pampas grassland (now dedicated to agricultural crops, predominantly soy) to forest areas.
- The real estate speculation associated with the urban development of the suburban belt of big and small cities surrounded by forests.
- Lack of environmental awareness that values forest services, mainly in the communities associated with them.
- Natural and anthropogenic forest fires. Between 2010 and 2016, an average of 102,041 ha of native forest were burned per year (SGAyDS 2018a).
- Precarious land tenure, with large fiscal areas occupied over generations without title to property and with permanent risk of eviction.
- Weakness of control and oversight policies and institutions, with low allocation of financial and human resources.

1.4   Official Trade Statistics for Timber and Non-timber Products of Native Tree Species

The official statistics of the National State represent the best source of information to weigh the impact that economic interests have on the use of Argentina's natural forest ecosystems. However, it must be recognized that informality and a gray economy characterize the use and exploitation of the resources provided by native forests, leaving this significant economic circuit out of consideration of official data.

Forest products accounted for just 1% of Argentine exports in 2017, while they represented 2.1% of imports, both largely defined by the trade of paper and cardboard that are produced almost exclusively with cultivated introduced species. The country's trade balance of forest products expressed in values has been clearly in deficit in recent years, with an average negative balance from 2011 to 2017 of 854.6 million US $/year (SGAyDS 2018b).

Primary forest production with native species comes almost exclusively from the natural forest. Only the plantations with *Araucaria angustifolia* (and to a lesser extent with *Prosopis* sp.) have an appreciable productive volume in formal trade. In the 2015 statistics, only 588 m³ of *A. angustifolia* roundwood are recorded, which were produced in plantations in the Province of Misiones and destined for the plywood industry (MAyDS 2017b). However, in the medium term, an increase in the offer of wood from planted native species is expected as the plantations of the last decades with *Prosopis* sp. and *Cedrela* sp. reach the cutting period. In this sense, the planting of native species would relax the pressure of use of natural forests, at the same time that it would ensure the maintenance of markets that are reduced year by year by the depletion of the natural resource.

Annually the native forests produce 3,776,591 tons of wood products (average from 2010 to 2016) corresponding to logs, firewood as such, and firewood for

Fig. 1.4 Participation by forest ecoregion of primary wood production in native forests of Argentina, average 2010–2016 (SGAyDS 2018a)

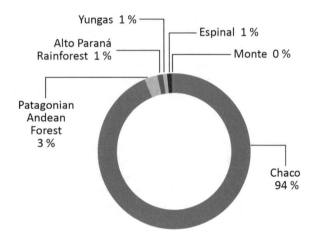

Yungas 1 %

Alto Paraná Rainforest 1 %

Espinal 1 %

Monte 0 %

Patagonian Andean Forest 3 %

Chaco 94 %

charcoal, poles, and other products (SGAyDS 2018a). Of this total amount, 94% comes from the Chaco (Fig. 1.4). In turn, 52.4% of the total corresponds to firewood to produce charcoal, 27.3% to firewood as such, and only 18.5% are logs, of which 35.3% are shredded for extraction of tannin (SGAyDS 2018a). The official statistics for 2016 indicate that of the 682,825 m^3 registered of roundwood of native species, 41% corresponded to *Schinopsis* spp., 16% to *Prosopis* spp., 15% to *Nothofagus pumilio*, and 8% to *Aspidosperma quebracho-blanco* (MAyDS 2018a). On the other hand, of the vast majority (88%) of the 991,206 tons registered of firewood extracted from natural forests, it was not stated to which species it belonged.

In addition to feeding the industries of sawing, plywood, charcoal, tannin, posts, and sleepers, the native forests of Argentina provide non-timber forest products to the formal commercial circuit. Among the most important, the following can be mentioned: wild honey, pine nuts from *Araucaria araucana*, algarrobo flour (from the pods of *Prosopis* spp.), edible fungi, ornamental fern fronds, poles of palm trees, mosses, lichens, and plants for ornamental purposes, canes for furniture, bromeliad fibers, seeds for afforestation, and fruits for fodder and human food.

1.5 Regulations for the Conservation and Restoration of Native Forests and Promotion of Afforestation with Native Species

With the persistence and aggravation of the processes of degradation and deforestation of the forest ecosystems of Argentina, the social demands of the large urban centers (in conjunction with international trends) increased, claiming for regulations that contributed to the conservation of native forests. These demands crystallized in the enactment in November 2007 of the National Law N° 26,331 on

Minimum Requirements for Environmental Protection of Native Forests, finally regulated and implemented in January 2009.

Law 26,331 regulates, promotes, and finances the conservation of native forests (including restoration) based on a territorial planning that divides the forests into three categories according to their conservation value: *very high* (symbolized with the color red), *moderate* (yellow), and *low* (green). Each of the categories has a series of activities enabled or restricted to be carried out in them. *Red areas* should not be transformed, but they can be the habitat of indigenous communities and the object of scientific research. The *yellow areas* can additionally be utilized for tourism, gathering, and sustainable use. *Green areas* can be transformed, radically changing land use, converting them into cultivation areas or urbanization, after approval of a change of use plan.

In keeping with the 1994 reform of the Argentine Constitution, which defined the provincial ownership of natural resources, it is the provinces that must carry out this territorial planning, establishing it by provincial law adhering to Law 26,331. Consequently, as of 2009, the 23 Argentine provinces, with unequal haste, began to perform the territorial planning of their forests with the criteria of Law 26,331 and to adhere to it (the last was the Province of Buenos Aires, which enacted its provincial adhesion law only in 2017). The sum of the areas subjected to the territorial planning by each of the provinces results in the surface occupied by native forests in Argentina according to the consideration of those who have jurisdiction over them (the provinces). This value is 53,654,545 ha, some 22 million hectares more than those reported as native forest by the 1998 national inventory. The difference is explained by the valuation of its own resources that each province made, considering ecosystems of the inventory category "other forest lands" as forests. Of this total area, 19% was classified as having a high conservation value, 61% as a moderate conservation value, and 20% as a low conservation value.

The Law creates a specific fund that must be assigned annually to the provinces according to their forest area in red and yellow categories, to compensate them for the conservation of their forests, as a kind of payment for the environmental services they provide for humanity. Forest land owners receive 70% of these funds in order to perform a conservation plan, including restoration programs or simple prevention of threats such as livestock grazing. The remaining 30% of the funds are available to the provinces for (a) institutional strengthening with the objective of supervising the conservation plans of the private properties, (b) monitoring the conservation status of their forests, and (c) assisting settlers and indigenous communities who live within the forests.

Ten years after its implementation, the Law has failed to suppress deforestation in Argentina. The prolonged discussion prior to its enactment and the delay in its regulation accelerated requests for land use change permits without the restrictions of the Law, causing the undesired effect of increasing the deforestation rate until the first years of its implementation (Fig. 1.3). Even a strong increase in forest fires has been identified during the transition stage in the implementation of the Law (Egolf 2017). In the period 2009–2011, that is, while the majority of the provinces were carrying out their territorial planning of forests, the number of forest fires doubled with respect to the records prior to the enactment of the law, going from 4 to 8 fires per every 100,000 ha of forest. The possible intentionality of these catastrophic events has been suggested, with the purpose of converting forests into agricultural land, anticipating use restrictions and the possible implementation of a rigorous control system. After a peak of 375,000 ha deforested in 2010, the deforestation rate fell sharply, but still maintaining alarming values. Presumably, forest degradation, even without deforestation, also continued at a sustained rate.

There is another national law with current effect on the restoration of degraded forests and afforestation with native species: the Law of Investments for Cultivated Forests N° 25,080 (promulgated in 1998 and extended in 2008 by Law N° 26,432 and in 2018 by Law N° 27,487). This law essentially subsidizes part of the costs of industrial forestry for productive purposes. However, it also includes a section for the enrichment of degraded native forests, which, in conjunction with Law 26,331, represents an additional benefit for restoration, although with the ultimate purpose of forest exploitation. With annual resolutions of the Ministry of Agriculture, the amounts and procedures are updated for each region of the country, and for the different species and silvicultural methods, distinguishing afforestation with irrigation or rainfed, and in blocks or enrichment strips. In recent years, some native species have been explicitly included with a specific cost calculation. These are *Araucaria angustifolia*, *Austrocedrus chilensis*, species of the *Prosopis* genus (mainly refers to *P. alba*, *P. nigra*, *P. flexuosa*, and *P. chilensis*), and species of the *Nothofagus* genus (mainly refers to *N. alpina* and *N. obliqua*). There are also some tax benefits granted by the provinces for the promotion of afforestation (both with native and introduced species). Also, some provinces are considering imposing the obligation to conserve (or recover) a minimum surface of each rural property covered with forests or afforestation. In this sense, the Province of Córdoba has enacted Law N° 10,467 in 2017, which requires agricultural properties to have a forest cover of at least 2% of its surface.

Finally, the National Seed Institute (INASE) should be mentioned, which is the institution that regulates the production and trade of seeds of forest species in Argentina, through Resolution 256/99. This Institute keeps a record of the basic propagation materials (seed sources) existing in Argentina, certifying the seed produced according to different defined types. There are still no restrictions on the trade in seeds according to their provenance and place of use, although there is an increase of 10% in the subsidy of law 25,080 if the material used is of certified origin and is at least of the "selected" category, that is, seed from a genetic improvement program. Currently there are 34 seed production areas of native species registered in INASE: 4 from *Araucaria angustifolia*, 3 from *Austrocedrus chilensis*, 1 from *Cedrela balansae*, 3 from *Nothofagus obliqua*, 1 from *Nothofagus pumilio*, 8 from *Prosopis alba*, 4 from *Prosopis chilensis*, 1 from the hybrid *Prosopis chilensis* X *Prosopis flexuosa*, 2 from *Prosopis flexuosa*, 1 from *Prosopis nigra*, and 5 from *Prosopis* sp. There are also one seed stand of *N. alpina* and three of *P. alba* (https:// www.argentina.gob.ar/sites/default/files/inase_guia_forestal.pdf).

1.6 International Commitments of Argentina, National Plan for the Restoration of Native Forests, and Certification of Sustainable Management of Forests and Plantations

Argentina is a signatory to the United Nations Framework Convention on Climate Change (UNFCCC) and the Kyoto protocol in which it participates through the Clean Development Mechanism. In 2016, through Law N° 27,270, the National State approved the subscription of Argentina to the Paris Agreement (COP21), committing not to exceed the net emission of 483 million tons of carbon dioxide equivalent ($MtCO_2eq$) in the year 2030 (advancing even to a conditional commitment of 369 $MtCO_2eq$), which represents a reduction of 18.5% of the emissions projected for that date. This goal involves mitigation measures that include the REDD + strategy, that is, avoidance of deforestation and degradation of native forests, and afforestation, reforestation, and restoration of degraded forest ecosystems. It has been estimated that in 2014, 15.6% (57.4 $MtCO_2eq$) of the total emission of greenhouse gases (GHG) in Argentina came from native forests, 89% of which were emitted by the conversion from forests to pastures or agricultural crops (PANByC 2017).

Starting in 2017, the Argentine Ministry of Environment and Sustainable Development (MAyDS) began to develop the National Action Plan on Forests and Climate Change (PANByCC), which aims to reduce emissions and increase the capture of GHGs related to the forestry sector. The capture of CO_2 through plantations can be faced with both native and exotic species. In this sense, fast-growing exotic species (i.e., pines, eucalypts, poplars and willows) seem to be a convenient alternative, since they have a higher CO_2 fixation rate. However, according to one of the UNFCCC Cancun safeguards (COP16), this increase in fixation cannot be detrimental to natural forests, and therefore its replacement by more efficient carbon fixation systems cannot be considered. On the other hand, Argentina is a state party to the Convention on Biological Diversity, which forces it to attend to the conservation of native species and ecosystems. It is also one of the 12 member states of the Montreal Process (MAGyP 2015), an international convention that sets criteria and indicators for the sustainable management of forests, obviously restricting the conversion of natural forests to exotic plantations. Therefore, planting on forest land should be considered through programs to restore or enrich the natural forest; and mass afforestation with introduced species should be limited to agricultural areas.

Strategic operational axes defined by the PANByCC include avoided deforestation, sustainable management of native forests, and restoration of degraded forests. Regarding this last axis, after the development of some pilot projects, in 2018 the MAyDS launched the National Plan for the Restoration of Native Forests (MAyDS 2018b), calling in 2018 and 2019 for the open application of projects.

Most of the approved restoration projects (still ongoing) have proposed active restoration strategies, that is, planting of foundational forest species from the degraded ecosystems to be restored. This implies the choice of the genetic materials to be used, which should be those adapted to the implantation site, with the dual purpose of avoiding maladaptation processes of the implanted individuals and preventing genetic contamination of the forest patches that remain within or surrounding the area under restoration (e.g., Vander Mijnsbrugge et al. 2010, and see Chap. 18). Some of these projects have explicitly considered the genetic origin of the seed for the production of the seedlings used. However, this is an unusual consideration in Argentina, since, for native species, commercial nurseries rarely record the provenance of the seed they are using. Technicians still need to be formed and trained on the importance of considering the genetic factor in the industrial production of seedlings of native species. Equally necessary is the generation of knowledge about the gene pools of native species, at least in the main species of each forest ecoregion. In this sense, this book hopes to contribute to the dissemination of current knowledge in this regard.

The certification of the sustainable use of native forests by independent and internationally recognized institutions is a safeguard that society is increasingly imposing to prevent the loss or degradation of natural forest ecosystems. This mode of action through non-governmental organizations is an indirect strategy of the organized civil society that resort to the "market" as a non-punitive regulator.

Since 2002, the Forest Stewardship Council (FSC) has certified in Argentina the sustainable management of forests, including all the links in the chain of custody of the transformation of the raw material to a final product. In the third quarter of 2019, FSC reported 117 chains of custody certificates in Argentina and a certified forest area of 467,585 ha distributed over 11 sustainable management certificates (https:// fsc.org/es/page/datos-y-cifras).

In 2010, the Argentine Forest Certification System (CERFOAR) was created in Argentina to promote good practices in the forest and ensure that both wood and non-timber forest products are produced respecting the highest ecological, social, and ethical standards. In 2014 CERFOAR was approved in the Program for the Endorsement of Forest Certification (PEFC) scheme, which is a global scheme that establishes requirements for forest certification, including the maintenance or improvement of biodiversity and the prohibition of forest conversion (plantations resulting from the replacement of natural forests cannot be certified). In the third quarter of 2019, CERFOAR reported the certification of forest plantations of two of the largest forestry companies in Argentina, and the chains of custody of 14 companies that use raw materials of forest origin (https://www.pefc.org.ar/index.php/emprecertificad/menuempregfs).

1.7 Domestication of Forest Tree Species

In the agricultural sciences, the domestication of wild species involves selection processes that lead to the constitution of highly productive gene pools that, at the same time, are dependent on artificial management systems. The ecologically advantageous qualities for the perpetuation of the species in a natural way are replaced by desirable characteristics for human being (i.e., high productivity of a particular organ, outstanding yield of a certain compound, resistance to pests or diseases inherent to monoculture) under a highly subsidized and artificial production system. The modification of the gene pool of the cultivar variety becomes so profound that sometimes without human intervention it would not be able to persist in successive generations.

In forest science, the concept of domestication is necessarily less strict, mainly due to the longevity of trees. Whereas in species such as cereals, 20 years imply 20 generations, in trees they can represent only one generation or even less. For this reason, when talking about forest species, domestication has not only genetic aspects but also of plantation silviculture, such as the adjustment of seedling production technologies in the nursery, planting in the field, and plantation management until the conformation of an adult population. Therefore, we can define the domestication of forest trees as the development of technologies to take this species to plantation on an industrial scale, both for productive or conservation purposes. From a strictly genetic point of view, tree domestication is limited to elemental selection that rules out gene pools that are not adapted to adjusted artificial management. This selection commonly involves a single generation, and only in some species, it has reached three or four generations.

The domesticated gene pool of a forest species is not far from the wild gene pool, so plantations maintain the ability to turning feral, and thus artificial populations could be perpetuated without the need for recurrent human intervention. This quality is not only a restriction imposed by the long generational times but also a sought-after characteristic due to the longevity of the trees. It is necessary to have a high genetic diversity in the growing populations, since the trees must be adapted to the environmental conditions at the time of planting but also need to have a high adaptability to future conditions. The environment of the future is unknown, but the longer the life cycle (or plantation rotation) of the species considered, the more likely the environment toward the end of the cycle will be different from the current one. This leads to sacrifice selection intensity for the sake of gaining adaptability. Only in highly productive species with very short cycles, the intensity of selection is maximized to the point of clonal forestry, that is, the cultivation of a few select clones.

Historically, since the beginning of civilizations, in species of food relevance and with an annual or very short life cycle, domestication has represented a cultural rather than a technological process. Just in the last century, domestication and genetic improvement have resulted from a planned and directed technical process (large-scale forest tree breeding programs began in the 1950s, White et al. 2007). Currently, a genetic improvement program commonly begins with the generation of information on the genetic variation patterns of the species in question, that is, its level of genetic variation, in general, and how it is distributed among its natural populations. This basic information is essential to outline a selection as well as a conservation strategy.

The genetic characterization of the natural populations of a species can be approached with genetic markers in laboratory studies or with quantitative characters in plant growth chamber, nursery, or field genetic tests. The former (e.g., isoenzymes, microsatellites, AFLP, SNPs; see Box 1.1) are genes or portions of the genome that do not have expression on the phenotype and therefore are not suscep- tible to selection; therefore they are defined as selectively neutral. The latter, on the other hand, are phenotypic characters (i.e., morphological, physiological, pheno- logical, growth; see Box 1.2) and therefore potentially adaptive, measured in prov- enance, progeny, or clonal trials. The two sources of information are valuable and complementary. Genetic markers will allow us to identify differences between and within populations due to past demo-stochastic processes (fundamentally genetic drift and gene flow processes modeled by demographic fluctuations in populations), while variation in quantitative traits of individuals with known kindship relation- ships will allow us to recognize differences due to processes of adaptation to current or modern conditions.

Box 1.1: Characteristics of the Main Genetic Markers Used in the Study of Forest Trees

A genetic marker is a gene or a fragment of DNA sequence with a specific location on a chromosome that is associated with a gene or a trait, shows polymorphism, and allows the distinction of different individuals. There are three conceptually different classes of genetic markers: protein variants (isozymes), DNA sequence polymorphism, and DNA repeat variation (Schlötterer 2004). Isozymes were the first markers used in studies of forest trees, back in the 1970s (Bergmann 1971). With the advent of techniques that allowed the screening of the DNA molecule, more powerful DNA markers became available which allowed detecting higher levels of variability. Some of the most used markers in forest tree species are described below:

Isozymes: Are isoforms of an enzyme, encoded by one or several genes, hav- ing the same function, which can be separated through gel electrophoresis. Variants that are coded by alleles at the same locus are called allozymes. Their main advantage as genetic markers relies upon their Mendelian inheritance with codominant expression of the alleles. Given the degener- ated condition of the genetic code and the fact that some mutations result in the same net charge of the molecule, many mutations are not detectable. The level of polymorphism is therefore moderated to low.

RAPD (random amplified polymorphic DNA): Is a PCR-based technic that uses aleatory primers to amplify non-specific DNA fragments. It is cost- effective and produces highly polymorphic markers, but the main disad- vantages are its dominant expression and its low repeatability.

AFLP (amplified fragment length polymorphisms): Is a DNA fingerprinting method that uses restriction enzyme digestion followed by selective ampli- fication of a subset of fragments. The main advantage is the hypervariable level of diversity; the main disadvantage is the dominant expression.

Nuclear microsatellites or SSRs (simple sequence repeats): Microsatellites are tandemly repeated DNA motifs (from one to six nucleotides), distributed throughout the genome. Their main advantages are the codominant expression and the high level of polymorphism detected. The need to know the flanked sequences to design the primers was the main disadvantage for their use, although with the advance of massive sequencing methods their development became easier and cheaper.

Organelle DNA markers (chloroplast and mitochondrial DNA): They consist in the amplification and sequencing of non-coding regions of the organelle genomes. Due to the uniparental and clonal inheritance of these genomes, they show moderate to low levels of polymorphism, allowing phylogeographic and phylogenetic reconstructions.

SNPs (single nucleotide polymorphisms): Are loci with alleles that differ in a single base, with the rarer allele having a frequency of at least 1% in a set of individuals. SNPs can occur in coding or non-coding regions and are highly frequent in the genomes of plants and animals and codominantly expressed.

Box 1.2: The Quantitative Perspective of Genetic Variation Patterns
Because the phenotype is the expression of the genotype modulated by environmental conditions, the analysis of the variation between individuals in phenotypic characters allows us to gain knowledge about the gene pools that determine them. For this, it is necessary to set up a trial in which the environmental effect can be controlled. Essentially, this trial seeks to raise the different genetic entities (i.e., provenances, families, clones) under the same environmental condition (common garden experiment), with the expectation that the differences that appear in the phenotype are the reflection of the genetic variations. These trials will have an experimental design defined by the environmental conditions and the entities to be tested, and their conclusions will be valid for the trial conditions.

The variables to consider can be of different kinds, some measured simply with a ruler (e.g., seedling height) and others with sophisticated devices such as a spectrophotometer (e.g., chlorophyll concentration). Some variables are of greater productive importance, while others will provide information particularly on adaptability. In this sense, juvenile characters may be the most relevant when the main interest is to ensure post-plantation survival.

The main statistical method to analyze the variation in quantitative characters is the analysis of variance (ANOVA), which can be applied to additive linear models of fixed effects, random effects, or mixed effects. A typical test in a low-intensity breeding program is the comparison of different provenances of a species and, within them, open-pollinated families. Such a trial can be set up by collecting seeds from 10 to 30 trees in each of about 20 populations of interest. All plants obtained with the seeds of one of the trees will be an open-pollinated family, and the kindship relationship between them will be at least half siblings. It is commonly of interest to test the difference between the means of provenances assayed (considering therefore that this factor is of fixed effects) and if the variability of the open-pollinated families has a significant influence on the character of interest (and consequently this factor is considered to have random effects). These considerations define a mixed linear model, which in the case of having an experimental design in blocks with single tree plots, its symbolic expression would be the following:

$$y_{ijk} = \mu + \rho_i + \varphi_{j(\rho_i)} + \beta_k + \varepsilon_{ijk}$$

where y_{ijk} is an observation of the variable of the seedling from the jth family within the ith population, located in the kth block, μ is the overall mean of the variable, ρ_i is the effect (fixed) of the ith population, $\varphi_j(\rho_i)$ is the effect (random) of the jth family within the ith population, β_k is the effect (random) of the kth block, and ε_{ijk} is the general error $\sim NID(0, \sigma^2)$.

Knowing the kinship relationships among the individuals of the trial allows estimating various genetic parameters such as additive genetic variance (V_A), narrow sense heritability (h^2), coefficient of additive genetic variance (CV_A), or quantitative differentiation (Q_{ST}). With them, it is possible to characterize the patterns of genetic variation of a species (inter- and intra-population variation) analogously to that achieved with genetic markers, although including the adaptive portion of the genome. In addition, they allow selecting provenances, families, or individuals in an improvement program, through the estimation of breeding values, the prediction of genetic gains, and finally the establishment of genetic rankings.

After the first steps in gaining baseline information on the path to domestication, pilot afforestation trials continue on an "almost-industrial" scale. Indeed, these assays overlap chronologically with the genetic trials, at least partially. For both, it will be necessary to adjust previously the technology for the production of seedlings in the nursery, from the harvesting and stocking of seeds to the development of sowing, irrigation, and fertilization protocols. The development of this technology requires knowledge of the autoecology of the species, its reproductive system, and its physiology, especially up to the seedling ontogenetic stage.

1.8 Low-Intensity Breeding of Forest Tree Species

Low-intensity (or low input) breeding is characterized by being cheap, simple, robust, locally projected, undemanding in record keeping and central control, capable of being continued after periods of neglect, and not dependent on high technology or highly complex specializations, in short, possible to be sustained on a small and uncertain budget (Lindgren and Wei 2006). In return, the genetic gains achieved are of course modest (Namkoong et al. 1980). However, the low-intensity breeding strategy turns out to be the most effective for species with a low annual planting rate (50–100 ha; White et al. 2007) and in particular for those used in active forest restoration programs, in which the interest for a high adaptability (and therefore high genetic variability) prevails over the productive ones. Many times the main objective is the establishment of basic propagation materials (i.e., seed-producing areas, seed stands, seed orchards) to ensure the provision of seeds for seedling production (Kjaer et al. 2006), and only a low degree of genetic gain that simple ensures good health and a not-inferior-to-the-average quality in the traits of interest.

In a low-intensity strategy, the improvement cycle is shorter, and the tools used are fewer and more direct in effect than those applied in high input programs, which use more complex technologies (complexity that depends on each species). Recurrent mass selection and open pollination are the basis, and instead controlled crosses and vegetative propagation are rarely used techniques. The selection of individuals for their phenotype can be carried out in natural populations or in existing plantations if the species has begun to be domesticated. In the first case, individual selection is recommended by the baseline method (Ledig 1974), while for the second, what is more effective is selecting by comparison with the neighbors.

In a low input program, the same physical population usually represents different populations of the improvement cycle. The selected population (set of selected individuals in the base population) usually coincides with the breeding population (the set of individuals that cross freely to start a new improvement cycle, thus regenerating genetic variability). In turn, it can also coincide with the propagation population, generally in the form of a seed orchard. Usually the breeding program does not pass this stage, petering out in the first cycle. An evaluation of the progeny of the selected population, even in the productive plantations, can lead to a genetic purification of the orchard already installed, achieving a 1.5 generation orchard.

Any breeding program should begin by identifying the breeding zones based on environmental criteria (i.e., edaphoclimatic). Each breeding zone requires a particular program, which produces the propagation material to be used within it (White et al. 2007). All accessible populations (natural or artificial) included within the breeding area will represent the program's base population. The next step will be the identification and evaluation of provenances, which will require a wide and adequate sampling for the good representation of each one. In a low-intensity program, a provenance trial will not only test the provenances best suited for the breeding zone but may also constitute the base population for the next generation. In this case, starting from trees that have already demonstrated their adaptation to artificial management might be an advantage, and additionally the selection can be based on the comparison between neighboring trees. However, conforming the selected population from these selections requires avoiding open pollination when propagating selected trees (i.e., controlled pollination or vegetative propagation should be used), since otherwise the progeny of the selected trees would include genes from unselected individuals and provenances.

Once the selection of provenances has been performed, progress can be made to define seed-producing areas (SPA), which usually represent the most elementary basic propagation material. SPA are natural stands or plantations, with good accessibility to the trees and their crowns, high productivity and low periodicity in seed production, high specific purity in the case of possible inter-specific hybridization, and with a mode of trees of good phenotype. The objective of the SPA is the provision of seeds from a known provenance. A thinning that eliminates undesirable individuals (i.e., unhealthy, phenotypically diminished) will lead to the formation of a seed stand (SS). This new basic propagation material already involves genetic gain, even when it is based on light phenotypic selection.

The next step is the establishment of seed orchards (SOs), which are plantations made with selected genetic material and with the specific purpose of producing seeds for the plantation program. Clonal SOs are often the core product of the first cycle of a high-intensity breeding program and require mastering the vegetative propagation technique by grafting or rooting of cuttings. Progeny SOs allow dispensing with vegetative propagation, which is important to simplify the process and occasionally crucial since in some species the percentage of success is very low (or null) or depends on the genotype. In a low-intensity strategy, these progenies will be the product of open pollination of the selected mothers. Thus, progeny orchards will have greater genetic diversity than clonal orchards, which may be a particularly sought-after effect if the goal is to increase adaptability rather than productivity. However, open pollination involves the inclusion of many undesirable genes, leading to the need to test individuals in the orchard. In this way, the same group of trees is a progeny trial and, simultaneously, the future seed orchard, which will only be established when the genetic thinning prescribed by the trial evaluation is carried out.

However, monitoring and evaluating a progeny trial often exceed the possibilities of a low-intensity breeding strategy. Gene source plantations (GSP) have been proposed as an alternative to progeny SO (Lindgren 2000). These are plantations established with seedlings from open pollinated selected trees, usually mass selected in previous plantations or directly in the natural forest. They may be established in a similar fashion to regular plantations and may serve concurrently several functions, such as production of desirable commercial products, conservation of the genetic diversity of the species, and as a local demonstration plot or as seed source for plantations, while retaining options to initiate a more regular improvement program. If it will effectively function as a seed source, it should be thinned aiming to eliminate undesirable individuals. The original material should include the seed of a high number of mothers, since an intense thinning of their progeny could leave a limited number of families in the GSP, thus favoring inbreeding.

Summarizing, recurrent selection cycles with recombination and infusions of new material typical of high-intensity breeding are replaced, in low-intensity breeding, by linear chains leading to the development of a program with a predicted end (Fig. 1.5). This does not mean that if the demand for seeds for productive or restoration purposes increases more than expected, the initial low-intensity program cannot be reformulated toward a new improvement cycle, moving in this case to a classic high-intensity program.

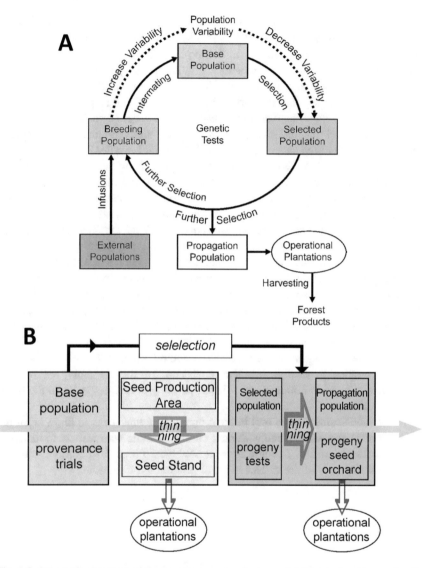

Fig. 1.5 Schematic diagrams of the breeding process for trees. (**a**) Classic breeding cycle of a high-intensity improvement program (from White 1987). (**b**) One possible breeding chain of a low-intensity improvement program. Different colors imply different physical populations

1.9 Aim and Scope of the Present Contribution

The development of forestry in Argentina has been based on the cultivation of fast-growing introduced species, mainly from the *Pinus*, *Pseudotsuga*, *Eucalyptus*, *Populus*, and *Salix* genera, for which both private companies and the National State through INTA have developed high-intensity genetic improvement programs.

However, in the late 1990s, three complementary reasons gave impetus to the development of a domestication program at INTA: (i) a certain social consensus (first international and finally national) on the importance of conserving and restoring natural forest ecosystems, (ii) sympathy for native species per se, and (iii) the baseline information that had been generated in previous years on the genetic resources of native forest species. Initial efforts were unstructured and responded to different projects with varying objectives, but finally a formal program started in 2006.

In addition to INTA's initiative to domesticate potentially productive native species using low-intensity breeding strategies (see Marcó and Llavallol 2016), efforts were also made by other institutions that had also started studying the genetic resources of forest species in Argentina. These actions make up the Argentine experience in the matter, at times systematic and well-structured and sometimes dispersed and with a varied level of depth. It is the intention of this book to summarize the most relevant experiences, trying to make visible the effort of many technicians and researchers who have managed to chart a path that, despite the years, has only just begun.

The book is organized through the main forest ecosystems in Argentina. In each one, the advances referred to the knowledge of the genetic resources of the most studied species are presented, together with the implementation of actions for their domestication and genetic improvement. The most prominent results are reported, avoiding accounting for failures or low impact results. Toward the end of the book, we present the latest advances in the development of tools by means of high-throughput sequencing in native species of Argentina that will be the basis of the next technological generation for their genetic improvement. Finally, we conclude with essential considerations inherent in the planting of native species in the frame of climate change.

We are addressing a wide and international public that seeks to know what has been achieved and what is being worked on in Argentina with respect to its native forest genetic resources. The interested reader should look for the most scientific details in the articles of the extensive list of bibliographic citations.

References

Bergmann F (1971) Genetische Untersuchungen bei *Picea abies* mit Hilfe der Isoenzym-Identifizierung.I. Möglichkeiten für genetische Zertifizierung von Forstsaatgut. Allg Forst Jagdztg 142:278–280

BUR (2017) Segundo Informe Bienal de Actualización de la República Argentina a la Convención Marco de las Naciones Unidas Sobre el Cambio Climático. http://eurocli-maplus.org/intranet/_documentos/repositorio/02%20Bienal%20Convenci%C3%B3n%20 ONU%20cambio%20clim%C3%A1tico_2017.pdf

BUR (2019) Tercer Informe Bienal de Actualización de la República Argentina a la Convención Marco de las Naciones Unidas Sobre el Cambio Climático. https://www. argentina.gob.ar/ambiente/cambioclimatico/informespais/primera-comunicacion/ tercer-informe-bienal-de-actualizacion

Cabrera AL, Willink A (1973) Biogeografía de América Latina. OEA, Washington, DC, 120 pp

Egolf P (2017) Estudio econométrico sobre incendios forestales e incentivos económicos a partir de la Ley de Bosques de Argentina. MSc thesis, Universidad del Centro de Estudios

Macroeconómicos de Argentina (UCEMA), Buenos Aires, 40 pp. https://inta.gob.ar/sites/default/files/inta_cicpes_instdeeconomia_egolf_p_tesina_mae_ucema.pdf

Kjaer ED, Dhakal LP, Lilleso J-PB, Graudal L (2006) Application of low input breeding strategies for tree improvement in Nepal. In: Proceedings of IUFRO conference "Low input breeding and genetic conservation of forest tree species", 9–13 October 2006, Anatalya

Ledig FT (1974) An analysis of methods for the selection of trees from wild stands. For Sci 20:2–16

Lindgren D (2000) Low-intensity breeding. In: Lundkvist K (ed) Rapid generation turnover in the breeding population and low-intensity breeding, Research Notes, vol 55. Department of Forest Genetics, Uppsala, pp 37–48

Lindgren D, Wei R-P (2006) Low-input tree breeding strategies. In: Proceedings of IUFRO conference "Low input breeding and genetic conservation of forest tree species", 9–13 October 2006, Anatalya

MAGyP (2015) Segundo Reporte de Argentina al Proceso de Montreal. Buenos Aires, 185 pp. https://drive.google.com/file/d/0Bzc7dDyL0NwLeVg1UXUyOW5pdkE/view

Marcó MA, Llavallol CI (eds) (2016) Domesticación y Mejora de Especies Forestales. Ministerio de Agroindustria, Buenos Aires. 199 pp. http://forestoindustria.magyp.gob.ar/archivos/biblioteca-forestal/domesticacion-y-mejoramiento-de-especies-forestales.pdf

Marcó MA, Gallo LA, Verga AR (2016) Introducción. In: Marcó MA, Llavallol CI (eds) Domesticación y Mejora de Especies Forestales. Ministerio de Agroindustria, Buenos Aires. 199 pp. http://forestoindustria.magyp.gob.ar/archivos/biblioteca-forestal/domesticacion-y-mejoramiento-de-especies-forestales.pdf

MAyDS (2017a) Plan de Acción Nacional de Bosques y Cambio Climático (V1). Buenos Aires, 56 pp. https://redd.unfccc.int/files/4849_1_plan_de_accion_nacional_de_bosques_y_cambio_climatico_-_argentina.pdf

MAyDS (2017b) Anuario de Estadística Forestal – Especies Nativas 2015. https://www.argentina.gob.ar/manejo-sustentable-de-bosques/programa-nacional-de-estadistica-forestal

MAyDS (2018a) Anuario de Estadística Forestal – Especies Nativas 2016. https://www.argentina.gob.ar/sites/default/files/2016_-_anuario_de_estadistica_forestal_de_especies_nativas.pdf

MAyDS (2018b) Plan Nacional de Restauración de Bosques Nativos. Buenos Aires, 31 pp. https://www.argentina.gob.ar/sites/default/files/resumen_pnrbn_final_0.pdf

Montenegro C, Bono J, Parmuchi MG, Strada M, Manghi E, Gasparri I (2005) La deforestación y degradación de los bosques nativos en Argentina. Ediciones INTA, IDIA XXI 8:276–279

Namkoong G, Barnes RD, Burley J (1980) A philosophy of breeding strategy for tropical forest trees, Tropical forestry papers 16. Commonwealth Forestry Institute. University of Oxford, 67 pp

Peri PL, Monelos L, Díaz B, Mattenet F, Huertas L, Bahamonde H et al (2019) Bosques Nativos de Lenga, Siempreverdes y Mixtos en Santa Cruz: base para su conservación y manejo. Consejo Agrario Provincial de Santa Cruz, Río Gallegos

Saidman BO (1985) Estudio de la variación alozímica en el género *Prosopis*. Doctoral thesis, Facultad de Ciencias Exactas y Naturales. Universidad de Buenos Aires. http://digital.bl.fcen.uba.ar/Download/Tesis/Tesis_1893_Saidman.pdf

SAyDS (2005) Primer Inventario Nacional de Bosques Nativos. Secretaría de Ambiente y Desarrollo Sustentable, Ministerio de Salud y Ambiente de la Nación. http://www.ambiente-forestalnoa.org.ar/userfiles/nodo/informenacionalpinbn.pdf

Schlötterer C (2004) The evolution of molecular markers – just a matter of fashion? Nat Rev Genet 5:63–69

SGAyDS (2018a) Series Estadísticas Forestales 2010–2016. https://www.argentina.gob.ar/sites/default/files/series_2010-2016_1.pdf

SGAyDS (2018b) Comercio Exterior de Productos Forestales 2017. Buenos Aires, 34 pp. https://www.argentina.gob.ar/sites/default/files/comercio_exterior_2017_pdf

Vander Mijnsbrugge K, Bischoff A, Smith B (2010) A question of origin: where and how to collect seed for ecological restoration. Basic Appl Ecol 11:300–311

White TL (1987) A conceptual framework for tree improvement programs. New For (4):325–342

White TL, Adams WT, Neale DB (2007) Forest genetics. CABI Publishing, Cambridge, 682 pp

Part I
Temperate Subantarctic Forests

Chapter 2
Temperate Subantarctic Forests: A Huge Natural Laboratory

Paula Marchelli, Mario J. Pastorino, and Leonardo A. Gallo

2.1 Generalities and Particularities of the Andean-Patagonian Forests

In the southern tip of South America, temperate subantarctic forests develop on both sides of the Andes mountain range. Known in Argentina as Andean-Patagonian forests, these ecosystems cover a narrow but long latitudinal strip, from 35° S (at Maule River, in Chile) to 55° S (in the southern extreme of the Tierra del Fuego archipelago, in Chile and Argentina). Their current geographic discontinuity from the other forests of South America is a remarkable feature, for which they have been considered a biogeographic island (Armesto et al. 1995). The sclerophyllous scrub and the high-Andean-steppe border the Subantarctic forests to the north; the Patagonian steppe, in Argentina, confines its development to the east, and the ocean marks its western and southern edges. This geographic isolation dates back to the Oligocene (about 23–33 My ago), when South America began to drift northward; before that, the continents of the southern hemisphere were connected as Gondwanaland (Markgraf et al. 1996). Since its origin in the late Cretaceous (ca. 90 My ago; Dettmann et al. 1990) till the separation of the continents, floristic interchange endured. This explains the relationships with other southern forests like those of Tasmania and New Zealand (i.e., disjunct distribution of genus like *Araucaria*, *Aristotelia*, *Blechnum*, *Discaria*, *Lomatia*, *Nothofagus*, *Podocarpus*, among others; Veblen et al. 1996). In addition, abundant neotropical elements characteristics of the "Yungas" in NE Argentina (e.g., *Azara*, *Chusquea*, *Crinodendron*, *Drimys*, *Escallonia* among others) reflect a history of recurrent connections with ecosystems of lower latitudes (Arroyo et al. 1995).

P. Marchelli (✉) · M. J. Pastorino · L. A. Gallo
Instituto de Investigaciones Forestales y Agropecuarias Bariloche (IFAB) INTA–CONICET, Bariloche, Argentina
e-mail: marchelli.paula@inta.gob.ar

© Springer Nature Switzerland AG 2021
M. J. Pastorino, P. Marchelli (eds.), *Low Intensity Breeding of Native Forest Trees in Argentina*, https://doi.org/10.1007/978-3-030-56462-9_2

The climate and topography along which these forests develop determine a wide variety of forest formations, soil types, and disturbance regimes, which greatly influence in species distribution as well as in their patterns of genetic variation. Then, the region is characterized by a high heterogeneity imposed by deep environmental gradients. First, the extension of these forests encompassing about 20° of latitude generate a gradient in the photoperiod with earlier start and later end of the daylight during the growing season toward the south, thus increasing day length. Also light quality and quantity differ across latitude due to seasonal variation in the distribution of semidarkness (twilight and moonlight) (Mills 2008) and the amount of solar ultraviolet radiation that decreases toward the poles. In addition, air temperature has a poleward decrease. The latitudinal mean annual temperature (MAT) gradient in the Southern Hemisphere is of −0.57 °C per degree of latitude (while in the Northern Hemisphere is of −0.73 °C), although many factors influence the local temperature making this relation less linear (i.e., vegetation, topography, winds, cloud, and snow cover) (De Frenne et al. 2013). For species with geographic ranges that span these gradients, local adaptation is often observed, with variation among populations in growth and reproductive traits such as phenology, as a response to light and temperature (Savolainen et al. 2007). Intraspecific variation in other key life history traits like leaf N:P ratio and seed mass was found to decrease with latitude in natural populations of vascular plants (De Frenne et al. 2013).

The amount of precipitation is also a major determinant of the distribution of vegetation. At latitudes between 35° S and 43° S in Chile, the Coastal range runs parallel to the Andes, separated by the Central Depression. While the Andes reach altitudes above 3000 m asl, the Coastal mountain range only reaches 2000 m asl at its maximum heights (at 38° S; Armesto et al. 1995). South of 43° S, the Andes gradually diminish in mean elevation. These two mountain ranges cause a rain shadow effect that concentrates rainfall on the western slopes. This effect is more pronounced in the Andes where precipitation decreases from more than 2000 mm to under 200 mm over a west-to-east distance of 80 km (Roig and Villalba 2008). In addition, precipitation increases from north (1500 mm at 35° S) to south (4500 mm at 47° S) (Di Castri and Hajek 1976), decreasing again toward the continental extreme (1400 mm at 55° S) (Arroyo et al. 1995). Not only does the precipitation volume change from north to south but also its seasonality. For example, Valparaiso (33° S) and Punta Arenas (53° S) have a quite similar annual mean, but in the first case, almost 80% of the rainfall is concentrated in the winter months (May to August), while in the second city, precipitation is homogeneously distributed over the whole year (Donoso 1992). In addition, at mid-latitudes (42° − 44° S) north-south-oriented mountain chains run parallel to the Andes, imposing another restriction to rains.

The precipitation gradient is associated with an increase in the daily thermal amplitude as the forests give rise to the steppe (Hadad et al. 2019). These climatic conditions enhance the likelihood of occurrence of frosts. Actually, the annual frost-free period is of only 90 days in eastern North Patagonia (December to March), with records of late frost during the first days of December and early frost at middle of March (Bustos 2001). Frost events occurring during the growing season can affect meristematic tissues and even leave injuries that can be recorded in tree rings as anatomical anomalies. Frost rings were reported in some native conifers with late-spring frosts having more incidence (Rojas-Badilla et al. 2017; Hadad et al. 2019).

The marked patterns in the annual precipitation, together with the gradual decrease in the mean annual temperatures from north to south, determine the occurrence of different forest formations. At northern latitudes, between 35° S and 37° 20′ S in Chile occurs the district of the deciduous warm temperate *Nothofagus* forests, mainly at the coastal range. These forests are the transition between the evergreen sclerophyllous shrublands of Northern Chile and the southern temperate forests and are dominated by deciduous species like *Nothofagus glauca* and *Nothofagus obliqua* (San Martin and Donoso 1995). Temperate Valdivian rain forests prevail from 37° 45′ S to 43° 20′ S and are followed by the North Patagonian temperate humid forest, from about 43° 20′ S to 47° 30′ S. Finally, the Magellanic humid forest covers the surfaces to the southern tip of the continent (Veblen et al. 1995b). The main forestland occurs in Chile, but smaller extensions enter Argentina in the wetter sections.

Biodiversity declines from north to south, accompanying the decrease in temperature and growing season length. Accordingly, the Valdivian rain forest represents the hotspot of species richness, including evergreen broadleaved trees (e.g., *Laurelia sempervirens*), evergreen conifers (e.g., *Austrocedrus chilensis*, *Fitzroya cupressoides*, *Saxegothaea conspicua*) and the deciduous *Nothofagus alpina* (= *N. nervosa*) and *N. obliqua* (Veblen et al. 1981; Donoso 1993). The evergreen species *Nothofagus dombeyi* and *Nothofagus nitida* prevail in the North Patagonian temperate humid forests in association with a few broad-leaved and conifer species. The Magellanic humid forest is characterized by the presence of *N. betuloides*, *Pilgerodendron uviferum*, and *Drimys winteri*. Finally, the deciduous *N. pumilio* and *N. antarctica* are the emblematic tree species of the Subantarctic forests, since they occur within all the Andean forest districts mentioned above (Veblen et al. 1995b; Chap. 5).

On the other hand, the west-east sharp precipitation gradient promotes a vegetation change from temperate humid forests, through cool temperate *Nothofagus* forests, to open forests and shrublands in the ecotone with the Patagonian steppe (Fig. 2.1). These cool temperate forests occur more extensively in the eastern side of the Andes (Argentina), from 37° 30′ S to 55° S, and in their north portion include the eastern drier forests characterized by the conifers *Austrocedrus chilensis* and *Araucaria araucana* (Veblen et al. 1995a).

Finally, an altitudinal gradient develops along the slopes of the mountains, particularly at northern latitudes where the Andes display higher heights. Elevational gradients impose strong adaptive challenges for plant species because of the marked changes in environmental conditions over relatively short distances (i.e., decreasing temperature, frost, shorter growing season). At around 40° of latitude, *Nothofagus pumilio* grows from 1000 to 1700 m asl, where it conforms the tree line. Along this elevation range, the species shows a marked variation in the growth habit distinguishing three morphotypes: arboreal, shrubby, and crawling. A genetic base for these differences was analyzed based on genetic markers and quantitative traits

Fig. 2.1 Vegetation change along the deep precipitation gradient in North Patagonia. (**a**) *Nothofagus* forests; (**b**) mix forests of *Araucaria araucana* and *Austrocedrus chilensis*; (**c**) *N. obliqua* in its extreme eastern range; (**d**) *Austrocedrus chilensis* forest patches at xeric sites. (Photos **a**, **c** and **d**, by M. Pastorino, **b** by M. Millerón)

(Premoli 2003; Premoli et al. 2007; Mondino et al. 2019; Soliani and Aparicio, 2020; see Chap. 5).

The distribution of species in mountainous areas is typically restricted to narrow and well- delimited altitudinal zones (Jump et al. 2009). In the Andes, the decrease in temperature toward higher elevations determines specific niches for the main three species that characterize the temperate *Nothofagus* mixed forest of the Lanín National Park in Argentina, at 40° S. Thus, although some overlap in their altitudinal distributions exists, *N. obliqua* dominates the lower altitudes between 650 and 800 m asl, *N. alpina* does so between 800 and 1000 m asl, and *N. pumilio* is found from 1000 m till 1700 m asl usually forming the tree line (Donoso 2006).

These particular settings of the Subantarctic forests, occupying different environments along temperature, precipitation, and day length gradients, suggest the occurrence of local adaptations of populations to their home environmental conditions. Thus, these gradients represent a kind of "natural laboratory" since the same species can be found living under contrasting environmental conditions (Fig. 2.2). Furthermore, in some of these gradients, sharp contrasts occur in a matter of few kilometers (i.e., precipitation) or even meters (i.e., altitude, slope aspect), and consequently, populations not geographically isolated (i.e., within possible gene flow distances) may, nevertheless, show signs of local adaptation.

Several recent studies are taken advantage of this "natural laboratory" condition of the Andean-Patagonian forests to shade light about adaptive behavior of different species. For example, adaptation processes along elevation gradients were studied in different *Nothofagus* species (e.g., in *N. pumilio* (Premoli 2003; Premoli et al. 2007; Mondino et al. 2019), in *N. pumilio*, *N. alpina*, and *N. obliqua* (Arana et al. 2016)). While the effect of the sharp precipitation regime on genetic patterns was evaluated, for example, in *A. chilensis* (Pastorino et al. 2010, 2012), and in *N. pumilio*, *N. alpina*, and *N. obliqua* (Soliani et al. 2020). Finally, genetic differences in morphological and phenological traits at contrasting latitudes (e.g., changing photoperiod) was investigated in *N. obliqua* (e.g., Barbero 2014), in *N. alpina* (e.g., Duboscq-Carra et al. 2018), and in *N. pumilio* (e.g., Torres et al. 2017). Recently, the current extent and bases of local adaptation in *N. pumilio* along the Andes are under study to disentangle the interactions between genotype, phenotype, and environmental conditions (Sekely et al. 2020).

Disturbances of different origins and at different scales have influenced and still do so in the persistence of the Subantarctic forests. The main influencing factors are those of geological origin given the huge volcanic and tectonic activity of the region. The tectonic history of the Patagonian region is exposed by very active and extensive volcanism, almost recurrently from the Late Permian (ca. 250 My ago) until today, with hundreds of active volcanoes along the Andes (Rabassa 2008). Denuded sites are suitable for the establishment of shade-intolerant species such as some *Nothofagus* spp. (e.g., *N. dombeyi* and *N. antarctica*) and *Fitzroya cupressoides* (Veblen et al. 1981). Also moraines are usually colonized by *Nothofagus* spp. (Veblen et al. 1989; Garibotti and Villalba 2017). Actually, coarse-scale disturbance is required for *Nothofagus* regeneration, particularly at lower elevations (Pollmann and Veblen 2004). In addition, the strong winds frequent in Patagonia play a role in

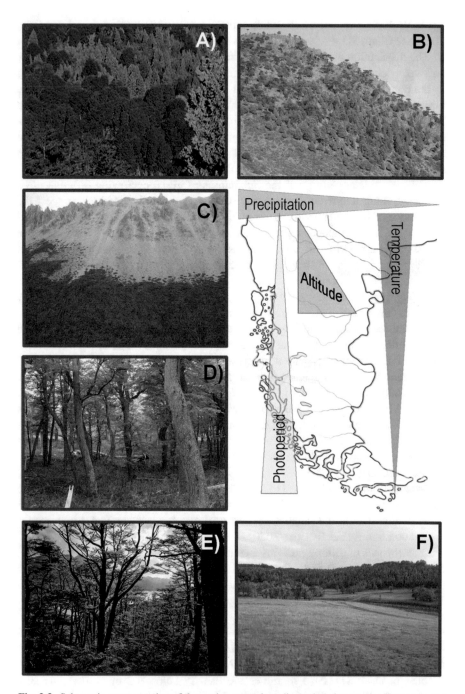

Fig. 2.2 Schematic representation of the environmental gradients that characterize Patagonia and pictures of the typical forests along these gradients. (**a** and **b**) Change in the forest types of *Austrocedrus chilensis* along the precipitation gradient in north Patagonia. (**c** and **d**) High and low altitude forests of *Nothofagus pumilio* in North Patagonia. (**e** and **f**) Humid and xeric forests of *Nothofagus* in Tierra del Fuego Island. (Photos **a–c** and **f** by M. Pastorino; **d** by C. Soliani and **e** by J. Sekely)

the generation of gaps that facilitate the regeneration dynamics (Rebertus and Veblen 1993).

Fire is another disturbance affecting these forests and the most important since European colonization (Veblen and Lorenz 1988). After human settlement, most fires are of anthropic origin, although natural fires due to lightning bolts also occur, particularly in the north. Some species have clear traits that indicate adaptations to fire, for example, the thick bark of *Araucaria araucana* and *Fitzroya cupressoides*, or the resprouting ability of *Nothofagus antarctica, N. obliqua,* and *N. alpina* (Burns 1993). Fires in Patagonia showed a strong climatic influence at an annual scale. Years of extreme fire occurrence are associated with dry winter-springs of La Niña events and with warm summers after El Niño events (Veblen et al. 1999). The greater propensity of shrublands to burn together with an increase in anthropogenic ignitions accelerates the replacement of forests by shrublands (Mermoz et al. 2005).

Certainly, the disturbances that affected the region on a broad scale were the climate changes during the Neogene (ca. 23–2.6 My BP). Like in all regions of the planet, climatic conditions in Patagonia were highly variable during the Cenozoic (that includes the Neogene and the Quaternary), particularly since the Miocene, with the occurrence of multiple cold-warm climatic cycles (Rabassa 2008). Continental and alpine-type glaciations are very well represented in Patagonia and Tierra del Fuego, with the occurrence of glaciations already at the end of the Miocene (ca. 5–6 My; Rabassa 2008). The Great Patagonian Glaciation was dated about 1 My before present (Singer et al. 2004) being the glaciation with the maximum expansion of ice in extra-Andean Patagonia (Flint and Fidalgo 1969; Rabassa et al. 2005). The Last Glacial Maximum (LGM) occurred about 18,000–20,000 years ago (Porter 1981), when the ice covered the Patagonian Region beyond the mountain ranges and toward the plains (Rabassa and Clapperton 1990; Glasser et al. 2008). However, pollen diagrams suggested that floristic elements of modern Patagonian forests were present during the LGM in low abundance and in a discontinuous pattern along both slopes of the Andes (e.g., Heusser et al. 1999; Henríquez et al. 2017; Moreno et al. 2019) indicating that forest species persisted at multiple micro-climatically favorable refugia (Markgraf et al. 1995). Some of these refugia become the center of expansion of the biota after the ice retreated (e.g., Pastorino and Gallo 2002; Marchelli and Gallo 2006; Azpilicueta et al. 2009; Marchelli et al. 2010; Mathiasen and Premoli 2010; Premoli et al. 2010). These complex scenario left an imprint in the vegetation patterns and, particularly, in the genetic structure of populations within species (see details for each species in the different chapters of this section of the book).

2.2 Social Aspects

Patagonia is one of the main tourist regions from Argentina, highly appreciated both within the country and abroad. Its main feature relies on the exotic and wild natural landscapes, relatively little affected by humans. With a population density of only

2.2 inhab/km^2, wild nature can be found on the outskirts of any of its populated cities. Actually, much of the economy of the region depends directly or indirectly on tourism. Therefore, this clearly highlights the economic importance of the conservation of natural resources, which in recent times received much attention and social valuation. Thus, at present, a large proportion of the Subantarctic forests is under some type of protection either national (within the National Parks), provincial (reserves managed by the different states), or municipal (under the cities jurisdiction). However, this was not always the case, and these forests have a long history of unsustainable exploitation and use.

The exploitation of the forests begun with the arrival of white men to these lands during the seventeenth century. Before the colonization, Patagonia has a known history of pre-Hispanic human occupation that dates back approximately 7000 to 9200 years ago. Ethno-historical knowledge reveals intense contacts and exchanges between different groups: Puelches, Pewenches, and Mapuches from both sides of the Andes Mountains, at least since the sixteenth century (Carpinetti 2007). These original folks lived within the forest (or in its margins) making a rationale use of it. For example, in the Araucania region, the Pewenches considered *Araucaria araucana* a sacred tree and used the seeds (pine nuts) for their own consumption (Chap. 7A). Another example is the medicinal use of tree species like *Salix humboldtiana*, whose bark contains salicin, phenols, and oxalates that have febrifuge, analgesic, sedative, tonic, astringent, and antispasmodic properties for medicinal and veterinary use (Chap. 7B). However, with the colonization, exploitation begun. Most of the native tree species hold an outstanding wood quality, highly appreciated both in regional and international markets. A clear example of overexploitation is the case of *Fitzroya cupressoides*, species that was logged to be used as tiles due to its prominent outdoors resistance (Fig. 2.3a). Historical buildings like churches made of wood in Chiloé Island (Chile), declared as UNESCO World Heritage, are still standing after 300 years. Another overexploited species, particularly west of the Andes, was *Austrocedrus chilensis* also used for construction (Fig. 2.3b) and even for exporting to the Virreinato del Perú (Donoso et al. 2006a). However, the most substantial impact on the native forest came with the European colonization at the end of the nineteenth century, after Chile and Argentina became independent nations. After the military campaigns that decimated the original folks, the new settlers initiated a non-sustainable use of the forests, not only for wood extraction but also provoking fires to transform the forests in land for cattle raising (Rothkugel 1913; Donoso and Lara 1995). After the subjugation of the Mapuche Nation in 1881, the almost untouched region between the rivers Biobío and Valdivia was decimated, to open up agricultural land. The erosion of these 300,000 ha can be observed until today. At the beginning of twentieth century, 580,000 ha were burnt between Arauco and Llanquihue for livestock. Similarly, the colonization of Aisén Province of Chile was based on the conversion of forests to pasturelands: more than 2 million hectares were burnt in a couple of decades since 1920 (Donoso and Lara 1995). Lara et al. (1999) estimated that about 40% of the original Valdivian rain forest was lost. The overexploitation and destruction of the native forests also occurred in Argentina, although at a lesser extent because the main agro-business was centered in the

Fig. 2.3 Buildings made with native species. (**a**) Typical house in the south of Chile made with *Fitzroya cupressoides* tills (see inset with details of the tills). (**b**) "Centro Cívico" of Bariloche, Argentina, where the city council functions, made with wood of *Austrocedrus chilensis*, and closer look to a roof in the inset. (Photos **a** downloaded from www.tripadvisor.com, **b** by L. Gallo)

Pampas region. In addition, the large afforestation with pinus species with industrial purposes from the second half of the twentieth century contributed to the destruction of the forests. In Chile, more than 2.5 million ha of *Pinus radiata* and *Eucalyptus* sp. were planted in former Subantarctic forests. In Argentina, the impact is lower since only 100,000 ha were planted and not all of them in forests lands.

Although overexploitation continued during the twentieth century, a new conception of the forest emerged, and public policies aimed at the conservation of the Subantarctic forest begun to take shape. In 1922, Argentina creates its first National Park, initially named "Parque Nacional del Sur," being since 1934 Nahuel Huapi National Park. This park occupies a surface of 717,000 ha. From then on, a national conservation policy was institutionalized, which was later imitated by provincial and municipal jurisdictions, based on which numerous protected natural areas were created. Until recent times, the aim was not focus on the preservation of vulnerable species or threatened genetic resources but on landscapes of tourist attraction or species of social interest. As an example, conservation of *Nothofagus alpina*, appreciated for its wood quality, was the main aim for the creation of Lanín National Park in 1937. This park covers almost the whole natural range of the species in Argentina (Sabatier et al. 2011). In addition, other six National Parks were subsequently created with the aim of protecting the Subantarctic forests: "Los Arrayanes," "Lago Puelo," "Los Alerces," "Perito Moreno," "Los Glaciares," and "Tierra del Fuego." All these National Parks cover the more humid area of the Andean Patagonian forests, where lush vegetation grows, leaving aside the ecotone zones and the steppe. Paradoxically, these left out areas harbor genetic hotspots of some of the most emblematic species of Patagonia such as *A. chilensis* (Pastorino and Gallo 2002; Arana et al. 2010), *A. araucana* (Gallo et al. 2004; Marchelli et al. 2010), and *Fitzroya cupressoides* (Premoli et al. 2000a).

However, forest conservation should not be restricted to preserving natural areas but to also promoting its sustainable management. The traditional exploitation implemented during the first half of the twentieth century consisted in a high grading sivicultural management, i.e., selection and cutting of all mature good-quality trees from a forestry point of view. This method changed to modern conceptions where promoting recruitment of new individuals through natural regeneration is of primary concern. Then, for example, the shelterwood system consists of successive regeneration cuttings (preparatory cuts in immature stands and seed cuts in mature stands) which retain a forest cover of around 30–40%, until completing the regeneration phase (ca. 20 years) (Chauchard et al. 2012). This management system is being used for the *Nothofagus* mixed forest within Lanín National Park (Sola et al. 2020). More recently, other methods also stimulating natural regeneration by opening the canopy are being implemented. Variable retention, for example, has been designed to conserve the original biodiversity and to maintain some heterogeneity in old-growth forests and is used for management of *N. pumilio* forests in Tierra del Fuego island (Martínez Pastur et al. 2011). Nowadays, forestry in the Andean-Patagonian forests focuses on the provinces of Tierra del Fuego and Chubut, where the main species under use is *N. pumilio*. At a lower scale, silvicultural management of *A. chilensis* in being done, as well as of *N. alpina* and *N. obliqua* in Lanín National Park as mentioned before.

In spite of the protection of large areas of the Subantarctic forests under national or provincial parks, two main menaces attempt against its preservation. Firstly, forest fires, mainly of anthropic origin that usually light in the outskirts of the main cities. Between 1993 and 1999, forest fires affected about 57,900 ha of native forests in the Provinces of Río Negro and Chubut. In addition, 44,257 ha of native forests burnt in 2015 in the Provinces of Neuquén, Río Negro, and Chubut, from which 27,101 ha belonged to a single forest fire at Las Horquetas – Cholila (Mohr-Bell 2015). Climate change predictions foreseen an increase in the frequency of fires (Veblen et al. 2011) causing concern for the future of these forests.

The other menace against the conservation of the Andean-Patagonian forests is the progress of urbanizations. In a region with a population growth rate doubling that of the rest of the country (INDEC 2012), there is genuine need to have land for housing construction but also a pressure of the real estate speculation. Most Andean cities are surrounded by forests; therefore, their growth necessarily advances toward them. In this regard, real estate interests take steps toward more flexible regulations for the conversion of forests into urban land. Then, causal relationships are suspected between this real estate pressure and some of the fires that occurred. This type of pressure is also observed in the periodic reviews of the territory planning of native forests that must be carried out in accordance with National Law 26,331 on Environmental Protection of Native Forests. Thus, the forests initially categorized as having a high or medium conservation value are reason for strong disputes on the part of those who seek to lower them to the category of "low conservation value," making it possible to transform them into urban lands. Finding the balance between the conservation of these peri-urban forests and the progress of the urbanizations is a great challenge.

In addition to the abovementioned real estate pressure, the original folks also claim for their ancestral rights, demanding the restitution of the lands that they occupied before the colonization. These are communities of the Mapuche-Tehuelche Nation, militarily subdued at the end of the nineteenth century, after 300 years of resistance. Although largely they have been integrated into the general society, they are also organized in community property reserves, which are mostly located in steppe or ecotone areas, where they practice a family livestock production economy. However, in some cases, its relationship with the natural forest is closer. Such is the case of several communities located in the *A. araucana* forests, which include in their economy the collection of the seeds (pine nuts). This non-timber forest product is consumed by the communities, used as fodder for their livestock, and even integrated into a regional commercial circuit (see Chap. 7A).

A remarkable case of the relationship between the forest and the native people is that of the Curruhuinca community, located within the Lanín National Park, near San Martín de los Andes. Being the original owners of the land, they were the first community (in 1888) to get the permission of the Argentine State to occupy and use part of what had been theirs (Curruhuinca and Roux 1993). Later, with the creation of the National Park in 1937, this concession came into conflict. Finally, in 1989, the Argentinean State recognized by National Law 23.750/89, the Community ownership of 10,500 ha within the Lanín National Park. Years later, promoted by an institutional crisis in the National Parks Administration (NPA) and a change of authorities in the national government, the Mapuche Neuquén Confederation proposed to the NPA the realization of a workshop called "protected indigenous territory" in order to discuss legislation, territoriality, and natural resources management. This workshop, held in May 2000, consolidated the co-management plan between the Curruhuinca community and the National Parks Administration. It is understood as the distribution of responsibilities and competences, as well as the definition of public authority and guidelines for the use and management of resources arising from the ancestral knowledge of the communities (Carpinetti 2007). This co-management plan constitutes a paradigm shift in the management of the Lanín National Park and, in short, the practical implementation of the Convention on Biological Diversity ratified by the Argentine State. In addition, it led to the creation

Fig. 2.4 Livestock in the native forests. (Photos by L. Gallo)

of the Committee of Intercultural Management in the Lanín National Park. This Committee is an essential tool to develop the new relationship that must be based on the mutual recognition of the rights of the parties (Carpinetti 2007). In these 20 years, the social and political experience has been strengthened. The internal organization of the communities and the implementation of intercultural environmental education programs have been promoted (M. González Peñalba, personal communication). However, tensions still exist with opposing views to the relative weight that is given to the conservation of natural resources and their use, mainly with regard to productive practices such as extensive livestock farming (Fig. 2.4). Unmanaged livestock is highly degrading for the natural regeneration of the forest in the National Park (Monjeau et al. 2007). In a context of increasing poverty, the population growth leads to a greater use of the forest for wood and/or firewood. Although these aspects still need to be improved to achieve a better state of forest conservation, the continuity of the co-management plan over 20 years and the paradigm shift that still means today are remarkable.

2.3 The Main Tree Species of the Andean-Patagonian Forests of Argentina

The Andean-Patagonian forests of Argentina are characterized by the predominance of angiosperms, accompanied by some long-lived gymnosperms. *Araucaria araucana*, which inhabits at northern latitudes, is perhaps the most iconic of them (see Chap. 7A). There are also three monotypic genus of the Cupressaceae family: *Austrocedrus chilensis*, *Pilgerodendron uviferum*, and *Fitzroya cupressoides*. *Austrocedrus chilensis* is the most abundant and widespread native conifer in Argentina and is fully described in Chap. 6.

Pilgerodendron uviferum is restricted to a few small areas in Argentina, usually peat bogs (Lara et al. 2006), but along a wide latitudinal range that extends over 1600 km (41° 00′ 03″ to 50° 32′ 51″ S) and at elevations from 250 to 1000 m asl

(Rovere et al. 2002). It usually occurs in wet, poorly drained sites, with little slope (Allnutt et al. 2003). Along its wide latitudinal range, it is found in association with *N. pumilio* at southern latitudes and with *F. cupressoides* and/or *A. chilensis* in the north but always as small and fragmented populations (Rovere et al. 2002). A study of genetic variation using isozymes revealed a reduced within-population variation with a very low expected heterozygosity at the species level (He = 0.035) and a differentiation between populations measured by an F_{ST} of 0.16 (Premoli et al. 2001). Similar results were obtained when using RAPD markers (Allnutt et al. 2003). The genetic differentiation was larger, and the diversity lower than that registered for the other native Cupressaceae: He = 0.07 and F_{ST} = 0.078 for *F. cupressoides* (Premoli et al. 2000b) and He = 0.14 and F_{ST} = 0.066 for *A. chilensis* (Pastorino and Gallo 2002). This genetic structure of *P. uviferum* would be related with its wide geographic distribution and the isolation among populations (Premoli et al. 2001).

Fitzroya cupressoides occurs in Argentina as disjunct populations in humid habitats between 41° S and 42° 43′ S, while in Chile, it grows as discontinuous populations along the Coastal Mountains, the Central Depression, and the Andes Mountains between 39° 50′ and 42° 45′ (Veblen et al. 1995a). It is usually found in nutrient-poor soils, at places where annual precipitations range between 2000 and 4000 mm, and at elevations between 100 and 1200 m asl (Veblen et al. 1995a). This conifer is one of the longest-lived tree species of the world, with a dendrochronological record extending back 3620 years (Lara and Villalba 1993). Because of its valuable wood, it was heavily logged, particularly in Chile, where most populations from accessible areas were decimated. In addition to the former exploitation, the change of land use from 1999 to 2011 caused the loss of 46% of potential habitats for *F. cupressoides* in Chile (Rodriguez et al. 2015). Since 1973, it is listed in CITES (Convention on International Trade in Endangered Species of Wild Fauna and Flora; 1984) which prohibits its commercialization and international trade. In Argentina, over 80% of its forests occur within protected areas (Kitzberger et al. 2000), so its logging never had the scale reached in Chile. Seedling establishment is scarce or nil after intense logging (clear-cutting), but abundant regeneration is found in exploited areas if remnant trees were left nearby and grazing was not intense (Donoso et al. 1993). Gap-phased regeneration in *F. cupressoides* varies according to site conditions and type of forests and depends mainly on light availability and the existence of a substrate free of debris (Donoso et al. 2004). Field trial experiences in Chile showed a survival over 70% with mean annual increment (MAI) in height of 20–30 cm in an arboretum of Valdivia (Gerding and Rivas 2006), while site conditions were determinant for a better establishment: western slopes, lowlands, and shadow protection (Donoso et al. 2000).

Genetic variation among populations of *F. cupressoides* was studied both with RAPD markers (Allnutt et al. 1999) and isozymes (Premoli et al. 2000a). A significant degree of variation within populations was observed, although lower than other conifers around the world (e.g., for isozymes, He = 0.077 for *F. cupressoides* vs. 0.155 for other conifers; Premoli et al. 2000b). Spatial analysis of the genetic variation showed three main clusters within Chilean populations (northern Coastal Range, southern Coastal Range, and Central Depression), while Argentinean populations were significantly different (Allnutt et al. 1999). The populations from eastern slopes of the Andes (Argentinean populations) were more variable than those of the west and suggested an origin from different glacial refugia (Premoli et al. 2000b).

Another group of tree conifers inhabiting these forests belong to the Podocarpaceae family: *Podocarpus nubigenus*, *Saxegothaea conspicua*, and *Prumnopitys andina*. All these species are more abundant in Chile and occur in very humid sites in Argentina, near the border. *Podocarpus nubigenus* has a widespread latitudinal range (between 39° 50′ and 50° 23′ S) but occurs in small and isolated populations with low density, under climatic conditions of high precipitation. While in Chile it grows at the Coastal Range, the Andes, and the Central Valley, in Argentina, it is restricted to the region of the Valdivian rainforests (Donoso et al. 2006b). It usually forms mixed forests with other conifers like *S. conspicua*, *F. cupressoides*, and *P. uviferum*, as well as some broadleaf species like *Nothofagus dombeyi*, *Drimys winteri*, or *Embothrium coccineum*. Genetic diversity was studied in this species through isozyme markers (Quiroga and Premoli 2010). A moderate genetic diversity was detected within populations ($H_T = 0.275$) and a significant genetic differentiation ($F_{ST} = 0.18$). Cluster analysis showed latitudinal divergence at around 43° S associated to a paleobasin that would have impeded gene flow among northern and southern populations (Quiroga and Premoli 2010). *Saxegothaea conspicua* is found between 36° and 49° S mainly in Chile, at low altitudes on poorly drained marine deposits in the Coastal and Andes Mountains and also at mid to high altitudes (400–950 m asl) on shallow soils (Veblen et al. 1981). *Prumnopitys andina* has a restricted distribution along the Andes Range between 35° 52′ and 39° 30′, with a high degree of fragmentation and degradation (Hechenleitner et al. 2005). A genetic analysis of few populations using AFLP markers revealed low levels of diversity ($He = 0.011$) and very high differentiation ($F_{ST} = 0.608$) (Hernández et al. 2012 cit. in González 2019).

Among the angiosperms, *Nothofagus* is the highly dominant genus conforming about 80% of the Andean-Patagonian forests. This genus has a disjunct distribution with more than 40 species in the world occurring in New Zealand, Australia, New Guinea, New Caledonia, and South America (Wardle 1984). In South American temperate forests, we can find ten species and three recognized hybrids, with six of these species occurring in Argentina *Nothofagus obliqua*, *N. alpina* (= *N. nervosa*), *N. pumilio*, *N. antarctica*, *N. dombeyi*, and *N. betuloides*. The most relevant of these *Nothofagus* species are described in detail in the specific chapters of this section of the book (Chaps. 3, 4 and 5).

Some small trees of the Proteaceae family can also be mentioned: *Lomatia hirsuta*, remarkable due to its fine wood used in carpentry and joinery work; *Gevuina avellana*, cultivated (not in Argentina) for the production of its edible fruits; and *Embothrium coccineum* (Fig. 2.5), whose abundant and colorful flowering causes it to be frequently planted with ornamental purposes. *Gevuina avellana* grows mainly on the western slopes of the Andes between 35° and 44° S, from sea level till 700 m asl, as an understory species usually associated with *N. obliqua* and *N. alpina* (Donoso 1978). Molecular markers (EST-SSRs) were developed and used to

Fig. 2.5 Some of the emblematic species of the Andean Patagonian forests that are not included in domestication programs. (**a**) *Luma apiculata* with its particular bright and cinnamon bark. (**b**) The showy flower of *Embothrium coccineum*. (**c**) Tall and impressive *Nothofagus dombeyi* individuals. (Photos **a** and **c** by M. Serati; **b** by M. Pastorino)

characterize Chilean populations of this species (Díaz 2010). A higher genetic diversity was found in northern populations, with a moderate differentiation between them ($F_{ST} = 0.118$; Díaz 2010). Similarly, chloroplast DNA markers revealed higher diversity in the north range of the species (Ferrada 2011). The range of *Embothrium coccineum* covers a huge latitudinal gradient of about 2200 km, from sea level to 1200 m asl (Tortorelli 1956). It is a little tree or shrub that occurs as naturally fragmented populations in ecologically distinct habitats (Vidal-Russell et al. 2011), from warm- to cold-temperate rainforest, or in the ecotone with the steppe. A low genetic diversity (He = 0.170) and moderate differentiation between populations (F_{ST} = 0.202) were detected through isozyme markers (Souto and Premoli 2007), while a phylogeographic study with chloroplast DNA markers suggests the existence of multiple glacial refugia for this species (Vidal-Russell et al. 2011). *Maytenus boaria* is also noteworthy, particularly in the ecotone with the steppe, being a source of shade and fodder for livestock (its evergreen leaves are palatable). It can reach 20 m in height, and its wood is usually used as firewood (Santos Biloni 1990).

Along Patagonian rivers occurs the only native willow species, *Salix humboldtiana*, a species that was decimated in Patagonia and just few remnant populations and isolated individuals exist. A domestication program is starting, and details are described in Chap. 7B. Another well-known tree species of the Andean Patagonian forest is *Luma apiculata* (Fig. 2.5), from the Myrtaceae family. It grows in riparian areas, and its ornamental value due to its bright and cinnamon bark, notable among the usual colors of the forest, is highly appreciated. A pure old-growth forest of *L. apiculata*, unusually large, was the main reason for the creation of Los Arrayanes National Park, near San Carlos de Bariloche, which is a site of particular interest for tourism. Seedlings of this species can be bought in almost every commercial nursery of northwest Argentine Patagonia. Genetic diversity in some Argentinean populations showed a low diversity (He = 0.129) with evidences of inbreeding (Caldiz and Premoli 2005).

2.4   Shared Patterns in the Distribution of the Genetic Variation

Many of the abovementioned species grow along the environmental gradients that characterized the Patagonian region, and their genetic variation is, in some cases, associated with these gradients (e.g., Pastorino and Gallo 2002; Soliani et al. 2020). In addition, all of them were affected by the Neogene climatic changes that generated shifts in their distributions. The retractions and expansions of the forests after these disturbs are still imprinted in the genetic structure of the populations. Moreover, different genetic studies have analyzed the distribution of the genetic variation and shared patterns between distant taxa became evident (see review in Sérsic et al. 2011).

The first studies used biochemical markers (i.e., isozymes) to find the more diverse populations and infer cryptic refugia. For example in *Austrocedrus chilensis* (Gallo and Geburek 1994; Pastorino and Gallo 2002; Pastorino et al. 2004), *Nothofagus obliqua* (Azpilicueta and Gallo 2009), *N. alpina* (Marchelli and Gallo 2001, 2004), *N. antarctica* (Pastorino et al. 2009), *Fitzroya cupressoides* (Premoli et al. 2000b), *Pilgerodendron uviferum* (Premoli et al. 2002), *Podocarpus nubigenus* (Quiroga and Premoli 2010), and *Embothrium coccineum* (Souto and Premoli 2007). The development of new and more specific molecular markers, able to trace recent demographic histories of populations (e.g., chloroplast and mitochondrial DNA polymorphisms), gave rise to phylogeographic studies. Then, different species were studied, for example, several *Nothofagus* spp. (e.g., Marchelli et al. 1998; Marchelli and Gallo 2006; Millerón et al. 2008; Azpilicueta et al. 2009; Acosta and Premoli 2010; Mathiasen and Premoli 2010; Soliani et al. 2012; Acosta et al. 2014), *Embothrium coccineum* (Vidal-Russell et al. 2011) and *Araucaria araucana* (Marchelli et al. 2010). Most of these studies agreed in the persistence of the different species in multiple and scattered refugia. A joint analysis of animal and different plant taxa suggested the existence of "lowland," "peripheral," and "valley" refugia, located at six stable areas: east of the Patagonian Andes, northern Chiloé Island, northern Chilean Coast, Atlantic Coastline, northern Patagonia, and South of Santa Cruz Province and Tierra del Fuego (Sérsic et al. 2011). An interesting shared pattern is the inferred potential lowland refugia at high latitudes (Tierra del Fuego; 54° S), both in the Andes and the Steppe (Jakob et al. 2009; Tremetsberger et al. 2009; Cosacov et al. 2010; Mathiasen and Premoli 2010; Soliani et al. 2012). The ability of populations to endure glacial times and persist in situ at these high latitudes would have been facilitated by the particular glaciation setting of the Andes which was notably different than the Northern Hemisphere (Markgraf et al. 1995), as described in the first section of this chapter.

A general trend observed in many tree species of the Subantarctic forests is a latitudinal divergence of their genetic variation. Some phylogeographical breaks were detected along the Patagonian Andes at 35° S, 37.5° S, 40.5° S, 43° S, and 50° S (reviewed by Sérsic et al. 2011). For example two *Nothofagus* species (*N. alpina* and *N. obliqua*) showed a genetic discontinuity at around 40° S (Marchelli et al. 1998; Azpilicueta et al. 2009) probably associated with the recent volcanic activity of the Villarrica-Lanín volcanic chain (ca. 2700 years BP; Lara et al. 2004) that might have avoided contact between northern and southern colonizing routes (Millerón et al. 2008). The widely distributed *N. pumilio* and *N. antarctica* are characterized by a deep divergence in maternal lineages at around 43° S with two disjunct groups having different evolutionary histories (Mathiasen and Premoli 2010; Soliani et al. 2012). Also among some conifers, latitudinal trends in the genetic diversity were detected (e.g., *A. chilensis*, (Pastorino et al. 2004), *A. araucana* (Bekessy et al. 2002)). It was proposed that the Andean breaks would be more ancient (associated with Pre-Quaternary processes like Andean orogeny and paleobasins) than the stepparian breaks (related with Quaternary events like glaciations) (Sérsic et al. 2011).

Another shared genetic pattern is the higher diversity at eastern and more isolated populations from the ecotone between the forests and the steppe. Eastern populations of, for example, *Austrocedrus chilensis* (Pastorino and Gallo 2002; Arana et al. 2010), *Fitzroya cupressoides* (Premoli et al. 2000a), *Nothofagus obliqua* (Azpilicueta et al. 2013; Soliani et al. 2020), and *Araucaria araucana* (Gallo et al. 2004; Marchelli et al. 2010) were found to be more genetically variable than larger and continuous populations from western locations. Divergent gene pools in the steppe together with high among population differentiation and high number of private alleles in some populations of *A. chilensis* suggested they would be ancient refugia from pre-Holocene times (Arana et al. 2010).

Concerning the genetic variation in adaptive traits, several species of the Subantarctic forests have been studied either by traditional provenance and progeny field tests or under controlled conditions in the nursery. The use of common garden experiments, where different genetic entities (provenances or progenies) are evaluated in the same environment, allows inferring the genetic bases of the observed phenotypic variation. These trials can be temporary installed in the nursery (inside or outside the greenhouse) for evaluating early traits under more or less controlled conditions, or permanently in the field for long-term evaluation, in this case usually as networks across environmental gradients. For example, early growth traits were evaluated in 24 provenances of *N. pumilio* (Mondino 2014, see Chap. 5) to estimate geographic and genetic variation, in 116 families from eight provenances of *N. obliqua* (Barbero 2014; see Chap. 4), and in 177 families from 10 provenances of *Austrocedrus chilensis* (Aparicio et al. 2010; Pastorino et al. 2014; see Chap. 6). Likewise, phenological traits were analyzed in 65 open pollinated families from 8 provenances of *N. alpina* (Duboscq-Carra et al. 2020) and in 12 provenances of *N. pumilio* (Torres et al. 2017).

Field trials are especially important for long-lived, slow-growing species like forest trees, because they allow evaluating later life history traits, which might show variation not evident in younger plants (O'Brien et al. 2007). Morphological traits like growth, branching, tree form, as well as phenological and reproductive traits are relevant both for breeding programs and for studies of adaptation and can only be measured in long-establish trials. These trials demand a huge effort for establishing and maintenance, and, currently, only few species from the Patagonian forests are being evaluated in this way. The oldest progeny test of a forest tree species of the Andean-Patagonian forest in Argentina was planted for *Nothofagus alpina* in 1997, at Trevelin Forest Station of INTA (Instituto Nacional de Tecnología Agropecuaria), much further south of its natural distribution. Since then, several trials (both provenance and progeny) were installed (mainly in INTA experimental stations) for the species with the highest domestication potential, with the aims of analyzing their genetic variation and evaluating their performance for breeding programs. Thus, around 20-year-old field trials are already available and under evaluation for *N. alpina*, *N. obliqua*, *A. chilensis*, and *A. araucana*, and more recently, long-term trials for *N. pumilio* and *N. antarctica* were also installed.

Several studies attempted to evaluate adaptive responses of populations to ecological gradients, like altitude, where the most important environmental variable is the temperature (e.g., (Premoli 2003; Premoli and Brewer 2007; Arana et al. 2016), and longitude, where the precipitation changes drastically (e.g., Pastorino et al. 2012; Soliani et al. 2020). In those studies, natural populations along the particular gradient are evaluated aiming to find local adaptations to their home environments. Using reciprocal transplants and/or common garden trials, morphological, ecophysiological, and phenological traits were measured to elucidate genetic vs. plasticity responses.

2.5   The Choice of Species for Breeding

So far, no native species have been planted for commercial purposes in the Patagonian region. However, for about 20 years, there has been increasing interest in planting for ecological restoration purposes, usually in response to the recurrent and catastrophic forest fires. Active restorations were carried out after several forests fires (Table 2.1; Fig. 2.6).

In any case, commercial afforestation with native species represent a certain potential. Although there are only few recent planting experiences, and technological adjustment is still far from complete, there are many incentives. The first is the preservation of the original landscapes and ecosystems, which are the sustainment of the tourist activity. In addition, the quality of the native species' wood is superior to that of the fast-growing exotic species grown in the region, essentially *Pinus ponderosa* and *Pseudotsuga menziesii*. This difference in the quality of wood is reflected in market prices (Table 2.2), where all the native woods of the region have values that double and even quadruple that of the wood of the most widely planted exotic species (*P. ponderosa*). Although native wood so far comes from natural forest, this difference is expected to become more acute in the future, maintaining the relative values even when native wood comes from cultivated forests.

On the other hand, the technology for production of forest seedlings of native species has been adjusted very efficiently in the last two decades, developing effective fertigation protocols. In particular, *Nothofagus* species respond very well to fertigation, allowing their growth and development to be manipulated, based on the same protocol with slight variants (Schinelli Casares 2013).

The advances and characteristics mentioned led to the initiation, in 2006, of a formal domestication program of native species of the Patagonian region as a part of the National Project developed by INTA (see Chap. 1). Initially the species *N. alpina* and *N. obliqua* were chosen for their excellent wood and growth rate. Then, *A. chilensis* was included for its wood quality and potential adaptation to drier environments, despite its lower growth rate. More recently, *N. pumilio* and *N. antarctica*, also of relatively rapid growth, were included. *Nothofagus pumilio* has a very good forest shape and, being a cold-tolerant species, allows its cultivation in cold places where other species could not be planted. The cultivation of *N. antarctica*, on the other hand, is specifically related to the management of the extensive silvopastoral systems most widely implemented in Patagonia (Peri et al. 2016).

This section of the book also includes two species for which genetic improvement programs are not being developed, although its domestication is in due course (Chaps. 7A and 7B). The first of these is the iconic *A. araucana*. Its growth rate is extremely slow, making it hardly possible to grow it for economic purposes. However, it is the species with the highest cultural value in Patagonia, to which the Mapuche Nation even attributes spiritual value. As mentioned, the collection of "pine nuts" in the natural forest is of great relevance in the economy of the related Mapuche communities. In turn, the restoration of degraded areas requires its

Table 2.1 Restauration experiences after forests fires occurred in Patagonia

Location	Province	Species used for restoration	Area covered (ha)	Year of the fire	References
Catedral Mountain (ski center)	Río Negro	*A. chilensis*	~10	1996	Oudkerk et al. (2003)
Loma del Medio	Río Negro	*A. chilensis*	~30	1999	Perdomo et al. (2009)
La Colisión	Chubut	*A. chilensis, N. pumilio, Maytenus boaria*	~7	2008	Urretavizcaya et al. (2011)
Lake Tromen	Neuquén	*A. araucana*	~3	2009	https://www.lmneuquen.com/reponen-araucarias-el-tromen-el-incendio-n30192
Cholila	Chubut	*A. chilensis, N. pumilio, N. dombeyi*	~90	2015	https://es.wikipedia.org/wiki/Incendios_forestales_en_Chubut_de_2015
Paso Bebán	Tierra del Fuego	*N. pumilio, N. betuloides*	~200	2012	Mestre et al. (2018)
Otto Mountain	Río Negro	*A. chilensis, N. pumilio, N. antarctica, N. dombeyi, M. boaria*	20	1995 and 2013	Pastorino et al. (2018)

Fig. 2.6 Examples of postfire restauration activities. (**a**) Planting of *Araucaria araucana* seedlings in 2009 and 2010, at Lake Tromen. (**b** and **c**) Restauration at Otto Mountain (Bariloche) in 2018 and 2019, with seedlings of *Austrocedrus chilensis* (**b**) and *Nothofagus dombeyi* (**c**). (**d**) *Austrocedrus chilensis* young tree of 15 years of age planted 3 years after a fire in Catedral Mountain (Bariloche). (Photos: **a** taken from Río Negro newspaper, 5-06-2009; **b–d** by M. Pastorino)

domestication. Despite the fact that its seeds are recalcitrant, the production of plants in the nursery is simple, and its survival in the field plantation is high.

The other species is *Salix humboldtiana*, which although it has a continental distribution, and in Argentina it crosses various ecosystems, it is in the marginal populations of the Patagonian steppe that its conservation acquires a greater value. To avoid the extinction of these populations, an active restoration strategy becomes necessary, that is, to domesticate the species. Its vegetative propagation by rooting of cuttings has been shown to be not only feasible but very simple.

Table 2.2 Absolute and relative prices (in US dollars) per board foot of native and exotic woods produced in Argentinean Patagonia

Species	Price U$S/bd.ft	Relative price
Pinus ponderosa	0.84	1
Pseudotsuga menziesii	1.23	1.47
Austrocedrus chilensis	1.44	1.72
Nothofagus dombeyi	1.73	2.06
Nothofagus obliqua	1.95	2.32
Nothofagus alpina	3.25	3.87
Nothofagus pumilio	3.47	4.13

From INTA-MinAgro (2018)

In the different chapters of this section of the book, the degree of progress of the domestication program for each of the mentioned species will be presented, as well as general knowledge about their genetic variation.

References

Acosta MC, Premoli AC (2010) Evidence of chloroplast capture in South American *Nothofagus* (subgenus Nothofagus, Nothofagaceae). Mol Phylogenet Evol 54:235–242

Acosta MC, Mathiasen P, Premoli AC (2014) Retracing the evolutionary history of *Nothofagus* in its geo-climatic context: new developments in the emerging field of phylogeology. Geobiology 12:497–510

Allnutt TR, Newton AC, Lara A, Premoli AC, Armesto JJ, Vergara S, Gardner M (1999) Genetic variation in *Fitzroya cupressoides* (alerce), a threatened South American conifer. Mol Ecol 8:975–987

Allnutt TR, Newton AC, Premoli A, Lara A (2003) Genetic variation in the threatened South American conifer *Pilgerodendron uviferum* (Cupressaceae), detected using RAPD markers. Biol Conserv 114:245–253

Aparicio AG, Pastorino MJ, Gallo LA (2010) Genetic variation of early height growth traits at the xeric limits of *Austrocedrus chilensis* (Cupressaceae). Austral Ecol 35:825–836

Arana MV, Gallo LA, Vendramin GG, Pastorino MJ, Sebastiani F, Marchelli P (2010) High genetic variation in marginal fragmented populations at extreme climatic conditions of the Patagonian Cypress *Austrocedrus chilensis*. Mol Phylogenet Evol 54:941–949

Arana MV, Gonzalez-Polo M, Martinez-Meier A, Gallo LA, Benech-Arnold RL, Sánchez RA, Batlla D (2016) Seed dormancy responses to temperature relate to *Nothofagus* species distribution and determine temporal patterns of germination across altitudes in Patagonia. New Phytol 209:507–520

Armesto JJ, León-Lobos P, Kalin Arroyo M (1995) Los bosques templados del sur de Chile y Argentina: una isla biogoegráfica. In: Armesto JJ, Villagran C, Kalin Arroyo M (eds) Ecología de los bosques nativos de Chile. Editorial Universitaria, Santiago de Chile, pp 23–28

Arroyo KM, Cavieres L, Peñaloza A, Riveros M, Faggi AM (1995) Relaciones fitogeográficas y patrones regionales de riqueza de especies en la flora del bosque lluvioso templado de sudamérica. In: Armesto JJ, Villagran C, Kalin Arroyo M (eds) Ecología de los bosques nativos de Chile. Editorial Universitaria, Santiago de Chile, pp 71–100

Azpilicueta MM, Gallo LA (2009) Shaping forces modelling genetic variation patterns in the naturally fragmented forests of a South-American Beech. Biochem Syst Ecol 37:290–297

Azpilicueta MM, Marchelli P, Gallo LA (2009) The effects of Quaternary glaciations in Patagonia as evidenced by chloroplast DNA phylogeography of southern beech *Nothofagus obliqua*. Tree Genet Genomes 5:561–571

Azpilicueta MM, Gallo LA, van Zonneveld M, Thomas E, Moreno C, Marchelli P (2013) Management of *Nothofagus* genetic resources: definition of genetic zones based on a combination of nuclear and chloroplast marker data. For Ecol Manag 302:414–424

Barbero F (2014) Variación genética de poblaciones naturales argentinas de *Nothofagus obliqua* ("Roble Pellín") en caracteres adaptativos tempranos relevantes para domesticación. Tesis Doctoral-Facultad de Agronomía – Universidad de Buenos Aires

Bekessy SA, Allnutt TR, Premoli AC, Lara A, Ennos RA, Burgman MA et al (2002) Genetic variation in the monkey puzzle tree (*Araucaria araucana* (Molina) K.Koch), detected using RAPDs. Heredity 88:243–249

Burns BR (1993) Fire-induced dynamics of *Araucaria araucana-Nothofagus antarctica* forest in the Southern Andes. J Biogeogr 20:669

Bustos C (2001) Heladas en el sector precordillerano nordpatagónico. Serie Técnica INTA N° 22

Caldiz MS, Premoli AC (2005) Isozyme diversity in large and isolated populations of *Luma apiculata* (Myrtaceae) in North-Western Patagonia, Argentina. Aust J Bot 53:781

Carpinetti B (2007) Una experiencia intercultural de co-manejo entre el Estado y las Comunidades Mapuches en el Parque Nacional Lanin, Argentina. Programa FAO/OAPN Fortalecimiento del manejo sostenible de los recursos naturales en las áreas protegidas de América Latina, 26 pp

Chauchard L, Bava JO, Castañeda S, Laclau P, Loguercio GA, Pantaenius P, Rusch VE (2012) Manual para las buenas prácticas forestales e bosques nativos de Nordpatagonia. Ministerio de Agricultura, Ganadería y Pesca, Presidencia de la Nación Argentina

Cosacov A, Sérsic AN, Sosa V, Johnson LA, Cocucci AA (2010) Multiple periglacial refugia in the Patagonian steppe and post-glacial colonization of the Andes: the phylogeography of *Calceolaria polyrhiza*. J Biogeogr 37:1463–1477

Curruhuinca C, Roux L (1993) Las Matanzas del Neuquén. Crónicas Mapuches. Ed. Plus Ultra, Buenos Aires

De Frenne P, Graae BJ, Rodríguez-Sánchez F, Kolb A, Chabrerie O, Decocq G et al (2013) Latitudinal gradients as natural laboratories to infer species' responses to temperature. J Ecol 101:784–795

Dettmann ME, Pocknall DT, Romero EJ, Zamaloa MC (1990) Nothofagidites Erdtman ex Potoni, 1960: a catalogue of species with notes on the paleogeographic distribution of *Nothofagus* Bl. (Southern Beech). N Z Geol Surv Paleontol Bull 60: 1–79

Di Castri F, Hajek ER (1976) Bioclimatología de Chile. Vicerrectoría Académica. Universidad Católica de Chile, Santiago

Díaz L (2010) Determinación de la diversidad y estructura genética en poblaciones de avellano *Gevuina avellana* Mol. mediante el uso de marcadores moleculares tipo microsatelites de secuencias expresadas (EST- SSR). Tesis de Magister, Facultad de Ciencias Agrarias, Universidad Austral de Chile, 113 pp

Donoso C (1978) Avance de investigación: antecedentes sobre producción de avellanas. Bosque 2:105–108

Donoso C (1992) Ecología Forestal, El Bosque y su Medio Ambiente, vol 368. Editorial Universitaria

Donoso C (1993) Bosques templados de Chile y Argentina. Variación, Estructura y Dinámica. Ecología Forestal. Editorial Universitaria, Chile, 484 pp

Donoso C (2006) Las especies arbóreas de los bosques templados de Chile y Argentina. Autoecología. Marisa Cuneo Ediciones, Valdivia

Donoso C, Lara A (1995) Utilización de los bosques nativos en Chile: pasado, presente y futuro. In: Armesto JJ, Villagran C, Arroyo KM (eds) Ecología de los Bosques nativos de Chile. Editorial Universitaria, Santiago, pp 363–388

Donoso C, Sandoval V, Grez R, Rodriguez J (1993) Dynamics of *Fitzroya cupressoides* forests in Southern Chile. J Veg Sci 4:303

Donoso C, Escobar B, Castro H, Zúñiga A, Gres R (2000) Sobrevivencia y crecimiento de alerce (*Fitzroya cupressoides* Mol. (Johnston)) en plantaciones experimentales en la Cordillera de la Costa de Valdivia. Bosque 21:13–24

Donoso C, Lara A, Escobar B, Premoli AC, Souto CP (2004) *Fitzroya cupressoides* (Molina) I.M. Johnst. Alerce, Lahuén, Lahuán Familia: Cupressaceae. In: Donoso C, Premoli AC, Gallo LA, Ipinza R (eds) Variación intraespecífica en especies arboreas de los bosques templados de Chile y Argentina. Editorial Universitaria, Santiago

Donoso C, Escobar B, Pastorino MJ, Gallo LA, Aguayo J (2006a) *Austrocedrus chilensis* (D.Don) Pic. Ser. et Bizzarri (Ciprés de la Cordillera, Len). In: Donoso C (ed) Las Especies Arbóreas de los Bosques Templados de Chile y Argentina, Autoecología, vol 678. Marisa Cuneo Ediciones, Valdivia

Donoso C, Utreras F, Zúñiga A (2006b) *Podocarpus nubigena* Lindl. In: Donoso Zegers C (ed) Las especies arbóreas del los bosques templados de Chile y Argentina. Marisa Cuneo, Valdivia, pp 92–101

Duboscq-Carra VG, Letourneau FJ, Pastorino MJ (2018) Looking at the forest from below: the role of seedling root traits in the adaptation to climate change of two *Nothofagus* species in Argentina. New For 49:613–635

Duboscq-Carra VG, Arias-Rios JA, El Mujtar VA, Marchelli P, Pastorino MJ (2020) Differentiation in phenology among and within natural populations of a South American *Nothofagus* revealed by a two-year evaluation in a common garden trial. For Ecol Manag 460. https://doi.org/10.1016/j.foreco.2019.117858

Ferrada PV (2011) Determinación de polimorfismo e identificación de haplotipos en *Gevuina avellana* y *Embothrium coccineum* (Magnoliópsida: Proteaceae), especies nativas de Chile, utilizando marcadores moleculares en cpDNA. Tesis Facultad de Ciencias, Universidad Austral de Chile

Flint RF, Fidalgo F (1969) Glacial drift in the Eastern Argentine Andes between latitude 41° 10′ S and latitude 43° 10′S. Geol Soc Am Bull 80:1043–1052

Gallo LA, Geburek T (1994) A short note: genetics of isozyme variants in *Austrocedrus chilensis* (Don.) Florin et Boutelje. Phyton (B Aires) 34:103–107

Gallo L, Izquierdo F, Sanguinetti LJ (2004) *Araucaria araucana* forest genetic resources in Argentina. In: Vinceti B, Amaral W, Meilleur M (eds) Challenges in managing forest genetic resources for livelihoods: examples from Argentina and Brazil. IPGRI, Rome, pp 105–131

Garibotti IA, Villalba R (2017) Colonization of mid- and late-Holocene moraines by lichens and trees in the Magellanic sub-Antarctic province. Polar Biol 40:1739–1753

Gerding V, Rivas TY (2006) Desarrollo de plantaciones experimentales jóvenes de *Fitzroya cupressoides* establecidas en el arboreto de la Universidad Austral de Chile, Valdivia. Bosque 27:155–162

Glasser NF, Jansson KN, Harrison S, Kleman J (2008) The glacial geomorphology and Pleistocene history of South America between 38° S and 56° S. Quat Sci Rev 27:365–390

González J (2019) Diversidad genética neutral y adaptativa, una simple explicación. Rev Cienc e Investig For 25:81–98

Hadad MA, Arco Molina J, Roig Juñent FA, Amoroso MM, Müller G, Araneo D, Tardif JC (2019) Frost record in tree rings linked to atmospheric circulation in northern Patagonia. Palaeogeogr Palaeoclimatol Palaeoecol 524:201–211

Hechenleitner P, Gardner N, Thomas P, Echeverría C, Escobar B, Brownless P, Martínez C (2005) Plantas Amenazadas del Centro-Sur de Chile. Distribución, Conservación y Propagación. Universidad Austral de Chile y Real Jardín Botánico de Edimburgo, Valdivia y Canbridge

Henríquez WI, Villa-Martínez R, Vilanova I, De Pol-Holz R, Moreno PI (2017) The last glacial termination on the eastern flank of the central Patagonian Andes (47 ° S). Clim Past 13:879–895

Heusser CJ, Heusser LE, Lowell TV (1999) Paleoecology of the southern Chilean lake district-Isla Grande de Chilo, during middle-late Llanquihue glaciation and deglaciation. Geogr Ann 81 A:231–284

INDEC (2012) Censo Nacional de Población, Hogares y Viviendas 2010. Argentina

INTA-MinAgro (2018) Boletín N°4 de costos y precios forestales Patagonia Andina – mayo 2018. INTA, Subsecretaría de Desarrollo Foresto Industrial – Ministerio de Agroindustria de la Nación

Jakob SS, Martinez-Meyer E, Blattner FR (2009) Phylogeographic analyses and paleodistribution modeling indicate Pleistocene in situ survival of *Hordeum* species (Poaceae) in Southern Patagonia without genetic or spatial restriction. Mol Biol Evol 26:907–923

Jump AS, Mátyás C, Peñuelas J (2009) The altitude-for-latitude disparity in the range retractions of woody species. Trends Ecol Evol 24:694–701

Kitzberger T, Perez A, Iglesias G, Premoli AC, Veblen TT (2000) Distribución y estado de conservación del alerce (*Fitzroya cupressoides* (Mol.) Johnst.) en Argentina. Bosque 21:79–89

Lara A, Villalba R (1993) A 3620-year temperature record from *Fitzroya cupressoides* tree rings in southern South America. Science 260:1104–1106

Lara A, Rutherford P, Montory C, Bran D, Perez A, Clayton S et al (1999) Mapeo de la Ecorregión de los bosques Valdivianos. Fundación Vida Silvestre Argentina, Buenos Aires. Boletín Técnico la Fund Vida Silv Argentina 51:1–27

Lara LE, Naranjo JA, Moreno H (2004) Lanín volcano (39.5°S), Southern Andes: geology and morphostructural evolution. Rev Geol Chile 31:241–257

Lara A, Bannister J, Donoso C, Rovere A, Soto D, Escobar B, Premoli AC (2006) *Pilgerodendron uviferum* (D. Don) Florin. In: Donoso C (ed) Las especies arbóreas de los bosques templados de Chile y Argentina. Autoecología, pp 82–91

Marchelli P, Gallo LA (2001) Genetic diversity and differentiation in a southern beech subjected to introgressive hybridization. Heredity 87:284–293

Marchelli P, Gallo LA (2004) The combined role of glaciation and hybridization in shaping the distribution of genetic variation in a Patagonian southern beech. J Biogeogr 31:451–460

Marchelli P, Gallo L (2006) Multiple ice-age refugia in a southern beech from southern South America as revealed by chloroplast DNA markers. Conserv Genet 7:591–603

Marchelli P, Gallo L, Scholz F, Ziegenhagen B (1998) Chloroplast DNA markers revealed a geographical divide across Argentinean southern beech *Nothofagus nervosa* (Phil.) Dim. et Mil. distribution area. Theor Appl Genet 97:642–646

Marchelli P, Baier C, Mengel C, Ziegenhagen B, Gallo LA (2010) Biogeographic history of the threatened species *Araucaria araucana* (Molina) K. Koch and implications for conservation: a case study with organelle DNA markers. Conserv Genet 11:951–963

Markgraf V, McGlone M, Hope G (1995) Neogene paleoenvironmental and paleoclimatic change in southern temperate ecosystems – a southern perspective. Trends Ecol Evol 10:143–147

Markgraf V, Romero EJ, Villagran C (1996) History and paleoecology of South American *Nothofagus* forests. In: Veblen TT, Hill RS, Read J (eds) The ecology and biogeography of *Nothofagus* forests. Yale University Press, New Haven/London, pp 354–386

Martínez Pastur GJ, Cellini JM, Lencinas MV, Barrera M, Peri PL (2011) Environmental variables influencing regeneration of *Nothofagus pumilio* in a system with combined aggregated and dispersed retention. For Ecol Manag 261:178–186

Mathiasen P, Premoli AC (2010) Out in the cold: genetic variation of *Nothofagus pumilio* (Nothofagaceae) provides evidence for latitudinally distinct evolutionary histories in austral South America. Mol Ecol 19:371–385

Mermoz M, Kitzberger T, Veblen TT (2005) Landscape influences on occurrence and spread of wildfires in Patagonian forests and shrublands. Ecology 86:2705–2715

Mestre L, Fernández-Génova L, Turi L, Collado L (2018) "Soy Parte del Bosque Fueguino", cinco años de restauración participativa. Patagon For 23:42–45

Millerón M, Gallo L, Marchelli P (2008) The effect of volcanism on postglacial migration and seed dispersal. A case study in southern South America. Tree Genet Genomes 4:435–443

Mills AM (2008) Latitudinal gradients of biologically useful semi-darkness. Ecography (Cop) 31:578–582

Mohr-Bell FD (2015) Superficies afectadas por incendios en la región bosque Andino Patagónico (BAP) durante los veranos de 2013–2014 y 2014–2015. Patagon For 21:34–41

Mondino V (2014) Variación geográfica y genética en caracteres adaptativos iniciales de *Nothofagus pumilio* (Poepp. et Endl.) Krasser en una zona de alta heterogeneidad ambiental. Tesis de Doctorado. Universidad de Buenos Aires

Mondino VA, Pastorino MJ, Gallo LA (2019) Altitudinal variation of phenological characters and initial growth under controlled conditions among *Nothofagus pumilio* populations from center-west Chubut, Argentina. Bosque 40:87–94

Monjeau A, Nazar Anchorena S, Montoni V, Marquez J, Alcalde D, D'Iorio A et al (2007) Biodiversidad, amenazas a la conservación y prioridades de inversion. Parque Nacional Lanín. In: Monjeau A, Pauquet S (eds) Estado de conservación, amenazas y prioridades de inversión en áreas protegidas andino-patagónicas. Ediciones Atlántida, pp 17–46

Moreno PI, Simi E, Villa-Martínez RP, Vilanova I (2019) Early arboreal colonization, postglacial resilience of deciduous *Nothofagus* forests, and the Southern Westerly Wind influence in central-east Andean Patagonia. Quat Sci Rev 218:61–74

O'Brien EK, Mazanec RA, Krauss SL (2007) Provenance variation of ecologically important traits of forest trees: implications for restoration. J Appl Ecol 44:583–593

Oudkerk L, Pastorino MJ, Gallo LA (2003) Siete años de experiencia en la restauración postincendio de un bosque de Ciprés de la Cordillera. Patagon For 9:4–7

Pastorino MJ, Gallo LA (2002) Quaternary evolutionary history of *Austrocedrus chilensis*, a cypress native to the Andean-Patagonian forest. J Biogeogr 29:1167–1178

Pastorino MJ, Gallo LA, Hattemer HH (2004) Genetic variation in natural populations of *Austrocedrus chilensis*, a cypress of the Andean-Patagonian Forest. Biochem Syst Ecol 32:993–1008

Pastorino MJ, Marchelli P, Milleron M, Soliani C, Gallo LA (2009) The effect of different glaciation patterns over the current genetic structure of the southern beech *Nothofagus antarctica*. Genetica 136:79–88

Pastorino MJ, Ghirardi S, Grosfeld J, Gallo LA, Puntieri JG (2010) Genetic variation in architectural seedling traits of Patagonian cypress natural populations from the extremes of a precipitation range. Ann For Sci 67:508

Pastorino MJ, Aparicio AG, Marchelli P, Gallo LA (2012) Genetic variation in seedling water-use efficiency of Patagonian Cypress populations from contrasting precipitation regimes assessed through carbon isotope discrimination. For Syst 21:189–198

Pastorino MJ, Sá MS, Aparicio AG, Gallo LA (2014) Variability in seedling emergence traits of Patagonian Cypress marginal steppe populations. New For 45:119–129

Pastorino MJ, Aparicio AG, Azpilicueta MM, Rusch V (2018) Restauración del bosque quemado del C° Otto, Bariloche: un compromiso de hoy con las generaciones futuras. Presencia (INTA) 70:14–17

Perdomo M, Andenmatten E, Basil G, Letourneau F (2009) La gestión de la reserve forestal Loma del Medio – Río Azul (INTA-SFA). Presencia 54:23–27

Peri PL, Dube F, Costa Varella A (2016) Silvopastoral systems in Southern South America. Advances in agroforestry. Springer Series, 276 pp

Pollmann W, Veblen TT (2004) *Nothofagus* regeneration dynamics in South-Central Chile: a test of a general model. Ecol Monogr 74:615–634

Porter SC (1981) Pleistocene glaciation in the southern Lake District of Chile. Quat Res 16:263–292

Premoli AC (2003) Isozyme polymorphisms provide evidence of clinal variation with elevation in *Nothofagus pumilio*. J Hered 94:218–226

Premoli AC, Brewer CA (2007) Environmental v. genetically driven variation in ecophysiological traits of *Nothofagus pumilio* from contrasting elevations. Aust J Bot 55:585

Premoli AC, Kitzberger T, Veblen TT (2000a) Conservation genetics of the endangered conifer *Fitzroya cupressoides* in Chile and Argentina. Conserv Genet 1:57–66

Premoli AC, Kitzberger T, Veblen TT (2000b) Isozyme variation and recent biogeographical history of the long-lived conifer *Fitzroya cupressoides*. J Biogeogr 27:251–260

Premoli AC, Souto CP, Allnutt TR, Newton AC (2001) Effects of population disjunction on isozyme variation in the widespread *Pilgerodendron uviferum*. Heredity 87:337–343

Premoli AC, Souto CP, Rovere AE, Allnutt TR, Newton AC (2002) Patterns of isozyme variation as indicators of biogeographic history in *Pilgerodendron uviferum* (D.Don) Florin. Divers Distrib 8:57–66

Premoli AC, Raffaele E, Mathiasen P (2007) Morphological and phenological differences in *Nothofagus pumilio* from contrasting elevations: evidence from a common garden. Austral Ecol 32:515–523

Premoli AC, Mathiasen P, Kitzberger T (2010) Southern-most *Nothofagus* trees enduring ice ages: genetic evidence and ecological niche retrodiction reveal high latitude (54°S) glacial refugia. Palaeogeogr Palaeoclimatol Palaeoecol 298:247–256

Quiroga MP, Premoli AC (2010) Genetic structure of *Podocarpus nubigena* (Podocarpaceae) provides evidence of Quaternary and ancient historical events. Palaeogeogr Palaeoclimatol Palaeoecol 285:186–193

Rabassa J (2008) Late Cenozoic Glaciations in Patagonia and Tierra del Fuego. In: Rabassa J (ed) Developments in quaternary sciences. Elsevier, pp 151–204

Rabassa J, Clapperton CM (1990) Quaternary glaciations in the southern Andes. Quat Sci Rev 9:153–174

Rabassa J, Coronato AM, Salemme M (2005) Chronology of the Late Cenozoic Patagonian glaciations and their correlation with biostratigraphic units of the Pampean region (Argentina). J S Am Earth Sci 20:81–103

Rebertus AJ, Veblen TT (1993) Structure and tree-fall gap dynamics of old-growth *Nothofagus* forests in Tierra del Fuego, Argentina. J Veg Sci 4:641–654

Rodriguez J, Echeverria C, Nahuelhual L (2015) Impacts of anthropogenic land-use change on populations of the Endangered Patagonian cypress *Fitzroya cupressoides* in southern Chile: Implications for its conservation Photo-gallery and Map of Vulnerable Socio-ecological Systems. www.sev-project.org

Roig FA, Villalba R (2008) Understanding climate from Patagonian tree rings. Dev Quat Sci 11:411–435

Rojas-Badilla M, Álvarez C, Velásquez-Álvarez G, Hadad M, Le QC, Christie DA (2017) Anomalías anatómicas en anillos de crecimiento anuales de *Austrocedrus chilensis* (D. Don) Pic.-Serm. et Bizzarri en el norte de su rango de distribución. Gayana Bot 74:269–281

Rothkugel M (1913) Los incendios en los Andes Patagónicos. Boletín No 3. Buenos Aires, Argentina

Rovere AE, Premoli AC, Newton AC (2002) Estado de conservación del ciprés de las Guaitecas (*Pilgerodendron uviferum* (Don) Florín) en Argentina. Bosque 23:11–19

Sabatier Y, Azpilicueta MM, Marchelli P, Gonzalez Peñalba M, Lozano L, García L et al (2011) Distribución natural de *Nothofagus alpina* y *Nothofagus obliqua* (*Nothofagaceae*) en Argentina, dos especies de primera importancia forestal de los bosques templados norpatagónicos. Bol la Soc Argentina Bot 46:131–138

San Martin J, Donoso C (1995) Estructura florística e impacto antrópico en el bosque Maulino de Chile. In: Armesto JJ, Villagran C, Kalin Arroyo M (eds) Ecología de los bosques nativos de Chile. Editorial Universitaria, Santiago de Chile, pp 153–168

Santos Biloni J (1990) Árboles autóctonos argentinos. Tipográfica Editora Argentina, Buenos Aires

Savolainen O, Pyhäjärvi T, Knürr T (2007) Gene flow and local adaptation in trees. Annu Rev Ecol Evol Syst 38:595–619

Schinelli Casares T (2013) Producción de *Nothofagus* bajo condiciones controladas. Ediciones INTA, Bariloche

Sekely J, Arana MV, Dalla Salda G, Heer K, Marchelli P, Martínez Meier A et al (2020) GenTree upside down: local adaptation in *Nothofagus pumilio* along environmental gradients in the Andes. In: Genetics to the rescue: managing forests sustainably in a changing world, Avignon

Sérsic AN, Cosacov A, Cocucci AA, Johnson LA, Pozner R, Avila LJ et al (2011) Emerging phylogeographical patterns of plants and terrestrial vertebrates from Patagonia. Biol J Linn Soc 103:475–494

Singer BS, Ackert RP, Guillou H (2004) 40Ar/39Ar and K-Ar chronology of Pleistocene glaciations in Patagonia. Bull Geol Soc Am 116:434–450

Sola G, El Mujtar V, Attis Beltrán H, Chauchard L, Gallo LA (2020) Mixed *Nothofagus* forest management: a crucial link between regeneration, site and microsite conditions. New For 51:435–452

Soliani C, Gallo L, Marchelli P (2012) Phylogeography of two hybridizing southern beeches (*Nothofagus* spp.) with different adaptive abilities. Tree Genet Genomes 8:659–673

Soliani C, Azpilicueta MM, Arana MV, Marchelli P (2020) Clinal variation along precipitation gradients in Patagonian temperate forests: unravelling demographic and selection signatures in three *Nothofagus* spp. Ann For Sci 77:4. https://doi.org/10.1007/s13595-019-0908-x

Soliani C, Aparicio AG (2020) Evidence of genetic determination in the growth habit of *Nothofagus pumilio* (Poepp. & Endl.) Krasser at the extremes of an elevation gradient. Scand J For Res 35:5–6, 211–220

Souto CP, Premoli AC (2007) Genetic variation in the widespread *Embothrium coccineum* (Proteaceae) endemic to Patagonia: effects of phylogeny and historical events. Aust J Bot 55:809

Torres AD, Aparicio AG, Mondino V, Schinelli Casares T, Paredes M, Pastorino MJ (2017) Variación en caracteres fenológicos entre poblaciones naturales de *Nothofagus pumilio* a lo largo de su distribución latitudinal. In: 6° Seminario de *Nothofagus*: Silvicultura, manejo y conservación. Fac. Cs. Agrs. Ftales, UNLP, La Plata

Tortorelli LA (1956) Maderas y bosques argentinos. Editorial Acme, SACI: Buenos Aires, Buenos Aires, Argentina

Tremetsberger K, Urtubey E, Terrab A, Baeza CM, Ortiz MA, Talavera M et al (2009) Pleistocene refugia and polytopic replacement of diploids by tetraploids in the Patagonian and Subantarctic plant *Hypochaeris incana* (Asteraceae, Cichorieae). Mol Ecol 18:3668–3682

Urretavizcaya MF, Oyharcabal MF, Deccechis F (2011) Análisis de área afectada por incendio: incendio La Colisión (Chubut Argentina), actividades realizadas. Patagon For 17:13–16

Veblen TT, Lorenz DC (1988) Recent vegetation changes along the forest/steppe ecotone of Northern Patagonia. Ann Assoc Am Geogr 78:93–111

Veblen TT, Donoso C, Schlegel FM, Escobar B (1981) Forest dynamics in South-Central Chile. J Biogeogr 8:211

Veblen TT, Ashton DH, Rubulis S, Lorenz DC, Cortes M (1989) *Nothofagus* stand development on in-transit moraines, Casa Pangue Glacier, Chile. Arct Alp Res 21:144–155

Veblen TT, Burns BR, Kitzberger T, Lara A, Villalba R, Enright NJ, Hill RS (1995a) The ecology of the conifers of southern South America. In: Ecology of the southern conifers. Melbourne University Press, Melbourne, pp 120–155

Veblen TT, Kitzberger T, Burns BR, Rebertus AJ (1995b) Perturbaciones y dinámica de regeneración en bosques andinos del sur de Chile y Argentina. In: Armesto JJ, Villagran C, Kalin Arroyo M (eds) Ecología de los bosques nativos de Chile. Editorial Universitaria, Santiago de Chile, pp 169–198

Veblen TT, Donoso C, Kitzberger T, Rebertus AJ (1996) Ecology of southern Chilean and Argentinean *Nothofagus* forests. In: Veblen TT, Hill RS, Read J (eds) The ecology and biogeography of *Nothofagus* forest. Yale University, New Haven/London, pp 293–353

Veblen TT, Kitzberger T, Villalba R, Donnegan J (1999) Fire history in northern Patagonia: The roles of humans and climatic variation. Ecol Monogr 69:47–67

Veblen TT, Holz A, Paritsis J, Raffaele E, Ktizberger T, Blackhall M (2011) Adapting to global environmental change in Patagonia: what role for disturbance ecology? Austral Ecol 36:891–903

Vidal-Russell R, Souto CP, Premoli AC (2011) Multiple Pleistocene refugia in the widespread Patagonian tree *Embothrium coccineum* (Proteaceae). Aust J Bot 59:299

Wardle J (1984) The New Zealand beeches. New Zealand Forest Service, New Zealand

Chapter 3
Raulí (*Nothofagus alpina* = *N. nervosa*): The Best Quality Hardwood in Patagonia

Paula Marchelli, Mario J. Pastorino, María Marta Azpilicueta, Virginia Duboscq-Carra, Georgina Sola, Verónica El Mujtar, Verónica Arana, Jorge Arias-Rios, Natalia Fernández, Sonia Fontenla, Marcelo González Peñalba, and Leonardo A. Gallo

3.1 General Features of *Nothofagus alpina*

Nothofagus alpina (Poepp. and Endl.) Oerst. (= *N. nervosa* (Phil.) Dim. et Mil.), raulí, is an iconic forest tree species of the South American temperate forests due to its hardwood quality. With a pinkish colour and light red tints, raulí's wood is very appreciated in the regional and international market for its excellent physical and mechanical characteristics. The texture is fine and homogeneous, presenting a very smooth veining with linear fibres that allow obtaining even excellent roof tiles and a density of approximately 0.55 g/cm^3 at 12% moisture content. Easy to dry, once dry it is dimensionally stable, easy to saw, brush and turn, achieving excellent terminations and therefore very suitable for door and window frames, furniture and fine woodwork (Diaz-Vaz 1987). Owing to this superb wood quality, raulí has been

P. Marchelli (✉) · M. J. Pastorino · M. M. Azpilicueta · V. Duboscq-Carra · V. El Mujtar
V. Arana · J. Arias-Rios · L. A. Gallo
Instituto de Investigaciones Forestales y Agropecuarias Bariloche (IFAB) INTA-CONICET, Bariloche, Argentina
e-mail: marchelli.paula@inta.gob.ar

G. Sola
Instituto de Investigaciones Forestales y Agropecuarias Bariloche (IFAB) INTA-CONICET, Bariloche, Argentina

AUSMA, Universidad Nacional del Comahue, San Martín de los Andes, Neuquen, Argentina

N. Fernández · S. Fontenla
IPATEC, Centro Regional Universitario Bariloche, Universidad Nacional del Comahue – CONICET, Bariloche, Argentina

M. González Peñalba
Parque Nacional Lanín, Administración de Parques Nacionales, San Martín de los Andes, Neuquén, Argentina

© Springer Nature Switzerland AG 2021
M. J. Pastorino, P. Marchelli (eds.), *Low Intensity Breeding of Native Forest Trees in Argentina*, https://doi.org/10.1007/978-3-030-56462-9_3

extensively logged, mainly during the first half of the twentieth century, causing a dramatic reduction of its forests (Donoso and Lara 1995; Gonzalez Peñalba et al. 1997).

Raulí is endemic to the temperate forests of Chile and Argentina. It occurs in Chile between 35° and 41° 30′ S in both, the Andes and Coastal Mountains (Donoso 1993). In Argentina, it has a very small range along the Andes Mountains (39° 25′ – 40° 35′S), which covers only 55,000 hectares in a narrow fringe of c. 120 km in length and c. 40 km in maximum width, following the numerous west-east lake basins of glacial origin (Fig. 3.1; Sabatier et al. 2011). It grows between 800 and 1000 m asl, although it can go down following humid micro-basins to 650 m asl, for example, in the Lake Lácar basin, where it enters the predominant area of *N. obliqua*, as also up to 1250 m asl at some locations where *N. pumilio* is the dominant species. Along the Andes Mountains, the species develops on soils of volcanic origin, while in the Coastal Range, in Chile, it does so on granitic soils (Donoso 1993). The natural range in Argentina expands from an average annual rainfall of around 1200 mm in the east towards more than 3000 mm to the west.

Raulí is a large tree that reaches heights between 25 m and 30 m, sometimes even 40 m (Fig. 3.2a) (Donoso 1993), whose trunk can exceptionally exceed 2 m in diameter. It has a deciduous habit, turning the leaves reddish during the fall (Fig. 3.2b). It is a monoecious species, with anemophilous and anemochorous dispersion. The male flowers are pedunculated, solitary or in clusters of two or three, with numerous stamens. The female flowers are axillary, in number of three, wrapped by a short pedunculated pubescent involucre: bifid style in the central flower and trifid in the lateral ones (Dimitri and Milano 1950). The pollination takes place between the months of October and November, and seeds fall between February and April. Each involucre contains three seeds (achene fruits; Fig. 3.2c): flat the one of the middle and triquetrous the other two.

The forests of raulí in Argentina are mature, with individual ages between 100 and 150 years, being able to reach 300 years (Dimitri and Milano 1950). These forests are one of the most productive in terms of commercial biomass, with growths of 12 m³/ha/year in some places (Chauchard et al. 1997, 2012). Fire represents one of the most important factors in the dynamics of these forests, and large recurrent low-frequency fires cause pulses of post-fire regeneration at the landscape scale (Gallo et al. 2000). Raulí requires moderate open spaces for natural regeneration, since it does not tolerate total exposure; therefore, regeneration occurs in protected sites with partial shadow of other vegetation.

3.2   Genetic Structure: Patterns of Genetic Variation at Neutral Markers

3.2.1   The Legacy of the Glacial Ages

Past climatic changes induced vegetation shifts in different areas around the globe, including the southern Patagonian Andes (e.g. Folguera et al. 2011). However, glaciations affected the two hemispheres in different ways, having a lower impact in

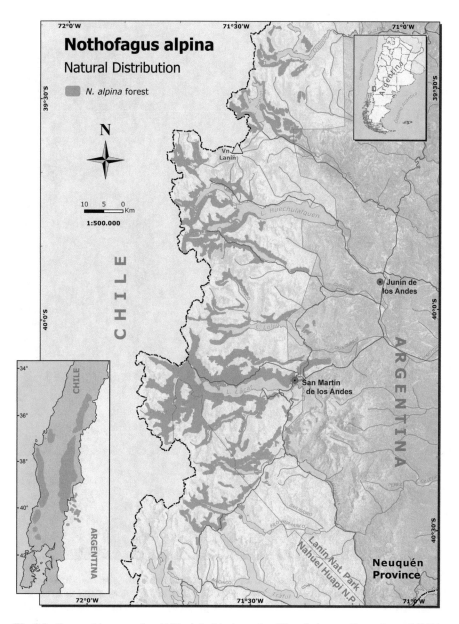

Fig. 3.1 Geographic range of raulí (*N. alpina*) in Argentina. The whole area (Argentina and Chile) is shown at the bottom-left. (Modified from Sabatier et al. 2011)

Fig. 3.2 Morphological traits of raulí (*N. alpina*): (**a**) whole tree in early spring, (**b**) leaves, (**c**) seeds (achene fruits), (**d**) mature forest, (**e**) view of the bark and (**f**) view of forests in autumn. (Photos: (**a**) by M. Pastorino; (**b** and **c**) by M. M. Azpilicueta; (**d** and **e**) by M. Gonzalez Peñalba and (**f**) by L. A. Gallo)

the south because of its greater oceanity and smaller land masses (Markgraf et al. 1995). Accordingly, several ice-free areas remained (Markgraf et al. 1995), probably even during pre-Pleistocene glaciations, constituting ancient and/or recent refugia for vegetation, some of which became the centre of expansion of the biota after the ice retreated (e.g. Marchelli et al. 1998, 2010; Premoli et al. 2002; Marchelli and Gallo 2006; Azpilicueta et al. 2009).

In southern South America, the greatest registered glaciation occurred in the Early Pleistocene, at about 1.2–0.7 My BP (Kodama et al. 1986), with a maximum expansion of ice in extra-Andean Patagonia (Flint and Fidalgo 1964). Last glacial maximum occurred about 20,000 years BP (Porter 1981), and the glacier complex that developed over the Andes was largely confined to the mountains in the region north of 41° S (i.e. the distribution range of raulí), both in Chile and Argentina (Glasser et al. 2008). The scarceness and discontinuity of paleoecological records in Patagonia (Iglesias et al. 2014), and the inability to distinguish between tree species with similar pollen morphology (Heusser 1984), complicate detailed reconstructions of past species distributions. In this sense, intraspecific molecular phylogeographies introduced a powerful tool to make inferences about the location of putative refugia (Petit et al. 1997) by studying patterns of genetic variation in a geographical context through gene trees.

In long-leaved species, like trees, the current genetic structure can still be imprinted by postglacial migration (Newton et al. 1999). Since migration occurs via seeds, maternal inherited plastids are the most suitable markers to describe past recolonization routes, in a way unaffected by posterior pollen movements (Petit et al. 2003). The low mutation rates of the chloroplast genome (i.e. about 10^{-9} substitutions/site/year; Clegg et al. 1991), together with its uniparental inheritance (i.e. reduced gene flow), promote that differences in the chloroplast sequence can persist allowing for population differentiation and the reconstruction of post-glacial migration routes. The predictions indicate that populations surviving in or near refugia should be highly divergent because of prolonged isolation and that intraspecific diversity should decline away from refugia because of successive founder events (Petit et al. 2003).

In order to unravel the evolutionary processes undergone by the populations of *N. alpina* during and after the glacial period, the intraspecific genetic variation at the chloroplast DNA level was evaluated (Marchelli et al. 1998; Marchelli and Gallo 2006). Five haplotypes were detected between 26 populations distributed along the complete natural range in Chile and Argentina (Fig. 3.3a). One of the main features of this phylogeographic study was the confirmation of the location of glacial refugia along the Coastal Range, as suggested by palynological records (Villagran 1991) evidenced by the presence of exclusive haplotypes (Fig. 3.3a). Moreover, Andean refugia were suggested, and even cryptic refugia located towards the eastern side of the Andes were proposed after analysing the distribution of haplotypes. The eastern cryptic refugia were supported later on by ecological niche modelling (Marchelli et al. 2017) and recently confirmed by pollen records (Moreno et al. 2019). As in other tree species (e.g. Premoli et al. 2000; Bekessy et al. 2002; Azpilicueta et al. 2009; Mathiasen and Premoli 2010; Soliani et al. 2012), the existence of multiple refugia in raulí was suggested by the distribution of the cpDNA variation. Also, as in other Patagonian species (reviewed by Sérsic et al. 2011), phylogeographic breaks shared by different species indicate a common history and a strong latitudinal division between populations, probably related to the different glaciation patterns (Flint and Fidalgo 1964; Glasser et al. 2008). Contrary to other species where admixture of colonization routes was detected (e.g. in *Nothofagus pumilio*; Soliani

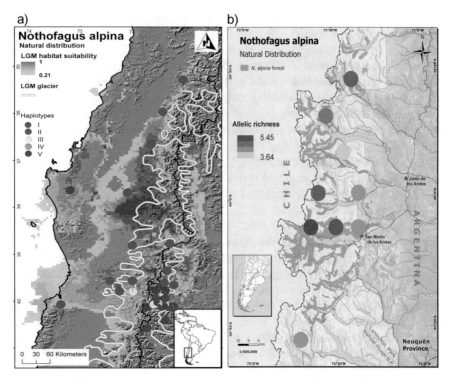

Fig. 3.3 Distribution of the genetic variation among populations of *N. alpina*. (**a**) Distribution of chloroplast haplotypes in the whole range (Chile and Argentina) in combination with the Last Glacial Maximum suitability distribution of *N. alpina*. (From Marchelli et al. 2017); (**b**) distribution of allelic richness in populations from Argentina for the combination of three markers (cpDNA, isozymes and nuclear microsatellites)

et al. 2015), the two main haplotype lineages did not enter into contact in the populations of raulí. The eruption of Lanín volcano, belonging to a volcanoes chain, contemporary to the moment of forests expansion (i.e. about 3000 y BP) probably avoided the mixing of northern and southern migratory routes (Millerón et al. 2008). Similar results were encountered in *N. obliqua*, highlighting the impact of volcanism on the evolutionary history of these species.

3.2.2 Recent Processes Shaping the Genetic Structure: Identifying Hotspots of Diversity

Phylogeographic studies revealed major patterns in the genetic structure of raulí populations that allowed the detection of two main groups with a north-south distribution in Argentina. In order to search for the most variable populations and strengthen this structure, it is necessary to look into recent processes (i.e. gene flow) by inspecting regions of the genome with a higher mutation rate. Therefore, the genetic diversity detected with the chloroplast DNA markers, with information provided by nuclear microsatellites and isozyme markers, was combined.

A first approach through the screening of the genetic variation with isozyme markers in 20 Argentine populations revealed higher levels of diversity in two populations from western areas (Hua Hum at Lake Lácar basin and Boquete at Lake Lolog basin) and moderate but significant population differentiation ($\delta = 0.047$, $F_{ST} = 0.052$; Marchelli and Gallo 2004). Then, species-specific nuclear microsatellites (Azpilicueta et al. 2004; Marchelli et al. 2008) were used to genotype individuals from 14 populations confirming the high diversity of the western populations (Azpilicueta et al. 2013).

For the identification of hotspots of genetic diversity and to prioritize conservation activities, it is ideal to measure allelic richness (Petit et al. 1998) which is highly sensitive to past demographic changes. Rarefaction provides unbiased estimates of allelic richness with great precision and statistical power (Leberg 2002). Accordingly, nine populations that were genotyped with the three markers (chloroplast DNA, isozymes and microsatellites) were selected, and the allelic richness (R_g) in each population after rarefaction to the lowest sample size (g) for all three markers was estimated. For prioritization and identification of the most diverse populations (hotspots of genetic diversity), a standardized allelic richness (R_{gst}) was calculated as described in Marchelli et al. (2017). Population Hua Hum at the west of Lake Lácar holds the highest genetic diversity (Fig. 3.3b), and accordingly, conservation actions were taken (Gallo et al. 2009). Another interesting result is the decreasing trend in allelic richness from west to east detected in the populations of Lake Lácar basin, which suggests a possible migration route (Fig. 3.3a, b) (Marchelli et al. 2017).

The mating system and the extent of pollen flow are two of the most important genetic features that determine the genetic structure of plant populations, and both are crucial for the design of conservation and management strategies. The mating system is both cause and effect of the pollen flow dynamics, defining the distribution of genetic variation within and among populations (e.g. Holsinger 2000). Thus, it is of considerable relevance to many areas of evolutionary and applied genetics including management (Neale and Adams 1985) and conservation practices (e.g. Burczyk et al. 1996). The mating system of raulí was characterized by high estimates of both single-locus ($ts = 0.952$) and multilocus ($tm = 0.995$) outcrossing rates, and the multilocus rate was not significantly different from unity (Marchelli et al. 2012). The estimated rate of biparental inbreeding ($tm-ts = 0.043$; SD = 0.026) was very low, and the correlated paternity rate was $rp = 0.101$ (SD = 0.022), translating into a number of effective fathers per mother tree of $Nep = 9.9$. The estimated average distance of pollen dispersal calculated with the indirect methods TwoGener (Smouse et al. 2001) and Kindist (Robledo-Arnuncio et al. 2006), in one population at the Lake Tromen basin was very short ($\delta = 34$ m), but the dispersal kernel was fat-tailed, and therefore long-distance events are expected (Marchelli et al. 2012). Notwithstanding, the cumulative fraction of pollination suggests that vast majority of pollination comes from within a 50 m radius (99.8%) indicating that pollination is mostly local.

In a more recent study (Sola et al. 2020b), gene dispersal distance of *N. alpina* and *N. obliqua* was indirectly estimated based on fine-scale spatial genetic structure (Vekemans and Hardy 2004). Additionally, the same study made direct estimates of seed and pollen dispersal based on spatially explicit mating models (Chybicki and Burczyk 2010) for both species in a plot that had been silviculturally managed 20 years ago (2.85 ha). Significant genetic correlation at the first distance class (<40 m) was detected for both species. Accordingly, restricted gene flow (≤45 m) was estimated, along with contemporary restricted seed (≤28 m) and pollen dispersal (≤46 m). Significant seed and pollen immigration (up to 57%) was detected; these migration events would correspond, however, to short dispersal distance as the study plot/area belongs to a region of continuous *Nothofagus* forests. This interpretation agrees with the fitting of seed and pollen dispersal kernels to exponential functions ($b = 1$), and the significant fine-scale spatial genetic structure was consistent with the restricted seed and pollen dispersal. Despite the existence of groups of related individuals, no significant inbreeding coefficients (selfing (s) = 0 for both species) were detected, a result that agrees with the high degree of self-incompatibility previously reported for *Nothofagus* (Riveros et al. 1995). On the other hand, size of trees (DBH and height) was significantly associated with female and/or male fertility, highlighting the importance of vigorous trees to increase fecundity and dispersion maintaining forest genetic diversity. Therefore, the local movement of genes has several implications both for understanding the evolution of the species and for defining conservation and management actions.

3.3 Adaptive Genetic Variation: Patterns of Variation Among Populations at Quantitative Traits

Neutral markers cannot unravel genetic variation resulting from adaptive processes. Instead, classic provenance and progeny trials are appropriate to analyse variation in quantitative traits, which due to their expression on the phenotype are potentially adaptive. The seedling and sapling phases of plant development are the most important and acute selective stages in the life span of a tree species (Moles and Westoby 2004). Therefore, knowing selection drivers at these phases is crucial to understand the natural processes of extinction or persistence and to take management decisions in afforestation programmes.

In order to study genetic variation at juvenile stages, a nursery trial (Trial 1, Table 3.1) was conducted with potted seedlings of 74 open-pollinated families corresponding to 7 Argentine natural populations (Duboscq-Carra et al. 2018). Several shoot and root traits were measured at the end of the first growing season (Table 3.1), and mixed linear models were applied to perform variance analysis on these traits. Intra-population variation resulted to be quite variable among populations and traits

Table 3.1 Genetic variation of Argentinean *Nothofagus alpina* populations in quantitative traits measured in two trials (see description above) (from Duboscq-Carra et al. (2018) and Duboscq-Carra (2019))

Trait	Mean (SD)	Mean h^2 (SE)	CV_A	Q_{ST}
Trial 1. 41° 7′ 21.17″ S, 71°14′ 56.95″ W; 795 m asl; 7 pops; 74 OP fam; 592 seedlings				
Shoot height (cm)	54 (1.2)	0.27 (0.12)	7.11	0.47
Collar diameter (mm)	2.04 (0.51)	0.26 (0.12)	9.66	0.35
Root length (cm)	29.8 (8.8)	0.13 (0.06)	7.07	0.30
Stem dry biomass (g)	0.091 (0.05)	0.25 (0.12)	20.52	0.36
Root dry biomass (g)	0.351 (0.24)	0.19 (0.01)	21.91	0.37
R/S	3.77 (1.37)	0.20 (0.10)	14.27	–
Specific root length	1.24 (0.90)	0.16 (0.07)	20.5	0.63
Root/aboveground	1.62 (0.58)	0.17 (0.09)	16.12	0.43
Trial 2. 41° 59′ 53″ S; 71° 31′ 34″ W; 408 m asl; 8 pops; 86 OP fam; 1290 trees				
Height 4th year (cm)	154.7 (59.5)	0.20 (0.10)	14.4	–
Height 6th year (cm)	272.4 (78.4)	0.11 (0.07)	6.3	–
Collar diam. 4th year (mm)	16.76 (6.93)	0.18 (0.09)	13.4	–
Collar diam. 6th year (mm)	31.80 (9.81)	0.08 (0.06)	5.3	–
Daily height growth 4th year (cm)	0.24 (0.33)	<0.01 (0.02)	4.9	–
Daily diam. growth 4th year (mm)	0.028 (0.041)	<0.01 (0.02)	1.5	
Height growth initiation 4th year (DOY)	278 (14)	0.13 (0.07)	1.2	0.15
Height growth initiation 5th year (DOY)	301 (12)	0.19 (0.10)	1.0	0.17
Height growth cessation 4th year (DOY)	353 (24)	0.14 (0.09)	2.2	–
Height growth cessation 5th year (DOY)	362 (21)	0.11 (0.07)	1.3	0.09
Marcescent leaves 4th year (%)	44.87	0.38 (0.18)	–	–
Marcescent leaves 6th year (%)	47.24	0.22 (0.11	–	–
Bud burst (DOY)	287.4 (8.4)	0.60 (0.20)	1.9	0.15
Leaf neoformation initiation (DOY)	335.6 (13.7)	0.10 (0.08)	0.9	–
Senescence initiation (DOY)	426.0 (14.7)	0.13 (0.08)	7.6	–
Leaves duration (days)	185.2 (26.8)	0.08 (0.06)	2.7	–
Specific area of preformed leaves (cm²/g)	112.5 (15.3)	0.25 (0.15)	5.6	0.07
Length/width ratio of preformed leaves	2.29 (0.35)	0.22 (0.13)	6.6	0.09
Stomatal density of pref. leaves (n/mm²)	250.1 (60.4)	0.21 (0.03)	7.9	–
Specific area of neoformed leaves (cm²/g)	112.4 (12.8)	0.42 (0.21)	5.2	1.00
Length/width ratio of neoformed leaves	1.81 (0.26)	0.32 (0.20)	6.9	–
Stomatal density of neoformed leaves (n/mm²)	290.9 (84.8)	0.44 (0.07)	16.9	–
Photosynthetic ratio (mmol CO²/m² seg)	3.05 (2.89)	0.25 (0.16)	32.95	–
Transpiration ratio (mmol H₂O/m² seg)	0.55 (0.47)	0.19 (0.15)	44.77	–

Mean values (standard deviation), mean narrow sense heritability (standard error), additive genetic coefficient of variation and quantitative differentiation

but, in general, tended to be low (mean heritability and mean additive genetic coefficient of variation of all populations for all variables: $h^2 = 0.20$ and $CV_A = 14.65$). CV_A is a proxy for evolvability (Houle 1992), namely, the ability of an organism to evolve, chiefly to adapt. Therefore, in general, the analysed populations have a low capacity for adaptation to eventual environmental changes, and for selective breeding. However, each population should be considered separately for each trait.

On the other hand, differences between populations were significant for all variables except for the ratio between root and stem dry biomasses. Quantitative differentiation between populations was moderate on average (Q_{STmean} = 36%) and higher than neutral differentiation (F_{ST} between 5 and 6%, Marchelli and Gallo 2001; Azpilicueta et al. 2013), thus indicating the likely occurrence of an adaptation process to current environmental conditions. According to post hoc comparisons, the northernmost of the analysed populations (Tromen) was the most different; a result that is consistent with previous studies based on chloroplast DNA markers (Marchelli et al. 1998; Marchelli and Gallo 2006), on SSRs (Azpilicueta et al. 2013) and on seed traits (Marchelli and Gallo 1999).

Still thinking on juvenile characters, a high-density field trial (Trial 2, Table 3.1) was installed in 2011 in San Martín Forest Station of INTA (Instituto Nacional de Tecnología Agropecuaria) for early evaluation. It consisted of 1290 2-year-old seedlings from 86 open-pollinated families, corresponding to 8 Argentinian natural populations, arranged in a randomized complete block design (Table 3.1). Four and five years later, during the growing seasons, the total height of each sapling was measured repeatedly once a week the first year and every other week the second year (Duboscq-Carra 2019). The genetic variation for daily growth rate (mean of the whole trial was 2.4 mm/d and 3.1 mm/d for the first and the second seasons, respectively) was analysed by means of mixed linear models. The effect of "population" and "family" factors were not significant in any of the 2 years. Subsequently, individual growth curves were regressed (Boltzmann model) for each sapling, and the genetic variation of some essential parameters of the adjusted curves was analysed. On average, growth initiated (t_{10}: time to reach 10% of the season growth) in the 278 day of the year (DOY) (October 4) and in the 301 DOY (October 27) in the fourth and the fifth seasons, respectively. On the other hand, growth ceased (t_{90}: time to reach 90% of the season growth) in the 353 DOY (December 18) and 363 DOY (December 27) in each season, respectively. Differences were observed among populations for t_{10} in both seasons, but only in the second season for t_{90}. Differentiation was estimated as moderate for t_{10} in the first (Q_{ST} = 15%) and in the second (Q_{ST} = 17%) seasons but low for t_{90} in the second season (Q_{ST} = 9%; Table 3.1). The family factor was not significant in any of the seasons for any of these two traits.

Among the adaptive traits, the phenological ones are probably the most affected by global climate change (Bertin 2008). In trees of temperate zones, phenology of processes such as bud burst and bud set, or leaf senescence, responds to a balance between survival and productivity. Since the risk of frost damage (necrosis by freezing of primary meristems) is more frequent in autumn, trees adapted to colder climates complete their growth early in the season, forming precociously the winter resistance structures (buds) (Howe et al. 2003). The cessation of growth, the formation of buds and the subsequent senescence of leaves occur mainly by changes in the photoperiod (longer nights) and lower temperatures (colder nights). On the other hand, the opening of buds, once the increasing photoperiod signal has occurred (about June 22 in the Southern Hemisphere), is largely influenced by the temperature (Chuine and Cour 1999; Howe et al. 2003), being the most accurate predictor for bud burst the thermal time in the late winter or the spring months (cumulative degree days; Bertin 2008).

The previous provenance and progeny test was used to study variation in phenological traits (Duboscq-Carra 2019). At the age of 7 years, terminal buds of each sapling were observed every 3 days from the beginning of the growing season (September 20) until the end of the bud expansion, registering their phenophases according to a scale of five categories (Fig. 3.4). Then, at the end of the growing season (from March 1 to May 26), the whole canopy of each sapling was observed every 3 days again, and the beginning (S_{10}: 10% of the leaves with yellowish or reddish colours) and the end (S_{90}: 90% of the leaves with autumnal colours) of foliar senescence were registered. On average for the whole trial, buds reached the third phenophase (semi-open bud) in the 287 DOY (October 13). Differences among populations were observed, and a moderate differentiation was estimated (Q_{ST} = 14.9%; Table 3.1). Regarding the intra-population variation, the family factor was significant and explained 16.3% of the total variance, and according to a mean

Fig. 3.4 Main three phenophases in the field trial and description of phenophases of the bud burst phenology, from closed to completely open buds. Third phenophase (semi-open bud) was the criterion to date the occurrence of the process. (Photos by J. Arias-Rios (1st), M. Pastorino (2nd and 3rd) and V. Duboscq-Carra)

narrow sense, heritability of $h^2 = 0.61$ must be considered as high. These results are similar to those of studies with other forest tree species (Gallo and Geburek 1991; Vitasse et al. 2009; Alberto et al. 2011; Premoli and Mathiasen 2011) and are evidence of a moderate adaptation to the local conditions as well as a high ability to adapt to eventual environmental changes. From a productive point of view, the high heritability reveals the possibility of selecting individuals that sprout early or late in the season, depending on the selective pressure of the occurrence of late frosts (spring) of the implantation site.

Foliar senescence started, on average, 426 days after the beginning of the previous year (March 1) and ended on April 18 (474 days), without significant differences among populations. On the other hand, the family factor resulted significant but only explained 3.3% and 1.1% of the total variance of S_{10} and S_{90}, respectively, with a low intra-population variation as estimated through the narrow sense heritabilities ($h^2_{S10} = 0.13$; $h^2_{S90} = 0.07$; Table 3.1). A subset (65 open-pollinated families, 373 individuals) of the above evaluated trees was measured again for the same traits three seasons later in a climatically contrasting year (Duboscq-Carra et al. 2020). Bud burst date (BBD), foliar senescence (FL), growing season length (GSL) and relative height growth (RHG) were analysed. Growing degree days (GDD) and chilling hours (CH) until bud burst were also calculated, with two possible basal temperatures (5 °C and 7 °C) to evaluate their role in BBD. Significant differences among populations and years in BBD and GSL were found by means of linear mixed models, with the family factor explaining around 30% and 12% of the total variance, respectively. In addition, differences between years but not among populations were detected for FL and for RHG. A tight relationship between GDD and CH with BBD was found, and significant differences among populations and years were found for both variables. Also, a high and positive correlation between the altitude of natural populations and the mean BBD and GDD suggests that probably altitude is conditioning thermal requirements of bud burst. Altogether these results revealed the genetic control of bud burst and foliar senescence, but also the plasticity of these traits, and that GDD and CH are implicated in BBD, suggesting good perspectives to face the climate change scenario.

Stomatal density variation of preformed leaves was also analysed in this trial (Duboscq-Carra 2019). Three leaves were measured per sapling from a subsample of the trial (five saplings per family). Nail polish impressions from dehydrated leaves (Brewer 1992) were taken on the abaxial surface of each leaf. Stomata were counted under an optical microscope (Leica® DM500, Leica Microsystems) on a field of view of 0.05 mm² with a magnification of 500x. The mean stomatal density of the whole trial was 250 s/mm², and significant differences among populations cannot be shown. However, the family factor was significant, explaining 5% of the total variance. The intra-population variation in this trait resulted moderate according to an estimated narrow sense heritability of $h^2 = 0.21$.

Several other characters were analysed in the same trial (Trial 2, Table 3.1), including architectural, physiological, phenological, and growing traits. The general pattern for the species in Argentina in quantitative traits resulting from these analyses (Duboscq-Carra 2019) is low differentiation and low intra-population variation, with some significant exceptions (being bud burst phenology the main of them, as mentioned above). This pattern matches with the restricted distribution area of the species in Argentina and especially with its lack of large environmental ranges. In this regard, some phenological traits were found to be correlated with altitude and mean July temperature of the provenances, but no trait was correlated with mean annual precipitation or UNEP aridity index. The low genetic variation found within populations means a worrisome risk with respect to the global climate change, since the species might be not capable to adapt, and its local persistence could depend exclusively on phenotypic plasticity.

3.4 The Underground Diversity: Relevance of Mycorrhizas and Other Soil Fungi Associated to Raulí

In temperate and boreal forests, the establishment and growth of different forestry species that occupy large areas worldwide are usually dependent on the mycorrhizas present in their roots. Mycorrhizas are a symbiotic association between plants and soil fungi that usually benefit host plants by enhancing water and nutrient uptake, in addition to increasing host resistance to pathogens and other biotic and abiotic stresses. Consequently, mycorrhizas have a direct bearing on plant community structure and productivity (Smith and Read 2008). All *Nothofagus* species are extensively colonized by ectomycorrhizas (EcM), thus suggesting a strong dependence on this symbiosis (Diehl et al. 2008; Fernández et al. 2013, 2015). Given the key role of EcM in raulí's life cycle, it is important to consider this symbiosis to describe the ecology of the species but also to ponder its inclusion in nursery technology during seedling production.

3.4.1 Intraspecific Variation of Ectomycorrhizal Fungal Communities in Raulí

In Argentina, *N. alpina* is not commonly inoculated with EcM fungi during its cultivation in nurseries, in spite of the deep impact that this practice can presumably have on the industrial seedling production. In order to analyse the mycorrhizal status of raulí seedlings produced under standard local protocols, a 2-year experiment was carried out (Fernández et al. 2013). Seedlings were produced in containers with ferti-irrigation according to the procedures adjusted for *Nothofagus* species (Schinelli Casares 2013). When seedlings were 1-year-old, half of them were outplanted to the nursery soil (outside the greenhouse). After the rustification stage (nutrient supply and irrigation frequency were significantly reduced), between 6 and 12 months after germination, the containerized seedlings in the greenhouse were naturally colonized by EcM fungi present in the cultivation system. Meanwhile, seedlings transplanted to the nursery soil had lower colonization rates than those that remained growing in the greenhouse, which was probably related to transplant stress and changes in the environmental conditions. However, these seedlings had higher EcM richness (six species) with respect to those that were in the greenhouse (two species), indicating that after being exposed to a natural soil inoculum they were colonized by other fungi. These results indicated that mycorrhization of *N. alpina* occurs spontaneously under the standard protocols generally used in local nurseries, and that the abundance and diversity of EcM in domesticated raulí seedlings depends on the cultivation technique (Fernández et al. 2013).

In order to obtain information regarding the colonization dynamics of EcM in *N. alpina* outside the greenhouse, the abundance and diversity of EcM in plants of different age (seedlings, young plants and adults) naturally established in the native forest, and in young specimens cultivated in the nursery and then implanted in the same forest (10-year-old provenance trial), were analysed (Fernández et al. 2015). This study was conducted in a forest located in the Lake Lácar basin (Lanín National Park). All the specimens were highly colonized by EcM (>90%). The highest colonization values corresponded to the adults and the lowest to the seedlings. A total of 25 fungal species were identified (Fig. 3.5a), being most of them Basidiomycetes (88%). Richness and diversity of EcM were higher in adults and young specimens than in seedlings and were also higher in cultivated specimens growing in the native forest for 10 years than in naturally established plants. This indicates that the passage of the plants through the nursery in addition to the management practices associated with the establishment of the provenance trial (i.e. fences against livestock, understory removed before plantation, removal of fallen branches and trees) might influence EcM fungal communities in *N. alpina* roots. Additionally, the associated changes are still evident even several years after the trial establishment. It was also observed that the number of fungal species forming EcM in specimens located in the forest was significantly higher than in nursery seedlings. Besides, the composition of EcM fungal communities was completely different, and no common fungal species was found between the native forest (Fernández et al. 2015) and the nursery (Fernández et al. 2013). According to these findings, plant age, substrate inoculum (artificial greenhouse substrate vs forest soil) cultivation techniques and forest management seem to be important factors in determining mycorrhizal development and diversity in raulí.

It is also interesting to mention that, in natural ecosystems, there is a high level of similarity in the EcM fungal communities associated with different *Nothofagus* species (Fernández et al. 2013, 2015; Nouhra et al. 2013; Truong et al. 2019; Barroetaveña et al. 2019; Fioroni 2020). In fact, it has been recently described that one of these common species (*Ruhlandiella patagonica*, Fig. 3.5b; Kraisitudomsook et al. 2019) might improve plant growth in recently established *N. obliqua* seedlings (Fioroni 2020). This information is relevant because it gives us clues regarding the fungal species that could be used for simultaneous inoculation of different *Nothofagus* species during their cultivation in the nursery.

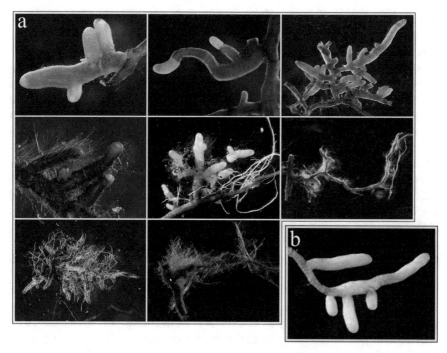

Fig. 3.5 Ectomycorrhizas associated with *N. alpina* in a native forest from Patagonia, Argentina (**a**) (Pictures by N. Fernández). The ectomycorrhizal fungus *Ruhlandiella patagonica* colonizing roots of *N. obliqua* (**b**) (Photo by F. Fioroni)

3.4.2 Other Soil Fungi Associated with Raulí and Its Ectomycorrhizas

Increasing our knowledge about microbial community structure and composition is important for improving our conceptual and projective understanding of soil-ecosystem processes, functions and management. In this context, soil fungi have a prominent importance due to their abundance and the significance of the processes they perform (i.e. decomposition, nutrient cycling, biological interactions such as mycorrhizas). In addition, fungi are excellent bioindicators (organisms highly sensitive to environmental shifts), so it is possible to use them for monitoring ecological changes and the effects of human activities on environmental health, including soil quality and fertility (Rai and Varma 2011).

Afforestation with fast-growing exotic conifers has been promoted as an important economic activity in Patagonia. However, the establishment of plantations of exotic species causes severe disturbances at ecosystem level. In a recent study (Fernández et al. 2020), temperature gradient gel electrophoresis (TGGE, molecular fingerprint method) was utilized for evaluating the structure and composition of different soil fungal communities (Ascomycetes, Basidiomycetes and total fungi) associated with the root system of raulí specimens from populations with different genetic diversity (i.e. low, intermediate and high). Plants installed in two contrasting ecosystems were analysed: a natural *Nothofagus* forest and a *Pinus ponderosa* plantation. It was clearly observed that Ascomycetes, Basidiomycetes and total fungi communities were completely different between the natural forest and the pine plantation, regardless of *N. alpina* provenance. Based on species richness, it was determined that Basidiomycetes was the dominating fungal group in the natural forest, as has been previously described (Nouhra et al. 2013; Mestre et al. 2014; Fernández et al. 2015; Truong et al. 2019; Barroetaveña et al. 2019), and Ascomycetes in the pine plantation. Indeed, richness of these fungal groups was negatively correlated. The fact that the Basidiomycetes to Ascomycetes richness ratio changed significantly in the plantation suggests that these fungal groups might be used as bioindicators of environmental changes reflected in the soil (e.g. forest management, anthropogenic disturbs). Factors such as soil physicochemical characteristics, quantity and quality of organic matter and the different EcM fungi associated with each forestry species seemed to be the main factors determining the structure of these soil fungal communities. Some tendencies relating the intraspecific genetic diversity of *N. alpina* to the richness and composition of soil fungal communities were also noticed. For example, specimens corresponding to a provenance with intermediate genetic diversity presented higher richness of Basidiomycetes and total fungi than other provenances with lower and higher genetic diversity, respectively. Consequently, this study provides evidence that the genetic diversity of the population from which seeds are collected could influence the composition of the fungal communities in the soil, which in turn could affect the plant performance and different ecosystem processes.

3.4.3   Concluding Remarks on the Relevance of Fungal Communities for Conservation and Domestication Programmes

As described previously, there is a significant intraspecific variation regarding the fungal communities associated with *N. alpina* roots, and it depends on different factors, such as cultivation technique, plant age, plant genetic diversity, management practices and changes in soil use. This information is relevant for setting down management guidelines that include the employment of EcM fungi for cultivation during domestication processes. It also highlights the potential of applying soil fungi as bioindicators and contributes to the understanding of how these microorganisms respond to changes related to land use and anthropogenic disturbances such as the establishment of exotic plantations. Finally, these considerations are relevant not only for a better understanding of ecological processes but also for their application in the production of seedlings ulteriorly used in afforestation for economic or restoration purposes.

3.5 Domestication of Raulí: A Potentiality with a Long History

The availability of seeds constitutes a bottleneck for the production of raulí seedlings and may become a problem for the massive production of plants for commercial purposes. Until present, all seeds utilized come from natural forests, where a considerable annual variation in seed production was observed at every stand, with mast years and years of virtually zero production, in cycles of unpredictable duration yet. A marked variation among stands for the same year also occurs (Marchelli and Gallo 1999), and even genetic diversity significantly varied in the same population between good and bad seed-producing years (Marchelli 2002).

Moreover, not only productivity seems to vary from site to site and year to year but also the percentage of empty seeds. The genus is capable of developing fruits through parthenogenesis, that is, without fertilized ovules mediation (Poole 1951). This happens every year in a very high proportion but with temporal and spatial variation. In a single-year collection of seeds in eight Argentinean populations, Barbero (2014) measured mean values of empty seeds per population ranging from 38% to 79%, with individuals producing only empty seeds. This author also showed that the proportion of filled seeds was higher in those of flat shape than in the triquetrous ones (Fig. 3.6). Just as the collection of seeds in nature can be a bottleneck, it was also observed (at least preliminary) that plantations with direct insolation begin to produce seeds after 12 years, reaching regular production almost every year at age 15, with a low relative proportion of empty seeds. These observations encourage the establishment of seed orchards.

Viable seeds are needed to produce seedlings at industrial scale, but also the knowledge to manage the natural process of germination is necessary. Because seed germination is probably the most important decision in plant life history, it is strongly environmentally regulated. This regulation provides advantages to the organism because it restricts seedling emergence to the conditions that are likely to be favourable for the success of the new individual, promoting fitness. However, the existence of specific requirements for germination may constitute a problem and a real bottleneck for the massive production of plants for commercial purposes.

The regulation of germination by environmental cues is usually related to the existence of seed dormancy, i.e. when a viable seed is not able to germinate even when exposed to favourable conditions of water, concentration of gases and temperature (Bewley et al. 2013). In several species, seeds use to be highly dormant at the moment of dispersal, and dormancy is alleviated in a process defined as after-ripening, which is expressed by an increment in the permissive ranges of environmental conditions in which germination occurs (Batlla and Benech-Arnold 2015).

Temperature is one of the main environmental factors that regulate seed behaviour of *N. alpina* seeds, having major effects both in dormancy and in the rate of germination of non-dormant seeds. The use of thermal time models provides a conceptual framework for distinguishing these two effects of temperature on seed behaviour (Batlla and Benech-Arnold 2015) and establishes that a certain quantity

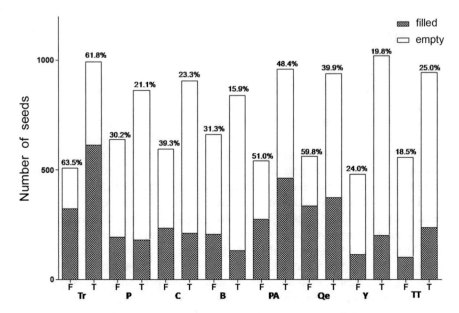

Fig. 3.6 Number of filled and empty seeds (proportions of filled seeds indicated at the top of the bars) in eight natural populations of raulí from Argentina, distinguishing flat (F) and triquetrous (T) seeds. Populations: *Tr* Tromen, *P* Paimún, *C* Currhué, *B* Boquete, *PA* Puerto Arturo, *Q* Queñi, *Y* Yuco, *TT* Tren Tren

of thermal time (degree days or hours) is required to be "accumulated" to complete a given phenological stage (i.e. germination) (Garcia-Huidobro et al. 1982). Through this approach, it is possible to predict the moment of germination in a complex thermal environment (Batlla and Benech-Arnold 2003). The mathematical model that best fitted with *N. alpina* seed behaviour in relation with environmental temperature establishes that the thermal time needed to complete germination is accumulated above a base temperature (Tb) when the environmental temperature is higher than a lower temperature threshold (Tl). Therefore, Tl is the minimum limit temperature permissive for germination (Arana et al. 2016). Experiments under controlled conditions show that warm temperatures, in the range between 22 and 15 °C, promote the germination of a large proportion (more than 75%) of *N. alpina* freshly harvested seeds. However, just a small fraction of the population (less than 1%) is able to germinate at 12 °C. Therefore, *N. alpina* seeds express dormancy at low temperatures (Arana et al. 2016).

The exposure of the seeds to a treatment of cold stratification, this means the incubation of imbibed seeds in a range between 0.5 °C and 6 °C, alleviates the degree of dormancy. The decrease of dormancy with the cold stratification is evident up to 100 days of treatment. After this period, the mean lower limit temperature for germination remains constant in the range between 1.5 °C and 4 °C (Fig. 3.7). From an ecological point of view, the alleviation of dormancy by low temperatures allows *N. alpina* seeds, which are dispersed during fall with relatively high levels of

Fig. 3.7 Calculated values of the lower limit temperature (Tl) for germination in *N. alpina* seed population at different stratification periods. References indicate the time, in days (*d*) of the duration of the stratification treatment. Equations used for the estimation of Tl are described in Arana et al. (2016)

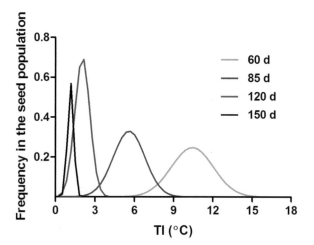

dormancy, to germinate during early spring after spending long periods in the soil during winter. This is possible because Tl of the seeds decreases due to the natural chilling of the seeds during winter and becomes lower than the daily mean soil temperature, a condition that allows germination (Fig. 3.8) (Arana et al. 2016). This behaviour places the development of germinated seeds in a temporal window characterized by an environment with high soil moisture and warm spring temperatures that favours the subsequent growth and survival of the seedlings. Incubation of imbibed seeds during a period of 60 days at a temperature range between 0.5 °C and 6 °C allows germination in the greenhouse in a wide range of temperatures (1–22 °C), although at temperatures below 8 °C, germination rates of the population are very low. This last condition is related to the fact that temperature not only regulates the level of dormancy but also has major effects on the rate of germination of non-dormant seeds.

Seedling production protocols have been very well adjusted in the last years. Very good seedlings (~30 cm tall; 4 mm minimum collar diameter) can be ready to be planted in the field in 8 months when they are produced under greenhouse conditions, by means of a ferti-irrigation regime, in 250 cm³ containers filled with inert substrate (volcanic sand and sphagnum peat in equal proportions) (Schinelli Casares 2013).

Interest on cultivating raulí for timber production has first arisen in Great Britain, where it was introduced in 1910 (Mason et al. 2018). Despite those early attempts, raulí timber is provided exclusively by the natural forest until today. However, the availability of this source has decreased deeply in both Chile and Argentina: in Chile due to a severe change of land use, whereas in Argentina due to the depletion of the existing small productive forests restricted to protected areas. This fact has promoted in Argentina the interest to domesticate the species to produce timber outside its natural range, thus additionally diminishing the pressure over the natural forests. Some silviculture experiences and assays have demonstrated the potential of raulí to be cultivated in Argentina. For example, a trial established in 2005 in a

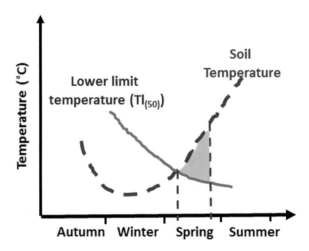

Fig. 3.8 Graphical representation of the dynamics of the daily soil temperature and mean lower limit temperature for germination ($Tl_{(50)}$) in natural conditions. Germination will occur in the period indicated between dashed lines, when $Tl_{(50)}$ is lower than the daily mean soil temperature. The accumulation of thermal time to complete germination will be established by the difference between $Tl_{(50)}$ and the daily soil mean temperature, represented in the green area of the graphic

first category forestry site near to Lake Puelo (42° 1′ 50.72″ S; 71° 33′ 12.49″ O; 380 m asl) showed a 95% survival 2 years later, and at the age of 11 years, the trees reached a mean height of 6.0 m ± 0.3 and a mean DBH of 5.1 cm ± 0.2, without further reductions in the survival percentage (Urretavizcaya et al. 2018).

3.6   Building a Breeding Strategy for Raulí in Argentina

The promising performance of the initial plantations in Scotland and Wales, with fast-growing trees of excellent stem form and good self-pruning, promoted in the 1970s the planting of the species more systematically in Great Britain (Mason et al. 2018). A series of provenance trials were established in 1979 at 17 forests across Britain and included 16 Chilean and 1 Argentine (Lake Quillén basin) provenances and 2 British seed sources. The effect of the provenance was evident, presenting the poorer performance those coming from northern sites of Chile. This result was in agreement with the studies of Murray et al. (1986) who found that the northern Chilean provenances were the most frost sensitive. Due to the recognized vulnerability of the species to frost and winter cold, the next attempt was carried out with eight cold hardy Chilean origins from elevations not lower than 1000 m asl (Ortega and Danby 2000). These new trials were established at the end of the 1990s in south-west Wales and central Highlands and included a local seed source and an Argentinean unknown provenance (Mason et al. 2018). Now the survival was higher, confirming the relevance of choosing the appropriate provenances for a better adaptability, as was also revealed by the phenological studies presented above.

In South America, breeding of raulí started in the 1990s, simultaneously but separately in Chile (Ipinza 1999) and Argentina (Gallo et al. 2005). The first progeny tests of Argentina were established in 1997, thus beginning the breeding programme conducted by INTA until today. At the same time, a temporary provenance trial was installed at the nursery representing 22 Argentinean natural populations of the species ($N = 1200$). Since then, ten provenance and progeny tests were installed in the field to gain knowledge about morphological, functional and adaptation traits (Table 3.2). Five of them are located more than 300 km south from the southernmost limit of the natural distribution area of the species, in Trevelin and San Martín Forest Stations of INTA, thus providing valuable information for the possible strategy of assisted migration in the frame of global climate change. Others were installed under the protection of pine plantations or within the native forest. The site conditions are variable, and, accordingly, preliminary estimated age-age correlations varied between sites (between $r = 0.37–0.56$). In the southern site (Trevelin), which has very good conditions (Table 3.2), the highest growth was registered with individuals above 5 m height at the seventh year. The influence of site conditions played an important role also in the diameter growth. DBH measures were compared at age 13 between provenance tests located at different site conditions, and the best site (Trevelin, DBH = 85 mm) was almost double that of one of the average sites (Yuco Alto, DBH = 45 mm) and almost three times higher than that of the poorer site (Meliquina, DBH = 30 mm) (Gallo, unpublished data).

Since 97% of the lands with presence of raulí in Argentina are under the protection of Lanín and Nahuel Huapi national parks (Sabatier et al. 2011), an agreement was established in 1994 between the National Park Administration and INTA for the use of the Argentine genetic resources of the species. In the frame of this agreement, both institutions have been collaborating for seed collection with scientific and deployment purposes. Thus, natural stands with good seed productivity and accessibility were identified. Moreover, in 2008 a pure stand of 3 ha, situated in a place of Lanín National Park locally known as Yuco Alto (40° 08' 40.01" S; 71° 29' 56.31" W; 1010 m asl), was registered with the National Seed Institute (INASE) as a Seed Production Area (SPA).

Following the first experiences in the management of the species for cultivation and with the preliminary results of the provenance and progeny tests, in 2006 INTA formally initiated in Argentina a low intensity breeding programme for raulí. The initial strategy was to establish clonal seed orchards with individuals selected from the natural forest or precociously from young trials. Since natural forests of raulí in Argentina are not even-aged, mass selection is not expected to achieve large genetic gains due to the impossibility to compare candidate trees with controls of the same age in their neighbourhood (Ledig 1974). This is not an issue in the case of selection in field trials, but their young age is, namely, trees would be selected by juvenile traits. However, the low intensity scale explicitly proposed for the breeding programme of raulí endorsed this strategy.

Since preliminary experiences revealed that adult material was not adequate for vegetative propagation, the resprouting ability of the species from tree stumps was taken as advantage to get rejuvenated material for vegetative propagation. Thus, trees were phenotypically selected in the forest under simple criteria: dominant

Table 3.2 Location and description of the provenance and progeny trials from the domestication programme of *N. alpina* from INTA

Location	Type of site	Lat. (S)	Long. (W)	Alt. (m)	Annual mean T (°C)	Annual pption (mm)	Site conditions	Type of trial and year of installation	N° prov.	N° fam.	Design	Observations
Trevelin Forest Station INTA	Outside natural range	43° 07'	71° 33'	460	9.7	1013	Deep holophane soil, shading management by successive thinning of wild vegetation	Progenies 1997	1	19	RCBD	
								Progenies 1999	4	26	RCBD STP	Thinning (2009–2011) selected seed stand
								Provenances 1999	18	–	RCBD	Thinning (2009–2011) selected seed stand
								Progenies 2000	1	23	RCBD STP	
								Provenances 2000	22	–	RCBD STP	
S. Martin Forest Station INTA	Outside natural range	41° 59'	71° 31'	410	10.08	908	Clay soil, side shading by pine windbreak	Progenies 2011	8	86	RCBD STP	Selective thinning (2017) Progeny seed orchard
Yuco Alto Lanín National Park	Native forest	40° 09'	71° 34'	930	7.7	1900	Halophane soil, open irregular shading	Progenies 1999	2	12	RCBD	
								Provenances 1999	20	–	CRD	
								Progenies 2000	1	17	CRD	
								Provenances 2000	21	–	CRD	
Meliquina Private property	Under *Pinus* plantation	40° 23'	71° 15'	966	7.9	1100	Shallow soil, continuous cenital shading	Provenances 2002	15	–	RCBD	

Lat latitude, *Long* longitude, *Alt* altitude, *N° prov* number of provenances, *N° fam* number of families within provenances, *RCBD* ramdomized complete block design, *CRD* completely ramdomized design, *STP* single-tree plot trees, good shaped and with good diametric growth. They were cut down and their stumps resprouted profusely in the following year. Spring cuttings of semi-woody sprouts of the current year from those resprouted stumps and also from branches of 3 and 12-year-old trees of ongoing trials were attempted to root in warm sand beds with irrigation provided by a fogger system. The best propagation protocol resulted from 8 cm long cuttings with half leaf and its axillary bud above the surface and two buds under the surface, treated with rapid immersion in an IBA alcoholic solution at 7000 ppm. With this methodology and a total sampling of 1008 cuttings from 12 genotypes, rooting was achieved in only 7.9% of the cuttings assayed, with variations among genotypes from 1.2 to 23.8% (Pastorino et al. 2016). Terminal bud grafting was also assayed, using 3-year-old saplings as rootstock. Scions were taken from the same sources as for rooting. Grafts were mostly tested in winter time, though a test was also performed with spring material. Success was achieved only in 27% of the grafts, which is an insufficient proportion considering the efforts of the technique, which was also dependent of the genotype (Pastorino et al. 2016).

The poor performance of the vegetative propagation and its dependence on the genotype led to change the breeding strategy for raulí from clonal seed orchards to progeny seed orchards. Additionally, a research result supports this decision. The Chilean breeding programme of raulí established a provenance and progeny trial in 2000 near to Lake Maihue (40° 12′ S; 72° 2′ W; 180 m asl). It included the progeny of 114 not selected open-pollinated mothers sampled in the natural forests (14 sampling points), 4 control treatments and the progeny of 8 mothers belonging to a clonal seed orchard established in 1989 by Corporación Forestal Nacional (CONAF) with 25 clones (Ipinza et al. 2015). At 14 years of age, this trial presented a survival of 72.2%, with an average height of 14.7 m and a DBH of 15.4 cm. These mean values are quite promising, but the point to highlight is that the progenies coming from the seed orchard mothers had mean height and diameter below the average of the trial. That is, progenies from not selected mothers were better than those coming from crossbreeding of mass-selected trees.

According to the new strategy of installing progeny seed orchards, the already described progeny test of the San Martín Forest Station (INTA) containing 86 open-pollinated families from 8 Argentine origins was transformed in a progeny seed orchard (0.3 ha; Table 3.2). From the original 1290 seedlings planted in 2011 at high density (1 × 1 m) with a single-tree plot design, 622 trees remained in the seed orchard in 2017 after mortality and a preliminary systematic thinning, with a mean height of 272 cm (maximum: 491 cm). Seed production is expected to begin in 2023, at which time a new evaluation will be performed to prescribe a genetic thinning.

Similarly, two trials installed in 1999 in Trevelin Forest Station (INTA) were transformed in a selected seed stand (SSS; Table 3.2) that was registered with INASE in 2017. Within these two trials, a study performed at 13 years of age in the progeny test revealed significant differences in total height and basal diameter among families (Puntieri et al. 2013). In the provenance trial, an interesting association was found between genetic diversity and growth. The best provenance according to DBH at 12 years of age was Hua-Hum (Lake Lácar basin; Gallo unpublished data), which is also the population with the highest genetic diversity determined with different molecular markers (Marchelli and Gallo 2004, 2006; Azpilicueta et al. 2013; Marchelli et al. 2017). More genetic diversity implies not only more importance for conservation purposes but also for growth performance. Both trials together occupied an area of 0.8 ha and originally contained 1188 trees. Best forestry phenotypes were mass-selected regardless of the trial at the ages of 10 and 12 years, and two thinning interventions were subsequently carried out in 2009 and 2011, leaving almost 400 trees. Seed production begun in 2016. Aiming to perform a genetic selection, the trees were genotyped by means of 11 microsatellite markers in order to build a matrix of genomic relationships (Pastorino et al. 2016). A preliminary analysis showed in many cases a relationship closer than the assumed between half-sibs among trees of the same family (data not published). This analysis also allowed detecting the presence of five natural hybrids among the selected trees, which must be cut down to maintain the purity of the stand.

3.7 Definition of Genetic Zones Based on Genetic Markers

Appropriate measures for forest management and conservation demand the identification of genetically homogeneous zones across the distribution of each species to steer germplasm transfer within and between areas, for example, in restauration and reforestation activities (Newton et al. 1999). Genetic zones are defined as genetically homogeneous regions within which propagation material can be transferred, minimizing the risk of introducing changes in the genetic structure (Honjo et al. 2009; Pastorino and Gallo 2009). Therefore, the identification of genetic zones is relevant for the definition of management units, because they reflect a population of interbreeding trees adapted to geographically restricted areas (Azpilicueta et al. 2013).

A first regional approach for *N. alpina* was done by Vergara (2000), who defined 14 provenance regions for the entire natural range (Chile and Argentina). This definition was predominantly based on climatic and geographic information and only to minor extent on morphological characterization of the individuals of several natural populations. The 11 provenance regions identified for the Chilean forests have a clear practical importance for management and conservation programmes (Gutiérrez 2003). By contrast, the three main provenance regions identified for the Argentinean forests are less satisfactory, considering the high fragmentation and spatial heterogeneity of these forests. Therefore, a more specific definition was necessary.

Molecular markers are useful tools for defining genetic zones, with better definitions being obtained by combining different genetic markers (e.g. Bucci et al. 2007). Therefore, a definition of genetic zones for *N. alpina* was done by combining genetic diversity at maternal inherited chloroplast DNA (Marchelli and Gallo 2006) with nuclear genetic variation detected with isozymes (Marchelli and Gallo 2004) and microsatellites (Azpilicueta et al. 2013). Bayesian clustering analyses (BAPS, Corander et al. 2008) identified five groups among Argentinean populations (Azpilicueta et al. 2013): three of them compound each by a unique peripheral population, another represented by marginal populations towards the east and northwest and the largest one composed by the central area of the species distribution range in Argentina, where the most genetically diverse populations are located (Fig. 3.9). These five genetic zones constitute the first step towards the definition of provenance regions and contribute in promoting a rational use of the native genetic resources (Azpilicueta et al. 2016).

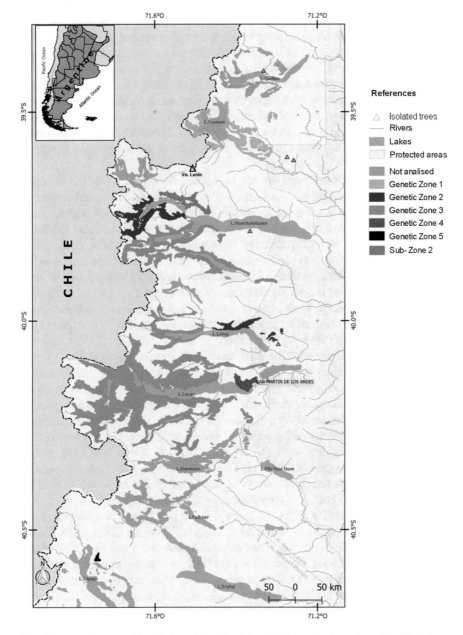

Fig. 3.9 Genetic zones identified and described for *N. alpina* in Argentina. (Modified from Azpilicueta et al. 2016)

3.8   Conservation and Management of Raulí Forests in Argentina

In Argentina, *N. alpina* has been protected since the creation of the National Park Lanín in 1937, and, although devastated in some areas in the past because of its valuable wood, now it is not threatened by logging. The current tendency in Argentine *Nothofagus* mixed forest (*N. alpina*, *N. obliqua* and *N. dombeyi*) is to extract every time less wood from the native forests due to social pressures. Actually, the main threat for this species is climate change. As has been shown by Marchelli et al. (2017; see Chap. 18 for details), according to the more plausible climate change scenarios for the mountainous zone of North Patagonia, the distribution of raulí will be seriously reduced. This species has very specific autoecological requirements for germination and establishment of regeneration, among which humidity and shadow protection counts as the most important. Therefore, forest management in the coming decades should consider the effects of global climate change in *N. alpina* forests.

Assisted migration activities are often proposed for mitigation of the effects of climate change in endangered species (Vitt et al. 2009; Williams and Dumroese 2013). In this sense, large areas of suitable habitat are predicted to appear in the future at southern latitudes, outside of the species' current distribution ranges, as well as locally but at higher altitudes in western sites (Marchelli et al. 2017). However, assisted migration can have unintended and unpredictable consequences (Fady et al. 2016), and caution is in place before introducing a species to areas where is considered as exotic. To maintain the native forest with all its production and ecosystem functions, in situ conservation strategies are the best choice. These strategies might be enhanced by promoting ecological connectivity to expedite natural migration of high diversity populations to new suitable areas. Also management practices should promote the persistence of ecosystem functionality.

Silvicultural practices that involve natural regeneration, regulation of species mixtures, thinning and harvesting operations could have a strong influence on the genetic variation of the populations (Finkeldey and Ziehe 2004), which is essential to preserve the adaptation capacity (Lefèvre et al. 2013). Silvicultural management of *Nothofagus* mixed forests (*N. dombeyi*, *N. alpina* and *N. obliqua*) has been carried out within Lanín National Park (40° 9′ S, 71° 21′ W) since late 1980, mainly following the shelterwood system. This system consists in successive regeneration cuttings that retain partial forest cover (canopy cover of around 30–40%) until the regeneration phase is complete (approximately 20 years, Chauchard 1989).

In order to evaluate the effects of a seed cut (cut performed with the purpose of seed dissemination) on the regeneration dynamics, Sola et al. (2015, 2016, 2020a) studied the genetic diversity and gene flow at different spatial scales in *Nothofagus* mixed forests after about 20 years of seed cut implementation. At microsite scale (a 2.85 ha plot), an intensive sampling of pre- (mature trees) and post- (regeneration) harvest individuals of all three species (*N. alpina*, *N. obliqua* and *N. dombeyi*) was combined with microsatellite genotyping (>2000 individuals and 15 microsatellite markers), together with characterization of micro-environmental conditions. At stand scale, the influence of site condition, altitude and post-harvest stand structure on regeneration composition and establishment was studied in two sites located along Lácar watershed, where mean annual precipitation level has a drastic variation.

Globally, these studies indicated that canopy cover after intervention is a key factor modelling regeneration dynamics in mixed forest, but other factors such as site and forest structure are also relevant and should be taken into consideration (Sola et al. 2020a). The pioneer and widely distributed species *N. dombeyi* has been favoured in all managed areas, while *N. obliqua* was associated to low altitude and open microsites (reduced basal area of mature trees and low competition of understory species) and *N. alpina* was restricted to shady conditions with low water stress and high altitude (Sola et al. 2020a). In this way, *N. alpina* resulted underrepresented within the total regeneration. A similar trend of species composition change has been reported after the implementation of the group selection cutting silvicultural system (Dezzotti et al. 2003). Weinberger and Ramirez (1999) showed that in sites with low precipitation levels, juvenile stages of *N. alpina* became established only in shady conditions (with cover >40%). Thus, reproductive strategies do not depend only on light but on the interaction with other environmental factors, such as temperature and water economy.

The modification of relative abundance of species in the post-harvest population that followed the implemented management was found altering the global genetic diversity of the mixed forest; however, no impact was detected at species level (Sola et al. 2016). The use of species-specific markers also allowed to determine that the level of introgressive hybridization between *N. alpina* and *N. obliqua* was not changed by management (Sola et al. 2016). Therefore, the potential negative genetic impact of increasing interindividual distances resulting from harvesting could be counterbalanced by high gene flow, suggesting that evolutionary processes were maintained at pre-harvest levels or that derived changes had compensatory effects on genetic diversity at species level (Fageria and Rajora 2013). Accordingly, the comparison of historical (indirect) and contemporary (direct) gene dispersal estimates (see before in this Chapter) revealed that the recent history (20 years ago) of logging activities within the study area has not significantly affected the patterns of gene dispersal distance (Sola et al. 2020b). The residual tree density maintaining species composition and the homogeneous spatial distribution of trees (followed by the applied silvicultural system) possibly allowed the maintenance of gene dispersal.

Concluding, these studies provide important information to plan management, conservation and restoration of *Nothofagus* mixed forests based on improved knowledge of the regeneration dynamics and the dispersal pattern. Sola et al. (2020a) proposed a set of recommendations depending on the site conditions that contributes in a practical way to the sustainable use of this forest and can be summarized as: (i) in sites with stressful conditions, a first harvest of low intensity could be applied (residual canopy cover greater than 60%), to facilitate seedling establishment; (ii) in mixed stands of the three species, a larger proportion of *N. dombeyi* individuals could be harvested, favouring higher seed availability of the other species, in order to maintain the original composition in the new generation; (iii) in areas under management, where the *N. obliqua* and, mainly, *N. alpina* regeneration processes are affected, regeneration management of *N. dombeyi* saplings (thinning) should be implemented to promote the release and establishment of *N. obliqua* and *N. alpina* individuals; (iv) in sites with low sapling density, topsoil removal and/or planting of the most affected species could be implemented.

References

Alberto F, Bouffier L, Louvet JM, Lamy JB, Delzon S, Kremer A (2011) Adaptive responses for seed and leaf phenology in natural populations of sessile oak along an altitudinal gradient. J Evol Biol 24:1442–1454

Arana MV, Gonzalez-Polo M, Martinez-Meier A, Gallo LA, Benech-Arnold RL, Sánchez RA, Batlla D (2016) Seed dormancy responses to temperature relate to *Nothofagus* species distribution and determine temporal patterns of germination across altitudes in Patagonia. New Phytol 209:507–520

Azpilicueta MM, Caron H, Bodenes C, Gallo L (2004) SSR markers for analysing South American *Nothofagus* species. Silivae Genet 53:240–243

Azpilicueta MM, Marchelli P, Gallo LA (2009) The effects of Quaternary glaciations in Patagonia as evidenced by chloroplast DNA phylogeography of southern beech *Nothofagus obliqua*. Tree Genet Genomes 5:561–571

Azpilicueta MM, Gallo LA, van Zonneveld M, Thomas E, Moreno C, Marchelli P (2013) Management of *Nothofagus* genetic resources: definition of genetic zones based on a combination of nuclear and chloroplast marker data. For Ecol Manag 302:414–424

Azpilicueta MM, Marchelli P, Gallo LA, Umaña F, Thomas E, van Zonneveld M et al (2016) Zonas genéticas de raulí y roble pellín en Argentina: herramientas para la conservación y el manejo de la diversidad genética. Ediciones INTA

Barbero F (2014) Variación genética de poblaciones naturales argentinas de *Nothofagus obliqua* ("Roble Pellín") en caracteres adaptativos tempranos relevantes para domesticación. Tesis Doctoral-Facultad de Agronomía – Universidad de Buenos Aires

Barroetaveña C, Salomón MES, Bassani V (2019) Rescuing the ectomycorrhizal biodiversity associated with South American Nothofagaceae forest, from the 19th century naturalists up to molecular biogeography. For Int J For Res 92:500–511

Batlla D, Benech-Arnold RL (2003) A quantitative analysis of dormancy loss dynamics in *Polygonum aviculare* L. seeds: development of a thermal time model based on changes in seed population thermal parameters. Seed Sci Res 13:55–68

Batlla D, Benech-Arnold RL (2015) A framework for the interpretation of temperature effects on dormancy and germination in seed populations showing dormancy. Seed Sci Res 25:147–158

Bekessy SA, Allnutt TR, Premoli AC, Lara A, Ennos RA, Burgman MA et al (2002) Genetic variation in the monkey puzzle tree (*Araucaria araucana* (Molina) K.Koch), detected using RAPDs. Heredity 88:243–249

Bertin RI (2008) Plant phenology and distribution in relation to recent climate change. J Torrey Bot Soc 135:126–146

Bewley JD, Bradford KJ, Hilhorst HWM, Nonogaki H (2013) Dormancy and the control of germination. In: Seeds. Springer, New York, pp 247–297

Brewer CA (1992) Responses by stomata on leaves to microenvironmental conditions. In: Glase JC (ed) Tested studies for laboratory teaching. Proceedings of the 13th workshop/conference of the Association for Biology Laboratory Education (ABLE), pp 67–77

Bucci G, González-Martínez SC, Le Provost G, Plomion C, Ribeiro MM, Sebastiani F et al (2007) Range-wide phylogeography and gene zones in *Pinus pinaster* Ait. revealed by chloroplast microsatellite markers. Mol Ecol 16:2137–2153

Burczyk J, Adams WT, Shimizu JY (1996) Mating patterns and pollen dispersal in a natural knob-cone pine (*Pinus attenuata* Lemmon.) stand. Heredity 77:251–260

Chauchard L (1989) Plan de ordenación Quilanlahue. In: Municipalidad de Junjn de los Andes-Administración de Parques Nacionales. San Martjn de los Andes, Neuquén

Chauchard L, Sbrancia R, Gonzales Peñalba M, Rabino A, Maresca L, Blacher C (1997) Dinámica y manejo del bosque de Nothofagus. Universidad Nacional del Comahue-Administración de Parques Nacionales-Instituto Nacional de Tecnología Agropecuaria

Chauchard L, Bava JO, Castañeda S, Laclau P, Loguercio GA, Pantaenius P, Rusch VE (2012) Manual para las buenas prácticas forestales e bosques nativos de Nordpatagonia. Ministerio de Agricultura, Ganadería y Pesca, Presidencia de la Nación Argentina

Chuine I, Cour P (1999) Climatic determinants of budburst seasonality in four temperate-zone tree species. New Phytol 143:339–349

Chybicki IJ, Burczyk J (2010) NM+: software implementing parentage-based models for estimating gene dispersal and mating patterns in plants. Mol Ecol Resour 10:1071–1075

Clegg MT, Learn GH, Golenberg EM, Selander RS, Clarck AG, Whittam TS (1991) Molecular evolution of chloroplast DNA. In: Evolution at the molecular level. Sinauer Associates, Sunderland, pp 135–149

Corander J, Marttinen P, Siren J, Tang J (2008) Enhanced Bayesian modelling in BAPS software for learning genetic structures of populations. BMC Bioinformatics 9:539

Dezzotti A, Sbrancia R, Rodríguez-Arias M, Roat D, Parisi A (2003) Regeneración de un bosque mixto de *Nothofagus* (Nothofagaceae) después de una corta selectiva. Rev Chil Hist Nat 76:591–602

Diaz-Vaz JE (1987) Anatomia de madera de *Nothofagus alpina* (Poepp. et Endl.) Oerstedt. Bosque 8:143–145

Diehl P, Mazzarino MJ, Fontenla S (2008) Plant limiting nutrients in Andean-Patagonian woody species: effects of interannual rainfall variation, soil fertility and mycorrhizal infection. For Ecol Manag 255:2973–2980

Dimitri MJ, Milano VA (1950) El nombre botanico del "Rauli". Bol la Soc Argentina Bot 3:85–87

Donoso C (1993) Bosques templados de Chile y Argentina. Variación, Estructura y Dinámica. Ecología Forestal. Editorial Universitaria, Chile

Donoso C, Lara A (1995) Utilización de los bosques nativos en Chile: pasado, presente y futuro. In: Armesto JJ, Villagran C, Arroyo KM (eds) Ecología de los Bosques nativos de Chile. Editorial Universitaria, Santiago, pp 363–388

Duboscq-Carra VG (2019) Variación y diferenciación genética de *Nothofagus alpina*: proyecciones para su domesticación y análisis de sus estrategias evolutivas en el contexto del Cambio Climático Global. Doctoral thesis, Universidad Nacional del Comahue, Bariloche

Duboscq-Carra VG, Letourneau FJ, Pastorino MJ (2018) Looking at the forest from below: the role of seedling root traits in the adaptation to climate change of two *Nothofagus* species in Argentina. New For 49:613–635

Duboscq-Carra VG, Arias-Rios JA, El Mujtar VA, Marchelli P, Pastorino MJ (2020) Differentiation in phenology among and within natural populations of a South American *Nothofagus* revealed by a two-year evaluation in a common garden trial. For Ecol Manag 460. https://doi.org/10.1016/j.foreco.2019.117858

Fady B, Cottrell J, Ackzell L, Alía R, Muys B, Prada A, González-Martínez SC (2016) Forests and global change: what can genetics contribute to the major forest management and policy challenges of the twenty-first century? Reg Environ Chang 16:927–939

Fageria MS, Rajora OP (2013) Effects of harvesting of increasing intensities on genetic diversity and population structure of white spruce. Evol Appl 6:778–794

Fernández NV, Marchelli P, Fontenla SB (2013) Ectomycorrhizas naturally established in *Nothofagus nervosa* seedlings under different cultivation practices in a forest nursery. Microb Ecol 66:581–592

Fernández NV, Marchelli P, Gherghel F, Kost G, Fontenla SB (2015) Ectomycorrhizal fungal communities in *Nothofagus nervosa* (Raulí): a comparison between domesticated and naturally established specimens in a native forest of Patagonia, Argentina. Fungal Ecol 18:36–47

Fernández NV, Marchelli P, Tenreiro R, Chaves S, Fontenla SB (2020) Are the rhizosphere fungal communities of *Nothofagus alpina* established in two different environments influenced by plant genetic diversity? For Ecol Manage 473:118269. https://doi.org/10.1016/j.foreco.2020.118269

Finkeldey R, Ziehe M (2004) Genetic implications of silvicultural regimes. In: Forest ecology and management. Elsevier, pp 231–244

Fioroni F (2020) Efecto de distintas estrategias de cultivo sobre las comunidades fúngicas radicales y el desarrollo vegetal de *Nothofagus obliqua* y *Pinus ponderosa*. Tesis de Licenciatura en Biología, Centro Regional Universitario Bariloche, Universidad Nacional del Comahue

Flint RF, Fidalgo F (1964) Glacial geology of the East flank of the Argentine Andes between latitude 39° 10′S and latitude 41° 20′S. Geol Soc Am Bull 75:335–352

Folguera A, Orts D, Spagnuolo M, Vera ER, Litvak V, Sagripanti L et al (2011) A review of Late Cretaceous to Quaternary palaeogeography of the Southern Andes. Biol J Linn Soc 103:250–268

Gallo LA, Geburek T (1991) Genetics of isozyme variants in *Populus tremula, P. tremuloides* and their hybrids. Euphytica 53:225–233

Gallo LA, Marchelli P, Crego P, Oudkerk L, Izquierdo F, Breitembücher A et al (2000) Distribución y variación genética en características seminales y adaptativas de poblaciones y progenies de raulí en Argentina. In: Domesticación y Mejora Genética de raulí y roble. Universidad Austral de Chile-Instituto Forestal, Valdivia, pp 133–156

Gallo L, Marchelli P, Pastorino M, Izquierdo F, Azpilicueta MM (2005) Programa de conservación y utilización de los recursos genéticos de especies forestales nativas patagónicas. Idia XXI:152–158

Gallo LA, Marchelli P, Chauchard L, Gonzalez Peñalba M (2009) Knowing and doing: research leading to action in the conservation of forest genetic diversity of Patagonian temperate forests. Conserv Biol 23:895–898

Garcia-Huidobro J, Monteith JL, Squire GR (1982) Time, temperature and germination of pearl millet (*Pennisetum typhoides* S. H.) I. Constant temperature. J Exp Bot 33:288–296

Glasser NF, Jansson KN, Harrison S, Kleman J (2008) The glacial geomorphology and Pleistocene history of South America between 38°S and 56°S. Quat Sci Rev 27:365–390

Gonzalez Peñalba M, Chauchard L, Maresca L, Perez A, Iglesias G, Gross M et al (1997) Plan de gestión ambiental de la cuenca del Lago Lolog. Relevamiento forestal de la cuenca del Lago Lolog en el área de Reserva Nacional. Primera Etapa. Municipalidad de San Martín de los Andes. Administración de Parques Nacionales, San Martín de los Andes

Gutiérrez B (2003) Mejoramiento genético y conservación de recursos forestales nativos en Chile. Inv Agra Sist Rec For 12(145):153

Heusser CJ (1984) Late-glacial-Holocene climate of Lake District of Chile. Quat Res 22:77–90

Holsinger KE (2000) Reproductive systems and evolution in vascular plants. Proc Natl Acad Sci U S A 97:7037–7042

Honjo M, Kitamoto N, Ueno S, Tsumura Y, Washitani I, Ohsawa R (2009) Management units of the endangered herb *Primula sieboldii* based on microsatellite variation among and within populations throughout Japan. Conserv Genet 10:257–267

Houle D (1992) Comparing evolvability and variability of quantitative traits. Genetics 130:195–204

Howe GT, Aitken SN, Neale DB, Jermstad KD, Wheeler NC, Chen THH (2003) From genotype to phenotype: unraveling the complexities of cold adaptation in forest trees. Can J Bot 81:1247–1266

Iglesias V, Whitlock C, Markgraf V, Bianchi MM (2014) Postglacial history of the Patagonian forest/steppe ecotone (41 – 43 S). Quat Sci Rev 94:120–135

Ipinza R (1999) Ensayos de progenie y procedencias en Chile, análisis espacial (ASReml) en *Nothofagus alpina*, un estudio de caso. In: Kleinn C, Köhl M (eds) Long-term observations and research in forestry. CATIE, Turrialba, pp 187–203

Ipinza R, Gutiérrez B, Molina MP (2015) Evaluación genética a los 9 y 14 años de un ensayo de progenies y procedencias de raulí (*Nothofagus alpina*) en Arquilhue, Región de los Ríos, Chile. In: Gutierrez B, Ipinza R, Barros S (eds) Conservación de Recursos Genéticos Forestales, principios y práctica. Instituto Forestal, Santiago, p 320

Kodama KP, Rabassa J, Evenson EB, Clinch JM (1986) Paleomagnetismo y edad relativa del drift Pichileufu en su área tipo, San Carlos de Bariloche, Río Negro. Rev Asoc Geológ Arg XLI:165–178

Kraisitudomsook N, Healy RA, Mujic AB, Pfister DH, Nouhra ER, Smith ME (2019) Systematic study of truffles in the genus Ruhlandiella, with the description of two new species from Patagonia. Mycologia 111:477–492

Leberg PL (2002) Estimating allelic richness: effects of sample size and bottlenecks. Mol Ecol 11:2445–2449

Ledig FT (1974) An analysis of methods for the selection of trees from wild stands. For Sci 20:2–16

Lefèvre F, Koskela J, Hubert J, Kraigher H, Longauer R, Olrik DC et al (2013) Dynamic conservation of forest genetic resources in 33 European countries. Conserv Biol 27:373–384

Marchelli P (2002) Variabilidad genética en Raulí (*Nothofagus nervosa* (Phil.) Dim. et Mil.) su relación con procesos evolutivos y su importancia en la conservación y utilización de sus recursos genéticos. Tesis para optar al grado de Doctor en Ciencias Biológicas, Centro Regional Universitario Bariloche, Universidad Nacional del Comahue

Marchelli P, Gallo LA (1999) Annual and geographic variation in seed traits of Argentinean populations of southern beech *Nothofagus nervosa* (Phil.) Dim. et Mil. For Ecol Manag 121:239–250

Marchelli P, Gallo LA (2001) Genetic diversity and differentiation in a southern beech subjected to introgressive hybridization. Heredity 87:284–293

Marchelli P, Gallo LA (2004) The combined role of glaciation and hybridization in shaping the distribution of genetic variation in a Patagonian southern beech. J Biogeogr 31:451–460

Marchelli P, Gallo L (2006) Multiple ice-age refugia in a southern beech from southern South America as revealed by chloroplast DNA markers. Conserv Genet 7:591–603

Marchelli P, Gallo L, Scholz F, Ziegenhagen B (1998) Chloroplast DNA markers revealed a geographical divide across Argentinean southern beech *Nothofagus nervosa* (Phil.) Dim. et Mil. distribution area. Theor Appl Genet 97:642–646

Marchelli P, Caron H, Azpilicueta MM, Gallo LA (2008) Primer note: a new set of highly polymorphic nuclear microsatellite markers for *Nothofagus nervosa* and related south American species. Silvae Genet 57:82–85

Marchelli P, Baier C, Mengel C, Ziegenhagen B, Gallo LA (2010) Biogeographic history of the threatened species *Araucaria araucana* (Molina) K. Koch and implications for conservation: a case study with organelle DNA markers. Conserv Genet 11:951–963

Marchelli P, Smouse PE, Gallo LA (2012) Short-distance pollen dispersal for an outcrossed, wind-pollinated southern beech (*Nothofagus nervosa* (Phil.) Dim. et Mil.). Tree Genet Genomes 8:1123–1134

Marchelli P, Thomas E, Azpilicueta MM, van Zonneveld M, Gallo LA (2017) Integrating genetics and suitability modelling to bolster climate change adaptation planning in Patagonian *Nothofagus* forests. Tree Genet Genomes 13:119

Markgraf V, McGlone M, Hope G (1995) Neogene paleoenvironmental and paleoclimatic change in southern temperate ecosystems – a southern perspective. Trends Ecol Evol 10:143–147

Mason B, Jinks R, Savill P, Wilson SMG (2018) Southern beeches (*Nothofagus* species). Q J For 112:30–43

Mathiasen P, Premoli AC (2010) Out in the cold: genetic variation of *Nothofagus pumilio* (Nothofagaceae) provides evidence for latitudinally distinct evolutionary histories in austral South America. Mol Ecol 19:371–385

Mestre MC, Fontenla S, Rosa CA (2014) Ecology of cultivable yeasts in pristine forests in northern Patagonia (Argentina) influenced by different environmental factors. Can J Microbiol 60:371–382

Millerón M, Gallo L, Marchelli P (2008) The effect of volcanism on postglacial migration and seed dispersal. A case study in southern South America. Tree Genet Genomes 4:435–443

Moles AT, Westoby M (2004) Seedling survival and seed size: a synthesis of the literature. J Ecol 92:372–383

Moreno PI, Simi E, Villa-Martínez RP, Vilanova I (2019) Early arboreal colonization, postglacial resilience of deciduous *Nothofagus* forests, and the Southern Westerly Wind influence in central-east Andean Patagonia. Quat Sci Rev 218:61–74

Murray MB, Cannell MGR, Sheppard LJ, Lines R (1986) Frost hardiness of *Nothofagus procera* and *Nothofagus obliqua* in Britain. For Int J For Res 59:209–222

Neale DB, Adams J (1985) The mating system in natural and shelterwood stands of Douglas-fir. Theor Appl Genet 71:201–207

Newton AC, Allnutt TR, Gillies ACM, Lowe AJ, Ennos RA (1999) Molecular phylogeography, intraspecific variation and the conservation of tree species. Trends Ecol Evol 14:140–145

Nouhra E, Urcelay C, Longo S, Tedersoo L (2013) Ectomycorrhizal fungal communities associated to *Nothofagus* species in Northern Patagonia. Mycorrhiza 23:487–496

Ortega A, Danby NP (2000) *Nothofagus* en Gran Bretaña. In: Ipinza R, Gutierrez B, Emhart V (eds) Domesticación y Mejora Genética de raulí y roble. Universidad Austral de Chile, Santiago, pp 53–62

Pastorino MJ, Gallo LA (2009) Preliminary operational genetic management units of a highly fragmented forest tree species of southern South America. For Ecol Manag 257:2350–2358

Pastorino MJ, El Mujtar VA, Azpilicueta MM, Aparicio AG, Marchelli P, Mondino V et al (2016) Subprograma Nothofagus. In: Marcó M, Llavallol C (eds) Domesticación y Mejoramiento de Especies Forestales. Min. Agroindustria, UCAR, Buenos Aires, 422 pp

Petit JR, Pineau E, Demesure B, Bacilieri R, Ducousso A, Kremer A (1997) Chloroplast DNA footprints of postglacial recolonization by oaks. Proc Natl Acad Sci U S A 94:9996–10001

Petit JR, El Mousadik A, Pons O (1998) Identifying populations for conservation on the basis of genetic markers. Conserv Biol 12:844–855

Petit RJ, Aguinagalde I, de Beaulieu JL, Bittkau C, Brewer S, Cheddadi R et al (2003) Glacial refugia: hotspots but not melting pots of genetic diversity. Science 300:1563–1565

Porter SC (1981) Pleistocene glaciation in the southern Lake District of Chile. Quat Res 16:263–292

Premoli AC, Mathiasen P (2011) Respuestas ecofisiológicas adaptativas y plásticas en ambientes secos de montaña: *Nothofagus pumilio*, el árbol que acaparó los Andes australes. Ecol Austral 021:251–269

Premoli AC, Kitzberger T, Veblen TT (2000) Isozyme variation and recent biogeographical history of the long-lived conifer *Fitzroya cupressoides*. J Biogeogr 27:251–260

Premoli AC, Souto CP, Rovere AE, Allnutt TR, Newton AC (2002) Patterns of isozyme variation as indicators of biogeographic history in *Pilgerodendron uviferum* (D.Don) Florin. Divers Distrib 8:57–66

Poole AL (1951) The development of *Nothofagus* seed (Incuiding a preliminary account of the embryogeny,etc). Trans R Soc New Zeal 80:207–212

Puntieri J, Grosfeld J, Tejera L, Mondino V, Gallo LA (2013) Within-population variability in architectural traits and suitability to forestry conditions in *Nothofagus nervosa* (= *N. alpina*; Nothofagaceae). Ann For Sci 70:471–479

Rai M, Varma A (2011) Diversity and biotechnology of Ectomycorrhizae. Springer, Berlin

Riveros M, Parades MA, Rosas MT, Cardenas E, Armesto JJ, Arroyo KM, Palma B (1995) Reproductive biology in species of the genus *Nothofagus*. Environ Exp Bot 35:519–524

Robledo-Arnuncio JJ, Austerlitz F, Smouse PE (2006) A new method of estimating the pollen dispersal curve independently of effective density. Genetics 173:1033–1045

Sabatier Y, Azpilicueta MM, Marchelli P, Gonzalez Peñalba M, Lozano L, García L et al (2011) Distribución natural de *Nothofagus alpina* y *Nothofagus obliqua* (Nothofagaceae) en Argentina, dos especies de primera importancia forestal de los bosques templados norpatagónicos. Bol la Soc Argentina Bot 46:131–138

Schinelli Casares T (2013) Producción de *Nothofagus* bajo condiciones controladas. Ediciones INTA, Bariloche, 56 pp

Sérsic AN, Cosacov A, Cocucci AA, Johnson LA, Pozner R, Avila LJ et al (2011) Emerging phylogeographical patterns of plants and terrestrial vertebrates from Patagonia. Biol J Linn Soc 103:475–494

Smith SE, Read DJ (2008) Micorrhizal symbiosis. Academic, London

Smouse PE, Dyer RJ, Westfall RD, Sork VL (2001) Two-generation analysis of pollen flow across a landscape. I. Male gamete heterogeneity among females. Evolution 55:260–271

Sola G, Attis Beltran H, Chauchard L, Gallo LA (2015) Efecto del manejo silvicultural sobre la regeneración de un bosque de *Nothofagus dombeyi, N. alpina* y *N. obliqua* en la Reserva Nacional Lanín (Argentina). Bosque 36:113–120

Sola G, El Mujtar V, Tsuda Y, Vendramin GG, Gallo LA (2016) The effect of silvicultural management on the genetic diversity of a mixed *Nothofagus* forest in Lanín Natural Reserve, Argentina. For Ecol Manag 363:11–20

Sola G, El Mujtar V, Attis Beltrán H, Chauchard L, Gallo LA (2020a) Mixed *Nothofagus* forest management: a crucial link between regeneration, site and microsite conditions. New For 51:435–452

Sola G, El Mujtar VA, Gallo LA, Vendramin GG, Marchelli P (2020b) Staying close: short local dispersal distances on a managed forest of two Patagonian *Nothofagus* species. Forestry 93:652–661

Soliani C, Gallo L, Marchelli P (2012) Phylogeography of two hybridizing southern beeches (*Nothofagus* spp.) with different adaptive abilities. Tree Genet Genomes 8:659–673

Soliani C, Tsuda Y, Bagnoli F, Gallo LA, Vendramin GG, Marchelli P (2015) Halfway encounters: meeting points of colonization routes among the southern beeches *Nothofagus pumilio* and *N. antarctica*. Mol Phylogenet Evol 85:197–207

Truong C, Gabbarini LA, Corrales A, Mujic AB, Escobar JM, Moretto A, Smith ME (2019) Ectomycorrhizal fungi and soil enzymes exhibit contrasting patterns along elevation gradients in southern Patagonia. New Phytol 222:1936–1950

Urretavizcaya MF, Caselli MM, Contardi LT, Loguercio G, Defossé G (2018) Enriquecimiento de Bosques Degradados de ciprés de la cordillera con especies nativas de alto valor forestal. Informe Final Proyecto PIA 14067. UCAR-CIEFAP

Vekemans X, Hardy OJ (2004) New insights from fine-scale spatial genetic structure analyses in plant populations. Mol Ecol 13:921–935

Vergara R (2000) Regiones de procedencia de *N. alpina* y *N. obliqua*. In: Ipinza R, Gutierrez B, Emhart V (eds) Domesticación y Mejora Genética de raulí y roble. Universidad Austral de Chile-Instituto Forestal, Valdivia

Villagran C (1991) Historia de los bosques templados del sur de Chile durante el Tardiglacial y Postglacial. Rev Chil Hist Nat 64:447–460

Vitasse Y, Delzon S, Bresson CC, Michalet R, Kremer A (2009) Altitudinal differentiation in growth and phenology among populations of temperate-zone tree species growing in a common garden. Can J For Res 39:1259–1269

Vitt P, Havens K, Kramer AT, Sollenberger D, Yates E (2009) Assisted migration of plants: changes in latitudes, changes in attitudes. Biol Conserv 143:18–27

Weinberger P, Ramirez C (1999) Sinecología de la regeneración natural del Raulí (*Nothofagus alpina*) Fagaceae, Magnoliopsida. Rev Chil Hist Nat 72:337–351

Williams MI, Dumroese RK (2013) Preparing for climate change: forestry and assisted migration. J For 111:287–297

Chapter 4
Roble pellín (*Nothofagus obliqua*): A Southern Beech with a Restricted Distribution Area But a Wide Environmental Range in Argentina

María Marta Azpilicueta, Paula Marchelli, Alejandro G. Aparicio, Mario J. Pastorino, Verónica El Mujtar, Cristian Daniel Torres, Javier Guido Puntieri, Marina Stecconi, Fernando Barbero, Liliana Lozano, and Leonardo A. Gallo

4.1 *Nothofagus obliqua* Forests in Argentina

Nothofagus obliqua (Mirb.) Oerst. (Nothofagaceae, Hill and Jordan 1993), known as 'roble', 'pellín', 'roble pellín' or 'hualle', is a tree species endemic to South American temperate forests. It is under protection in Argentina, mostly within national and provincial protected areas. However, because of its high wood quality, similar to *Nothofagus alpina* (= *N. nervosa*) and *Nothofagus pumilio* and in order to supply the timber trade, it is harvested in some of these protected areas under the regulation of forest management plans. Roble pellín's wood is pink yellowish in the sapwood and brown to reddish brown in the heartwood, with a fine texture, homogeneous and right grain (Dimitri et al. 1997) and high density (1.220 kg/dm³ for green wood and 0.720 kg/dm³ for dry wood; INTI-CITEMA 2003). It is characterised by high mechanical resistance, durability and low putrefaction due to the high content of tannins.

M. M. Azpilicueta (✉) · P. Marchelli · A. G. Aparicio · M. J. Pastorino · V. El Mujtar
F. Barbero · L. A. Gallo
Instituto de Investigaciones Forestales y Agropecuarias Bariloche (IFAB) INTA-CONICET, Bariloche, Argentina
e-mail: azpilicueta.maria@inta.gob.ar

C. D. Torres · M. Stecconi
INIBIOMA (UNCo-CONICET), Bariloche, Argentina

J. G. Puntieri
IRNAD (UNRN-CONICET), Bariloche, Argentina

L. Lozano
Parque Nacional Lanín (Administración de Parques Nacionales), San Martìn de los Andes, Argentina

© Springer Nature Switzerland AG 2021
M. J. Pastorino, P. Marchelli (eds.), *Low Intensity Breeding of Native Forest Trees in Argentina*, https://doi.org/10.1007/978-3-030-56462-9_4

Roble pellín wood is used for outdoor construction (sleepers, shipbuilding, poles), as well as carpentry, furniture and musical instruments, but it requires careful drying to avoid crack problems. It reaches in the regional market almost three times the price of that of ponderosa pine, the most planted exotic species in the region (MinAgro 2018). The wood quality of roble pellín encouraged the over-exploitation of its Argentinean forests in the late nineteenth and the early twentieth centuries (González Peñalba et al. 2008), until the creation of Lanín National Park in 1937. Nowadays, within National Park jurisdiction, some roble pellín forests are managed for sustainable timber production.

Nothofagus obliqua extends over a wide precipitation range (about 850–2500 mm/year), expressing tolerance to drought and some degree of plasticity (Varela et al. 2010). It occupies the lowest altitude locations in the mountain, which is also indicative of its better adaptation to the warmest niches. Accordingly, future climate models predict a wider distribution since current warm and dry conditions will spread over a larger area (Marchelli et al. 2017). These ecological characteristics together with its high wood quality make *N. obliqua* a remarkable species for domestication and breeding.

4.1.1 Species Characteristics and Natural Distribution Range

Nothofagus is a genus with disjunct distribution represented by more than 40 species in the world. In South American temperate forests, we can find ten species and three recognized hybrids, with seven species occurring in Argentina. The genus is subdivided in four subgenera, being *N. obliqua* classified in the subgenus *Lophozonia* along with other six species (Hill and Read 1991; Martin and Dowd 1993; Heenan and Smissen 2013), three of which are also South American: *Nothofagus alpina* (Poepp. & Endl.) Oerst., *Nothofagus glauca* (Phil.) Krasser (Veblen et al. 1996) and *Nothofagus macrocarpa* (A. DC.) F. M. Vázquez and R. Rodr. (Vázquez and Rodríguez 1999). Recently, Heenan and Smissen (2013) raised this classification at genus level, meaning that the Latin name of roble pellín would be *Lophozonia obliqua*, but controversies arose (Hill et al. 2015), and former names are currently accepted (Darwinion Institute http://www.darwin.edu.ar/Proyectos/FloraArgentina/fa.htm).

Nothofagus obliqua is a deciduous tree species; leaf blades are lanceolate to ovate, with asymmetric base (hence the name obliqua) and double-serrated edge. Adult individuals present straight stems that can reach up to 40 m in height, characterised by grey bark on plates. The species is monoecious, wind-pollinated, preferentially anemochorous and exhibits protandria. Fruits are winged achenes, subtended on woody cupules (Donoso et al. 2006a) (Fig. 4.1) and fall between March and the beginning of May, depending on the climatic conditions of the year and site features (Veblen et al. 1996). Seed production is characterised by cycles of 2–3 years (Donoso 1993), although according to our field observations, cycles seem longer in

Fig. 4.1 View of a marginal *N. obliqua* forest (**a**), botanical characteristics of the species (**b, c, d**), example of collecting activities (**e, f**) and seedling production (**g, h**)

recent times. This species often develops high percentages of empty seeds (Azpilicueta et al. 2010).

The distribution range of *N. obliqua* extends between 33° and 41° S both west (Chile) and east (Argentina) of the Andes Mountains, along the Pacific Coastal Range, and in the Longitudinal Valley between the Coastal Range and the Andes, covering more than 1.2 million hectares (Araya and Oyarzún 2000). A marked human impact can be observed in the west side of the Andes (Chile) (Donoso 1993), while naturally fragmented forests characterise the eastern distribution (Argentina) (Gallo et al. 2000). In Argentina, this species occupies ca. 33,800 ha along a heterogeneous environmental landscape, from 36° 49′ S to 40° 11′ S at five lake basins (Lácar, Quillén, Ñorquinco, Moquehue and Epulauquen) and a river margin (Aluminé river) (Azpilicueta et al. 2016a; Sabatier et al. 2011; Fig. 4.2). The northernmost population of *N. obliqua* in Argentina (Epulauquen) conforms a particular forest growing at high altitude (1500–1700 m asl) and geographically isolated. More than 220 km separate this population from the nearest con-specific forests in Argentina (Fig. 4.2), ~ 100 km is the distance to the closest Chilean population, on the western side of the Andes. Trees at these fragmented forests show particularities at biochemical, molecular, morphological and architectural levels that make them different from con-specific populations. Based on pollen records (Villagrán 1991) and molecular tools (Azpilicueta et al. 2014), some studies suggested that this population could have originated as an introgression from the western side of the Andean

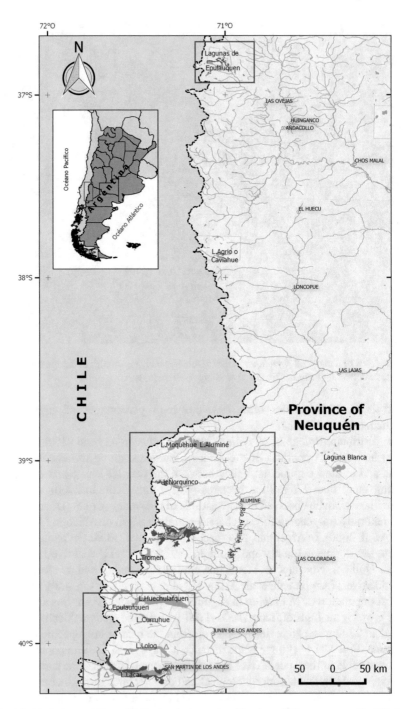

Fig. 4.2 *Nothofagus obliqua* distribution in Argentina. The three different geographical (latitudinal) distribution areas are highlighted (from Azpilicueta et al. 2016a)

Cordillera during postglacial recolonisation. Hybridisation and introgression processes with taxa such as *N. macrocarpa* and *N. leonii* from Chilean populations could not be discarded according to this hypothesis (Azpilicueta et al. 2016b). Further studies are necessary to detect highly divergent molecular markers among these taxa and finally achieve a higher degree of accuracy in the taxonomic identification of robles from Epulauquen population.

Although *N. obliqua* populations grow mostly within a precipitation range of 1200–2500 mm/year in Argentina, the easternmost population, Pilo lil, is located in more xeric conditions, with a precipitation regime of ca. 700 mm/year (Fig. 4.2). Here, the species conforms mixed forests with the native conifer *Austrocedrus chilensis*. Although ecophysiological behaviour of Pilo lil seedlings subject to drought stress experiments could not be differentiated from that of more humid populations (Varela et al. 2010), adaptation processes might be expected at this extreme and xeric environment.

More than 90% of roble pellín forests in Argentina stand in protected areas (Sabatier et al. 2011). However, historical over-exploitation and the restricted area that this species occupy in Argentina highlight the need of a deeply discussion about their present and future conservation and management.

4.1.2 Hybridisation Processes

Natural hybridisation is common between species of the *Nothofagus* genus. *Nothofagus* × *leonii* Espinosa (Donoso and Landrum 1979; Grant and Clement 2004) is recognised as the hybrid between *N. obliqua* and *N. glauca*. Also, hybridisation between *N. obliqua* and *N. macrocarpa* has been suggested (Vázquez and Rodríguez 1999). Likewise, natural hybridisation between *N. obliqua* and *N. alpina* has been reported, firstly based on morphological traits (Donoso et al. 1990) and later confirmed by genetic studies (Gallo et al. 1997), emphasising the evolutionary importance of this genetic process for the long-term maintenance of intra-generic variation. Even more, Vergara et al. (2014) postulated the hybridisation of *N. obliqua* with both *N. alpina* and *N. glauca* in northern Chilean populations in order to explain the admixture found while analysing molecular data of these three sympatric species.

In Argentina, *N. obliqua* hybridises with *N. alpina* (Donoso et al. 1990; Gallo et al. 1997; Marchelli and Gallo 2001; Gallo 2004) in sympatric areas such as the basins of the lakes Lácar and Quillén. Gallo (2002) proposed a conceptual model of introgressive hybridisation for these species, which is characterized by hybridisation asymmetry (*N. obliqua* acting as pollen donor) and bidirectionality of introgression and low fitness of F1 hybrids at low temperatures. The latter feature is based on the results of a progeny trial including both pure and hybrid juvenile individuals, where 15% of the hybrids exhibited higher performance in growth characters but showed a high rate of mortality (~ 40%), probably related to susceptibility to low temperatures (Crego 1999; Gallo 2004). Gallo (2002) also highlighted the

impact of abundance of parental species, directionality of gene flow, gamete competition and hybrid fitness in the introgression dynamics on a local scale. His model
suggests that introgression may be structured along ecological gradients due to
changes in the relevance of different reproductive barriers.

In 2017, El Mujtar et al. begun to test the model by studying the pattern of introgression in a natural hybrid zone of these species at Lake Lácar basin. This study
focused on two plots 280 m apart in altitude (ca. 1.9 °C difference in mean temperature) and two subplots which captured microsite variation (abundance and spatial
distribution of species and predominance of wind direction) within each plot. Based
on intensive sampling of individuals (2055, including adults and regeneration) and
molecular genotyping with six highly species-specific nuclear microsatellites, the
study revealed that introgressive hybridisation occurs at a global rate of 7.8% and
that different types of reproductive isolation barriers could be acting along the altitudinal gradient. At low altitude, introgression occurs at a rate of ca. 11% with a
unimodal distribution of hybrid genotypes, suggesting weak reproductive barriers.
In contrast, at high altitude, introgression occurs at a rate of ca. 6% with an asymmetric bimodal distribution, suggesting stronger reproductive barriers. Moreover,
F1 hybrids were detected at a global frequency of 3.8% and are fertile, according to
the detection of first- and late-generation hybrids. Further studies are necessary to
determine the relative contribution of pre- and post-zygotic reproductive isolation
barriers to the introgression pattern detected along the altitudinal gradient (i.e. sur-
veys of phenological variation, genotyping of seed samples, parentage analysis,
evaluation of relative hybrid fitness).

El Mujtar et al. (2017) corroborated some predictions of Gallo's conceptual model
(Gallo 2002) and contributed with new elements related to the direction and generation
(early and late) of introgression along the altitudinal gradient. Other predictions such as
the asymmetry of hybridisation could not be evaluated as divergent cytoplasmic markers are not available. Other hybrid zones, including altitudinal and rainfall gradients,
should be studied in order to understand *N. alpina* and *N. obliqua* introgressive hybridisation dynamics at a more generalised level and the role of this evolutionary process in
the response of *Nothofagus* species to climate change. Furthermore, such studies,
together with long-term field trials, might contribute to consider the potential relevance
of hybrids as valuable entities for breeding these two species.

4.2   Genetic Variation Patterns Based on Neutral
Genetic Markers

4.2.1   *Inference of Past Demographic Changes: The Impact*
of the Pleistocene Glaciations

Climatic oscillations during the quaternary strongly affected the distribution of
warm-temperate ecosystems, with repeated glaciations occurring since 1.8 million
years BP, from the Pleistocene to the Holocene. These processes affected the

present distribution patterns and the genetic structure of southern South American temperate forests (Veit and Garleff 1996), by their retraction and posterior expansion from glacial refuges, colonisation routes and convergence points.

To evaluate the impact of these changes on the genetic structure of *N. obliqua* populations, a phylogeographical analysis based on chloroplast DNA was carried out including the whole range of this species (Azpilicueta et al. 2009). Three intergenic chloroplast DNA regions were analysed on 27 natural populations of Chile and Argentina. Evidence of glacial refuges in the Chilean Coastal Range were shown (this region harbours high diversity together with old and private haplotypes), which is in agreement with previous studies based on pollen records (Villagrán 1991, 2001). Notwithstanding, refuges were also inferred along the Andes Cordillera and the Longitudinal Valley thus revealing a survival pattern of multiple glacial refuges during the Last Glacial Maximum, ca. 18,000–20,000 years BP. Therefore, and considering only the restricted Argentinean range of *N. obliqua*, two maternal lineages were identified (north and south of Lanín volcano at ~39°40′ S) which seems to be the latitudinal limit, inferring at least two different glacial origins.

In a subsequent study (Azpilicueta and Gallo 2009), 14 Argentinean natural populations were subjected to genetic analysis using isozymes. The genetic pattern found suggested distinct glacial origins for northern and southern populations, as shown with cpDNA markers. Evenmore, and based on allelic richness, relict areas could be postulated within each of the identified latitudinal groups. An additional study based on the analysis of seven nuclear microsatellite markers was carried out in ten Argentinean *N. obliqua* populations (Azpilicueta et al. 2013). Higher mutation rates of microsatellites in comparison with isozymes allowed higher sensitivity to polymorphism detection. Once again, higher allelic richness values were found in the surroundings of areas identified as potential glacial refuges, what suggests that the distribution pattern of allelic richness is closely associated with the history of *N. obliqua* in this region during the glaciations. Likewise, along the Chilean distribution area, Vergara et al. (2014) analysed 20 populations of this species with 7 microsatellites. Three latitudinal regions were detected (at north and south of 36° S and south of 38°30′ S, respectively), allowing the reconstruction of the area affected by the most recent ice age based on the genetic structure of *N. obliqua* and suggesting several centres of genetic diversity in agreement with the multiple refugia hypothesis.

The combined analysis of cpDNA intergenic regions, isozymes and microsatellites allowed the recognition of potential refuges east of the Andes Cordillera at Epulauquen and the likely convergence points of postglacial migratory routes. Rucañancu (39°33′ S) is postulated as a potential refuge at the Longitudinal Valley in Chile with a migratory route coming from the north also confluencing in this area. In Argentina, Epulauquen forests could have probably conformed an Eastern refuge, with the arrival of a Western introgression after ice-retraction (Azpilicueta and Gallo 2009; Azpilicueta et al. 2009, 2013). Biochemical and molecular neutral markers proved to be valuable tools for the study of past demographic changes and distribution levels of genetic variation in order to retrace the evolutionary history of *N. obliqua* forests, allowing their historical reconstruction (Azpilicueta and Gallo 2009; Azpilicueta et al. 2009, 2013; Vergara et al. 2014).

4.2.2 Level of Genetic Diversity as an Indicator for Conservation

An analysis on ten Argentinean *N. obliqua* populations based on information provided by three types of genetic markers (isozymes, chloroplast DNA and microsatellites) (Table 4.1) allowed the identification of the most diverse populations: the so-called hotspots of genetic diversity. Allelic richness is considered the best parameter for defining conservation actions (Petit et al. 1998). In order to combine this parameter for maternal (haploid, based on cpDNA) and biparental (diploid, based on isozymes and microsatellites) markers, a parameter was standardized as described by Marchelli et al. (2017). The constructed multiple-markers parameter is calculated as follows: allelic richness at population level was obtained for each marker and subsequently divided by the average value of the species for the same marker type, and finally all standardized allelic richness values were summed up across markers. Since cpDNA has a lower mutation rate, but polymorphism revealed ancient lineages, its contribution was doubled to the overall richness score. Figure 4.3 shows with different coloured dots the diversity level observed at each *N. obliqua* Argentinean population based on the combined analysis of the three markers (Azpilicueta et al. 2013, 2016a).

The highest genetic variation was observed at the north of its range (Fig. 4.3a, b, Table 4.1). Within the southern forests, eastern populations exhibited higher genetic diversity, suggesting that this area could also have been a glacial refuge for *N. obliqua* (Azpilicueta et al. 2013) (Fig. 4.3b, Table 4.1). Although the information is

Table 4.1 Geographical location and corresponding genetic zone of the analysed *Nothofagus obliqua* Argentinean populations

Watershed	Pop	ID	Lat (S)	Long (W)	Alt (m asl)	Pption (mm/ year)	Genetic diversity level	Genetic zone
Lácar	Bandurrias	L1	40°09′00″	71°21′00‴	850	1,400	Intermediate	2
	Yuco	L2	40°09′07″	71°30′39″	930	1,900	Low	3
	Quilanlahue	L3	40°08′18″	71°28′04″	910	2,000	High (SSRs)	3
	Nonthué	L4	40°08′46″	71°37′03″	680	2,450	Intermediate	3
	Hua Hum	L5	40°07′55″	71°40′02″	670	2,500	Low	3
	Quila Quina	L6	40°10′40″	71°26′37″	980	1,300	High	3 (sz)
	Catritre	L7	40°10′26″	71°24′10″	700	1,300	-	3
	Pío Protto	L8	40°06′48″	71°14′31″	840	900	-	3
Quillén	Corral Bueyes	Q1	39°22′16″	71°17′31″	1,140	1,500	Intermediate	1
Ñorquinco	Seccional	Ñ1	39°09′11″	71°15′03″	1,070	1,500	High	1
Aluminé	Pilo lil	P1	39°30′05″	70°57′44″	835	700	Low	1
Epulauquen	Epulauquen	E1	36°49′09″	71°04′07″	1,500	1,500	High	1 (sz)

Genetic diversity level estimated by means of cpDNA, isozyme and microsatellite markers (except for Quilanlahue population)
References: sz (subzone)

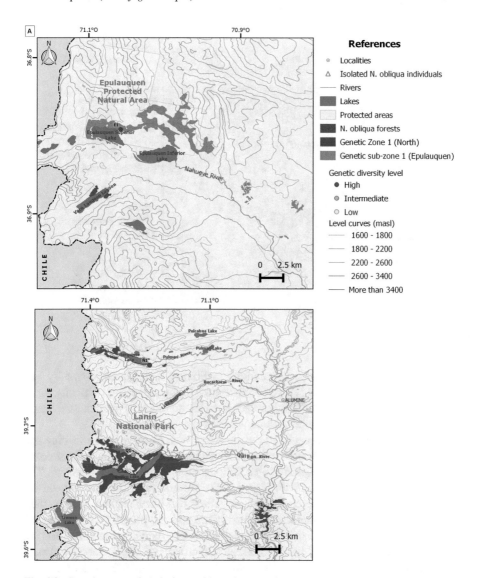

Fig. 4.3 Genetic zones of *Nothofagus obliqua* in Argentina at North (**a**) and South (**b**) of its distribution. Level of genetic diversity is defined with different colour dots: red, high genetic diversity; orange, intermediate genetic diversity; and yellow, low genetic diversity

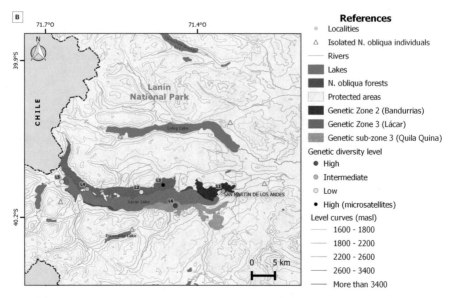

Fig. 4.3 (continued)

provided by neutral markers, populations exhibiting high genetic diversity and assumed as probable glacial refuges likely have a higher adaptation capacity, since within refugial areas, populations have survived for hundreds of generations, facing different environmental scenarios along which the species evolved.

To face climate change predicted for the Patagonian region – with warming and decreasing annual precipitation – information concerning genetic diversity at population level should be considered as an additional tool for seed source selection. The identification of populations with high priority for conservation purposes should then rely on their genetic diversity level, together with other features such as their risk of extinction, their general ecological value and the presence of endemisms harboured by them. For assisted migration programs under progressive climate change, the identification of genetic zones together with knowledge about hotspots of genetic diversity should constitute a relevant input for planning the activities.

4.2.3 Genetic Zones as Seed Sources for Breeding and Restoration

Development of appropriate forest conservation and management measures for a given tree species includes the identification of distinct genetically homogeneous zones across its range on the basis of information on molecular and phenotypic variation within and between populations. Such management units are pivotal for steering the decision-making process concerning germplasm transfer within and between areas, for example, in reforestation and/or restoration activities (Newton et al. 1999).

Vergara (2000) defined provenance regions for the total range of *N. obliqua* in Chile and Argentina based on climatic information, together with vegetation maps, phytogeographic limits and genecological variation (Donoso 1979). Focusing on the restricted and fragmented distribution of this species in Argentina, Azpilicueta et al. (2013, 2016a) defined genetic zones for *N. obliqua* at the eastern side of the Andes based on the available neutral genetic information, thus giving a deeper support for the identification of genetically homogeneous units (Table 4.1, Fig. 4.3). The movement of propagation material within genetic zones minimizes the chance of genetic structure changes. The identification of genetic zones contributes to the definition of genetic management units thus helping in breeding and conservation programs.

The four northernmost watersheds where *N. obliqua* occurs in Argentina (Epulauquen, Ñorquinco, Quillén and Aluminé) were clustered in one unique group (North genetic zone, Fig. 4.3a), whereas the forests at Lake Lácar were subdivided in two different clusters, one being composed of a single population (Bandurrias genetic zone) and the other by the rest of the populations (Lácar genetic zone) (Fig. 4.3b).

In summary, three genetic zones were defined for *N. obliqua* in Argentina (Table 4.1, Fig. 4.3). However, additional subdivisions must be recognized in two of them according to genetic particularities: the Epulauquen subzone in the North genetic zone and the Quila Quina subzone in the Lácar genetic zone. These subdivisions were mainly defined based on the presence of different variants at chloroplast DNA level coexisting with the shared genetic zone haplotype (Azpilicueta et al. 2013, 2016a). Future transfers of planting material for restoration activities and assisted natural regeneration of Argentinean *N. obliqua* forests have to be limited to the same genetic zone and subzone from where seeds or plants were collected. This information could also be used as a guiding tool for identifying the origin of seed or other plant material.

4.3   Quantitative Genetic Variation and Phenotypic Plasticity

Genetic variation estimated through the analysis of quantitative traits is crucial to study adaptive processes of the past and to forecast adaptive responses in case of eventual environmental changes. Classical provenance and progeny tests are a key tool for this purpose, and the design of multiple-site trials additionally allows studying phenotypic plasticity, which could be a decisive evolutionary strategy of the populations to persist in situ in the face of climate change (Aitken et al. 2008).

4.3.1 Seed and Germination Analyses

Germination is a vital process subjected to natural selection, and therefore the analysis of variation in related seed traits is relevant to understand its modulation. Aiming to unravel the effects of several seed traits on adaptation processes of roble pellín, a study on natural populations was done (Barbero 2014). Seeds were collected from 116 trees corresponding to eight populations of roble pellín, covering the entire distribution area of this species in Argentina. The proportion and weight of filled seeds were measured, and, subsequently, a germination assay was performed in germination chambers, adjusting a cumulative germination curve per mother tree by means of nonlinear regression (Gompertz function). Mean proportion of filled seeds per population varied between 4% and 57%. Differences among populations and among mother trees were significant. There were also large differences among populations regarding the weight of 100 filled seeds, which ranged from 0.82 to 2.68 g and was significantly affected by both population of origin and mother tree (Fig. 4.4). With regard to the test, mean germination capacity varied among the seven analysed populations from 89.3% to almost null (Fig. 4.4), resulting significant both the *population* and the *mother tree* factor (the latter explained 39.6% of the total variance). It must be highlighted that Epulauquen, the northernmost population, is the most different considering these three traits.

4.3.2 Survival, Growth and Morphological Seedling Traits

Genetic variation in seedlings height at the end of the first growing season was analysed by Barbero (2014), in a greenhouse provenance and progeny trial. The experimental design included 42 potted seedlings from each of 100 open-pollinated families corresponding to seven Argentinean natural populations. Mean height of

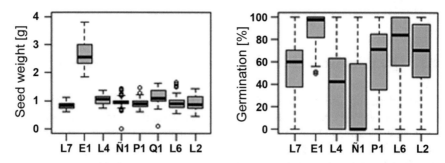

Fig. 4.4 Boxplots showing the median, the upper and lower quartiles and the minimum and maximum data values of seed weight and germination capacity of roble pellín studied populations (Barbero 2014). Populations ID are shown in Table 4.1

the entire trial was 53 cm, being Epulauquen the population with, on average, the smallest seedlings. The effects of both population and family were significant; family explained 5% of the total variance. Accordingly, intrapopulation variation was low (mean $h^2 = 0.19$), whereas differentiation resulted moderate ($Q_{ST} = 35\%$). It is noteworthy that the estimation of Q_{ST} after excluding the Epulauquen population fell down to 17%.

During 2004, a field trial including ten natural *N. obliqua* provenances was installed at Lake Lácar, Lanín National Park (40° 07′ 48″ S, 71° 28′ 48″ W, 920 m asl) with the aim of evaluating genetic variation at growth traits and survival, beneath a mixed forest of *Nothofagus dombeyi* (Mirb.) Oerst, *N. obliqua* and *N. alpina* (Azpilicueta et al. 2014). The site is characterised by volcanic soils and a mean precipitation of 2200 mm/year. Survival and height were measured at the first and fourth years from trial installation, and both measurements were used to calculate growth rate. Tree survival was high (94% and 91% at the first and fourth year after installation, respectively), according to the optimal environmental conditions of the trial site. Analysis of variance revealed that the *population* effect gave rise to significant differences for both years of measurements regarding total height and relative growth. According to Tukey's tests, mean plant height was significantly lower for Epulauquen than for all other analysed populations, except for two of them also belonging to the north of the species range. On the other hand, Epulauquen exhibited the highest relative growth rate, significantly different from all the other populations except for two northern populations belonging to the watersheds of Quillén and Ñorquinco.

Another field trial was installed in 2009 in Bariloche Experimental Station of INTA (Instituto Nacional de Tecnología Agropecuaria) (41° 07′ 21.17″ S; 71° 14′ 56.95″ W; 795 m asl) with the aim of analysing genetic variation in early traits (Barbero 2014). It was composed by 40 open-pollinated families from four environmentally contrasting natural populations of Argentina (Epulauquen, Pilo lil, Yuco, Catritre; 24 seedlings per family; N = 960). It had a randomised complete block design, with single-tree plots and seedlings distributed at high density, with a plantation frame of 25 × 25 cm, with the purpose of reducing the environmental noise.

In order to analyse growth traits, during the second growing season, the height of each sapling was measured every 10 days so as to build an individual growth curve for each sapling by means of nonlinear regressions (Boltzmann model). On average, growth initiated (t_{10}: time to reach 10% of the season growth) in the 332 day of the year (DOY) (November 27); whereas growth ceased (t_{90}: time to reach 90% of the season growth) in the 388 DOY (January 22). Differences for t_{10} were proved only for Epulauquen population, which was the last to initiate growth, thus evidencing a low interpopulation differentiation for this trait ($Q_{ST} = 11\%$). The significance of the *family* factor could not be proved. The variation for t_{90} was higher, but the pattern was similar: *family* was also not significant, and the most different population was Epulauquen which, on average, was the first to cease its growth. Differentiation in this case resulted moderate to high ($Q_{ST} = 47\%$).

Table 4.2 Genetic variation of four environmentally contrasting Argentinean *Nothofagus obliqua* populations in quantitative traits measured in a progeny trial installed in Bariloche (see description above)

Trait	Mean	Mean h^2	CV_A	Q_{ST}
Shoot height 1st year [cm]	61.0	0.22	8.96	
Shoot height 2nd year [cm]	97.1	0.25	14.03	–
Collar diameter 1st y. [mm]	8.1	0.07	5.16	–
Collar diameter 2nd y. [mm]	14.1	0.03	2.68	–
Height growth initiation 2nd y. [DOY]	332	0.02	3.47	0.11
Height growth cessation 2nd y. [DOY]	388	0.06	4.13	0.47
Branching index	4.5	0.09	19.62	–
Specific area of preformed leaves [cm²/g]	128.3	0.12	11.17	0.05
Marcescent leaves 2nd year [%]	37.2	0.52	6.49	0.18
Bud burst [DOY]	257.6	0.33	1.13	0.09
Bud set [DOY]	446.4	0.07	6.84	0.67
Vegetative period [days]	188.9	0.27	1.84	0.38

Mean value, mean narrow sense heritability, additive genetic variation coefficient and quantitative differentiation (from Barbero 2014)

4.3.3 Phenological Characters

Bud burst and bud set phenologies were evaluated during the second growing season in the last mentioned trial representing 40 open-pollinated families from four environmentally contrasting natural populations of Argentina (Barbero 2014). The phenophase of the terminal bud of each sapling was registered every other day, according to a scale of five categories (similar to that presented in Chap. 3 for *N. alpina*), from September 4 to October 28 (bud burst) and from March 14 to April 1 (bud set).

On average for the entire trial, buds were open (phenophase 3) in the 258 DOY (September 15) and closed in the 446 DOY (March 21). Both factors, *population* and *family*, were significant for bud burst, but only *population* had a significant effect on bud set. Similarly to growth initiation and finalization, Epulauquen was the most different population for these traits (it was, on average the last in bud bursting and the first in bud closing), and differentiation among populations was estimated as low for bud burst (Q_{ST} = 9%) but high for bud set (Q_{ST} = 67%). Intrapopulation variation presented the opposite pattern, that is, higher variation for bud burst ($h^2 = 0.33$) than for bud set ($h^2 = 0.07$). Mean bud and growth phenologies measured in Epulauquen population were consistent with its mean growth, since it was the population with the shortest vegetative period and the lowest height reached at the end of the first growing season, as observed in other studies (Azpilicueta et al. 2014). The results of this trial regarding the main traits analysed (those already described and some others) are presented in Table 4.2.

4.3.4   Resilience Against Eventual Damages

Seedling resprouting may have profound effects on population fitness, since it implies the survival of novel and recombined genotypes and thus may enhance opportunities for adaptive evolution under climate warming and associated changes in the frequency and severity of disturbances. With regard to domestication, resprouting might be an interesting trait to cope with wild and domestic animal browsing, a limiting factor for the cultivation of roble pellín.

In order to analyse genetic variation and plasticity in seedling resprouting ability, the previous trial (40 families from 4 populations) and a replica of it installed 400 m apart in altitude were analysed together after a simulated damage (Aparicio et al. 2015). Just before starting the third growing season (September 2015), all 1920 saplings of both replicas were cut at ground level removing their entire aerial biomass. Seven weeks later almost all the seedlings resprouted in both replicas (Fig. 4.5) but with varying profusion (RP: number of dominant and codominant shoots per plant) and vigor (RV: the percentage of pre-clipping seedling height reached by the dominant resprout). For RP, differentiation among populations was moderate in the low elevation replica ($Q_{ST} = 38.5\%$) and high in the high elevation replica ($Q_{ST} = 66.5\%$), while the intrapopulation variation resulted low (h^2 was not significant in both replicas). On the other hand, for RV, differences among populations were observed, but differentiation was estimated as very low, whereas variation within populations was also very low. The main contrasts between populations were found for pre-clipping seedling size and resprouting profusion, the latter trait showing a clear trade-off with resprouting vigour. Site × population interactions were due mainly to the behaviour of the highest altitude population (Epulauquen), suggesting its divergent adaptive trajectory and higher plasticity for resprouting traits.

Additionally, damage to resprouts was recorded for the highest (coldest) site after an episodic summer frost (SFD, the percentage of green tissue damaged) and

Fig. 4.5 Roble pellín resprouting trial. Seedling cut at the age of 3 years, 54 days after cutting and 5 months after cutting

seasonal autumn cold weather (AFD, dominant and/or co-dominant apices damaged or not, as a binary trait). The early-summer frost caused moderate to severe apex and leaf damage (SFD = 37.5% of tissues damaged, on average), being Epulauquen and Pilo lil populations the least affected, whereas the autumn cold spells caused apical damage in 64% of resprouted plants, being Pilo lil the least affected population. Both types of damage were correlated (r = 0.40; p < 0.001). The differentiation among populations for SFD was moderate (Q_{ST} = 36%) and significantly higher than the neutral differentiation estimated through molecular markers (Aparicio et al. 2015).

A further study of phenotypic plasticity was performed by means of a field trial installed in an altitudinal gradient at Lake Lácar basin (40° 08′ S, 71° 28′ W) (Barbero 2014). In this case only two natural provenances of contrasting altitude (650 and 890 m asl, respectively) were planted in three altitudinal levels (at 610, 850 and 1100 m asl). Survival of the 360 seedlings was registered at the ages of 2 and 3 years, showing differences among sites but not between the provenances for both ages, thus evidencing plasticity. Additionally, bud burst phenology was analysed with point observations, in which the phenophase of each seedling was registered. Again, differences among altitudinal sites were shown, but not between provenances, revealing the plasticity of this trait.

4.3.5 Some Conclusions Based on Quantitative Genetic Variation and Phenotypic Plasticity Studies

The traits analysed in these trials lead to the conclusion that there is evidence of local adaptation, whose modulation is directly related to environmental variations (Barbero 2014). Thus, the largest differences are found in the northernmost and highest-altitude origin (Epulauquen) and secondly in the origin of the eastern and dry extreme (Pilo lil), with fewer differences between the origins corresponding to less extreme conditions, located at the Lake Lácar basin.

4.4 On the Way to Domesticate Roble pellín

The first Argentinean plantations of roble pellín were established in the historical nursery of Nahuel Huapi National Park, in Isla Victoria. There are still about 50 *N. obliqua* trees that were planted in 1943 (Adolfo Moretti, pers. comm.). Likewise, a few trees from a small plantation established in 1956 at the Trevelin Forest Station of INTA are still standing. At the age of 50 years, this stand had a mean height of 31.9 m and a quadratic mean diameter of 44.8 cm. Moreover, the physical-mechanical features of the timber of some trees sampled at that age were found to be similar to those of older trees coming from the natural forest (Mondino and Tejera 2006). More recently, a trial established in 2005 in two plots at a first quality site near to Lago Puelo (42° 1′50″ S; 71°33′12″ O; 380 m asl) had a 90% survival 11 years later, with a mean height of 7.5 m ± 0.2 and 8.1 m ± 0.3, and a mean DBH of 10.6 cm ± 0.4 and 7.0 ± 0.3 in both plots, respectively (Urretavizcaya et al. 2018).

Likewise, early interest in planting South American *Nothofagus* species also existed abroad. Similarly to raulí, roble pellín has been assayed in Great Britain since the beginning of twentieth century (Mason et al. 2018), showing a sensitivity to spring and autumn frosts and severe winter cold. This suggests that its plantation should be confined to milder regions, often near the coast (Murray et al. 1986). There are about 40 ha of roble pellín in the Public Forest Estate in Britain, nearly all in England.

The technology for seedlings production in Argentina has been very well developed in the last decade (e.g. Azpilicueta et al. 2010). A ferti-irrigation protocol has been adjusted for the South American *Nothofagus*, which are very responsive to fertilization, making thus possible to get routinely 50-cm tall roble pellín seedlings in 8 months under greenhouse conditions (Schinelli Casares 2013). On the other hand, the regular provision of good quality seeds is a bottleneck for the industrial production of roble pellín seedlings. Not only the proportion of filled seeds is quite variable among populations (as mentioned above), but also seed production per tree varies sharply from year to year (Donoso 1993; Donoso et al. 2006a; Pastorino et al. 2016, Fig. 4.6). This uncertainty in the availability and quality of seeds from natural forests must be considered for the domestication of this species. In this regard, the existence of a 16-year-old roble pellín plantation producing seeds abundantly with a 60% proportion of filled seeds (Pastorino et al. 2016) encourages the establishment of seed orchards.

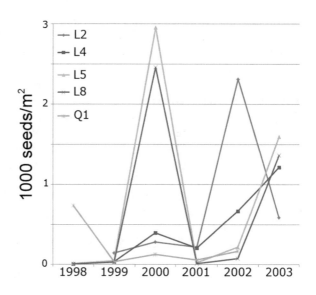

Fig. 4.6 Variation of seed production among years and among populations in natural forests, measured in thousands of seeds/m² under-canopy nets (from Pastorino et al. 2016)

Fortunately, roble pellín seeds are orthodox, and therefore their viability can be maintained at 4 °C for several years if they are stocked at a humidity lower than 10% (Escobar and Donoso 1996). Thus, mast years should be considered for collecting and stocking seeds.

4.5   First Steps Towards Breeding of Roble pellín in Argentina

The oldest provenance trials of *N. obliqua* were installed in Britain in 1979 and in 1981 but almost exclusively with Chilean provenances and local seed sources (Mason et al. 2018). Only one Argentinean provenance (Lake Lácar basin) was assayed and only in the two trials of 1981. This provenance resulted in the best in comparison with two provenances from Chile (100% survival and 6.1 m tall at 10 years in Yorkshire; 99% survival and 7.7 m tall at 13 years in Scotland). Meanwhile, in Argentina breeding of roble pellín has begun in the 2000s, with the first field trials installed in 2004. That year, INTA established a provenance trial in Lanín National Park (40° 07' 48''S, 71° 28' 48'' W, 920 m asl) representing ten Argentinean natural populations. The average survival four years later was 91%, with a mean height of 115.82 cm ±42.26 cm. Differences in trunk height were found among populations: one population from the north (Lake Quillén basin) included the tallest individuals, and the northernmost population of Argentina (Epulauquen) had the shortest ones (Azpilicueta et al. 2014), which is consistent with the greenhouse and field trials performed by Barbero (2014).

The distinct character of Epulauquen population has repeatedly been shown, both through genetic markers as through variation in quantitative traits. This distinctness is relevant for breeding since it is likely to be caused by adaptation to cold and/or high altitude environments (this population is located at 1500 m asl), which have an effect similar to that of drought (e.g. Yang et al. 2010). Thus, Epulauquen could be a key provenance for the cultivation of the species in harsh sites characterised by drought or out-of-season frosts. Based on these considerations, a Progeny Seed Orchard composed by 20 open-pollinated families exclusively of Epulauquen population was installed in the Trevelin Forest Station of INTA in 2004 (N = 480). In a first evaluation at 7 years of age (data not published), the mean height of this seed orchard was 199 ± 63 cm, with a survival of 96%. Differences among families were shown at this early stage, so further genetic evaluations will allow selecting the best genotypes and subsequently thinning the trial to finally turn it into an improved Progeny Seed Orchard. At the light of this preliminary results, a new progeny trial exclusively of the Epulauquen provenance has been installed in 2013, this time with 25 open-pollinated families (N = 600) and located in a property of CORFONE, the most important plantation company of Argentine Patagonia, near San Martín de los Andes (40° 4' 57'' S; 71°20'26'' W; 1125 m asl). At the age of three years, with a

94.7% survival, mean tree height in this trial was 142 ± 44 cm, and differences among families were found (data not published). Three years later, mean tree height more than doubled (294 ± 65 cm); total maximum height was 5 m; and survival rate was still high (81.3%). Differences among families continued to be significant and even more acute, with 3.50 m of mean tree height for the tallest family and 2.47 m for the shortest one (data not published), thus giving rise to genetic improvement.

The first trials and experiences led INTA to formally plan a low intensity breeding program for roble pellín in 2006. The main purpose was to guarantee the supply of seeds with identified origin and a genetic quality not lower than the average. Before that, seeds for seedling production were obtained exclusively from natural forests which, because of their cyclical seed production, are an uncertain source. An agreement between INTA and the National Park Administration allowed collecting seeds from Lanín National Park where 83% of the areas with roble pellín in Argentina are located (Sabatier et al. 2011). In the frame of this agreement, natural stands with good seed productivity and accessibility were identified. More experience in seed collection eventually led, in 2008, to register with the National Seed Institute (INASE) a pure 5 ha *N. obliqua* stand (40° 09′ 6.05″ S; 71° 30′ 41.54″ W; 870 m asl) as a Seed Production Area (SPA). Two years later a new natural stand, this time pertaining to a provincial reserve (Epulauquen, 36° 49′ 6.20″ S; 71° 04′ 14.5″ W; 1500 m asl), was also registered as a SPA. After several years of seed collection from the natural forest, it became clear that if a more regular provision of seeds was desired, seeds had to be collected from artificial stands. Thus, INTA decided to transform a 16-year-old plantation located in its Trevelin Forest Station (43° 07′ 3″ S; 71° 33′ 45″ W; 420 m asl) in a 2.35 ha SPA, which was registered with INASE in 2016 (it started to produce seeds 3 years before, in 2013). The origin of this SPA was unknown, so a subsequent study based on chloroplast DNA markers was performed, finally suggesting its belonging to the Lácar genetic zone (Azpilicueta et al. accepted).

Yet, seeds collected in these SPA lack genetic improvement. In this sense, the initial breeding strategy was to install Clonal Seed Orchards with trees selected in the natural forest (massal selection) and in young trials (genetic selection). However, similarly to raulí (see Chap. 3), the poor performance achieved in attempting vegetative propagation, both through rooting and through grafting, and its dependence on the genotype (Pastorino et al. 2016), led to change the breeding strategy to the installation of Progeny Seed Orchards. Consequently, INTA took advantage of the high density progeny trial composed by 40 open-pollinated families and its replica that were analysed by Barbero (2014) and Aparicio et al. (2015), to perform an early genetic selection in order to install a progeny seed orchard. Progenies were evaluated from three out of the four populations assayed (Catritre, Pilo Lil and Yuco). Thus, selection was carried out among 1440 trees according to 2 main criteria: height at 3-year age and cold resistance after an episodic summer frost and the seasonal autumn cold. All saplings were cut at ground level and then potted and grown in a nursery for 2 years. Finally, the selected trees were planted in San Martín Forest Station of INTA (42° 0′ 22.39″ S; 71° 32′ 50.78″ W; 285 m asl) in 2016, with a single-tree-plot design, keeping the saplings of the same family apart from each

Fig. 4.7 Progeny seed orchard of *Nothofagus obliqua* installed in 2016 in San Martín Forest Station of INTA. (Photos: M. Pastorino)

other. After the initial mortality, the progeny seed orchard was constituted by 358 trees from 10 open-pollinated families (Fig. 4.7). Since the saplings were planted at 7 years of age, they are expected to begin seed production in 2022.

4.6   Intraspecific and Interspecific Controlled Crossings of *N. obliqua*

Controlled crossings, both intra- and interspecific, constitute a valuable tool in forest improvement programs, allowing the production of progenies from selected parental trees. Here, we resume some experiences regarding the application of controlled crossings in *N. obliqua*. Flowers of *Nothofagus* species are little evident, especially to pistillate flowers, which are often very difficult to spot. Therefore, both the topology and the phenology of flowers must be thoroughly understood in order to carry out manual pollinations in these species.

4.6.1   *Flower Topology in* **N. obliqua**

Staminate flowers of *N. obliqua* are arranged in 1- to 3-merous inflorescences, whereas pistillate flowers are arranged invariably in inconspicuous, 3-merous inflorescences (Fig. 4.8a). Flowering shoots of *N. obliqua* mostly produce both staminate and pistillate inflorescences, although short shoots may produce only staminate inflorescences. In adult trees, both flower types may derive from shoots of all axis categories (main, secondary and short branches) except the trunk and in numbers

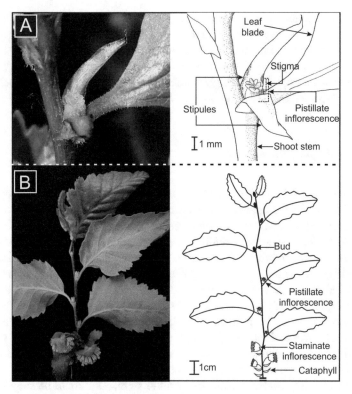

Fig. 4.8 Picture and schematic representation of (**a**) a pistillate inflorescence (3-merous) and (**b**) a typical annual flowering shoot of *N. obliqua.* (Adapted from Torres and Puntieri 2013)

directly proportional to shoot size; the number of pistillate inflorescences per shoot is more dependent upon shoot size than the number of staminate inflorescences (Torres et al. 2012). Within each flowering shoot, staminate inflorescences are invariably located in proximal nodes, corresponding to the axillary productions of cataphylls and proximal green leaves. On the other hand, pistillate inflorescences arise from nodes located distally respect to those with staminate inflorescences and are always subtended by green leaves (Puntieri et al. 2009; Fig. 4.8b). This general pattern equals that observed in other *Nothofagus* species.

Flowers of *N. obliqua* complete their development during the extension of their bearing flowering shoots. Due to inflorescence position, flowering phenology largely depends on the dynamics of primary shoot growth, since the proximal-to-distal sequence of internode elongation is paralleled by a similar sequence of flower deployment. Because of their proximal axillary position, staminate inflorescences expand during budbreak, whereas pistillate inflorescences are exposed later on, as their subtending leaves unfold. Even though the general pattern of distribution of staminate and pistillate inflorescences is similar in all *Nothofagus* species, some interspecific variations could be observed regarding the precise location of both inflorescence types. In *N. obliqua*, nodes bearing pistillate inflorescences are consecutive to those bearing staminate inflorescences, whereas both flower types are separated by one to several sterile nodes in the closely related species *N. alpina*. Therefore, at the shoot level, a higher probability of geitonogamous self-pollination would be expected in *N. obliqua* than in *N. alpina* (Puntieri et al. 2009).

4.6.2 Flower Phenology

The knowledge of the periods of pollen release and pistillate flowers receptivity is highly relevant for the application of technics of manual pollination as well as for understanding the processes conditioning seed production in natural populations. In several *Nothofagus* species, including *N. obliqua*, the production of great amounts of seedless fruits is common (Donoso et al. 2006a; Azpilicueta et al. 2010; Torres et al. 2019). Factors related to such a pattern are rather unknown and may include pollen limitation, self-interference and resource limitations, among others. Concerning self-interference, previous observations have suggested high degrees of overlapping between the periods of pollen release and stigma receptivity at the shoot level (Torres and Puntieri 2013). In the cited research, the period of stigmatic receptivity of *N. obliqua* was assessed through the application of manual sequential pollination in hydroponically grown shoots.

After evaluating pollen germination on the stigmas, it could be suggested that pistillate flowers may be receptive as soon as the expansion of their subtending leaves allows the exposure of the stigmas to air currents. Pollen germination on the stigmas was observed until 15 days after the anthesis of pistillate flowers, whereas maximum pollen germination occurred between 3 and 6 days after the anthesis of pistillate flowers. Preliminary results have shown that pollen release would be highly dependent upon air temperature and humidity (Torres 2013). Therefore, it could be hypothesized that the degree of temporal intra-individual overlapping between pollen release and stigma receptivity could be linked to climatic conditions during early shoot growth.

4.6.3 Test of Manual Pollinations

So far, there are few experimental populations of mature *Nothofagus* trees in Argentina, and *Nothofagus* seeds are mainly collected from natural populations. High interannual variations in flower and, therefore, seed production are typical in several *Nothofagus* species, including *N. obliqua*. Thus, the simultaneous occurrence of high flower production in selected trees is one of the main conditioning factors for controlled crossings.

Cycles of seed production vary even between closely related species. Previous observations in populations from Argentina and Chile suggest that high seed production occurs every 3 years in *N. obliqua* (Donoso et al. 2006a). For *N. alpina*, in contrast, cycles of high flower production would occur every 2 years (Donoso 1993; Marchelli and Gallo 1999; Donoso et al. 2006b). Nevertheless, at an individual scale, such periodicities are to be taken cautiously, since inter-individual variations in the periodicity of flower production have been observed even between neighbour con-specific trees (Torres et al. 2016). Years of massive seed production are characterised by both a high production of seeds per plant and a high proportion of seed-producing trees (Marchelli and Gallo 1999).

Phenological concurrence is a critical/important issue to be considered for controlled crossings: viable pollen from the selected pollen-donor tree must be available within the period of stigmatic receptivity of the pollen-receptor tree. In this regard, it has been demonstrated that, under natural conditions, pollen viability of *N. alpina* is drastically reduced 4 days after pollen release (García et al. 2015) so that both pollen viability and stigmatic receptivity provide a limited time-lapse available for pollination. This inconvenience may be sorted out by obtaining pollen from cut-flowering branches before the natural occurrence of anther dehiscence. Pollen release may be accelerated by keeping cut-flowering branches in hydroponic culture in a warm and dry environment (Torres and Puntieri 2013). To guarantee pollen collection, flowering branches should be cut after the unfolding of the perigonium of staminate flowers, when stamen lengthening begins.

Manual pollinations were applied to assess the degree of self-interference and the compatibility between *N. obliqua* and *N. alpina* in the two possible crossing directions (Torres and Puntieri 2013). In trees of both species, the following pollinations were applied: (I) cross-pollination, (II) self-pollination, (III) cross-pollination followed by 24 h-postponed self-pollination, (IV) self-pollination followed by 24 h-postponed cross-pollination, (V) simultaneous cross-pollination and self-pollination and (VI) interspecific pollination. Pollen germination on the stigmas and seed viability were assessed after each treatment. The obtained results suggested that the stigma would be the first barrier to self-fecundation (Figs. 4.9 and 4.10). In

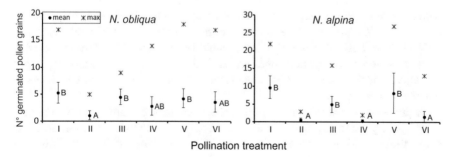

Fig. 4.9 Average (± SE) and maximum number (asterisks) of pollen grains germinated on the stigmas after (I) cross-pollination, (II) self-pollination, (III) cross-pollination followed by 24 h-postponed self-pollination, (IV) self-pollination followed by 24 h-postponed cross-pollination, (V) simultaneous cross-pollination and self-pollination and (VI) interspecific pollination. (Adapted from Torres and Puntieri 2013)

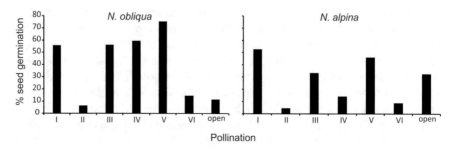

Fig. 4.10 Percentage of seed germination obtained from three *N. obliqua* and three *N. alpina* trees after the following pollination treatment: (I) cross-pollination, (II) self-pollination, (III) cross-pollination followed by 24 h-postponed self-pollination, (IV) self-pollination followed by 24 h-postponed cross-pollination, (V) simultaneous cross- and self-pollination, (VI) interspecific pollination and open pollination. (Adapted from Torres and Puntieri 2013)

this regard, self-pollinated flowers presented low pollen germination in the stigmas and produced few viable seeds. When mixed pollinations were applied, self-interference was confirmed for those cases in which self-pollination preceded cross-pollination; in such cases the probability of compatible pollen-tube formation was hindered, especially in *N. alpina* (Fig. 4.9). Consistently, self-pollination prior to compatible cross-pollination tended to reduce the production of viable seeds in this species (Fig. 4.10). These results suggest that self-interference could potentially reduce seed production in natural conditions in *N. alpina*.

Regarding interspecific crosses, certain degree of compatibility between *N. obliqua* and *N. alpina* was also confirmed in this study, although high inter-individual variations were observed. In natural populations, interspecific hybridisation always seems to occur in the sense *N. alpina* x *N. obliqua*, i.e. with *N. obliqua* acting as pollen donor (Gallo et al. 1997). Based on the results presented by Torres and Puntieri (2013), the probability of hybridisation in the opposite crossing direction (i.e. with *N. obliqua* acting as ovule donor) may not be totally discarded. Nevertheless, in natural conditions, budbreak of *N. obliqua* precedes that of *N. alpina*. Therefore, hybridisation through the pollination of *N. alpina* stigmas with *N. obliqua* pollen may seem more probable than that in the opposite sense, as no pollen of *N. alpina* would be airborne during the period of stigmatic receptivity of *N. obliqua*. These preliminary results would be important for future applications of intraspecific crossings in *N. obliqua*, as well as for interspecific crossings with *N. alpina*, which would represent a valuable tool for the selection of characters of interest of these species.

A preliminary test of manual pollinations has been performed in *N. alpina* x *N. obliqua* hybrid trees growing in a common garden. Such hybrid trees showed low compatibility with *N. alpina* pollen (tests with *N. obliqua* pollen were not performed), whereas one hybrid tree produced 42% of filled fruits after pollination with pollen from another hybrid tree. These results show that interspecific hybrids could produce filled fruits in a similar proportion to those observed in pure individuals (Torres 2013).

4.7 Concluding Remarks

Nothofagus obliqua wood is already positioned in the timber market in Chile, but in Argentina it has a more local market, mostly restricted to the surroundings of San Martín de los Andes city, where the Forest Department of Lanín National Park makes a sustainable use of its forests. The official statistics report a production of 1624 m³ of roble pellín timber in 2017 in Argentina, 65% originated within National Park jurisdiction (SGAyDS 2019). The knowledge generated for its domestication and improvement encourages its positioning as a breeding option aiming at diversifying forest production in Argentinean Patagonian region, which nowadays is focused in introduced fast-growing species such as ponderosa pine and Douglas fir. Within the current political scenario, with national laws promoting native species cultivation in Argentina, provincial states should deepen the implementation of instruments to support foresters. We hope cultivation of roble pellín will be installed in the region, generating a new productive alternative, offering high wood quality and indirectly contributing to the persistence of their beautiful natural forests.

References

Aitken SN, Yeaman S, Holliday JA, Wang T, Curtis-McLane S (2008) Adaptation, migration or extirpation: climate change outcomes for tree populations. Evol Appl 1:95–111

Aparicio AG, Zuki SM, Azpilicueta MM, Barbero FA, Pastorino MJ (2015) Genetic versus environmental contribution to variation in seedling resprouting in *Nothofagus obliqua*. Tree Genet Genomes 11:23

Araya LY, Oyarzún MV (2000) Descripción de los bosques de *N. alpina* y *N. obliqua* en Chile. In: Ipinza Carmona R, Gutiérrez Caro B, Emhart Schmidt V (eds) Domesticación y Mejora Genética de Raulí y Roble. UACH/INFOR, Valdivia, pp 25–42

Azpilicueta MM, Gallo LA (2009) Shaping forces modelling genetic variation patterns in the naturally fragmented forests of a South American Beech. Biochem Syst Ecol 37:290–297

Azpilicueta MM, Marchelli P, Gallo LA (2009) The effects of quaternary glaciations in Patagonia as evidenced by chloroplast DNA phylogeography of Southern beech *Nothofagus obliqua*. Tree Genet Genomes 5:561–571

Azpilicueta MM, Varela S, Martínez A, Gallo LA (2010) Manual de viverización, cultivo y plantación de Roble Pellín en el norte de la región andino patagónica. Ediciones INTA, San Carlos de Bariloche, 67 pp

Azpilicueta MM, Gallo LA, van Zonneveld M, Thomas E, Moreno C, Marchelli P (2013) Management of *Nothofagus* genetic resources: definition of genetic zones based on a combination of nuclear and chloroplast marker data. For Ecol Manag 302:414–424

Azpilicueta MM, Pastorino MJ, Puntieri J, Barbero F, Martínez-Meier A, Marchelli P, Gallo LA (2014) Robles in Lagunas de Epulauquen, Argentina: previous and recent evidence of their distinctive character. Rev Chil Hist Nat 87:24–35

Azpilicueta MM, Marchelli P, Gallo LA, Umaña F, Thomas E, van Zonneveld M, et al. (2016a) Zonas Genéticas de Raulí y Roble Pellín en Argentina. Herramientas para la conservación y el manejo de la diversidad genética (Azpilicueta MM, Marchelli P, eds). Ediciones INTA, Bariloche, 50 pp

Azpilicueta MM, El Mujtar V, Gallo LA (2016b) Searching for molecular insight on hybridisation in *Nothofagus* spp. forests at Lagunas de Epulauquen, Argentina. Bosque 37:591–601

Azpilicueta MM, Marchelli P, Aparicio AG, Pastorino MJ. Identificación del origen genético de un rodal semillero implantado de *Nothofagus obliqua* a través del análisis de dos regiones intergénicas de ADN de cloroplasto. RIA (accepted)

Barbero FA (2014) Variación genética de poblaciones naturales argentinas de *Nothofagus obliqua* en caracteres adaptativos tempranos relevantes para domesticación. Doctoral thesis, Universidad Nacional de Buenos Aires, Buenos Aires, 174 p

Crego MP (1999) Vraiación genética en el comportamiento fenológico y el crecimiento juvenil de progenies puras e híbridas de raulí, *Nothofagus nervosa* (Phil.) Dim. et Mil. Graduate thesis, Fac. Cs. Biológicas, Universidad Nacional del Comahue, 105 pp

Dimitri MJ, Rosario F, Leonardis J, Biloni JS (1997) El nuevo libro del árbol. Editorial El Ateneo

Donoso C (1979) Genecological differentiation in *Nothofagus obliqua* (Mirb.) Oerst. in Chile. For Ecol Manag 2:53–66

Donoso C (1993) Bosques templados de Chile y Argentina. Variación, estructura y dinámica (C Donoso ed). Editorial Universitaria, Santiago de Chile, p 484

Donoso C, Landrum LR (1979) *Nothofagus leonii* Espinosa, a natural hybrid between *Nothofagus obliqua* (Mirb.) Oerst. and *Nothofagus glauca* (Phil.) Krasser. N Z J Bot 17:353–360

Donoso C, Morales J, Romero M (1990) Hibridación natural entre roble (*Nothofagus obliqua* (Mirb,) Oerst.) y raulí (*N. alpina* (Poepp. et Endl.) Oerst,) en bosques del sur de Chile. Rev Chil Hist Nat 63:49–60

Donoso P, Donoso C, Gallo L, Azpilicueta MM, Baldini A, Escobar B (2006a) *Nothofagus obliqua* (Mirb.) Oerst. Roble, Pellín, Hualle. In: Donoso C (ed) Las especies arbóreas de los bosques templados de Chile y Argentina. Autoecología. Valdivia, Chile, pp 471–485

Donoso P, Donoso C, Marchelli P, Gallo L, Escobar B (2006b) *Nothofagus nervosa* (Phil.) Dim. et Mil., otros nombres científicos usados: *Nothofagus alpina*, *Nothofagus procera*, Raulí. In: Donoso C (ed) Las especies arbóreas de los bosques templados de Chile y Argentina. Autoecología. Valdivia, Chile, pp 448–461

El Mujtar V, Sola G, Aparicio A, Gallo LA (2017) Pattern of natural introgression in a *Nothofagus* hybrid zone from south American temperate forests. Tree Genet Genomes 13:49

Escobar B, Donoso C (1996) Resultados preliminares de almacenamiento en frío de semillas de coigüe (*Nothofagus dombeyi*), roble (*Nothofagus obliqua*) y raulí (*Nothofagus alpina*). Bosque 17:101–105

Gallo LA (2002) Conceptual and experimental elements to model natural inter-specific hybridisation between two mountain southern beeches (*Nothofagus* spp). In: Degen B, Loveless MD, Kremer A (eds) Modelling and experimental research on genetic processes in tropical and temperate forests. Embrapa-Silvolab-Guyane

Gallo LA (2004) Modelo conceptual sobre hibridación natural interespecífica entre *Nothofagus nervosa* y *Nothofagus obliqua*. In: Donoso C, Premoli A, Gallo L, Ipinza R (eds) Variación intraespecífica en las especies arbóreas de los bosques templados de Chile y Argentina. Editorial Universitaria, Santiago de Chile, pp 397–408

Gallo LA, Marchelli P, Breitembücher A (1997) Morphological and allozymic evidence of natural hybridization between two Southern beeches (*Nothofagus* spp) and its relation to heterozygosity and height growth. For Genet 4:15–23

Gallo L, Marchelli P, Crego P, Oudkerk L, Breitembücher A, Peñalba M et al (2000) Distribución y variación genética en características seminales y adaptativas de poblaciones y progenies de Raulí enArgentina. In: Ipinza Carmona R, Gutiérrez Caro B, Schmidt VE (eds) Domesticación y mejora genética de raulí y roble. UACH/INFOR, Valdivia, pp 133–156

García L, Riveros M, Droppelmann F (2015) Descripción morfológica y viabilidad del polen de *Nothofagus nervosa* (Nothofagaceae). Bosque 36:487–496

González Peñalba M, Lara A, Lozano L, Clerici C, Catalán M, Velásquez A et al (2008) Plan de Manejo Forestal Nonthué, Reserva Nacional Lanín. Parque Nacional Lanín. Provincia del Neuquén, San Martín de los Andes

Grant ML, Clement EJ (2004) Clarification of the name *Nothofagus alpina* and anew epithet for a *Nothofagus* hybrid. Bot J Linn Soc 146:447–451

Heenan PB, Smissen RD (2013) Revised circumscription of *Nothofagus* and recognition of the segregate genera Fuscospora, Lophozonia, and Trisyngyne (Nothofagaceae). Phytotaxa 146:1–31

Hill RS, Jordan GJ (1993) The evolutionary history of *Nothofagus* (Nothofagaceae). Aust Syst Bot 6:111–126

Hill RS, Read J (1991) A revised infrageneric classification of *Nothofagus* (Fagaceae). Bot J Linn Soc 105:37–72

Hill RS, Jordan GJ, Macphail MK (2015) Why we should retain *Nothofagus* sensu lato. Aust Syst Bot 28:190–193

INTI-CITEMA (2003). http://www.inti.gov.ar/maderas/pdf/densidad_cientifico.pdf

Marchelli P, Gallo L (1999) Annual and geographic variation in seed traits of Argentinean populations of southern beech *Nothofagus nervosa* (Phil.) Dim. et Mil. For Ecol Manag 121:239–250

Marchelli P, Gallo LA (2001) Genetic diversity and differentiation in a southern beech subjected to introgresive hybridization. Heredity 87:284–293

Marchelli P, Thomas E, Azpilicueta MM, van Zonneveld M, Gallo LA (2017) Integrating genetics and suitability modelling to bolster climate change adaptation planning in Patagonian *Nothofagus* forests. Tree Genet Genomes 13:119–132

Martin PG, Dowd JM (1993) Using sequences of rbcL to study phylogeny and biogeography of *Nothofagus* species. Aust Syst Bot 6:441–447

Mason B, Jinks R, Savill P, McG Wilson S (2018) Southern beeches (*Nothofagus* species). Quart J Forest 112:30–43

Minagro (2018) Boletín N°4 de costos y precios forestales Patagonia Andina. Ministerio de Agroindustria, Presidencia de la Nación. Subsecretaría de Desarrollo Foresto Industrial, 6 pp

Mondino VA, Tejera LE (2006) Plantaciones de Roble Pellín y Raulí. Boletín Forestal INTA Esquel 11: 47-50. https://inta.gob.ar/sites/default/files/script-tmp-inta_forestal11_ nothofagus.pdf

Murray MB, Cannell MGR, Sheppard LJ (1986) Frost hardiness of *Nothofagus procera* and *Nothofagus obliqua* in Britain. Forestry 59:209–222

Newton AC, Allnutt TR, Gillies ACM, Lowe AJ, Ennos RA (1999) Molecular phylogeography, intraspecific variation and the conservation of tree species. Trends Ecol Evol 14:140–145

Pastorino MJ, El Mujtar V, Azpilicueta MM, Aparicio AG, Marchelli P, Mondino VA et al (2016) Subprograma *Nothofagus*. In: Marcó M, Llavallol C (eds) Domesticación y Mejoramiento de Especies Forestales. Min. Agroindustria, UCAR, Buenos Aires, 422 pp

Petit RJ, El Mousadik A, Pons O (1998) Identifying populations for conservation on the basis of genetic markers. Conserv Biol 12:844–855

Puntieri JG, Grosfeld JE, Heuret P (2009) Preformation and distribution of staminate and pistillate flowers in growth units of *Nothofagus alpina* and *N. obliqua* (Nothofagaceae). Annal Bot 103:411–421

Sabatier Y, Azpilicueta MM, Marchelli P, González-Peñalba M, Lozano L, García L et al (2011) Distribución natural de *Nothofagus alpina* y *Nothofagus obliqua* (Nothofagaceae) en Argentina, dos especies de primera importancia forestal de los bosques templados norpatagónicos. Bol Soc Argent Bot 46:131–138

Schinelli Casares T (2013) Producción de *Nothofagus* bajo condiciones controladas. Ediciones INTA, 56 pp

SGAyDS (2019) Anuario de Estadística Forestal 2017–2018, Buenos Aires, 175 pp. https:// www.argentina.gob.ar/ambiente/tierra/bosques-suelos/manejo-sustentable-bosques/ programa-nacional-estadistica-forestal

Torres CD (2013) Biología reproductiva de *Nothofagus,* con especial referencia a *N. obliqua* (Mirb.) Oerst. (roble pellín) y *N. nervosa* (Phil.) Krasser (raulí). Doctoral thesis, Universidad Nacional del Comahue, Bariloche, Argentina, 168 pp

Torres CD, Puntieri JG (2013) Pollination and self-interference in *Nothofagus*. Flora 208:412–419

Torres CD, Puntieri JG, Stecconi M (2012) Flower and seed production as affected by axis category and shoot size in two Patagonian *Nothofagus* species. Botany 90:261–272

Torres CD, Magnin A, Stecconi M, Puntieri JG (2016) Testing individual inter-annual variations in flower production by means of retrospective analysis of meristem allocation in two tree species. Folia Geobot 51:361–371

Torres CD, Gallo LA, Puntieri JG (2019) Positional effects on fruit production and filling in two anemophilous *Nothofagus* species. Flora 262:151529

Urretavizcaya MF, Caselli MM, Contardi LT, Loguercio G, Defossé G (2018) Enriquecimiento de Bosques Degradados de ciprés de la cordillera con especies nativas de alto valor forestal. Informe Final Proyecto PIA 14067, UCAR-CIEFAP

Varela SA, Gyenge JE, Fernández MA, Schlichter T (2010) Seedling drought stress susceptibility in two deciduous *Nothofagus* species of NW Patagonia. Trees 24:443–453

Vázquez FM, Rodríguez RA (1999) A new subspecies and two new combinations of Nothofagus Blume (Nothofagaceae) from Chile. Bot J Linn Soc 129:75–83

Veblen TT, Donoso C, Kitzberger T, Rebertus AJ (1996) Ecology of southern Chilean and Argentinean Nothofagus forests. In: Veblen TT, Hill RS, Read J (eds) The ecology of biogeography of *Nothofagus* forests. Yale University Press, New Haven/London

Veit H, Garleff K (1996) Evolución del paisaje cuaternario y los suelos en Chile central-sur. In: Armesto JJ, Villagrán C, Arroyo MK (eds) Ecología de los bosques nativos de Chile. Editorial Universitaria, Santiago de Chile, Chile, pp 29–49

Vergara R (2000) Regiones de procedencia de *N. alpina* y *N. obliqua*. In: Ipinza R, Gutiérrez B, Emhart V (eds) Domesticación y Mejora de Raulí y Roble. Universidad Austral de Chile/ Instituto Forestal, Valdivia, pp 121–132

Vergara R, Gitzendanner MA, Soltis DE, Soltis PS (2014) Population genetic structure, genetic diversity, and natural history of the South American species of *Nothofagus* subgenus Lophozonia (Nothofagaceae) inferred from nuclear microsatellite data. Ecol Evol 4:2450–2471

Villagrán C (1991) Historia de los bosques templados del sur de Chile durante el Tardiglacial y Postglacial. Rev Chil Hist Nat 64:447–460

Villagrán C (2001) Un modelo de la historia de la vegetación de la Cordillera de la Costa de Chile central-sur: la hipótesis glacial deDarwin. Rev Chil Hist Nat 74:793–803

Yang F, Wang Y, Miao LF (2010) Comparative physiological and proteomic responses to drought stress in two poplar species originating from different altitudes. Physiol Plant 139:388–400

Chapter 5
Nothofagus pumilio and *N. antarctica*: The Most Widely Distributed and Cold-Tolerant Southern Beeches in Patagonia

Carolina Soliani, Paula Marchelli, Víctor A. Mondino,
Mario J. Pastorino, M. Gabriela Mattera, Leonardo A. Gallo,
Alejandro G. Aparicio, Ana D. Torres, Luis E. Tejera,
and Teresa Schinelli Casares

5.1 Main Characteristics, Ecological Features, and Distributional Ranges

Nothofagus pumilio (Poepp. & Endl.) Krasser, known as lenga, and *Nothofagus antarctica* (G. Forster) Oerster, commonly called ñire, are tree species of the temperate forests of southern South America, occurring in temperate-rainy forest districts, in the subalpine transition area, and in poorly drained sites (Veblen et al. 1996). Their natural distributions in Chile and Argentina mainly correspond to the Cordillera de los Andes, from 36° S to the south of Tierra del Fuego archipelago (55° S). In Chile, *N. pumilio* also inhabits the Coastal Range (Cordillera de Nahuelbuta) where it grows associated with *Araucaria araucana* forests above 1400 m asl. In turn, *N. antarctica* is distributed in the Central Depression of Chile, from Valdivia toward austral latitudes. Throughout their entire distribution in Argentina, lenga and ñire forests co-occur and overlap to a greater or lesser extent. This shared geographical distribution covers approximately 18° of latitude (2200 km of extension) and is the widest among the South American *Nothofagus*. Recently, the national forest inventory of Argentina (CIEFAP and MAyDS 2016) revealed a total of 1,595,661 ha and 864,148 ha of lenga and ñire forests, respectively. The National Park Administration of Argentina protects a portion of all this forest surface (34% and 15%, respectively), belonging the rest to provincial jurisdictions.

C. Soliani (✉) · P. Marchelli · M. J. Pastorino · M. G. Mattera · L. A. Gallo
A. G. Aparicio · A. D. Torres
Instituto de Investigaciones Forestales y Agropecuarias Bariloche (IFAB) INTA-CONICET, Bariloche, Argentina
e-mail: soliani.carolina@inta.gob.ar

V. A. Mondino · L. E. Tejera · T. Schinelli Casares
Instituto Nacional de Tecnología Agropecuaria (INTA) EEA Esquel, Esquel, Argentina

© Springer Nature Switzerland AG 2021   117
M. J. Pastorino, P. Marchelli (eds.), *Low Intensity Breeding of Native Forest Trees in Argentina*, https://doi.org/10.1007/978-3-030-56462-9_5

The two species conform an altitudinal ecological gradient along their natural distribution, with *N. pumilio* dominating at higher elevations and *N. antarctica* at lower sites. *Nothofagus pumilio* usually forms large masses of pure stands associated with climax forests (late-successional species) but also in co-dominance with *Nothofagus betuloides* (in the south) or *Araucaria araucana* (in the north). Individual trees can be taller than 35 m (Tortorelli 1956) (Fig. 5.1a) and grow in environments with deep, well-drained soils; while reaching the timberline, they tolerate low temperatures and frosts as a shrub. On the other hand, *N. antarctica* is the species with the widest ecological plasticity and phenotypic variation (Ramírez et al. 1997) among their South American congeners. It occurs in cold humid valleys with heavy clay soils, in peat bogs, and in rocky and xeric sites like the Patagonian steppe, forming monospecific masses of shrubs or small trees (Fig. 5.1b).

Nothofagus antarctica is considered a pioneer and resprouting species and has a great capacity for clonal reproduction (Premoli and Steinke 2008), while *N. pumilio* can only reproduce generatively. Clonal reproduction in ñire is suggested as an adaptation to recurrent disturbances, among which fires are the most common in the region.

Both are related species included in the subgenus *Nothofagus*, one of the four defined within the Nothofagaceae family (Hill and Jordan 1993). Phylogenetic analyses, including morphology, conserved DNA sequences, and fossils, agreed in a common ancestor among *Nothofagus betuloides* and *N. pumilio/N. antarctica* sister species (Manos 1997; Sauquet et al. 2012). Recently, it has been suggested that *N. pumilio* would be the ancestral species within the subgenus (Acosta and Premoli 2010). Extant species included in the clade shared the pollen type, nominated equivalently as "*N. fusca* Type b" (Manos 1997) or "*dombeyi*" (Villagrán et al. 1995; Heusser et al. 1999) all of them are cold-tolerant species.

As deciduous species, their dormant buds during winter determine a growing season restricted to spring and summer, whose length varies depending on available resources; particularly drought stress during summer is among the most critical factors. Pollen and seed dispersion is mediated by wind. Most commonly, seeds reach short distances from the mother tree (Rusch 1993). Seed production is not regular over consecutive seasons, being years of greater production associated with greater germination power and viability (*masting*) (Donoso 1993). Variability in seed production and seed quality was suggested to be linked to stand conditions, e.g., secondary *N. antarctica* forests have been observed at 51° S to have better seed quality (Soler Esteban et al. 2010), and also differences have been reported between pure and mixed *N. pumilio* forests (Toro Manríquez et al. 2016). Moreover, in these species, seeds have a stage of reduced viability and do not form persistent seed banks (Cuevas and Arroyo 1999). Seedling recruitment and survival are crucial stages for regeneration, varying in number of established individuals year after year, being tightly linked to the microclimate conditions within natural forests (e.g., Soler et al. 2013; Bahamonde et al. 2018). Regeneration dynamics is the result of large-scale disturbances, which cause the replacement of complete stands (e.g. mass removal on steep slopes), or small-scale disturbances, mainly the gap dynamics, which

Fig. 5.1 Species features. (**a**) *Nothofagus pumilio*, (**b**) *Nothofagus antarctica*; (1) seeds, (2) male flowers, (3) leaves, (4) arboreal morphotype, (5) view of a forest. (Photos: C. Soliani, M. J. Pastorino, V. Rusch; A2, B2 extracted from Giménez Gowland 2002)

consist of the natural falling of over-mature trees; both involve a massive recruitment related to canopy opening (Veblen et al. 1996; Donoso 2006).

Temporal overlapping of flower maturity during reproductive stages favors the occurrence of natural interspecific hybridization between these species, which is a widely reported phenomenon among South American *Nothofagus* (see Chaps. 3 and 4). The natural hybrid between *N. pumilio* and *N. antarctica* presents intermediate characteristics between both species (i.e., bark roughness and color, stem straightness, crown form). It has been suggested that *N. antarctica* acts predominantly as pollen receptor (Acosta and Premoli 2010). The introgression of the chloroplast genome from *N. pumilio* to *N. antarctica*, due to repeated interspecific crossings and/or hybrid-parent backcrossing, would have occurred more frequently during unfavorable climatic periods (e.g., glaciations, Palmé et al. 2004; Heuertz et al. 2006) or during forest recovering (e.g., postglacial recolonization), constituting an additional source of variation.

The climatic changes that occurred during the Quaternary modified the areas covered by forests in Patagonia and have strongly impacted in the population dynamics, possibly restricting gene flow, isolating them geographically, or even hindering their chances of regeneration. This regional disturbance left a trace on the distribution of genetic variation. At a narrower level, other processes could have shaped the population structure, such as the mating system or, more recently, anthropogenic disturbances. The genetic variability of these species, shaped along their evolutionary history, should be preserved in order to ensure their adaptability and, finally, their persistence.

5.2 Phylogeography: Inspecting *Nothofagus* Evolutionary History Through Chloroplast DNA

Climatic changes during the Quaternary imposed a great selection pressure along the distributional range of forest species. After ice advance, the remaining patches of forests withstood adverse climatic conditions. In Patagonia, glaciations occurred from the Late Miocene to the Pleistocene, being the Great Patagonian Glaciation (GPG; about 1 million years before present [M years BP]) the maximum expansion of ice in extra-Andean Patagonia (Flint and Fidalgo 1969; Rabassa et al. 2005). The Last Glacial Maximum (LGM) occurred about 18,000–20,000 years BP (Porter 1981), when the ice covered the Patagonian plains beyond the mountain range (Rabassa and Clapperton 1990; Glasser et al. 2008). However, several ice-free areas remained (Markgraf et al. 1995) and constituted refugia for vegetation, and some became the center of expansion of the biota after the ice retreated. Multiple evidence suggests a latitudinal trend in the type of glaciations during the LGM: valley-type glaciations characterized the north, whereas continuous ice layers covered the southern region (Glasser et al. 2008). A transitional zone at mid-latitudes (42° S–44° S) was established. The predominance of westerlies determined intermediate climatic conditions at these latitudes by the early Holocene (10,000–8500 years BP) (Markgraf et al. 2003), i.e., lack of seasonality and a drier and warmer environment (Manzini et al. 2008). Reinforcing this, three different zones based on a palynological reconstruction representing paleo-climates (Markgraf et al. 1996) were identified (north of 43° S, between 43° and 51° S, and south of 51° S).

The distribution of chloroplast genetic variation helps to thoroughly comprehend how forests were modeled after glaciation. Considering its slow evolution rate as well as the uniparental (maternal) inheritance, chloroplast DNA is a useful tool tracing the effective dispersal of seeds. Polymorphisms from cpDNA non-coding regions were screened in 40 Argentinean populations of *N. pumilio* and *N. antarctica* (Soliani et al. 2012; Table 5.1) and used to define haplotypes, coming from the following restricted regions: *trn*D-*trn*T/HinfI, *trn*C-*trn*D/taqI, and *atp*H-*atp*I/HinfI. Point mutations in the restriction site and indels (insertion/deletions) allowed identifying 9 and 13 haplotypes in *N. pumilio* and *N. antarctica*, respectively (based on different combinations of length variants). Eight haplotypes were shared among the species, with *N. antarctica* being the most variable with five unique haplotypes.

A genetic diversity trend along latitude, decreasing in both species from north to south, as well as a significant phylogeographic structure between the two main groups of populations and haplotypes (north and south of 42° S), evidenced regional footprints of glaciations (Fig. 5.2). The hypothesis of multiple glacial refugia is supported by these results (Premoli et al. 2000; Marchelli and Gallo 2006; Sérsic et al. 2011). In addition, a meeting area where migration routes could have encounter was proposed (ca. 42–43° S), which agrees with stratigraphic and palynological evidence. A geographical segregation of genetic lineages was identified (Mathiasen and Premoli 2010; Soliani et al. 2012), like in other widely distributed species of the region, such as *Austrocedrus chilensis* (Pastorino and Gallo 2002) and *Pilgerodendron uviferum* (Premoli et al. 2002). Alternatively, the great divergence between haplogroups was interpreted as isolated forest patches due to the settlement of pre-Quaternary depression areas (paleobasins) (Premoli et al. 2012).

Both species presented similar levels of average within-population gene diversity (h_s), total genetic diversity (h_t), and gene differentiation based on frequency (G_{ST}) and ordered alleles (N_{ST}) (Table 5.2). Cryptic refugia might be inferred from population allelic richness (A_R), a parameter independent from population size (Widmer and Lexer 2001) and with a significant value in conservation decisions (Petit et al. 1998). Northern populations (40° S) (4, IV) harbor the highest diversity in both species; at mid-latitudes (42–43° S), three populations of lenga (a, 12, 14) and two of ñire (XIII, XIV) showed a higher allelic richness. In southern Patagonia (54° S), one lenga population (XIX) was the most diverse. Trends of genetic diversity clearly follow the geographical latitude.

Haplotype sharing (cpDNA or mtDNA) among closely related species that hybridize naturally and occur in sympatry is very common (Rieseberg and Soltis 1991), signaling population variation. Introgression in this *Nothofagus* species complex (IG = 0.90; Soliani et al. 2012) could be occurring due to interspecific gene flow and backcrossing offspring. A similar geographical pattern for haplotype distribution in both species supports the idea of recent or at least postglacial

Table 5.1 Geographic location of coupled *Nothofagus pumilio* and *Nothofagus antarctica* sampled populations ordered from north to south

Population	Species[a]	Latitude (S)	Longitude (W)	N[b]	A_R[c]
Lagunas de Epulauquen	N. PUM [1]	36°49′39″	71°06′12″	5 (2,14)	1
	N. ANT [I]	36°49′30″	71°05′51″	5 (2,6)	1
Caviahue	N. PUM [2]	37°51′18″	71°05′02″	5 (1,3)	1
	N. ANT [II]	37°49′55″	71°1′5″	6 (1,2,3)	1.8
Tromen	N. PUM [3]	39°34′47″	71°27′35″	5 (3,5)	1
	N. ANT [III]	39°36′	71°27′	5 (3)	0
Quilanlahue	N. PUM [4]	40°7′59″	71°29′35″	5 (4,5,14)	2
	N. ANT [IV]	40°08′15″	71°28′1″	5 (3,4,5)	2
Challhuaco	N. PUM [5]	41°14′39″	71°17′9″	5 (3,5)	1
	N. ANT [V]	41°14′0″	71°17′27″	5 (3,5)	1
Northern group mean				5.1	1.2
Cholila	N. PUM [a]	42°40′36″	71°29′51″	9 (7,10)	1
	N. ANT [00]	42°31′32″	71°31′35″	10 (5,10)	0.5
Huemules	N. PUM [6]	42°50′14″	71°28′46″	10 (7,10)	0.8
	N. ANT [VI]	42°49′22″	71°27′53″	10 (7,10)	1
La Hoya	N. PUM [7]	42°50′27″	71°15′51″	10 (7,10)	0.8
	N. ANT [VII]	42°50′54″	71°15′33″	10 (7,10)	0.9
Nahuelpan	N. PUM [8]	42°58′59″	71°11′24″	10 (10,13)	0.9
	N. ANT [VIII]	42°59′24″	71°11′22″	10 (7,10)	0.9
Trevelin	N. PUM [9]	43°4′0″	71°34′44″	10 (7,10)	0.8
	N. ANT [IX]	43°4′5″	71°34′29″	11 (7,10,13)	1.6
Lago Guacho	N. PUM [11]	43°48′53″	71°29′41″	10 (10,13)	0.8
	N. ANT [XI]	43°49′38″	71°27′1″	10 (10,11)	0.8
J. San Martín	N. PUM [12]	43°49′40″	70°45′27″	10 (7,10)	1
	N. ANT [XII]	43°49′39″	70°45′10″	10 (7,10)	1
Lago Fontana	N. PUM [13]	44°50′26″	71°37′58″	10 (7,10)	0.8
	N. ANT [XIII]	44°50′35″	71°37′38″	10 (7,10,11,12)	1.9
Río Unión	N. PUM [14]	44°51′27″	71°39′11″	10 (7,10)	1
	N. ANT [XIV]	44°51′30″	71°39′24″	10 (7,8,9,10,12)	2.8
Arroyo Perdido	N. PUM [15]	44°50′17″	71°41′40″	10 (7,10)	0.5
	N. ANT [XIV]	44°50′12″	71°41′36″	10 (8,10,11,12)	1.8
Center group mean				10	1.1
Cancha Carrera	N. PUM [16]	51°13′19″	72°16′21″	5 (10)	0
	N. ANT [XVI]	51°13′21″	72°15′34″	5 (7,10)	1
Mina I	N. PUM [17]	51°31′17″	72°21′2″	5 (7,10)	1
	N. ANT [XVII]	51°31′48″	72°20′31″	5 (10)	0
Tierra del Fuego Norte	N. PUM [18]	54°4′30″	68°31′52″	5 (10)	0
	N. ANT [XVIII]	54°4′26″	68°31′26″	5 (7,10)	1

(continued)

Table 5.1 (continued)

Population	Species[a]	Latitude (S)	Longitude (W)	N[b]	$A_R{}^c$
Tierra del Fuego Centro	N. PUM [19]	54°22′28″	67°15′49″	6 (7,10,13)	1.8
	N. ANT [XIX]	54°22′14″	67°15′34″	6 (7,10)	1
Tierra del Fuego Este	N. PUM [20]	54°35′28″	66°37′13″	6 (7,10)	0.8
	N. ANT [XX]	54°34′21″	66°38′2″	6 (7,10)	1
Southern group mean				5.4	0.8

From Soliani et al. (2012)
[a]Populations' nomenclature is indicated between brackets; [b]*N*: sample size; chloroplast haplotypes are indicated between parentheses. [c]A_R: allelic richness per population (rarefaction method)

hybridization. Ecological features also support hybridization in our species complex: flowering phenology and pollen release overlap (González et al. 2006), although other pre- and even post-zygotic barriers can also occur. The predominance of westerlies in the region and the altitudinal ecological gradient formed by both species might favor *N. antarctica* acting as the mother tree. However, the lack of evidence supporting backcrosses of hybrids toward *N. antarctica* as the parental species and the scarce genetic information about F1 and F2 generations do not allow to thoroughly conclude about the directionality of the hybridization. Moreover, since individuals from both species coexist in some places, hybridization and directionality of the crosses (hybrids to parental species) could vary according to the relative abundance of the taxa (Lepais et al. 2009) or particular conditions of the site where they co-occur (e.g., Heuertz et al. 2006).

5.3   Genetic Structure at Nuclear Markers Across Species Ranges

5.3.1   *Latitudinal Trends, Species Admixture, and the Identification of a Contact Zone*

In addition to the phylogeographical analysis presented in the previous section, the same 20 pairs of populations of sympatric interspecific natural forests were studied with seven microsatellite loci in *N. pumilio* and six in *N. antarctica* (Soliani et al. 2010) (five shared by both species). Besides, one sample from a single *N. antarctica* site (42° S, Cholila (00) Table 5.1, Fig. 5.2) that showed admixture of maternal lineages (Pastorino et al. 2009; Soliani et al. 2012) was also analyzed. All genetic diversity parameters evaluated were significantly higher in *N. antarctica* than in *N. pumilio* (Wilcoxon paired test: $N_e = 3.55$ vs. $N_e = 3.16$, p = 0.022; $A_R = 4.80$ vs. $A_R = 4.45$, p = 0.015; $H_E = 0.656$ vs. $H_E = 0.598$, p = 0.029; $H_O = 0.520$ vs. $H_O = 0.450$, p = 0.017). The hierarchical AMOVA showed that the proportion of genetic variance partitioned between the two species was 15% for nSSRs ($\varphi_{RT} = 0.15$, p = 0.001), much greater than the percentage for cpDNA variation ($\varphi_{RT} = 0.012$, p = 1).

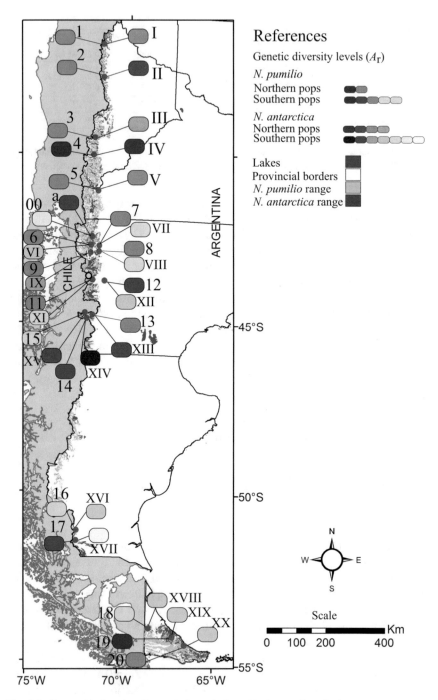

Fig. 5.2 Geographic location of sampled populations. Haplotype diversity is expressed by different colors in each species and with different tones for each population (darker colors represent higher diversity levels). Population codes have correspondence with Table 5.1. From Soliani et al. (2012)

Table 5.2 Average values of within-population gene diversity (h_s), total genetic diversity (h_t), and gene differentiation in all populations for unordered alleles (G_{st}) and for ordered alleles (N_{ST}), for the analyzed populations of each species

Genetic parameters	*Nothofagus pumilio*	*Nothofagus antarctica*
(h_s) (*s.d.*)	0.424 (0.0460)	0.488 (0.0506)
(h_t) (*s.d.*)	0.645 (0.0865)	0.761 (0.0536)
G_{ST} (*s.d.*)	0.344 (0.0589)	0.359 (0.0786)
N_{ST} (*s.d.*)	0.885 (0.0217)*	0.841 (0.0318)*

From Soliani et al. (2012)
s.d. standard deviation, * significant test, evidence of phylogeographic structure

Standardized genetic differentiation between the species was also higher for nSSRs ($G'_{ST} = 0.335$) with respect to cpDNA ($G'_{ST} = 0.061$) data.

A latitudinal trend was revealed in both species, which is probably related to the impact of past glaciations. Higher allelic richness and gene diversity values were found in *N. pumilio* populations located around 37°–42° S (populations 2, 3, 4, 6, and 8), whereas at about 40°–43° S for *N. antarctica* (V, 00, VIII, and IX). A decrease in A_R and H_E toward southern populations (southward 42° S) was more evident in *N. antarctica* than in *N. pumilio*, in agreement with previous results based on cpDNA data (Soliani et al. 2012). Genetic differentiation was significant in both taxa and slightly higher in *N. pumilio* ($F_{ST} = 0.094$, p = 0.001) than in *N. antarctica* ($F_{ST} = 0.083$, p = 0.001). The standardized differentiation (G'_{ST}) was higher but similar in both species ($G'_{ST} = 0.296$ and $G'_{ST} = 0.303$, respectively).

A clear separation between both species was found with a Bayesian clustering analysis (STRUCTURE; Pritchard et al. 2000) of the 41 populations ($\Delta K = 2$ indicated the optimal number of groups) (Fig. 5.3). Some populations presented a high level of admixture, suggesting hybridization and even introgression through backcrosses, e.g., populations 8 (42°59′ S, 71°11′ W) and 9 (43°04′ S, 71°35′ W) in *N. pumilio* and X (43°51′ S, 71°33′ W), IX (43°04′ S, 71°34′ W), XII (43°50′ S, 70°45′ W), and XIII (44°51′ S, 71°38′ W) in *N. antarctica*. Evidence of interspecific gene flow was also observed when cluster partitioning increased (e.g., at K = 4 and K = 5).

Through a coalescent model, the putative origin of population genetic variation of each species at intermediate latitudes was inferred (approximate Bayesian computation (ABC) in DIYABC v1.0.4.39; Cornuet et al. 2008, 2010). Species divergence was estimated around 302,500 years BP. Then, a recent species admixture could have occurred ~18,950 years BP (50 years/generation time) across individuals from the 8 (42° S) and 9 (43° S) populations (ABC1, scenario 1; Fig. 5.3). Although hybridization could have determined *N. pumilio* population variation, it seems not to be the most probable explanation of the *N. antarctica* sympatric populations. In this species, divergence from an ancestral population to around 26,700 BP could have been the origin of the current variation (ABC2, scenario 4; Fig. 5.3). The settlement of a hybrid zone at intermediate latitudes (42° ~ 43° S, 8–9 and VIII–IX populations) could have been facilitated by niche overlapping and a

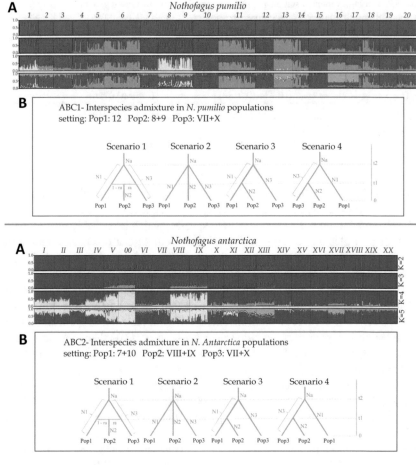

Fig. 5.3 Inferred genetic structure at the between-species level ($K = 2$ to $K = 5$, STRUCTURE, Pritchard et al. 2000) (**a**) and past demography scenarios tested with DIYABC (**b**). Populations are ordered from north to south and are coded according to Table 5.1 and Fig. 5.2 (except populations 10 and X that were added for the genotyping with microsatellite markers). Representation of clusters is indicated by the same colors in each species; figure was split in two for better comprehension and legibility, from Soliani et al. (2015)

flowering lag along altitude (Rusch 1993), an aspect that fosters hybridization in contact areas. Accordingly, the high proportion of admixed individuals, mainly detected around 42–44° S, suggests a tight correspondence between the frequency and the geographic location of hybridization.

Colonization from multiple refugia, as reported before, implied population bottlenecks, founding events, and admixture of genetic lineages. The convergence of isolated, independently evolving, intra-specific lineages during colonization might increase genetic diversity (Comps et al. 2001; Alberto et al. 2008; Durand et al. 2009) due to their admixture (e.g., Vendramin et al. 1998; Petit et al. 2003). The secondary contact zones that are therefore established constitute genetic reservoirs relevant for their conservation and might enrich sources of material in ex situ breeding programs (Petit et al. 2003; Grivet et al. 2008).

The identification of a contact zone where the northern lineages mixed with the southern lineages was inferred for *N. pumilio* and *N. antarctica* (Soliani et al. 2015). These results are in agreement with previously reported evidence of contact zones for Patagonian taxa, e.g., in fishes (Zemlak et al. 2008), forest trees (Pastorino et al. 2009), and herbs (Cosacov et al. 2010; Sérsic et al. 2011). The high allelic richness, low level of inbreeding, and a lack of evidence of genetic bottlenecks in populations around the contact zone add support to the meeting of colonization routes from northern and southern refugia (Soliani et al. 2015). In addition, the contact zone might be the result of immigrants from local or nearby refugia that remained in unglaciated areas at central latitudes. The described patterns of genetic variation were taken into consideration for the definition of preliminary operational genetic management units of *N. pumilio* and *N. antarctica* in Argentina (see next section). The identification of at least one population with high level of genetic diversity within each unit could be considered as base material for future breeding programs.

5.3.2 Impact of Selective Logging on Patterns of Genetic Diversity: A Case Study in Nothofagus pumilio

Logging is one of the human activities that has impacted on the natural evolution of the forest, by altering the genetic diversity and structure of main species like trees (Rajendra et al. 2014). In particular, the removing of trees and the impoverishment of a forest could lead to within-population changes in genetic variation and diversity, which is the key to adaptation (El-Kassaby et al. 2003; Finkeldey and Ziehe 2004). Signs of impact could be a decrease in allelic richness or modifications in heterozygote proportions, a reduction in allele frequencies, or loss of variants between the adult cohort and its regeneration (Cornuet and Liukart 1996; Rajora et al. 2000). Then, erosive forces (i.e., genetic drift, selection) might affect the remnant population. Logging could also affect the spatial genetic structure, i.e., the amount and distribution of genetic variation between and within local populations and individuals of a species, with consequences in regeneration recruitment.

In Patagonian natural forests, selective extraction in a high grading management system was implemented over many decades (Bava and Rechene 2004; Bava et al. 2006; González et al. 2006). The removal of best-featured individual trees (stem straightness and best sanitary conditions) was the most frequent technique employed, which may be expected to induce changes in allelic richness or modifications in the spatial distribution of alleles. Because of its excellent wood properties (high quality, long-time durability; González et al. 2006), *N. pumilio* has historically been one of the most exploited native species in Patagonia, threatening its populations.

Table 5.3 Location and genetic characterization of sampling sites of *Nothofagus pumilio* representing stands with selective extraction of individuals (LOGG) and natural forest (NONL)

Pop	Treat[a]	LA (°S)	LO (°W)	AC	N	A_R	Na$_{<5\%}$	H_O	H_E
Huemules (Hm)	LOGG ^	42°49′44″	71°27′41″	O	30	39	3	0.530	0.634
				A	30	37	3	0.468	0.589
				R	39	33	2	0.454	0.550
	NONL	42°50′14″	71°28′46″	A	35	32	2	0.553	0.600
				R	35	30	3	0.458	0.571
L. Guacho (G)	LOGG†	43°49′35″	71°27′41″	O	30	28	2	0.430	0.529
				A	30	31	1	0.488	0.577
				R	30	30	2	0.578	0.585
	NONL	43°48′53″	71°29′41″	A	30	31	3	0.463	0.539
				R	40	30	1	0.467	0.539
L. Engaño (Eg)	LOGG ^	43°51′25″	71°32′36″	O	30	24	2	0.439	0.547
				A	30	21	1	0.428	0.506
				R	30	32	2	0.483	0.518
	NONL	43°49′52″	71°35′04″	A	30	22	1	0.360	0.546
				R	30	30	2	0.458	0.513

From Soliani et al. (2016)

[a]Last date of registered management extraction in the 1990s ^ and 2004†. Pop, sampled populations; treat, management treatment; LA, latitude; LO, longitude; AC, age class; N, number of sampled individuals; A_R, allelic richness with rarefaction to a common sample size for each population; Na$_{<5\%}$, mean number of rare alleles across loci; H_O observed and H_E expected heterozygoses; O, over-mature; A, adult; R, regeneration

A comparison between logged and non-logged stands of *N. pumilio* was made through the estimation of genetic diversity of adults and their regeneration by means of nuclear microsatellites (Soliani et al. 2016). Three pairs of stands from areas around 42° S, which were reported as strongly intervened and degraded after a high grading logging (Bava et al. 2006), were sampled. A slight decrease in allelic richness (A_R) and a lower number of rare alleles (frequencies ≤5%) were observed in adults from the logged stand with respect to adults from the non-logged-stand in one population (Lago Guacho (11), Table 5.3). However, an unexpected result with the opposite trend was found in the other two studied populations (Huemules (6) and Lago Engaño (10)), with more alleles in the logged stand. A tendency toward a decrease in number or frequency of alleles was detected in old growth and remnant adults of managed forests in all populations. These results could be a sign of the impact of logging on the adult cohorts. The regeneration cohort from the logged sites had greater allelic richness than non-intervened forests, which could not be attributed unequivocally to logging but to other factors not contemplated by the study.

Even though a genetically impoverished population is expected after logging under a high grading system, no signs of recent bottlenecks were observed (*Bottleneck* software, Cornuet and Liukart 1996). The proportion of genetic variance partitioned between the two treatments (i.e., logged vs non-logged) was low but significant in only one population located at 42° 50′ S (5%, $F_{RT} = 0.048$; $p = 0.001$). Spatial autocorrelation and heterogeneity tests between management treatments were non-significant in the populations, making inviable to assess the true impact of management on spatial variation. The effects of management varied widely according to the type of treatment, having multiple effects (positive, negative, or neutral) on the genetic diversity and the mating system. Additional factors might be playing a role in the final determination of genetic variation, i.e., intensity of management, time elapsed since the last intervention, recent practices, and presence of livestock.

5.4 Genetic Zones: On How Molecular Tools Can Contribute to the Conservation and Management of Forest Resources

In widely distributed species conformed by hundreds of natural populations, as many tree species, it is unfeasible the management from a genetic perspective of each population separately. The definition of operational genetic management units (OGMU) overcomes the limitation of making decisions at the level of single populations. Knowledge on the genetic pool of a species gained by sampling Mendelian populations can contribute to the application of specific management actions to groups of populations. Those management decisions are commonly related to the conservation and use of genetic resources and involve actions such as the planning of reforestation or restoration programs. Properly designed strategies will focus on the preservation of the local provenance in order to avoid maladaptation and genetic contamination. In this regard, genetic zones' (GZs) delineation is the first step toward OGMUs' definition. GZs are known as genetically homogeneous regions (Bucci and Vendramin 2000) within which genetic material can be moved with minimum risk of altering the genetic constitution of the local and nearby populations (McKay et al. 2005). At the same time, GZs would represent discernible genetic pools that are desired to be conserved because of its distinctive genetic attributes. To accomplish this purpose, both a genetic inventory and a representation of the natural distributional range (mapped geographic area) of the species are crucial requirements. In order to classify the genetic information, a hierarchical clustering of sampled populations is useful to group populations sharing the same genetic background. Among the available methods, Bayesian clustering represents the most accurate and reliable.

Genetic analyses using molecular markers provide information based on the neutral evolution of the populations (i.e., not affected by selection) for identifying GZs. This type of markers could offer information about demographic history of a population and allow identifying the genetic structure of a set of populations modeled by historical processes. Still, adaptive traits could be evaluated by quantitative genetic studies and, combined with neutral marker analyses, might provide a complete assessment of the genetic resources. Thus, GZs constitute a first step toward the definition of provenance regions (PRs), which ensure not only the conservation of evolutionary significant variants related to its life history (Crandall et al. 2000; de Guia and Saitoh 2007) but also the ecological viability and local adaptation of populations. The final goal of OGMUs' definition is to conserve relevant ecological entities that represent short- and long-term genetic processes (Fraser and Bernatchez 2001).

Population genetic diversity of a species is shaped in terms of evolutionary time as the product of the interplay between enhancing forces (i.e., gene flow, generation of new mutations, hybridization) and erosive forces (i.e., genetic drift, selection). Highly diverse populations should be prioritized for conservation, since they probably expose a better response to environmental changes that might risk their persistence (e.g., extreme climatic events, biological invasions). Unique variants (i.e., infrequent genotypes, private alleles, or haplotypes) or geographically restricted variants are also relevant to face new and unpredictable environments. It is crucial to understand the distribution of the genetic diversity along the natural species range since it reveals the degree of interconnection among them. The variation level of each population might be related to past evolutionary footprints, the mating system, or current genetic processes. Whatever the cause, its identification could assist managers when collection of propagation material is needed.

Based on molecular markers, standardize allelic richness SAR (Marchelli et al. 2017) was obtained for *Nothofagus pumilio* and *N. antarctica* populations, and the more diverse were identified (1–5, 8, 12, 14, 16, and 19 and II, IV, IX, XIII, XIV, XV, and XX, respectively (Table 5.1); Soliani et al. 2017). Considering the emergent patterns of genetic diversity and structure, but also the prevailing topography and the geographic isolation of the populations, preliminary genetic zones were proposed for both species. Several GZs include large areas of the natural distribution of the target species, demanding the screening of more and new populations to refine the divisions. In *N. pumilio* case, a re-delineation of GZs has being carried out by adding 14 new populations to the previous set and thus covering almost completely the main distribution range of this species (Mattera et al., in press). As a result of this, 18 GZs were defined in the species within four main regions along Patagonia (36–42° S, 42–44° S, 44–51° S, and 55° S corresponding to populations included in the island of Tierra del Fuego). In *N. antarctica*, nine GZs were proposed (Fig. 5.4) based on the genetic data from the 21 populations presented in Table 5.1. From north to south, they are North, Tromen, Central, Chubut, Río Grande, LGM East, LGM West, South, and Tierra del Fuego.

5.5 Adaptive Genetic Variation of *N. pumilio*: Assessment of Juvenile Traits in Common Garden Experiments

Genetic diversity is a population intrinsic property, the basis on which the evolutionary force of natural selection operates. The geographic variation of quantitative traits displayed by individuals in situ can be associated with their distribution in environmental gradients and inferred as a result of adaptation

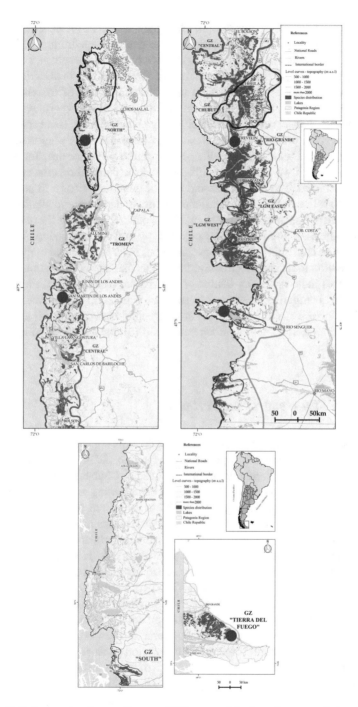

Fig. 5.4 *Nothofagus antarctica* genetic zones (GZ) proposed for Argentinean populations (Soliani et al. 2017). Colored dots represent highly diverse populations identified based on chloroplast and nuclear markers

processes, thus providing a first approach to the analysis of the genetic diversity of a species. A generalized interpretation of the genotype-environment relationship will make it possible to discern whether the type of variation is clinal, i.e., gradually changing characters, or ecotypical, that is, variation occurring in discrete changes of the character (it does not correspond to a gradual change of the environmental conditions). However, the establishment of common garden experiments is needed to distinguish between environmental and genetic effects. To accomplish this, breeding the bulk progeny of several natural populations in a common location is a further step on the way to unravel the genetic patterns of a species. Yet, a next step can be advanced if we keep the identification of the parental relationships of those progenies assayed. The analysis of variation of quantitative traits in progeny trials allows us to estimate the genetic variation within and between populations.

Out of the two species considered in this chapter, *N. pumilio* is the one with more breeding potential, due to its better productivity and forestry shape. Additionally, the interest on cultivating these species is higher in *N. pumilio* due to the fact that it lacks the ability to resprout from the stumps, thus making its cultivation for restoration purposes more necessary than in *N. antarctica*. Consequently, there are more advances in quantitative genetics of *N. pumilio*, which will be presented below following the three levels of analysis: (1) in situ geographic variation, (2) among populations' variation in common garden trials, and (3) genetic variation by means of progeny trials.

5.5.1   In Situ Geographic Variation

Variation in seed traits across environmental gradients could evidence adaptive processes. Aiming to analyze this possibility, Mondino (2014) collected seeds in 14 natural populations of *N. pumilio* in the Province of Chubut (a small portion of the wide Argentine distribution of the species), representing a latitudinal range of 2 degrees and a precipitation range from 400 to 1000 mm of annual average. The mean weight of 100 seeds was 1.36 g, and differences among populations were shown by means of an ANOVA for weight, width, length, and length/width ratio of the seeds (80.6% of the total variance was explained by *population* in the weight trait). However, it was not possible to recognize a consistent pattern associated with the two environmental characters considered (different interactions were shown between *precipitation* and *latitude* factors depending on the trait).

Based on a new seed collection in six populations, now representing three altitudinal levels in two sites (Mondino 2014), interaction between altitude and sites was shown for 100-seed weight (the highest population produced the heaviest seeds in one site, while the opposite was verified in the other). A different picture was found for seed-shape traits, where those of the highest altitude were the narrowest in both sites.

5.5.2 Variation Among Populations in Common Garden Trials

Annual growth synchronization with climate usually reflects a compromise between tolerance to cold and optimization of growth. In trees of temperate climates, the initiation and cessation of primary growth determine the duration of the stem elongation period and indicate the transition between resistant and frost-vulnerable stages. Therefore, the synchronization is critical for both optimal biomass production and fitness. Because of its wide latitudinal range, *N. pumilio* is expected to show deep phenological differences among populations.

Torres et al. (2017) carried out a population variation study concerning the bud burst process, including the entire latitudinal range of the species. They installed a high-density provenance trial in Trevelin Forest Station of INTA (43° 7' 17"S, 71° 33' 41"W, 390 m asl) representing 12 Argentine populations (4 blocks, 24 seedlings per block, N = 1152). Budburst phenology was registered in all seedlings every 3 days during 3 months at the beginning of their second growing season, according to five phenophases as described in Chap. 3 for *Nothofagus alpina*. The *population* factor had a significant effect on the day of the year to reach phenophase 3 (completely open buds) and explained 21% of the total variance. However, a geographically defined pattern was not found.

Similarly, Mondino (2014) studied the effect of the main environmental gradients on the phenology of the growth process. First, he installed a greenhouse provenance trial in Trevelin Forest Station with 27 potted seedlings (in a three-block design) from each of 11 Argentine populations from the Province of Chubut, representing a two degrees' latitudinal range and a 600 mm precipitation range. At the beginning of the second growing season, the height of each seedling was measured every 10 days in order to build individual growth curves fitting a Boltzmann sigmoidal model by regression. Growth initiation (t_{10}: time to reach 10% of the season growth) and cessation (t_{90}: time to reach 90% of the season growth) were estimated in days since the first changes in the budburst phenology (bud swelling). Interaction between *latitude* and *precipitation* factors, as well as significance of the precipitation level, was not detected. On the other hand, the *latitude* factor was significant for t_{90} and for the duration of the growing season (Dur = $t_{90} - t_{10}$) (northern populations ceased growing later and presented a longer growing period).

In a new study, Mondino et al. (2019) collected seeds from six populations from three altitudinal levels (200 m of difference between each other) in two sites of similar latitude and precipitation regime and installed a greenhouse provenance trial similar to the previous one. Again, the height of each seedling was measured every 10 days during the second growing season, and individual growth curves were regressed. The average final height of the entire trial was 29.2 cm. Interaction between altitude and site and differences between sites were not significant for any of the traits considered. On the contrary, significant differences among altitudinal levels were shown for several variables (t_{10} and Dur among them), and in all of them, the significance was due to the difference between the highest-level plants and the other two undifferentiated. Thus, in both sites, seedlings from the uppermost

Fig. 5.5 *Nothofagus pumilio* morphotypes coexisting at the maximum altitude reached by the species in a 50 m altitudinal range strip: arboreal, shrubby, and crawling. Bottom right box: effect of altitude of origin for the variables: (**A**) mean time (t_{50}), (**B**) form of the growth curve (S), (**C**) time of onset of growth (t_{10}), and (**D**) duration of the growth period (Dur). Different letters indicate significant differences with a $P < 0.05$. Altitudinal floors: low, white; medium, light gray; high, dark gray. (Photos: Víctor A. Mondino)

populations initiated growth later, had a shorter period of growth, and displayed a steeper growth curve than those from the other two altitudinal levels.

Nothofagus pumilio builds the treeline of the Subantarctic forests in Patagonia. It is possible to find different morphotypes coexisting at the maximum altitude reached by the species in a 50 m altitudinal range strip: arboreal, shrubby, and crawling (Fig. 5.5). In order to analyze whether these deep phenotypic differences are genetically determined or are the expression of the phenotypic plasticity according to the reaction norm of architectural and growing traits, two different studies were carried out. In a first case study (Mondino 2014), seeds from 20 trees corresponding to each of the three morphotypes were collected in two nearby sites at 42° 50′ S. The altitudinal ranges of both sites were around 1450 m asl and around 1550 m asl, respectively, that is, an intermediate altitude between the low-altitude good shaped forest and the high-altitude Krummholtz (Fig. 5.5). Seedlings were produced and a greenhouse trial was installed with potted plants arranged in a three-block design (9 seedlings per block; N = 162) to test morphotype and site differences by means of ANOVA. Plant growth rhythm was characterized by measuring the height of each seedling every 10 days during the second vegetative period and regressing

162 individual growth curves from these data (Boltzmann model). Then, several variables were derived: initial (H_0) height, time to growth initiation (t_{10}), cessation (t_{90}) and to reach 50% of growth (t_{50}), total length of the growing period (Dur), and the maximum growth rate through the slope of the curves (S) (Fig. 5.5). Plant architecture at the end of the second season was characterized by measuring collar diameter (d), total height (H), H/d ratio, number of buds in the main stem (NBu), apical meristem necrosis (AN, binary trait), forking (F, binary trait), first- and second-order branch length (BL1, BL2), number of branches of the first and second order (NB1, NB2), apical dominance (H/BL1), branchiness (H/NB1), internode length (H/NBu), and stem decumbent habit (Dh, binary trait). ANOVA showed differences between morphotypes only in the following traits: H, S, d, H/NBu, and H/d. This was surprising, particularly because the most vigorous seedlings corresponded to the crawling morphotype (Table 5.4). Thus, the different growth habits found in the nature among plants close to each other at that intermediate altitude seem to be a plastic response in most of the juvenile traits surveyed, especially in those that were expected to present a difference among the three morphotypes. Perhaps this evaluation has been too early to find a genetic cue.

In a second case study (Soliani and Aparicio, 2020), seedlings were produced with seeds collected from two stands of contrasting altitude located at 41° 15′ S: (L) low altitude, arboreal morphotype (1200 m asl), and (H) high altitude, shrubby morphotype (1560 m asl). A greenhouse common garden trial was established with

Table 5.4 Average values and variation coefficients of analyzed architectural variables in three different morphotypes of *Nothofagus pumilio* growing in a mountain slope: arboreal, shrubby, and crawling (see Fig. 5.5) (from Mondino 2014)

Variables	Arboreal		Shrubby		Crawling	
	Mean	CV	Mean	CV	Mean	CV
H [mm]	237.5	34	283.8	34	323.5	32
d [mm]	4.8	20	5.1	17	5.4	22
NBu	13.7	41	17.1	38	20.0	122
NB1	3.7	96	3.9	71	3.2	61
BL1 [mm]	92.3	66	120.5	56	138.7	48
NB2	0.1	693	0.0	0	0.0	0
BL2 [mm]	2.0	693	0.0	0	0.0	0
AN [%]	0.1	267	0.1	324	0.0	505
F	0.1	335	0.1	286	0.0	721
H/d	49.2	29	55.7	31	59.7	20
H/NBu	18.1	31	17.0	21	19.5	36
H/NB1	0.0	81	0.0	56	0.0	61
H/BL1	0.4	58	0.4	52	0.5	56

Height (H), diameter (d), number of buds in the main axis (NBu), number of first-order branches (NB1), length of the longest branch of first order (BL1), number of second-order branches (NB2), length of the longest branch of second order (BL2), percentage of apex loss (AN), forking (F), height/diameter ratio (H/d), internode length (H/NBu), apical dominance (H/BL1), branchiness (H/NB1)

~100 potted seedlings from each stand. Once seedlings had entered their first dormant period (June 15), their height (h_0) from the collar to the top of the most distal bud of the main axis (A1) and their diameter (d_0) at the base of the most distal bud were recorded. In the second year growing period, the cumulative length of the main axis was measured, six times along 172 days, finishing when the plants had reached asymptotic growth (December 12). With those data, individual growth curves were fitted using a sigmoidal equation (Boltzmann model), and traits were calculated as before t_{10}, t_{90}, Dur, and the maximum growth rate grmax ($mm.day^{-1}$). Budburst phenology of the main axis terminal bud (BB) was registered at the second and third growing seasons (considering four phenophases). Day of the year until phenophase 3 (at least one leaf is unfolded and spreading) was registered. Architectural traits were recorded during the second growing period: dominance (dom_1) of the main axis and delayed (dlb) and/or immediate (ilb) lateral branching of second-order branches (A2). A whole-plant size and form characterization was performed when plants were 6 years old, by obtaining the total height (h6), number of co-dominant axes (nAx) (which was also used to calculate the dominance of the main axis: (dom6)), length of the largest co-dominant axis (lAx), total number of secondary branches (nA2br), and length of the largest branch (lBr) carried by the largest co-dominant axis. With these measures, two continuous traits were constructed: branchiness (Br = nA2br/(lAx)/10) and slenderness (Sl = lAx/(lBr/nAx)). Together with h6, Br and Sl describe plant architecture within a continuum from shrubby, multi-stemmed to slender, single-stem growth forms.

The majority of the variables of phenology, growth rhythm, and architecture traits showed significantly different means between stands H and L. Although the individual age-age correlations for height (h_0, h_6) and dominance (dom_1, dom_6) of the main axis were not significant, the plants from stand H were on average consistently shorter and had lower dominance of the main axis. Besides, they had a later budburst phenology and growth initiation, i.e., an average temporal lag of 6 days, similar to the findings of Premoli et al. (2007), and were more branched and less slender than plants from stand L.

From the whole-plant analysis, three archetypes representing the juvenile growth form were retained: archetype 1 represents plants that were small and typically formed by several (2 to 4) co-dominant axes, relatively dense in lateral branches. On the other hand, archetype 3 reflects large single-stemmed, slender plants, with low branchiness. Archetype 2 represents highly branched plants, which, although in general had one main stem, was short (Fig. 5.6). The frequency of plants resembling each archetype (according to their nearest Euclidean distance) was not independent of the provenance stand. Within the group of plants from stand H, 74.5% were phenotypically closer to archetype 1, 15.5% to archetype 3, and 10% to archetype 2. In stand L, the proportions of plants resembling archetypes 1 and 3 were similar (48.7% and 50%), while those closer to archetype 2 were only 1.3%.

The juvenile growth habit of *N. pumilio* differed between the lower and higher extremes of a 360 m elevation gradient. This should be attributed to genetic determination and not solely to plastic responses to varying environmental constraints imposed by altitude. The significant temporal lag of ca. 6 days in budburst phenology

Traits	Archetype 1 (left)	Archetype 2 (center)	Archetype 3 (right)
h_6	30.00	33.81	95.43
Br	7.99	3.75	2.76
Sl	1.67	0.64	3.56

Fig. 5.6 Archetypal growth habits in 6-year-old plants of *Nothofagus pumilio* from the low and high extremes of an altitudinal gradient, according to three size and architecture traits: height (h6), branchiness (Br), and slenderness (Sl). From Soliani and Aparicio (2020)

between plants of high and low altitudes marked a shortening of the growing season that is perhaps the most relevant constraint for plants at high altitudes (Meloche and Diggle 2003; Cox 2005). Plants from stand H were shorter, more densely branched (immediate and delayed), and had a lower proportion of individuals whose main axes were dominant. At year 6, they presented a higher mean branchiness, and the majority (74.5%) of them clustered near archetype 1, which represents small multistemmed, densely branched phenotypes (Fig. 5.6), resembling the dominant architectural forms observed at high altitude. Instead, plants from stand L did not display one clearly prevailing growth habit. Thus, at high altitudes, *N. pumilio* seems to experience natural selection favoring late flushing and fast shoot extension to avoid frost damage. With increasing altitude, trees could be maximizing carbon gain by a more efficient photosynthetic performance (e.g., Premoli and Brewer 2007; Molina-Montenegro et al. 2012) and the efficient allocation of non-structural carbohydrates (e.g., Fajardo et al. 2013).

The dissimilar results thrown by the two last essays may be due to the different ages of evaluation, what was evidenced by the lack of age-to-age correlation in the second and longer trial. Alternatively, differences could be related to the sampling sites. In the first case, the contrasting morphotypes were vegetating close to each other, and consequently gene flow likely exists among them, thus restricting differentiation and consequently adaptation. On the contrary, in the second study, both morphotypes are separated by a relatively large distance, and gene flow is likely more limited than in the first case, and therefore adaptation is more probable.

Based on the previous results, altitudinal and latitudinal zonation should be considered in the definition of management units and/or in the delineation of provenance regions in *N. pumilio*, although not in a clinal way, since ecotypic variation seems to prevail. The seed sources for restoration programs, as well as the seed orchards for low-intensity breeding, should avoid the admixture of genetic materials from stands markedly separated in altitude or latitude. Observed gaps in bud sprouting and growth rhythm development between provenances in both gradients evidenced adaptation to local conditions. Particular site conditions should not be dismissed in active restoration. On the other hand, the amount of annual precipitation does not seem to make a difference among populations.

5.5.3   Genetic Variation by Means of Progeny Trials

A preliminary study of intra-population genetic variation (Mondino 2014) was conducted by means of a greenhouse progeny test that included the progeny of 68 open-pollinated mother trees corresponding to four natural populations of the Province of Chubut (N = 897). At the end of the first growing season, several architectural traits were measured in each potted seedling. The dispersion of data was large due to the low sampling level; however, the difference among populations and the significance of the *family* factor could be shown for some traits. Differences among populations were proved to be significant for height, slenderness index (height/diameter ratio), and branchiness index (height/number of branches ratio), for which a differentiation among populations of $Q_{ST} = 15\%$, $Q_{ST} = 16\%$, and $Q_{ST} = 17\%$ was estimated, respectively. It must be highlighted that the differentiation was due to only one population that was different from the other three: San Martin, which is in fact a marginal population of the steppe, completely isolated from the forest continuum (43° 49′ 47′ S, 70° 45′ 33′ W; 1350 m asl), and subjected to stressful precipitation conditions (annual average of 300 mm). The heritability estimated for the traits whose *family* factor resulted significant were mostly moderate, ranging from 0.15 to 0.57 (Table 5.5).

Table 5.5 Additive genetic variance (σ_A), variation coefficient ($CV_A\%$), and heritability (h^2) in five variables measured in *Nothofagus pumilio* seedlings

Populations Sample size		LH_{low} N = 12	Hm N = 26	LH_{high} N = 15	SM N = 15	
LA LO		42°50' 71°16'	42°49' 71°15'	42° 50' 71° 27'	43° 49' 70° 45'	
AL		1100	1450	1000	1450	Q_{ST}
Variables						
H	σ_A	0	0	790.74	544.83	0.15
	CVA%	0	0	55.23	44.31	
	h^2	0	0	0.57	0.40	
D	σ_A	0.34	0.12	0.49	0	–
	CVA%	11.65	5.74	14.25	0	
	h^2	0	0.15	0.54	0	
h/d	σ_A	0	0	226.35	146.59	0.16
	CVA%	0	0	35.56	39.27	
	h^2	0	0	0.43	0.48	
I	σ_A	0	0	5.18	3.97	0.17
	CVA%	0	0	18.43	19.67	
	h^2	0	0	0.54	0.43	
Lr/h	σ_A	0	0	0	0	–
	CVA%	114.76	0	0	0	
	h^2	1.16	0	0	0	

From Mondino (2014)
LA-LO, latitude-longitude (S – W); AL, altitude (m asl); Q_{ST}, genetic differentiation coefficients between populations; H, main axis length; D, collar diameter; h/d, height/diameter ratio; I, average internode length on the main axis; lr/h, ratio between length of the longest primary branch and height

5.6 Domestication and Low-Intensity Breeding Strategies

5.6.1 First Steps in N. antarctica

Although regeneration by sprouts (even after fire) is common in the species, viable seeds are produced regularly. In primary old growth stands of Tierra del Fuego (Soler Esteban et al. 2010), a one-season production of 1.85 million of viable seeds per hectare was estimated, which in fact represent a very low proportion of the 10.28 millions of seeds produced. These values surely vary among years and stands, but anyway, seeds are commonly available. This opens the possibility of seedling production and consequently active restoration (plantation) in the management of markedly degraded *ñirantales* (i.e., natural forests of the species), ensuring and accelerating the times of forest regeneration. Likewise, planting of *N. antarctica* for productive purposes is also possible, mainly in silvopastoral systems related to cattle raising (Hansen et al. 2004), in the forest-steppe ecotone. Seedlings suitable for planting can be obtained in one season by means of ferti-irrigation under greenhouse

Fig. 5.7 Greenhouse seedling production of *Nothofagus*. (**a**) 1-month-old plantlets of *Nothofagus antarctica*, (**b**) 8-month-old plants of *Nothofagus pumilio*, with an outstanding seedling in foreground. (Photos: Mario J. Pastorino)

conditions, as in the other *Nothofagus* of Argentina (Schinelli Casares 2012) (Fig. 5.7).

Nothofagus antarctica is a species with a great adaptability to dry and waterlogged environments and to different types of soils. Since 2003, 16 plots of ñire plantations have been established at different site conditions in the mountain ranges of the Province of Chubut (~42–44° S). Both survival and growth rate significantly changed depending on site conditions (Luis E. Tejera, personal communication). In humid sites, an average height of 3.5 m has been reached at 5 years of age, with tallest trees presenting 4.5 m of total height. Similar to other *Nothofagus*, the species usually presents apical death probably related to stressful conditions (i.e., peaks of high temperatures or drought), which causes stem damage in juvenile individuals. Among the main limitations for its establishment are damage by rodents such as tuco-tucos (*Ctenomys* spp.), European hares (*Lepus europaeus*), and rabbits (*Oryctolagus cuniculus*) (Contreras 1973; Vincon 2010) (Fig. 5.8). In addition, the attack of wood-boring insects (*Lautarus concinnus*, *Calydon submetallicum*, *Calydon globithorax*, *Phymantoderus bizonatus*, *Callisphyris semicaligatus*) on the trunks of living trees (Rizzuto 2003) is highly widespread, which greatly affects its timber (Fig. 5.8).

Breeding activities are barely initial for *N. antarctica*. The main subject related to the productive cultivation of the species is its poor forestry shape, with many stems, often tortuous, and with forking and coarse branches. However, there is much variation in growth form, and trees growing straight upright can be found in any forest, which gives the chance to think about selection. In order to test the genetic control of the forestry shape (Schinelli Casares et al. 2016), 39 trees were selected in a natural stand according to two categories: monopodial and straight trees (named as "plus") and common phenotypes (named as "general," including multi-stem, tortuous, or forked trees). Seeds were collected, seedlings were produced, and a

Fig. 5.8 (**a**) A specimen of tuco-tuco (*Ctenomys* spp.) and (**b**) the damage it causes in *N. antarctica* roots, (**c**) damage caused by wood-boring insects. (Photos: Luis E. Tejera)

greenhouse progeny trial was established with 60 seedlings per open-pollinated family randomly ordered in three repetitions (N = 2340). At the end of the first growing season, the total height (H) of each seedling was measured, and they were categorized in three shape types (S): straight, crooked, or decumbent. Differences between the progeny of plus and general mother trees were shown for the proportion of S (Fig. 5.9), but not for H. Plus trees had a progeny with a greater proportion of straight seedlings. On the contrary, when testing the *family* factor, it was not significant for the proportion of S but for H, and in fact a high heritability was estimated (h^2 = 82%). We still do not know the evolution of these traits at older ages, but it seems wise to select mother trees in the natural forest according to their shape. On the other hand, the selection according to the growth in height would also be quite convenient. However, this could hardly be done in the natural forest, since it is ignored if the higher relative height of a tree is due to a better growth or an older age.

After these preliminary results, a low-intensity breeding strategy was planned by INTA for *N. antarctica*, based on the establishment of progeny seed orchards with seedlings obtained from seeds of good forestry shape individuals selected in the natural forest. Thus, 40 trees with monopodial and straight stems were selected in two natural populations of the Province of Chubut (Trevelin and Lake Rosario). In 2009, open-pollinated seeds were collected from those trees, expecting a greater proportion of good shape seedlings in the nursery production, which was conducted by INTA. Finally, in 2011, two seed orchards were established with 2-year-old seedlings. The first one was settled in Trevelin Forest Station of INTA (43° 05′ 29″S, 71° 32′ 21″W, 450 m asl; Fig. 5.9), with 1750 seedlings and without irrigation, and the second in a nearby private property (43° 7′ 6″S, 71° 27′ 30″W, 360 m asl), with 400 seedlings and furrow irrigation. In order to perform a preliminary evaluation, the height of the saplings was measured in 2015 in both orchards, together with a

Fig. 5.9 Contrasting *Nothofagus antarctica* individual phenotypes photographed in 2019 in the progeny seed orchard installed in 2011 in Trevelin. (Photos: Luis E. Tejera). Central box: proportion of seedlings of the defined shape categories for each of the two types of progenitors selected as "plus" and "general" trees

characterization of their shape by means of three binary traits: mono-/sympodial growth, straight/tortuous stem, and presence/absence of forking. The mean height of each orchard was 113.5 ± 49 cm and 196 ± 48 cm, respectively. No differences could be shown between the populations in both orchards for any of the shape traits, nor for the height, and the *family* factor was significant only for the height in the irrigated orchard and in fact with a $h^2 = 0.24$. It may be that the environmental heterogeneity is too large in the non-irrigated orchard to detect differences so soon.

5.6.2 First Steps in N. pumilio

Seed production has been estimated in old growth natural forests of Tierra del Fuego (Martínez Pastur et al. 2008) and does not seem to be a limiting factor for seedling production (42.5 million of viable seeds per hectare in one season). However, in the experience of INTA in central and north Argentine Patagonia, seed provision represents a true bottleneck. Seed production fluctuates greatly from year to year in the natural forest, and the proportion of empty or nonviable seeds is rather large. It may take several years until an abundant collection of good quality seeds is achieved. Furthermore, the seeds of *N. pumilio* are almost recalcitrant, and, therefore, their viability decreases considerably in only 1 year, even when stocked at low temperature. Thus, the strategy of stocking seeds during mast years, as recommended for *N. alpina* and *N. obliqua*, does not apply in this case, at least till a specific stocking protocol can be adjusted for this species. With viable seeds, very good seedlings (more than 35 cm tall and 0.5 mm of collar diameter) can be produced in 8 months by means of ferti-irrigation under greenhouse conditions (Schinelli Casares 2012; Fig. 5.7).

Although the species has been planted profusely on a very low scale, even in Great Britain (Mason et al. 2018), Denmark, and Norway (Sondergaard 1997), forest plantations with *N. pumilio* (as with *N. antarctica*) are scarce. Just in the last decade, the species has been planted on a larger scale in the frame of active restoration programs related to extensive forest fires that occurred in Chile and Argentina. Good examples of these efforts are the restoration programs performed in Torres del Paine National Park (Chile) and in the Argentine provinces of Río Negro, Chubut, Santa Cruz, and Tierra del Fuego, where hundreds of thousands of lenga seedlings have been planted (Pastorino et al. 2018; Guzmán et al. 2019; Mattenet et al. 2019; Mestre et al. 2019; Paredes et al. 2019; Salinas et al. 2019). However, the vast majority of those seedlings were not produced in nurseries but collected from the regeneration banks of the natural forests. In any case, these experiences have boosted interest in the breeding of the species at both sides of the Andes Cordillera.

In Chile, a trial was installed in 2000 in the Forest Reserve of Coyhaique (Ipinza and Gutierrez 2015) with the open-pollinated progeny of 111 mother trees sampled in the natural forest of three provenance regions (N = 2466). Plus and not selected trees were included in the sample. The average survival of the trial was 58.5% at 11 years of age, with 2.14 m of mean height and 21.8 mm of mean collar diameter. The heritability of these two last traits was $h^2 = 0.18$ and $h^2 = 0.27$, respectively. It is remarkable that among the 20 better mothers, there was not any plus tree, and among the 50 better trees, there was not any progeny of the plus trees. This result leads to discard the strategy of selecting trees from the natural forest. Instead, the mass selection from planted trees, where the method of comparison with neighbor individuals is applied (Ledig 1974), or better yet, the genetic selection in progeny tests seems recommendable.

In Argentina, the first steps toward the genetic improvement of the species have been taken recently. A small provenance trial network has started to be established in 2017. Seeds were collected from 12 natural stands of the whole distribution area of *N. pumilio* in Argentina, from south of Tierra del Fuego to north of the Province of Neuquén. Seedlings were produced and four provenance trials were installed in (1) Río Turbio (Punta Gruesa Forest Reserve, 51° 32′ 44.7″S, 72° 07′ 22.2″W; 535 m asl; N = 675; first year survival = 92%), (2) Trevelin Forest Station of INTA (43° 05′ 29″S, 71° 32′ 21″W; 450 m asl; N = 900; first year survival = 97%), (3) Bariloche alto (private property, 41°13′38.86″S, 71°14′32.85″W; 1060 m asl; N = 480; second year survival = 51%), and (4) Bariloche Experimental Station of INTA (41° 7′ 21.17″ S, 71°14′ 56.95″ W; 795 m asl; N = 270; second year survival = 80%).

Finally, aiming to count with seeds of known origin for seedling production, in 2017, a production seed area was registered with the National Seed Institute

(INASE). It is a natural pure stand of 24 ha located in the Trevelin Forest Station of INTA (43° 3′ 49″S, 71° 34′ 33″W, 1100 m asl) with very good accessibility, seed productivity, and seed quality.

References

Acosta MC, Premoli AC (2010) Evidence of chloroplast capture in South American *Nothofagus* (subgenus Nothofagus, Nothofagaceae). Mol Phylogenet Evol 54:235–242

Alberto F, Massa S, Manent P, Diaz-Almela E, Arnaud-Haond S, Duarte et al (2008) Genetic differentiation and secondary contact zone in the seagrass *Cymodocea nodosa* across the Mediterranean-Atlantic transition region. J Biogeogr 35:1279–1294

Bahamonde HA, Lencinas MV, Martínez Pastur G, Monelos L, Soler R, Peri PL (2018) Ten years of seed production and establishment of regeneration measurements in *Nothofagus antarctica* forests under different crown cover and quality sites, in Southern Patagonia. Agrofor Syst 92:623–635

Bava J, Rechene C (2004) Dinámica de la regeneración de lenga (*Nothofagus pumilio* (Poepp. et Endl.) Krasser) como base para la aplicación de sistemas silvícolas. In: Arturi MF, Frangi JL, Goya JF (eds) Ecología y manejo de bosques nativos de Argentina. Edt. Universidad Nacional de La Plata, La Plata, 23pp

Bava J, Lencinas J, Haag A (2006) Determinación de la materia prima disponible para proyectos de inversión forestales en cuencas de la provincia del Chubut. Informe Final. Consejo Federal de Inversiones - Gobierno del Chubut. Fundación Para el Desarrollo Forestal Ambiental y del Ecoturismo Patagónico (FDFAEP), 139pp

Bucci G, Vendramin GG (2000) Delineation of genetic zones in the European Norway spruce natural range: preliminary evidence. Mol Ecol 9:923–934

CIEFAP, MAyDS (2016) Actualización de la Clasificación de Tipos Forestales y Cobertura del Suelo de la Región Bosque Andino Patagónico. Informe Final. CIEFAP. https://drive.google.com/open?id=0BxfNQUtfxxeaUHNCQm9lYmk5RnM

Comps B, Gömöry LJ, Thiébaut B, Petit RJ (2001) Diverging trends between heterozygosity and allelic richness during postglacial colonization in the European Beech. Genetics 157:389–397

Contreras J (1973) El tucu tuco y los problemas del suelo en la Argentina. INTA. IDIA 19

Cornuet JM, Liukart G (1996) Description and power analysis of two tests for detecting recent population bottlenecks from allele frequency data. Genetics 144:2001–2014

Cornuet JM, Santos F, Beaumont MA, Robert CP, Marin JM, Balding DJ et al (2008) Inferring population history with DIY ABC: a user-friendly approach to approximate Bayesian computation. Bioinformatics 24:2713–2719

Cornuet JM, Ravigné V, Estoup A (2010) Inference on population history and model checking using DNA sequence and microsatellite data with the software DIYABC (v1.0). BMC Bioinf 11:401

Cosacov A, Sérsic AN, Sosa V, Johnson LA, Cocucci AA (2010) Molecular evidence of ice-age refugia in the Patagonia steppe and post-glacial colonisation of the Andes slopes: insights from the endemic species *Calceolaria polyrhiza* (Calceolariaceae). J Biogeogr 37:1463–1477

Cox SE (2005) Elevational gradient of neoformed shoot growth in *Populus tremuloides*. Can J Bot 83:1340–1344

Crandall KA, Bininda-Emonds ORP, Mace GM, Wayne RK (2000) Considering evolutionary processes in conservation biology. Trends Ecol Evol 15:290–295

Cuevas J, Arroyo M (1999) Ausencia de banco de semillas persistente en *Nothofagus pumilio* (Fagaceae) en Tierra del Fuego, Chile. Rev Chil Hist Nat 72:73–82

de Guia APO, Saitoh T (2007) The gap between the concept and definitions in the evolutionarily significant unit: the need to integrate neutral genetic variation and adaptive variation. Ecol Res 22:604–612

Donoso C (1993) Bosques templados de Chile y Argentina. Variación, Estructura y Dinámica. Ecología Forestal Chile, Ed Univ, 484 pp

Donoso C (2006) Las especies arbóreas de los Bosques Templados de Chile y Argentina. Autoecología. Cuneo Ediciones Valdivia, p 678

Durand E, Jay F, Gaggiotti O, Francois O (2009) Spatial inference of admixture proportions and secondary contact zones. Mol Biol Evol 26:1963–1973

El-Kassaby Y, Dunsworth B, Krakowski J (2003) Genetic evaluation of alternative silvicultural systems in coastal montane forests: western hemlock and amabilis fir. Theor Appl Genet 107:598–610

Fajardo A, Piper FI, Hoch G (2013) Similar variation in carbon storage between deciduous and evergreen treeline species across elevational gradients. Ann Bot 112:623–631

Finkeldey R, Ziehe M (2004) Genetic implications of silvicultural regimes. For Ecol Manag 197:231–244

Flint RF, Fidalgo F (1969) Glacial drift in the eastern argentine Andes between latitude 41°10′S and latitude 43°10′S. Geol Soc Am Bull 80:1043–1052

Fraser DJ, Bernatchez L (2001) Adaptive evolutionary conservation: towards unified concept for defining conservation units. Mol Ecol 10:2741–2752

Giménez Gowland M (2002) Flora Nativa Norpatagónica Ilustrada, INTA-FVSA-APN. CD

Glasser NF, Jansson KN, Harrison S, Kleman J (2008) The glacial geomorphology and Pleistocene history of South America between 38ºS and 56ºS. Quat Sci Rev 27:365–390

González ME, Donoso C, Ovalle P, Martinez Pastur G (2006) *Nothofagus pumilio* (Poep. et Endl) Krasser. Lenga, roble blanco, leñar, roble de Tierra del Fuego. In: Zegers CD (ed) Las especies arbóreas de los Bosques Templados de Chile y Argentina. Autoecología. Cuneo Ediciones, Valdivia, p 678

Grivet D, Sork V, Westfall R, Davis F (2008) Conserving the evolutionary potential of valley oak (*Quercus lobata* Née) in California. Mol Ecol 17:139–156

Guzmán M, Roveta R, Urretavizcaya MF, Ríos-Campano F, Postler V, Antequera S et al (2019) Implementación del programa integral de manejo y restauración de las grandes áreas afectadas por los incendios forestales de la temporada 2014–2015 en la Provincia de Chubut. IV Jornadas Forestales de Patagonia Sur. Ushuaia, 22-26/04/2019

Hansen N, Tejera L, Fertig M (2004) Módulo 2. Desarrollo de sistemas silvopastoriles en bosques de *Nothofagus antarctica*. In: Alternativas de manejo sustentable para el manejo forestal integral de los bosques de la Patagonia. Capítulo 3. Sistemas silvopastoriles en Chubut. Informe PIARFON, pp 671–680

Heuertz M, Carnevale S, Fineschi S, Sebastiani F, Hausman JF, Paule L, Vendramin GG (2006) Chloroplast DNA phylogeography of European ashes, *Fraxinus* sp. (Oleaceae): roles of hybridization and life history traits. Mol Ecol 15:2131–2140

Heusser CJ, Heusser L, Lowell T (1999) Paleoecology of the Southern Chilean Lake District-Isla Grande de Chiloé during middle-late Llanquihue glaciation and deglaciation. Geografiska Annaler 81 A, 231–284

Hill RS, Jordan GJ (1993) The evolutionary history of *Nothofagus* (Nothofagaceae). Aust Syst Bot 6:111–126

Ipinza R, Gutierrez B (2015) Evaluación genética a los 8 y 11 años de un ensayo de progenies de lenga (*Nothofagus pumilio*) en la Reserva Nacional Coyhaique, Región de Aysén, Chile. In: Gutierrez B, Ipinza R, Barros S (eds) Conservación de Recursos Genéticos Forestales, principios y práctica. Instituto Forestal, Santiago, 320 p

Ledig F (1974) Analysis of methods for the selection of trees from wild stands. For Sci 20:2–16

Lepais O, Petit JR, Guichoux E, Lavabre JE, Alberto F, Kremer A, Gerber S (2009) Species relative abundance and direction of introgression in oaks. Mol Ecol 18:2228–2242

Manos P (1997) Systematics of *Nothofagus* (Nothofagaceae) based on rDNA spacer sequences (ITS): taxonomic congruence with morphology and plastid sequences. Am J Bot 84:1137–1155

Manzini MV, Prieto AR, Paez MM, Schäbitz F (2008) Late Quaternary vegetation and climate in Patagonia. In: Rabassa J (ed) Late Cenozoic Patagonia Tierra del Fuego. Elsevier, Amsterdam, pp 351–368

Marchelli P, Gallo LA (2006) Multiple ice-age refugia in a southern beech from southern South America as revealed by chloroplast DNA markers. Conserv Genet 7:591–603

Marchelli P, Thomas E, Azpilicueta MM, van Zonneveld M, Gallo L (2017) Integrating genetics and suitability modelling to bolster climate change adaptation planning in Patagonian *Nothofagus* forests. Tree Genet Genomes 13:119

Markgraf V, McGlone M, Hope G (1995) Neogene paleoenvironmental and paleoclimatic change in southern temperate ecosystems- a southern perspective. Trends Ecol Evol 10:143–147

Markgraf V, Romero EJ, Villagrán C (1996) History and paleoecology of South American *Nothofagus* forests. In: Veblen TT et al (eds) The ecology and biogeography of *Nothofagus* forests. Yale University Press, New Haven/London, pp 354–386

Markgraf V, Bradbury JP, Schwalb A, Burns SJ, Stern C, Ariztegui D et al (2003) Holocene palaeoclimates of southern Patagonia: limnological and environmental history of Lago Cardiel, Argentina. The Holocene 13:581–591

Martínez Pastur G, Lencinas MV, Peri PL, Cellini JM (2008) Flowering and seeding patterns in unmanaged and managed *Nothofagus pumilio* forests with a silvicultural variable retention system. Forstarchiv 79:60–65

Mason B, Jinks R, Savill P, Wilson SMG (2018) Southern beeches (*Nothofagus* species). Q J For 112:30–43

Mathiasen P, Premoli AC (2010) Out in the cold: genetic variation of *Nothofagus pumilio* (Nothofagaceae) provides evidence for latitudinally distinct evolutionary histories in austral South America. Mol Ecol 19:371–385

Mattenet F, Monelos L, Monaco M, Peri PL (2019) Restauración de bosque nativo en la zona de Río Turbio (Santa Cruz): una oportunidad de aprendizaje comunitario. IV Jornadas Forestales de Patagonia Sur. Ushuaia, 22-26/04/2019

Mattera MG, Pastorino MJ, Lantschner MV, Marchelli P, Soliani C (in press) Genetic diversity and population structure in *Nothofagus pumilio*, a foundation species of Patagonian Forests: defining priority conservation areas and management. Sci Rep

McKay JK, Christian CE, Harrison S, Rice KJ (2005) "How local is local?"—a review of practical and conceptual issues in the genetics of restoration. Restor Ecol 13:432–440

Meloche CG, Diggle PK (2003) The pattern of carbon allocation supporting growth of preformed shoot primordia in *Acomastylis rossii* (Rosaceae). Am J Bot 90:1313–1320

Mestre L, Fernández-Génova L, Turi L (2019) Soy parte del bosque fueguino: restauración participativa de un bosque sub-antártico afectado por incendio. IV Jornadas Forestales de Patagonia Sur. Ushuaia, 22–26/04/2019

Molina-Montenegro M, Gallardo-Cerda J, Flores TSM, Atala C (2012) The trade-off between cold resistance and growth determines the *Nothofagus pumilio* treeline. Plant Ecol 213:133–142

Mondino VA (2014) Variación geográfica y genética en caracteres adaptativos iniciales de *Nothofagus pumilio* (Poepp. et Endl.) Krasser en una zona de alta heterogeneidad ambiental. Doctoral Thesis, Universidad de Buenos Aires. Buenos Aires, 174 p

Mondino V, Pastorino M, Gallo L (2019) Altitudinal variation of phenological characters and initial growth under controlled conditions among *Nothofagus pumilio* populations from Center-West Chubut, Argentina. Bosque 40:87–94

Palmé AE, Su Q, Palsson S, Lascoux M (2004) Extensive sharing of chloroplast haplotypes among European birches indicates hybridization among *Betula pendula*, *B. pubescens* and *B. nana*. Mol Ecol 13:167–178

Paredes D, Parodi M, Ojeda J, Farina S, Trangoni F, Fagnani A, Quiroz D (2019) Restauración activa en Tierra del Fuego con plantas de lenga repicadas del bosque natural. IV Jornadas Forestales de Patagonia Sur. Ushuaia, 22-26/04/2019

Pastorino MJ, Gallo LA (2002) Quaternary evolutionary history of *Austrocedrus chilensis*, a cypress native to the Andean-Patagonian forest. J Biogeogr 29:1167–1178

Pastorino MJ, Marchelli P, Milleron M et al (2009) The effect of different glaciation patterns over the current genetic structure of the southern beech *Nothofagus antarctica*. Genetica 136:79–88

Pastorino MJ, Aparicio AG, Azpilicueta MM, Rusch V (2018) Restauración del bosque quemado del C° Otto, Bariloche: un compromiso de hoy con las generaciones futuras. Presencia 70:14–17

Petit RJ, El Mousadik A, Pons O (1998) Identifying populations for conservation on the basis of genetic markers. Conser Biol 12: 884–855

Petit J, Aguinagalde I, de Beaulieu J, Bittkau C, Brewer S, Cheddadi R et al (2003) Glacial refugia: hotspots but not melting pots of genetic diversity. Science 300:1563–1565

Porter SC (1981) Pleistocene glaciation in the southern Lake District of Chile. Quat Res 16:263–292

Premoli AC, Brewer CA (2007) Environmental v. genetically driven variation in ecophysiological traits of *Nothofagus pumilio* from contrasting elevations. Aust J Bot 55:585–591

Premoli AC, Steinke L (2008) Genetics of sprouting: effects of long-term persistence in fire-prone ecosystems. Mol Ecol 17:3827–3835

Premoli AC, Kitzberger T, Veblen TT (2000) Isozyme variation and recent biogeographical history of the long-lived conifer *Fitzroya cupressoides*. J Biogeogr 27:251–260

Premoli AC, Souto CP, Rovere AE, Allnut TR, Newton AC (2002) Patterns of isozyme variation as indicators of biogeographic history in *Pilgerodendron uviferum* (D. Don) Florin. Divers Distrib 8:57–66

Premoli AC, Raffaele E, Mathiasen P (2007) Morphological and phenological differences in *Nothofagus pumilio* from contrasting elevations: Evidence from a common garden. Austral Ecol 32:515–523

Premoli AC, Mathiasen P, Cristina Acosta M, Ramos VA (2012) Phylogeographically concordant chloroplast DNA divergence in sympatric *Nothofagus* s.s. How deep can it be? New Phytol 193:261–275

Pritchard JK, Stephens M, Donnelly P (2000) Inference of population structure using multilocus genotype data. Genetics 155:945–959

Rabassa J, Clapperton CM (1990) Quaternary glaciations of the Southern Andes. Quat Sci Rev 9:153–174

Rabassa J, Coronato AM, Salemme M (2005) Chronology of the Late Cenozoic Patagonian glaciations and their correlation with biostratigraphic units of the Pampean region (Argentina). J S Am Sci 20:81–103

Rajendra KC, Seifert S, Prinz K, Gailing O, Finkeldey R (2014) Subtle human impacts on neutral genetic diversity and spatial patterns of genetic variation in European beech (*Fagus sylvatica*). For Ecol Manag 319:138–149

Rajora O, Rahman M, Buchert G, Dancik B (2000) Microsatellite DNA analysis of genetic effects of harvesting in old-growth eastern white pine (*Pinus strobus*) in Ontario, Canada. Mol Ecol 9:339–348

Ramírez C, San Martín C, Oyarzún A, Figueroa H (1997) Morpho-ecological study on the South American species of the genus *Nothofagus*. Plant Ecol 130:101–109

Rieseberg LH, Soltis DE (1991) Phylogenetic consequences of cytoplasmic gene flow in plants. Evol Trends Plants 5:65–84

Rizzuto S (2003) Diversidad y grado de infestacion por insectos xilófagos en bosque de *Nothofagus antarctica* en Cerro Centinela, Chubut, Argentina. MSc thesis. Universidad Nacional de la Patagonia San Juan Bosco. Esquel, Argentina

Rusch V (1993) Altitudinal variation in the phenology of *Nothofagus pumilio* in Argentina. Rev Chil Hist Nat 66:131–141

Salinas P, Ruiz C, Larson J, Moncada N (2019) Estado de la plantación de *Nothofagus pumilio* (Poepp.&Endl.) en áreas afectadas por incendios forestales en el PN Torres del Paine. IV Jornadas Forestales de Patagonia Sur. Ushuaia, 22–26/04/2019

Sauquet H, Ho SYW, Gandolfo MA, Jordan GJ, Wilf P, Cantrill DJ et al (2012) Testing the impact of calibration on molecular divergence times using a fossil-rich group: the case of *Nothofagus* (Fagales). Syst Biol 61:289–313

Schinelli Casares T (2012) Producción de *Nothofagus* bajo condiciones controladas. Ediciones INTA, Esquel, 56 pp

Schinelli Casares T, Mondino VA, Paredes M, Pastorino MJ (2016, 9–13 November) Mejoramiento genético en ñire: selección temprana por forma forestal. Comisión producción forestal sustentable, Actas V Jornadas Forestales Patagónicas, Esquel, p 339

Sérsic A, Cosacov A, Cocucci AA et al (2011) Emerging phylogeographical patterns of plants and terrestrial vertebrates from Patagonia. Biol J Linn Soc 103:475–494

Soler Esteban R, Martínez Pastur G, Lencinas MV, Peri PL (2010) Flowering and seeding patterns in primary, secondary and silvopastoral managed *Nothofagus antarctica* forests in South Patagonia. N Z J Bot 48:63–73

Soler R, Pastur GM, Peri P et al (2013) Are silvopastoral systems compatible with forest regeneration? An integrative approach in southern Patagonia. Agrofor Syst 87:1213–1227

Soliani C, Sebastiani F, Marchelli P, Gallo LA, Vendramin GG (2010) Development of novel genomic microsatellite markers in the southern beech *Nothofagus pumilio* (Poepp. et Endl.) Krasser. In: Permanent genetic resources added to molecular ecology resources database 1 October 2009–30 November 2009. Mol Ecol Resour 10:404–408

Soliani C, Aparicio AG (2020) Evidence of genetic determination in the growth habit of *Nothofagus pumilio* (Poepp. & Endl.) Krasser at the extremes of an elevation gradient. Scand J For Res 35:5–6, 211–220. https://doi.org/10.1080/02827581.2020.1789208

Soliani C, Gallo L, Marchelli P (2012) Phylogeography of two hybridizing southern beeches (*Nothofagus* spp.) with different adaptive abilities. Tree Genet Genomes 8:659–673

Soliani C, Tsuda Y, Bagnoli F, Gallo LA, Vendramin GG, Marchelli P (2015) Halfway encounters: meeting points of colonization routes among the southern beeches *Nothofagus pumilio* and *N. antarctica*. Mol Phylogenet Evol 85:197–207

Soliani C, Vendramin GG, Gallo LA, Marchelli P (2016) Logging by selective extraction of best trees: does it change patterns of genetic diversity? The case of *Nothofagus pumilio*. For Ecol Manag 373:81–92

Soliani C, Umaña F, Mondino VA, Thomas E, Pastorino MJ, Gallo LA, Marchelli P (2017) Zonas genéticas de lenga y ñire en Argentina, y su aplicación en la conservación y manejo de los recursos forestales, Ediciones INTA. Bariloche, p55

Sondergaard P (1997) Experiences with *Nothofagus* in West-Norway and East-Denmark. Dansk. Dendrologisk Arsskrift 15:61–90

Toro Manríquez M, Mestre L, Lencinas MV, Promis A, Martínez-Pastur G, Soler R (2016) Flowering and seeding patterns in pure and mixed *Nothofagus* forests in Southern Patagonia. Ecol Process 5:21

Torres AD, Aparicio AG, Mondino VA, Schinelli Casares T, Paredes M, Pastorino MJ (2017) Variación en caracteres fenológicos entre poblaciones naturales de *Nothofagus pumilio* a lo largo de su distribución latitudinal. 6° Seminario de Nothofagus: Silvicultura, manejo y conservación. Fac. Cs. Agrs. Ftales., UNLP. La Plata, 10-11/08/2017

Tortorelli L (1956) Maderas y bosques argentino. Editorial ACME, Buenos Aires

Veblen T, Donoso C, Kitzberger T, Rebertus AJ (1996) Ecology of Southern Chilean and Argentinean *Nothofagus* forests. In: Veblen T, Hill RS, Read J (eds) The ecology and biogeography of *Nothofagus* forests. Yale University Press, New Haven, p 403

Vendramin G, Anzidei M, Madaghiele A, Bucci G (1998) Distribution of genetic diversity in *Pinus pinaster* Ait. as revealed by chloroplast microsatellites. Theor Appl Genet 97:456–463

Villagrán C, Moreno P, Villa R (1995) Antecedentes palinológicos acerca de la historia cuaternaria de los bosques chilenos. In: Armesto JJ (eds) Ecología de los bosques nativos de Chile. Universidad Santiago de Chile, Chile, p 51–70

Vincon SG (2010) UNPSJB. Conociendo los Tucu-Tucos (*Ctenomys* spp.). Estación Experimental Agroforestal Esquel (Chubut)

Widmer A, Lexer C (2001) Glacial refugia: sanctuaries for allelic richness, but not for gene diversity. Trends Ecol Evol 16:267–269

Zemlak TS, Habit EM, Walde SJ, Battini MA, Adams ED, Ruzzante DE (2008) Across the southern Andes on fin: glacial refugia, drainage reversals and a secondary contact zone revealed by the phylogeographical signal of *Galaxias platei* in Patagonia. Mol Ecol 17:5049–5061

Chapter 6
Patagonian Cypress (*Austrocedrus chilensis*): The Cedarwood of the Emblematic Architecture of North Patagonia

Alejandro G. Aparicio and Mario J. Pastorino

6.1 Characteristics of Patagonian Cypress and Traditional Uses for Its Wood

The Patagonian cypress *Austrocedrus chilensis* (D. Don) Pic. Serm. & Bizzarri, also known as Chilean cedar, is a dioecious, anemophilous, and anemochorous conifer (Cupressaceae), native to Argentina and Chile. In Argentina, it grows along a ca. 60 km wide strip that extends parallel to the Cordillera de Los Andes, from 37° 07′ to 43° 44′ S (Pastorino et al. 2006), within the Mediterranean climatic zone of North Patagonia (Fig. 6.1). A recent study compiling information on areas with presence of cypress, from dense compact forests to patches with scattered, isolated trees, yielded a total of 262,422 ha, of which 41.78% are under the jurisdiction and protection of the National Parks Administration (Pastorino et al. 2015).

One of the most remarkable auto-ecological aspects of the cypress is its occurrence across a severe west to east annual rainfall gradient that decreases from ca. 3000 to 400 mm or even less in the xeric eastern border of the species. In pure, compact forest patches, cypress averages 20–25 m in height, and we have measured individuals up to 44 m in certain highly productive stands. The average diameters in adult trees range between 30 and 40 cm, but individuals of more than 1 m can usually be found (Fig. 6.2). The boles are normally straight, somewhat conical, with a thin bark in young individuals and longitudinal cracks at maturity. The crowns are mostly pyramidal and compact.

The cypress could be classified as a slow-growing species. In pure compact stands, it can grow at mean annual increments of MAI = 4.2 m^3 ha^{-1}, but towards the west margin of its distribution, where it grows mixed with *Nothofagus dombeyi*, it can reach ca. MAI = 13 m^3 ha^{-1} $year^{-1}$ (Loguercio et al. 2018). Although the species

A. G. Aparicio (✉) · M. J. Pastorino
Instituto de Investigaciones Forestales y Agropecuarias Bariloche (IFAB) INTA-CONICET, Bariloche, Argentina
e-mail: aparicio.alejandro@inta.gob.ar

Fig. 6.1 Natural
geographic range of
Austrocedrus chilensis.
The green patches
represent areas with
presence of cypress.
Compiled by M. Pastorino
(Argentina) and Gustavo
Cruz M. (Chile)

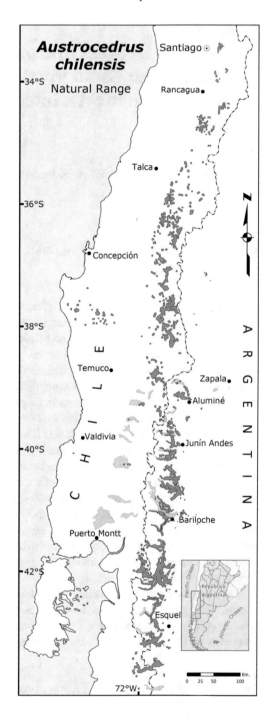

Fig. 6.2 Exceptional sawtimber tree of *Austrocedrus chilensis* at the southern shore of Lake Traful. (Photos: M. Pastorino)

has not yet been cultivated at an industrial forestry scale, the rotation length estimated from natural stands would be around 80 years. The cypress forests have been logged since the European settlement in the region at the end of nineteenth century, mostly without precise silvicultural systems and objectives. The species has been traditionally used in house building in rural areas and cities of the region, such as San Carlos de Bariloche, where once almost all the houses were built with cypress wood. This was due mainly to its moisture-resistant properties. With the passage of time, the use of cypress was diversified due to its aesthetic characteristics and workability. The wood is white-yellowish, durable, light, and aromatic, and although it is somewhat knotty, it is suitable for carpentry and furniture (Fig. 6.3). The advances towards the domestication with breeding purposes of the Patagonian cypress consist in a detailed proposal of provenance regions, the first for a tree species in Argentina (Pastorino et al. 2015). This proposal was based on a full distribution range description of the geographical patterns of neutral genetic variation, combined with the characterization of quantitative variation in juvenile stages of cypress's life.

Fig. 6.3 Aesthetic characteristics (upper panel, photo: A. Aparicio) and typical architecture design in North Patagonia using solid wood of *Austrocedrus chilensis* (lower panel; image kindly provided by Llao Llao Hotel, San Carlos de Bariloche)

6.2 Geographic Patterns of Neutral Genetic Variation

Selectively neutral genetic markers offer the possibility of reconstructing processes that have modeled the current distribution and patterns of genetic diversity of trees. Broadly speaking, the general agreement among studies on tree phylogeography and population genetics (complemented with other disciplines) in Patagonia is that the main modeling processes have been the glacial and interglacial cycles of the Quaternary, although volcanism, marine transgressions, and fires also played a role. Those processes can be understood as a drastic retraction of the distribution ranges of species up to the last glacial maximum, which dates back to ca. 25,000 years (Lowell et al. 1995; Flint and Fidalgo 1964, 1969), towards shelter areas located beyond the periphery of the ice, i.e., into the Patagonian steppe at the Argentinean side of the Andes. The subsequent recolonization of ice-free areas from multiple periglacial refuges is considered a major recent modeling process (e.g., Villagrán et al. 1996; Marchelli et al. 1998; Allnutt et al. 1999; Premoli et al. 2000; Pastorino and Gallo 2002; Azpilicueta et al. 2009; Cosacov et al. 2010), on which other effects of local scale can be superimposed.

The patterns of genetic variation of the cypress in Argentina have been characterized by means of isoenzymes (Pastorino 2001; Pastorino and Gallo 2002, 2009; Pastorino et al. 2004; Souto et al. 2012) and nuclear microsatellites (Arana et al. 2010). The general result of the studies with isozymes is that the intra-population genetic variation of cypress is relatively low in terms of allelic variants (number of alleles per locus A_L = 1.520; number of effective alleles Ae = 1.163) and moderate in terms of heterozygosity (Ho = 0.135) (Pastorino and Gallo 2009). With microsatellites, the number of allelic variants per locus was rather high (A_L = 12.08) (Arana et al. 2010). On the other hand, its degree of genetic differentiation between populations is low for both types of markers (Pastorino and Gallo 2009: F_{ST} = 0.06; Arana et al. 2010: R_{ST} = 0.08), that is, their populations differ little, even those that are subject to contrasting rainfall regimes or separated by hundreds of kilometers.

With respect to the structuring of the genetic variation of the species, all the studies agree in a latitudinal pattern, with the northern range of the distribution being on average more variable. This pattern has been interpreted as an effect of the last glaciation, with larger and/or more abundant refuges located in the north according to a lesser northwards development of the Andean glacial cover. In turn, another consistent finding in the studies of the different authors is a greater genetic variation in the steppe populations of the center and north of the species' range in Argentina. This is surprising, since these are mostly marginal forest patches of a few dozen trees and completely isolated by distance from the forest continuum. Low diversity values and the occurrence of inbreeding and genetic drift processes would be expected for these populations according to a center-periphery hypothesis (Pironon et al. 2016).

"Leading" and "rear" edges can be distinguished in the natural distribution of species (Hampe and Petit 2005). The first consists of relatively new marginal populations that act as a colonization front in the distributional adjustment that occurs following environmental changes, such as post-glacial climate change. The rear edge is the opposite margin of the distribution, which may end up being extinguished, and this will imply a displacement of the distribution area or otherwise persist (stable edge), implying this an expansion of the total distribution area. These stable populations, capable of surviving in situ the climatic oscillations of the Quaternary, would be relicts two or three orders of magnitude older than the rest of the populations and could be essential for the long-term conservation of genetic diversity and the evolutionary potential of species. In cypress, the results with neutral markers support the hypothesis that the steppe marginal forests represent the stable rear edge of the Argentine distribution of the species. They are located outside of the glaciated area, so they could have persisted during the last glacial maximum, and moreover, they could have been the glacial refuges from where the species initiated the post-glacial recolonization.

As for the presumption of inbreeding, we know that self-pollination is not possible in the Patagonian cypress, since it is a dioecious species, but it could still be verified as a cross between related individuals. Non-random pollination has been shown in a marginal population (Pastorino and Gallo 2006), as well as evidence of bi-parental inbreeding, although of a low degree, thanks to effective long-distance pollen flow (Colabella et al. 2014). Thus, it seems inbreeding does not compromise the population's levels of genetic diversity.

Likewise, the occurrence of genetic drift processes in these steppe populations has been shown by comparing the isozyme genetic pattern of neighboring forest patches, for which demographic, adaptive, and external genetic flow processes must be assumed to be the same (Pastorino and Gallo 2009). Not only a difference in the proportion of genetic variants was verified between neighboring populations but also the presence of exclusive allelic variants in one populations that are absent in the other (Pastorino and Gallo 2002; Arana et al. 2010). Despite the proven genetic drift, the species somehow manages to preserve a high genetic variation in these marginal forests, which still represents an unexpected result.

The pattern of neutral genetic variation described above was the basis for delineating a map of genetic zones (GZ) which can serve as preliminary operational genetic management units (OGMUs). We define a GZ as a group of natural populations with geographic continuity and genetic similarity shown with neutral genetic markers. Based on the isozyme genotypes of 746 adult trees corresponding to 27 Argentine populations, a structure analysis was performed (Pastorino and Gallo 2009) by means of different analytical approaches (UPGMA cluster analysis and PCoA main coordinates based on genetic distances; Bayesian inference with the BAPS program; hierarchical analysis of variance with the SAMOVA program) and also based on environmental features (i.e., precipitation, last glacial maximum, distance). Thus, five regional clusters of populations with similar genetic pools were identified:

1. North GZ: composed of populations north of 41° 30′ S
2. Central GZ: composed of populations between 41° 30′ S and 42° 30′ S
3. South GZ: composed of populations south of 42° 30′ S
4. Glacier Edge GZ: composed of populations located outside the glaciated area, close to the border of last glacial maximum, in the current ecotone, in the center of the Argentine distribution of the species (there is a certain continuity of these populations with the most conspicuous forest patches in the west)
5. Ancestral Distribution GZ: composed of populations located outside the glaciated area, extremely isolated, situated in the northern half of the Argentine distribution (there are several kilometers of discontinuity with the most conspicuous forest patches in the west)

This zonation is based not only on data from the analyzed, relatively few markers but rather on the inference of biological processes that affect the entire genome (e.g., population extinctions, recolonization after glaciations, etc.), which allows some extrapolation to populations not included in the sampling.

6.3 Quantitative Genetic Variation in Key Adaptive Juvenile Traits

As an introduction to the quantitative genetic variation of the Patagonian cypress, we summarize some key adaptations of the species that, together with the patterns of neutral variation, have guided us in the design of our study systems. During its juvenile life stage, *Austrocedrus* (together with *Libocedrus* and *Papuacedrus*) is distinguished from the other Cupressaceae by its aerial architecture. During the Paleogene, a context of change towards a warm and humid climate, and of transition from conifer-dominated to mixed evergreen forests, would have had a profound impact on the quality of the understory light. The flattening of the photosynthetic organs is a trait evolved in that context, which would have allowed *Austrocedrus* (its ancestor *Libocedrus*) to establish and compete with angiosperms (Hill and Brodribb 1999). This capacity would have been facilitated by the evolution of longevity, a characteristic that gives shade-tolerant species enough time to reach the canopy and to find suitable events to recruit seedlings. More recently, in the context of relative aridity of the Quaternary, the capacities to avoid desiccation and prevent xylem cavitation have been key adaptations for the persistence of many Cupressaceae (e.g., Brodribb and Cochard 2009). Under strong atmospheric demands, the cypress prevents cavitation by strict control of stomata closure; this adaptation allows it to persist in very arid environments. At the same time, cypress maintains physiological characteristics that are typical of shade-tolerant species, e.g., the ability to keep stomata open under low photosynthetic fluxes (Gyenge et al. 2007). At the end of the Pleistocene, while many tertiary species in the Northern Hemisphere became extinct during the glaciations (Sakai and Larcher 1987), in the Southern Hemisphere less rigorous conditions allowed the persistence of relicts, thus species evolving limited cold tolerance (Sakai 1971). The maximum cold tolerance of cypress would be between -6.7 and -12.1 °C (Bannister and Neuner 2001). In summary, Patagonian cypress has evolved adaptations to inhabit varied environments, from mixed rain forests to arid steppe environments (Fig. 6.4).

At the same time, it would be a species with moderate cold tolerance, which would explain its habitat location in deflated, protected rocky substrates and high slopes, at latitudes such as that of Bariloche (Anchorena and Cingolani 2002). This capacity for growing in ecologically contrasting situations suggests that there could be different genetic pools among populations, because natural selection may have "improved" certain functions in each environment. On the other hand, it could be postulated that this ability to inhabit contrasting environments is due to the development of a generalist strategy based on plasticity. In our study system, we intended to disentangle these questions at three scales of analysis: (1) at the whole regional scale, (2) between humid and arid extremes, and (3) along the arid border of the cypress' distribution.

The focus of our research was on juvenile quantitative trait variation. Natural selection, and therefore adaptation, intervenes along tree life, but seedling establishment is one of the most sensitive stages (Brubaker 1986; Green 2005). In addition,

Fig. 6.4 Ecological extremes of the east-west range of *Austrocedrus chilensis* in Argentina, from arid marginal. (Left panel: Pilcañeu, with 330 mm of mean annual precipitation; photo: A. Aparicio) to humid, mixed stands. (Right panel: Río Azul, where MAP is over 1500 mm; photo: M. Pastorino)

regeneration offers the highest levels of diploid genetic variation (with the exception of the embryo stage), and over this variation selection acts early. Fertility, which is defined as a parent's ability to produce successful gametes, can be assessed through the adaptability of the offspring, and the quantitative variation of early traits could be therefore used for low-intensity breeding programs.

6.3.1 Quantitative Genetic Variation at the Regional Scale

In a first broad-scale approach to the quantitative variation of cypress, we have installed a network of six provenance trials, with the general objective of describing the patterns of variation in survival and initial growth, which are the integrative result of multiple morphological, physiological, and biochemical responses to the environment. For this, we used seedlings with maternal identification from 14 populations of three forest types, which are in turn defined by their rainfall regimes: (1) marginal (MA), with up to 600 mm in mean annual precipitation (MAP); (2) mesic (ME), from 600 to 1000 mm in MAP; and (3) humid (H), with more than 1000 mm in MAP (Table 6.1). The sites were chosen at several locations within the distribution area of cypress in Argentina, covering latitude, altitude, and rainfall zones (Table 6.2). Under the hypothesis that cypress is adapted according to a generalist strategy based on the evolution of plasticity, we predicted similar behavior between the humid, mesic, and marginal rainfall zones (i.e., forest types), as well as a nonsignificant participation of the effect of population on the phenotypic variation of survival and growth. The details on the experimental design and statistical analyses are in Aparicio (2013).

Our predictions were confirmed throughout the trial network, with the exception of trial N° 6. That is, under the environmental conditions of most trials, the variation in survival and initial growth was largely attributable to residual variance, i.e., between individuals within the populations, nested within the forest types. Table 6.3

Table 6.1 Fourteen populations of *Austrocedrus chilensis* used in a network of common garden field trials

Forest type	Population	Latitude S	Longitude W	Altitude (m asl)	MAP
Marginal	**H**: Huingan-có	37°09′	70°37′	1500	604
Humid	**G**: Arroyo Pedregoso	40°37′	71°35′	900	2000
Humid	**B**: Puerto Blest	41°02′	71°42′	800	2650
Humid	**L**: Llao llao	41°03′	71°32′	800	1500
Mesic	**CL**: Cerro Leones	41°04′	71°08′	900	670
Marginal	**F**: La Fragua	41°05′	70°57′	1100	490
Humid	**C**: Cerro Catedral	41°08′	71°27′	850	1350
Mesic	**O**: Cerro Otto	41°08′	71°19′	900	1000
Marginal	**PN**: Pilcañeu Norte	41°13′	70°42′	1100	330
Marginal	**PS**: Pilcañeu Sur	41°14′	70°41′	1100	330
Mesic	**T**: Río Ternero	41°56′	71°22′	800	870
Marginal	**M**: El Maitén	42°02′	71°12′	750	490
Humid	**A**: Río Azul	42°06′	71°40′	300	1500
Mesic	**Co**: Corcovado	43°31′	71°27′	500	905

MAP mean annual precipitation (mm). The populations are ordered from north to south (bottom)

Table 6.2 Climatic description of a network of genetic field trials with 14 populations of *Austrocedrus chilensis*

Location	Trial (plantation year)	MAP	MAT	AI_M*	P_n	TM_n	P_{WS}/ MP_{WS}
Huingan-có[2]	N° 1 (2001)	604	12.9 (5.3–20.6)	32	928	no data	169/115
	N° 4 (2004)	604	12.9 (5.3–20.6)	32	428	no data	115/115
Meliquina[4]	N° 5 (2004)	1335	8.6 (2.3–14.7)	86	1423	8.3 (2.2–14.2)	331/324
El Bolsón[3]	N° 3 (2001)	950	9.6 (3.0–16.7)	58	954	no data	197/214
Trevelin[1]	N° 2 (2001)	1013	9.7 (3.5–15.9)	62	1240	9.4 (2.7–16.1)	313/291
	N° 6 (2004)	1013	9.7 (3.5–15.9)	62	1081	10.5 (4.4–15.6)	375/291

MAP mean annual precipitation (mm), *MAT* mean annual temperature (°C), AI_M aridity index (mm • °C^{-1}), P_n precipitation of the post-plantation annual period (mm), *TMn* mean temperature of the post-plantation year (°C), P_{WS}/MP_{WS} ratio between the accumulated precipitation in the first warm (growing) post-plantation season (October–March) and the historical mean for such period. The trials are arranged from north to south
Climate data from: [1]Estación Agroforestal INTA Trevelin (1970–2009); [2]AIC (1979–2006); [3]Subsecretaría de Recursos Hídricos de la Nación (1968–2009); [4]AIC (1997–2007). *De Martonne aridity index: $AI_M = 1.2 \, P/(T + 10)$, where P (mm) and T (°C) are the accumulated precipitation and the mean monthly temperature; lower AI_M indicates higher aridity

Table 6.3 Summary statistics for plant height (cm) at years one (h_0) and n (h_n) from plantation, relative height growth (%, relative to h_0) at year n (RG_n), and survival (%) at year n (S_n), in a network of field trials with 14 populations of *Austrocedrus chilensis*; *SD* standard deviation for the continuous traits

Trial	Period	Trait	mean ± SD	Population													
				A	B	C	CL	Co	F	G	H	L	M	O	PN	PS	T
N° 1	(–2002)	h_0	20.2 ± 5.2			19.9		19.1	17.7		22.7	21.8			19.4		
	(–2006)	h_4	25.5 ± 8.7			26.9		24.7	23.3		25.3	24.2			28.9		
	(2002–2006)	RG_4	28.4 ± 25.3			31.2		33.2	33.1		13.2	20.1			41.8		
	(2001–2002)	S_1	92.6			92.6		88.9	85.2		96.3	96.3			96.3		
	(2001–2006)	S_5	75.9			51.8		74.1	77.8		81.5	88.9			81.4		
N° 2	(–2002)	h_0	18.3 ± 5.8	17.9		17.1		16.7	18.9				21.3	19.9	17		
	(–2004)	h_2	31.9 ± 14	32.8		32.4		28.1	32.3				34.9	36.8	28.3		
	(2002–2004)	RG_2	72.5 ± 61	82.5		83.8		60.1	66				66.1	77.1	68.8		
	(2001–2002)	S_1	65.8	70		56		70	67				66	68	65		
	(2001–2004)	S_3	52.7	65		53		53	60				63	64	56		
N° 3	(–2002)	h_0	18 ± 5.9	15.9	16.9	16.8	15.4	15.4	18.2		22.5	19.7	19.2	18.9	16.5		19.5
	(–2004)	h_2	32.1 ± 12.8	29.8	29.3	33.8	32.5	28.1	26.2		31.1	32.7	33.6	37.2	34.3		38.9
	(2002–2004)	RG_2	75.8 ± 51.1	80.6	68.7	100.6	96.2	76.5	47.7		40	69.5	73.4	90.1	93.8		98.3
	(2001–2002)	S_1	69.3	56.2	75	77.1	58.3	62.5	79.2		77.1	62.5	83.3	66.7	68.8		64.6
	(2001–2004)	S_3	50.9	35.4	62.5	60.4	37.5	47.9	54.2		52.1	56.3	47.9	50	37.5		37.5
N° 4	(–2005)	h_0	37.1 ± 8.5	32.2	43.7	38.6	40.1	32	20.9	42.8	33.5	41.5	42.8	43.9	31.6	37.4	41.5
	(–2006)	h_1	37.4 ± 9	33	43.1	36.11	38.9	29.8	21.9	44.7	32.5	41.5	43.7	43.8	31.9	36.4	39.9
	(2005–2006)	RG_1	4.6 ± 6.8*	10.1	4.0	1.75	5.7	4.2	9.8	3.7	4.2	2.3	3.5	4.0	2.4	3.1	3.5
	(2004–2005)	S_1	78.3	92.6	96.3	74.1	85.2	66.7	92.6	59.3	74.1	85.2	88.9	74.1	85.2	66.7	55.6
	(2004–2006)	S_2	63.7	81.5	77.8	40.7	74.1	51.8	77.8	48.1	55.6	70.4	81.5	63	70.4	48.1	51.8

N° 5	(−2005)	h_0	33.7 ± 8.6	40.4	43	33.1	26	24.6	16.8	37.9	26.2	36.3	38.4	39.3	36.5	33.5	40.3
	(−2009)	h_4	50 ± 16.3	55.2	58	48.3	46.4	40	31.6	47	43.7	52.1	57.4	55.9	53.5	54.6	57
	(2004–2009)	RG_4	16.4 ± 14	14.7	15	15.2	20.4	15.3	14.7	9.8	17.5	15.8	19.2	16.6	17	21.1	16.7
	(2004–2009)	S_5	97.4	100	92.6	100	96.3	96.3	100	92.6	96.3	96.3	100	96.3	100	96.3	100
N° 6	(−2005)	h_0	35.7 ± 8.9	43.1	44.6	39.4	32.3	30.4	20.5	35	28.7	40.9	42.7	41.2	27.3	34.3	39.4
	(−2010)	h_5	94.6 ± 28.2	101.4	105.3	94.3	85.5	84	76.8	80.9	79.7	101.4	126.8	109	91.3	78.9	97.1
	(2005–2008)	RG_3	94.3 ± 50.5	88.6	69.8	78.8	99.6	110.4	145	67.8	96.4	76.5	102.7	102.8	131.6	66.8	72.3
	(2005–2010)	RG_5	167.2 ± 81.2	135.2	130.8	141.4	167.5	176.5	274.9	129.9	190.6	147.9	195.8	172.5	216.5	122.6	149.5
	(2004–2008)	S_4	87.6	96.3	92.6	100	96.2	81.5	81.5	77.8	81.5	51.8	100	96.3	96.3	85.2	88.9
	(2004–2010)	S_6	80.7	85.1	81.5	100	92.6	74.1	70.3	74.1	81.5	51.8	96.3	96.3	77.8	74.1	74.1

aThe SD is greater than the mean of RG_1 due to the excess of zeros in the data, i.e., plants that did not grow in the analyzed period

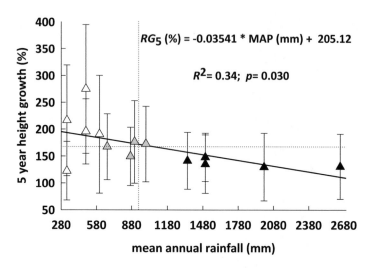

Fig. 6.5 Seedling height growth at year 5 (RG_5), relative to height at planting, in a common garden trial with 14 populations of *Austrocedrus chilensis*. The triangles represent the mean values for the xeric (white), mesic (grey), and humid (black) populations, with their standard deviations in vertical bars; the dotted vertical line is the mean annual precipitation (MAP) at the trial location; the dotted horizontal line is the mean of RG_5 in the trial (from Aparicio 2013)

shows the phenotypic statistics of the traits evaluated. In Trial N° 6, installed in a site of mesic climatic characteristics, the main result was that after 5 years in the field, the factor "forest type" showed significant differences in seedling growth: the populations from xeric origins had higher mean relative growth (and phenotypic variances) than those from mesic and humid environments (Fig. 6.5).

The lower growth and more homogeneous phenotypic variation of the humid populations could be interpreted as the effect of their local adaptation to favorable environmental conditions in terms of available resources, in particular humidity (Aparicio 2013). This adaptation would limit their ability to adequate under less favorable conditions, such as those of the experimental site. Instead, the xeric populations seemed to retain at least the same levels of intra-population variance than the mesic and humid ones but had also the ability to grow significantly more than the humid populations, given the favorable environment relative to their areas of origin. This could be evidence that marginal populations would not be simple relicts without the ability to adapt or adequate by means of plastic responses (and therefore, without ecological and evolutionary importance) and pointed out the importance of focusing our attention in the xeric marginal distribution of cypress. There is other evidence in that direction, some of which are addressed later. As a first conclusion, from the results of Trial N° 6, the hypothesis arises that there could be a differentiation between the ecological zones determined by aridity (Aparicio 2013). In the following section, we report some studies aimed at determining if the arid and humid extremes differ in traits that could be key for survival and initial growth.

6.3.2 Variation Between Humid and Xeric Populations

The architecture of the aerial part of trees describes the spatial location of photosynthesizing organs for light uptake, which is particularly delicate during recruitment. Filotaxis and branching are its most important traits and reflect the habitat preference of a species. Pastorino et al. (2010a) carried out a study, under controlled greenhouse conditions, of the genetic variation in seedling architecture, with the objective of determining the degree of differentiation between – and the variability within – populations contrasting in aridity (forest type). For this study, they used 840 1-year-old plants from 28 open-pollinated families corresponding to 2 xeric (Pilcañeu Norte, Pilcañeu Sur) and 2 humid (Cerro Catedral, Río Azul) populations of cypress (Table 6.1). Eight traits measuring size, vigor, and architectural balances between axes were measured and analyzed using linear mixed effect models, considering the genetic effects of population and half-sib family. The main result of this experiment was that the degree of genetic differentiation between populations was very low ($Q_{ST} = 0.088$ on average), virtually equal to the degree of neutral differentiation of the species (Pastorino et al. 2004). The additive variances and the heritability of the characters were moderate to high ($h^2 = 0.15$–0.53), but the genetic variability was not associated to the forest type.

In another experiment, Pastorino et al. (2010b) analyzed the variation of tolerance to a simulated drought in 3-year-old seedlings from two humid (Cerro Catedral, Río Azul) and six mesic (El Maitén, Cañada Molina) to xeric (Pilcañeu Norte, Pilcañeu Sur, Leleque, Chacabuco) Argentine populations (Table 6.4). For each of the two most arid populations (PN and PS), material from ten open-pollinated families was included in the design, in order to quantify their intra-population variance. The seedlings were grown in pots, to which irrigation was suppressed for 43 days at late spring. During that period, the functioning of the photosystems was evaluated weekly through the maximum quantum yield of the photosystem-II: Fv/Fm (Krause and Weis 1991), which indicates the degree of stress a plant is subjected to. Seedling survival was evaluated after a recovery period of 8 months. The authors have found significant differences between populations for Fv/Fm and survival, although these differences were not attributable to the type of population (rainfall regimes). The family effect within populations PN and PS was significant for survival and Fv/Fm, but for this last trait significance was largely influenced by the effect of one family, which showed an exceptional mean drought tolerance. The results of this experiment suggest that the regional variation of water stress tolerance (evaluated according to the Fv/Fm parameter) would not be modeled by divergent selection and that there could be some degree of heritable variation within xeric populations.

Long-term water-use efficiency (WUE, defined as the ratio of total plant dry matter produced to total water used over the same period) is a key physiological trait in trees, especially in semiarid Mediterranean climates. It can be estimated indirectly via carbon isotopic discrimination (Δ). Plants with C3 metabolism discriminate against the ^{13}C isotope because it is heavier than ^{12}C, which is then preferred by the carboxylation enzyme (Rubisco) and also has faster diffusion in the air (Farquhar

Table 6.4 Plant material used in a genetic trial with xeric populations of *Austrocedrus chilensis* under greenhouse conditions

Population		Latitude south	Longitude west	Altitude (m asl)	MAP (mm)	Open-pollinated families (number of seedlings)	
						Total number in the trial[1]	Growth rhythm study sample size
Mo	Cañada Molina	37°08′	70°36′	1450	604	19 (788)	11 (231)
Ra	Cañada Rahueco	37°10′	70°36′	1500	604	13 (543)	11 (231)
Ri	Riscos Bayos	37°59′	70°47′	1350	246	6 (252)	6 (126)
N	Catán Lil	39°21′	70°39′	1100	308	17 (713)	11 (231)
Y	Chacay	40°51′	70°59′	1250	488	19 (797)	13 (273)
CH	Chacabuco	40°39′	71°01′	900	450	27 (1118)	23 (483)
F	La Fragua[2]	41°05′	70°57′	1000	490	10 (393)	–
PN	Pilcañeu Norte	41°13′	70°42′	1100	330	35 (1446)	19 (399)
PS	Pilcañeu Sur	41°14′	70°42′	1100	330	30 (1225)	14 (294)
M	El Maitén	42°02′	71°12′	750	490	30 (1248)	20 (420)
Q	Leleque	42°20′	71°09′	850	353	24 (1002)	21 (441)

[a]All the plants were used for analyzing the first year branching degree
[b]Population F was not used in the studies of plantlet emergence and seedling annual growth rhythm
MAP mean annual precipitation

et al. 1982). When closing the stomata, the plant impedes the entrance of new air into the leaves and continues fixating carbon from the air of the intercellular spaces and sub-stomatal chambers. Consequently, the ^{13}C concentration starts to increase in the intercellular air spaces, and the discrimination in favor of ^{12}C diminishes. Thus, the enrichment of ^{12}C of the plant tissues decreases, making the proportion of both isotopes in the plant less distant from that of the atmospheric CO_2. Hence, plants showing higher stomatal control show lower carbon isotope discrimination when photosynthesis is not limited by non-stomatal factors. From the relation between the carbon isotope composition of the plant and that of the atmospheric CO_2, the carbon isotope discrimination parameter Δ can be derived, which is inversely correlated to WUE (Farquhar et al. 1989).

In order to know whether the Δ trait differs between the arid and humid extremes of the cypress range and also to determine the intra population genetic variance, Pastorino et al. (2012) sampled in a common garden field trial 246 5-year-old seedlings from 41 open-pollinated families corresponding to 1 humid (Río Azul) and 2 xeric, isolated populations from the steppe (Pilcañeu Norte and Pilcañeu Sur) (Table 6.1). The trial had been installed in the field 2 years before tissue sampling for the determination of their Δ values. Little twigs from each sampled seedling were dried, ground, and analyzed with a mass spectrometer to measure their stable carbon isotope composition ($^{13}C/^{12}C$). The main result of this work was that the arid and humid populations did not discriminate the carbon isotopes in a differential way, i.e., we found no evidence of adaptation. On the other hand, the two arid populations, very close to each other, proved to be very different in terms of genetic variation, which could be a consequence of genetic drift by isolation.

As a conclusion to this second stage of analysis, we found no evidence of genetic differentiation between populations from the humid and xeric extremes of cypress in traits linked to the early light capture and water stress tolerance. This is not a lesser information: the populations of the arid margin would be globally similar to the humid ones in their gene pool, but in some cases, high intra-population variance, as well as differences between them, indicates their evolutionary importance and potential for breeding in a context of climate change towards higher regional aridity.

6.3.3   Variation Along the Arid Marginal Edge

In trees that do not form seed banks, seed germination and plantlet emergence are precisely synchronized with the current environmental conditions. Modeling of its dynamics in common garden trials can be useful for assessing inter- and intra-population variance of key traits for breeding. We have analyzed the variation in cypress plantlet emergence, using seed samples from 177 trees randomly selected within 10 arid marginal populations (Table 6.4) (Pastorino et al. 2013). With the data of emergency accumulated over time, we fitted sigmoid curves (Gompertz model), from which we derived six parameters describing plantlet emergency capacity, energy, and timing. On average, for the emergency capacity, energy, and timing of cessation and duration of the process, the variance components analysis showed a larger effect of the mother tree ($V_F = 55.4\%$) than of population ($V_P = 22\%$), suggesting microevolutionary potential (although to clarify this, the additive genetic variance should be measured). For the initiation and timing to the maximum emergence rate (energy period), the variance component of population was larger than the family component ($V_P = 41.1\%$ vs. $V_F = 34.9\%$, on average), which may indicate differentiation between populations.

The genetic variation of seedling architecture between humid and arid edge populations has already been discussed above. In this stage, Aparicio (2013) deepens the analysis in the arid edge of cypress range (ten populations, Table 6.4), with focus in the immediate branching degree. This architectural trait expresses the number of branches of second order (A2) formed during the first year of life, in relation to the vigor of the main axis (A1). In this research, it was observed that the genetic differentiation between populations was low ($Q_{ST} = 0.08$), virtually equal to the degree of neutral differentiation. Despite the low Q_{ST} value, a linear trend ($r^2 = 0.53$) of ecotypic variation was detected, in which the mean population branching degree decreased with altitude. The intra-population genetic variance (additive variance coefficient: CV_A) of the branching degree was very high ($CV_A = 24.4\%$ on average), and its heritability was moderate ($h^2 = 0.26$ on average). This could reflect processes of heterogeneous selection within populations given high spatial variation of fine scale. Nevertheless, an alternative (or complementary) explanation would be that the estimates of the additive variance were inflated by maternal effects, which could cause in turn the Q_{ST} value to have been underestimated. Given the significant effect of population as an explanatory variable and association between the mean branching degree and population altitude, it seems reasonable to assume that there is a tendency for ecotypic regional structuring, modeled by ecological factors associated with geographic altitude. Probably, due to a higher probability of interspecific competition, populations of lower altitudes tend to a more branched mean phenotype.

The synchronization of the annual growth rhythm with the environment is critical in temperate trees, since it determines complex trade-offs between competitive ability and survival, in particular during seedling recruitment (Howe et al. 2003; Green 2005; Notivol et al. 2007). To understand how the genetic variation of the annual growth rhythm is structured along the arid edge of cypress range, Aparicio et al. (2010) carried out a genetic trial under greenhouse conditions (Table 6.4). By means of successive measurements of seedling's height during their second year of life, individual sigmoidal (Boltzmann) curves were fitted, from which the values of initial (h_0) and final (h_f) height; the timing points of initiation (t_{10}), cessation (t_{90}), and duration of the growth period (D), and the annual maximum growth rate (gr_{max}) were calculated (Fig. 6.6). With those calculated phenotypes, population genetic parameters were estimated.

One main result was that two traits, t_{10} and gr_{max}, showed significant and high to moderate genetic differentiation ($Q_{ST} \approx 1$ and $Q_{ST} = 0.29$, respectively) and their mean population phenotypes were negatively associated with altitude, suggesting altitudinal ecotypic patterns. Also, it was found that the additive genetic variances and trait heritabilities were on average higher at the northern half of the distribution range (Fig. 6.7). This suggests a special relevance of the northern region of the

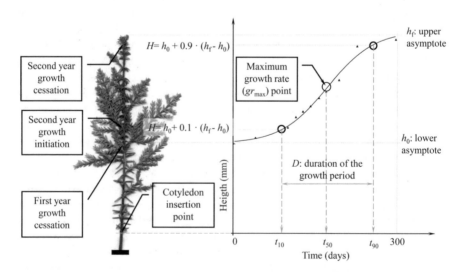

Fig. 6.6 Modeling of the annual height growth for estimating size and timing traits in 2-year-old seedlings of *Austrocedrus chilensis* (from Aparicio 2013)

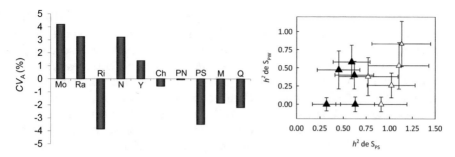

Fig. 6.7 Structure of the intra-population genetic variation in ten populations from the arid margin of *Austrocedrus chilensis*. In the left panel, the bars show the differences between the mean additive genetic coefficient of variation (CV_A) of six annual growth rhythm traits and the metapopulation overall mean CV_A. Populations are ordered in the x axis from north (left) to south (right) (from Aparicio et al. 2010). In the right panel, the triangles are the heritabilities (h^2) of survival after summer drought (S_{PS}) and winter extreme cold (S_{PW}), for populations to the south (in black) and north (white) of 40.5° S in latitude. (From Aparicio et al. 2012)

species for the conservation of its genetic resources but also points to a latitudinal break with regard to the structuring of the additive variance within cypress populations, at least for its arid edge. For the cessation and duration of the annual growth period, the majority of the marginal populations seemed to retain enough levels of additive variation to allow in situ adaptation, for example, in the face of changes in summer aridity due to altered patterns of seasonality.

Climate change implies increases in the amplitude and/or frequency of extreme weather events (Mitchell et al. 2006), whose impacts are expected to be stronger at the margins of the current ranges of trees. The survival of tree regeneration can be interpreted as the result of the expression of multiple functional traits to the environmental conditions, which with certain frequency can be extreme. In those circumstances, the acclimation capacities of seedlings are exceeded, thus leading to persistent effects on performance and eventually to death. Therefore, if variation exists in traits underlying survival, this could vary between or within populations. In the context of the study of the arid margins of cypress range, two progeny field trials were planted using the material presented in Table 6.4. Fortunately, in one of them ($N = 161$ open-pollinated families; $n = 2415$ plants), it was possible to evaluate the effects of one summer heat wave and drought and of one extreme winter cold. These events produced two clearly defined waves of seedling mortality, allowing a survival analysis under such infrequent conditions (Aparicio et al. 2012). One main result was that survival was not explained by the effect of population, suggesting that on a regional scale, cypress could be under homogenizing selection for drought and extreme cold tolerances. Within the populations, survival after summer drought (S_{PS}) was moderate to highly heritable. This suggests that the long-term persistence of the arid cypress populations could rely on maintaining high levels of genetic variation for drought tolerance, which could reflect heterogeneous, fine-scale spatial selection within populations. The heritability of survival after the extreme cold event (S_{PW}) was significant in most populations, although lower (with some cases of

$h^2 = 0$) than the heritability values obtained for post-summer survival. Another main result was an apparent north (mean $h^2 = 0.84$) to south (mean $h^2 = 0.28$) structure of the populations, according to their heritabilities of survival under summer drought. This structure seems consistent with previous results on the patterns of genetic variation of the initial growth rhythm (Fig. 6.7) and also coincident with the general patterns of selectively neutral variation of the species (Pastorino et al. 2004; Pastorino and Gallo 2009; Arana et al. 2010). This suggests a greater relative importance of the northern region of the cypress distribution in Argentina for the conservation of its genetic resources, focusing on both neutral and adaptive variation.

6.4   First Definition of Provenance Regions for a Forest Tree Species in Argentina

We mentioned above that based on a detailed mapping of the distributional range and information on selectively neutral genetic variation, Pastorino and Gallo (2009) have recognized five distinctive genetic clusters or zones (GZ) in Argentina. Based on them, at least five seed transfer zones were subsequently indicated as necessary to preserve the genetic identity of the Argentine natural populations of the Patagonian cypress (Pastorino 2012). With this pattern in mind, the information extracted from the studies on the species' quantitative variation was superimposed for delineating provenance regions (PR, Pastorino et al. 2015) by means of the agglomerative method, that is, the grouping of known forest patches of the species of interest, genetically and environmentally characterized (CTGREF 1976). PRs are groups of natural populations with geographic continuity, which belong to the same GZ, and for which similar adaptive responses are expected as verified in genetic trials and/or inferred given homogeneous environmental conditions.

One main conclusion of the works on the cypress' quantitative and neutral variation was that the overall differentiation between populations was relatively low. This hints that there would not be necessary too many PR for the conservation and breeding of the species. In turn, several results gave clues that the marginal xeric populations should be taken into account for the definition of operational genetic management units (OGMUs), not so much by criteria of differentiation with the rest of the species' forests but by its levels of intra-population variance and typical fragmentary and/or isolation conditions. As shown previously, most of the marginal populations seemed to retain at least the same levels of variation than the mesic and humid, but they showed also the ability to respond to less rigorous environments than those of their home habitats. Both experimental and empirical evidence also indicate that the xeric margin of the cypress should not be considered as depressed in genetic diversity and therefore adaptively unviable or functionless. For example, in the cited study on seedling emergence (Pastorino et al. 2012), the marginal xeric populations did not show loss of seed viability, a predictable problem for "relict" populations, reported in other Mediterranean Cupressaceae (e.g., Montesinos et al.

2007, in *Juniperus thurifera*) and frequently attributed to depression due to inbreeding (e.g., Ferriol et al. 2011). Moreover, relatively low levels of inbreeding have been found in xeric cypress populations quite small in size (Colabella et al. 2014). Indeed, such levels of inbreeding could be the result of grouped mating, related to the typical low tree density in the arid marginal forests (Pastorino and Gallo 2002) rather than to the effect of extreme reductions in genetic variance. Arana et al. (2010), in their study including 7 of the 11 marginal xeric populations reported in Table 6.4, did not find evidence of recent bottlenecks, using nuclear microsatellite markers under 2 different models. Other observations, such as the high production of normal pollen and seeds in trees of different cohorts, or the total survival of the seedlings produced in a large nursery trial reported by Aparicio (2013, and derived papers), are indicators of the viability of the xeric margins of the cypress. Nevertheless, based on population genetics theory and on several contrasting results between neighbor populations, both in neutral and selective variation, it must be considered that adaptation might not prevail against the predictable effects of genetic drift. This was considered for defining six OGMUs of local scope, named Restricted Area Provenance Regions (RAPR). This was done by joining individuals and small forests patches, sometimes separated by a few kilometers, assuming certain genetic unity between them and based on their differential genetic character. The six RAPR defined are Leleque, Pilcaniyeu, Cerro Los Pinos, Catán Lil, Riscos Bayos-Trolope, and Huingan-có (Fig. 6.7), and their detailed descriptions can be found in Pastorino et al. (2015).

Another conclusion of the quantitative studies was that the range of the arid margin north of 40.5° S presented higher levels of additive variation for several traits of the annual growth rhythm, linked to the cessation of the vegetative activity, which could be regulated by the annual evolution of aridity. This was also observed as a heritable variation of survival after strong summer drought. Added to the background on neutral variation, this information indicates that the northern range of the cypress should be a priority for the conservation of the species' genetic resources. For the definition of PR, we proposed that one main north-south grouping of populations should be considered, with a dividing zone at approximately 40.5 °S in latitude.

We have also found evidence of ecotypic variation associated to geographic altitude, in traits linked to initial light uptake (degree of sylleptic branching) and initiation of primary growth (a phenological trait) and its maximum annual rate. This could be attributed to adaptation to altitude (i.e., thermal regimes), which may in turn determine different probabilities of episodic winter heat events and different interspecific competition pressures (Aparicio 2013). Our results suggested that for long-term ex situ conservation or breeding, population samples from different altitudinal levels should not be mixed. We proposed therefore to separate the large North GZ into two PR, i.e., North High and North Low, taking into account the average altitude, which in the region correlates quite well with latitude. The limit between the North Low and the Central PR is defined by an abrupt change in mean altitude and a large discontinuity in the presence of cypress, which gives place to pure forests of *Nothofagus antarctica*. The Central PR coincides with the Central

GZ that extends from 41° 30′ to 42° 30′ S in latitude. The South PR fits the South GZ, excluding the population Leleque (RAPR), and is the less genetically variable region. Finally, the Eastern PR corresponds to the Glacier Edge GZ and is made up of some ecotone and marginal forest patches. This is the PR of smaller area and of more difficult delimitation because of large discontinuities between patches. Concluding, after having taken into account all the available genetic information and geographic and operational criteria, a total of 11 OGMUs were proposed. Five of them are provenance regions, with a clear north-south trend, and the other six, quite smaller in area and number of tree patches, are Restricted Area Provenance Regions. Their proposed borders and general delineation are presented in Fig. 6.8.

6.5   Domestication and Breeding Strategy

First attempts to cultivate Patagonian cypress were made in the 1950s, using plants collected in natural seedling banks. For example, there is a still standing 700-m^2 plantation installed in 1947 in Los Alerces National Park, in which a volumetric increment of 12.2 m^3 ha^{-1} year^{-1} was registered at 57 years of age (Loguercio et al. 2005). Since the last two decades, interest in using the Patagonian cypress in afforestation has focused on the ecological restoration of degraded forests (Urretavizcaya 2006). Several small- to medium-scale active restoration projects were carried out with the species (e.g., Oudkerk et al. 2003, ~7 ha; Perdomo et al. 2009, ~50 ha) mainly to restore forests devastated by extensive anthropogenic fires. However, its potential use in plantations for productive purposes is still in force. Although with a slow initial growth, it is the best adapted native tree for the semiarid ecotone between the North Patagonian temperate forest and the steppe.

Seed availability is not a limiting factor for cultivating the species due to frequent good production in the natural forests (approximately every other year) and their good viability after storage, that is, seeds are expected to conserve ca. 70% germination capacity for some 5 years if dried and stowed at −18 °C (Urretavizcaya et al. 2016). Also, seed viability is commonly high: a mean emergence capacity of 76.2% was measured in a nursery trial with 177 open-pollinated mother trees from 10 Argentine natural populations (Pastorino et al. 2013). There is a good experience in the collection of seeds from the natural forest, and practical recommendations have been given for this task (Pastorino et al. 2001).

The technology of seedling production has been improved in the last years, including the development of specific ferti-irrigation protocols, successively adjusted and optimized (Massone et al. 2018; Massone 2020). However, the species seems to be recalcitrant to fertilizer aggregate since it responds dimly during the entire first growing season. Only in the second season, with the root system already well developed and colonizing the whole container, seedlings react vigorously to the addition of fertilizers. Thus, two growing seasons are necessary to obtain a suitable plant for planting in the field (~25 cm high), and according to quality standards

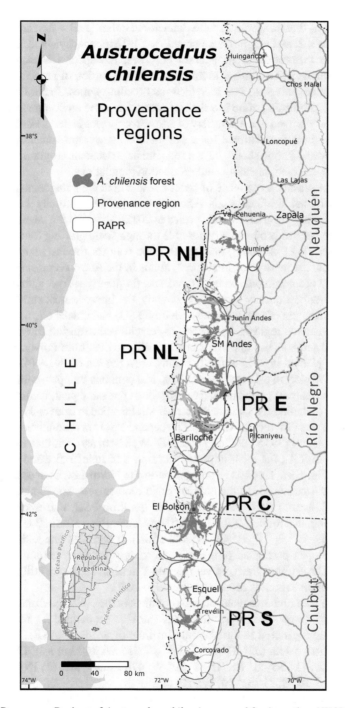

Fig. 6.8 Provenance Regions of *Austrocedrus chilensis* proposed for Argentina. *NH* North High; *NL* North Low; *E* east; *C* central; *S* south. Restricted Area Provenance Regions (RAPR) order from north to south: Huingan-có, Riscos Bayos-Trolope, Catán Lil, Cerro Los Pinos, Pilcaniyeu, and Leleque. (From Pastorino et al. 2015)

for the species, that is, seedlings with diameter at collar level > 2.5 mm, shoot/root biomass ratio < 2, and height/diameter ratio < 54 (Urretavizcaya et al. 2016).

Either for restoration or production purposes, the species is expected to be planted within its natural range. In this regard, the definition of provenance regions was taken in advantage to consider them as breeding zones. According to White et al. (2007), a breeding zone is a distinct area with some environmental variation for which an improved genetic variety of a forestry species is being developed. This critical decision is foundational for a well-planned tree improvement program. In fact, each breeding zone should have a separate improvement program with its own distinct base, selected, breeding, and production populations.

With a two-decade experience of genetic research in the Patagonian cypress, the Instituto Nacional de Tecnología Agropecuaria (INTA) formally began a low-intensity breeding program for the species in 2012. Although the vegetative propagation of the species is operatively possible through scion grafting (but not by stem rooting, Aparicio et al. 2009), thus opening the way for the installation of clonal seed orchards, the mass selection of individuals in the forest seems hardly feasible because they are commonly uneven-aged and frequently mixed with *Nothofagus dombeyi* in the best quality sites. Consequently, the improvement strategy focused on the establishment of progeny seed orchards (PSO) and indeed one for each of the defined provenance regions. So far, we have achieved to install three PSOs with cypress in Argentina. The first one, corresponding to the South Provenance Region (PR S), was planted in 2015 at the Trevelin Forest Station of INTA (43° 06′ 51″ S; 71° 33′ 53″ W; 455 m asl), occupies 2/3 ha, and contains 92 open-pollinated families from four natural populations (N = 1336). In the same year, a complementary progeny test including the same mother trees was installed in Epuyén Forest Reserve belonging to the Province of Chubut. The second PSO was established in 2017 in Junín de los Andes (39° 57′ 27″ S; 71° 05′ 09″ W; 810 m asl), within a land property of CORFONE S.A., the most important forestry and industrial wood company in Patagonia (Fig. 6.9). This PSO corresponds to the North Low Provenance Region (PR NL) and is composed of the progeny of 65 mother trees from 5 natural populations (N = 634; area = 1/3 ha). We installed the third PSO, corresponding to the Central Provenance Region (PR C), in 2018 at General San Martín Forest Station of INTA (41° 59′ 46″ S; 71° 31′ 45″ W; 400 m asl). It occupies one half hectare and contains the open-pollinated progeny of 66 trees from three natural populations (N = 644). The saplings have not yet reached 1 m high, so we are still far from the genetic selection, which is expected to begin in 2030.

While the seed orchards mature, seeds continue being collected from wild stands. In this regard, three seed production areas (SPA) were defined for the species in Argentina and registered in the National Institute of Seeds (INASE). They are (1) Cañada Molina, 50 ha (37° 07′ 41″ S; 70° 36′ 22″ W; 1400 m asl); (2) Loma del Medio, 32 ha (41° 56′ 51″ S; 71° 33′ 10″ W; 470 m asl); and (3) INTA Trevelin, 46 ha (43° 06′ 16″ S; 71° 31′ 50″ W; 430 m asl).

Finally, two provenance tests of the species were installed in 2015 and 2016. The first one, representing 23 Argentine provenances (N = 2292), was established in General San Martín Forest Station of INTA and the second, which represents 18 of

Fig. 6.9 Progeny seed orchard of *Austrocedrus chilensis* recently installed (2017) in Junín de los Andes. (Photo: M. Pastorino)

the previous provenances (N = 1395), in Trevelin Forest Station of INTA. This material will be useful not only as provenance tests for research purposes but also for future infusions in the developing seed orchards by eventually multiplying the best genotypes by means of scion grafting.

References

Allnutt T, Newton A, Lara A, Premoli A, Armesto J, Gardner M (1999) Genetic variation in *Fitzroya cupressoides* (alerce), a threatened South American conifer. Mol Ecol 8:975–987

Anchorena J, Cingolani A (2002) Identifying habitat types in a disturbed area of the forest-steppe ecotone of Patagonia. Plant Ecol 158:97–112

Aparicio AG (2013) Variación genética de la adaptación inicial del Ciprés de la Cordillera. Doctoral thesis. Universidad Nacional del Comahue (CRUB), San Carlos de Bariloche, 160 pp

Aparicio AG, Pastorino MJ, Martinez-Meier AG, Gallo LA (2009) Vegetative propagation of Patagonian cypress, a vulnerable species from the sub-Antarctic forest of South America. Bosque 30:18–26

Aparicio AG, Pastorino MJ, Gallo LA (2010) Genetic variation of early height growth traits at the xeric limits of *Austrocedrus chilensis* (Cupressaceae). Austral Ecol 35:825–836

Aparicio AG, Zuki SM, Pastorino MJ, Martinez-Meier AG, Gallo LA (2012) Heritable variation in the survival of seedlings from Patagonian cypress marginal xeric populations coping with drought and extreme cold. Tree Genet Genomes 8:801–810

Arana MV, Gallo LA, Vendramin G, Pastorino MJ, Sebastiani F, Marchelli P (2010) High genetic variation in marginal fragmented populations at extreme climatic conditions of the Patagonian cypress *Austrocedrus chilensis*. Mol Phylogenet Evol 54(3):941–949

Azpilicueta M, Marchelli P, Gallo L (2009) The effects of Quaternary glaciations in Patagonia as evidenced by chloroplast DNA phylogeography of southern beech *Nothofagus obliqua*. Tree Genet Genomes 5:561–571

Bannister P, Neuner G (2001) Frost resistance and the distribution of conifers. In: Bigras F, Colombo S (eds) Conifer cold hardiness. Kluwer Academic Publishers, Dordrecht, pp 3–21

Brodribb T, Cochard H (2009) Hydraulic failure defines the recovery and point of death in water-stressed conifers. Plant Physiol 149:575–584

Brubaker L (1986) Responses of tree populations to climatic change. Vegetatio 67:119–130

Colabella F, Gallo LA, Moreno AC, Marchelli P (2014) Extensive pollen flow in a natural fragmented population of Patagonian cypress *Austrocedrus chilensis*. Tree Genet Genomes 10:1519–1529

Cosacov A, Sérsic A, Sosa V, Johnson L, Cocucci A (2010) Multiple periglacial refugia in the Patagonian steppe and post-glacial colonization of the Andes: the phylogeography of Calceolaria polyrhiza. J Biogeogr 37:1463–1477

CTGREF (1976) Semences Forestières. Les régions de provenance d'épicéa commun (Picea abies Karst.), Note Technique 30. Ministère de l'Agriculture, Centre Technique du Génie Rural, des Eaux et des Forêts. Nogent-sur-Vernisson, p 51

Farquhar GD, O'Leary MH, Berry JA (1982) On the relationship between carbon isotope discrimination and the intercellular carbon dioxide concentration in leaves. Aust J Plant Physiol 9:121–137

Farquhar GD, O'Leary MH, Condon AG, Richards RA (1989) Carbon isotope fractionation and plant water-use efficiency. In: Rundel PW, Ehleringer JR, Nagy KA (eds) Stable isotopes in ecological research. Springer, Berlin, pp 21–40

Ferriol M, Pichot C, Lefevre F (2011) Variation of selfing rate and inbreeding depression among individuals and across generations within an admixed *Cedrus* population. Heredity 106:146–157

Flint R, Fidalgo F (1964) Glacial geology of the east flank of the Argentine Andes between latitude 39°10' S and latitude 41°20' S. Geol Soc Am Bull 75:335–352

Flint R, Fidalgo F (1969) Glacial drift in the eastern Argentine Andes between latitude 41°10' S and latitude 43°10' S. Geol Soc Am Bull 80:1043–1052

Green D (2005) Adaptive strategies in seedlings of three co-occurring ecologically distinct northern coniferous tree species across an elevational gradient. Can J For Res 35:910–917

Gyenge JE, Fernández ME, Schlichter T (2007) Influence of radiation and drought on gas exchange of *Austrocedrus chilensis* seedlings. Bosque 28:220–225

Hampe A, Petit RJ (2005) Conserving biodiversity under climate change: the rear edge matters. Ecol Lett 8:461–467

Hill R, Brodribb T (1999) Southern conifers in time and space. Aust J Bot 47:639–696

Howe G, Aitken S, Neale D, Jermstad K, Wheeler N, Chen T (2003) From genotype to phenotype: unraveling the complexities of cold adaptation in forest trees. Can J Bot 81:1247–1266

Krause GH, Weis E (1991) Chlorophyll fluorescence and photosynthesis: the basics. Ann Rev Plant Physiol Plant Mol Biol 42:313–349

Loguercio GA, Buduba C, La Manna L (2005) Plantación de ciprés de la cordillera de 57 años de edad: una experiencia en el Parque Nac. Los Alerces Patagonia Forestal 11(1):7–8

Loguercio GA, Urretavizcaya MF, Caselli M, Defossé GE (2018) Chapter 6: Propuestas silviculturales para el manejo de bosques de *Austrocedrus chilensis* sanos y afectados por el mal del ciprés de Argentina. In: Donoso PJ, Promis Á, Soto DP (eds) Silvicultura en bosques nativos.

Experiencias en silvicultura y restauración en Chile, Argentina y el oeste de Estados Unidos, pp 117–131

Lowell T, Heusser C, Andersen B, Moreno P, Hauser A, Huesser L, Schlüchter C, Marchant D, Denton G (1995) Interhemispheric correlation of Late Pleistocene glacial events. Science 269:1541–1549

Marchelli P, Gallo L, Scholz F, Ziegenhagen B (1998) Chloroplast DNA markers reveal a geographical divide across Argentinean southern beech *Nothofagus nervosa* (Phil.) Dim. et Mil. distribution area. Theor Appl Genet 97:642–646

Massone DS (2020) Estudio de las limitaciones ambientales y fisiológicas del crecimiento inicial de plántulas de *Austrocedrus chilensis* (D. Don) Pic. Ser. et Bizzarri "Ciprés de la Cordillera". Doctoral thesis, Fac. Cs. Nat., Universidad Nacional de La Plata

Massone DS, Bartoli CG, Pastorino MJ (2018) Efecto de la fertilización con distintas concentraciones de nitrógeno y potasio en el crecimiento de plantines de Ciprés de la Cordillera (*Austrocedrus chilensis*) en vivero. Bosque 39:375–384

Mitchell J, Lowe J, Wood R, Vellinga M (2006) Extreme events due to human-induced climate change. Philos Trans R Soc A 364:2117–2133

Montesinos D, Verdú M, García-Fayos P (2007) Moms are better nurses than dads: sex biased self-facilitation in a dioecious juniper tree. J Vegetation Sci 18:271–280

Notivol E, García-Gil M, Alía R, Savolainen O (2007) Genetic variation of growth rhythm traits in the limits of a latitudinal cline in scots pine. Can J For Res 37:540–541

Oudkerk L, Pastorino MJ, Gallo L (2003) Siete años de experiencia en la restauración postincendio de un bosque de Ciprés de la Cordillera. Patagonia Forestal 9:4–7

Pastorino MJ (2001) Genetic variation and reproduction system of *Austrocedrus chilensis* (D. Don) Florin et Boutelje, a cypress endemic to the Andean-Patagonian forest. PhD thesis, Cuvillier Verlag, Göttingen. ISBN 3-89873-033-6

Pastorino MJ (2012) How many seed transfer zones are necessary for the preservation of the genetic identity of *Austrocedrus chilensis* natural populations in Argentina? Restor Ecol 20:551–554

Pastorino MJ, Gallo L (2002) Genetic drift evidence in marginal isolated populations of Patagonian cypress (*Austrocedrus chilensis*). International conference: Dynamics and conservation of genetic diversity in forest ecosystems. Strasbourg 2–5 Dec. 2002

Pastorino MJ, Gallo LA (2006) Mating system in a low-density natural population of the dioecious wind-pollinated Patagonian cypress. Genetica 126:315–321

Pastorino MJ, Gallo LA (2009) Preliminary operational genetic management units of a highly fragmented forest tree species of southern South America. For Ecol Manag 257:2350–2358

Pastorino MJ, Gallo LA, Oudkerk L (2001) Aspectos genéticos a tener en cuenta en la cosecha comercial del "Ciprés de la Cordillera". Patagonia Forestal-CIEFAP 7:3–5

Pastorino MJ, Gallo LA, Hattemer H (2004) Genetic variation in natural populations of *Austrocedrus chilensis*, a cypress of the Andean-Patagonian forest. Biochem Syst Ecol 32:993–1008

Pastorino MJ, Fariña M, Bran D, Gallo L (2006) Extremos geográficos de la distribución natural de *Austrocedrus chilensis* (Cupressaceae). Bol Soc Arg Bot 41:307–311

Pastorino MJ, Ghirardi S, Grosfeld J, Gallo L, Puntieri J (2010a) Genetic variation in architectural seedling traits of Patagonian cypress natural populations from the extremes of a precipitation range. Ann Forest Sci 67:508–518

Pastorino MJ, Caballé G, Varela S, Gallo L (2010b) Variation in seedlings survival to drought stress of Patagonian cypress natural populations. Evoltree conference: Forest Ecosystem Genomics and Adaptation, San Lorenzo de El Escorial, Madrid, 9–11 Jun 2010

Pastorino MJ, Aparicio AG, Marchelli P, Gallo LA (2012) Genetic variation in seedling water-use efficiency of Patagonian Cypress populations from contrasting precipitation regimes assessed through carbon isotope discrimination. Forest Syst 21:189–198

Pastorino MJ, Sá MS, Aparicio AG, Gallo LA (2013) Variability in seedling emergence traits of Patagonian cypress marginal steppe populations. New For 45:119–129

Pastorino MJ, Aparicio AG, Azpilicueta MM (2015) Regiones de procedencia del ciprés de la cordillera y bases conceptuales para el manejo de sus recursos genéticos en Argentina. Ediciones INTA, 115 pp. https://inta.gob.ar/sites/default/files/script-tmp-inta_cipres_bariloche.pdf

Perdomo MJ, Andenmatten E, Basil JG, Letourneau FJ (2009) La gestión de la Reserva Forestal Loma del Medio-Río Azul (INTA-SFA). Presencia, Ediciones INTA 54:23–27

Pironon S, Papuga G, Villellas J, Angert AL, García MB, Thompson JD (2016) Geographic variation in genetic and demographic performance: new insights from an old biogeographical paradigm: the Centre-periphery hypothesis. Biol Rev 92:1877–1909

Premoli AC, Kitzberger T, Veblen TT (2000) Isozyme variation and recent biogeographical history of the long-lived conifer *Fitzroya cupressoides*. J Biogeogr 27:251–260

Sakai A (1971) Freezing resistance of relicts from the Arcto-Tertiary flora. New Phytol 70:1199–1205

Sakai A, Larcher W (1987) Frost survival of plants, Ecological studies 62. Springer, Berlin

Souto C, Heinemann K, Kitzberger T, Newton A, Premoli A (2012) Genetic diversity and structure in *Austrocedrus chilensis* populations: implications for dryland forest restoration. Restor Ecol 20:568–575

Urretavizcaya MF (2006) Ciprés de la Cordillera: plantación en bosques quemados y recomendaciones para su establecimiento. Patagonia Forestal 12:13–16

Urretavizcaya MF, Pastorino MJ, Mondino VA, Contardi L (2016) Chapter 12: La plantación con árboles nativos. In: Chauchard L, Frugoni MC, Nowak C (eds) Manual de Buenas Prácticas para el Manejo de Plantaciones Forestales en el Noroeste de la Patagonia, Buenos Aires, pp 335–368

Villagrán C, Moreno P, Villa R (1996) Antecedentes palinológicos acerca de la historia cuaternaria de los bosques chilenos. In: Armesto J, Villagrán C, Arroyo M (eds) Ecología de los bosques nativos de Chile. Editorial Universitaria, Santiago de Chile, pp 51–70

White TL, Adams WT, Neale DB (2007) Forest genetics. CABI Publishing, Cambridge, p 682

Chapter 7
Araucaria araucana and *Salix humboldtiana*: Two Species Highly Appreciated by the Society with Domestication Potential

7.1 *Araucaria araucana* (Pewen): The Sacred Tree of the Mapuche Nation

Paula Marchelli, Javier Sanguinetti, Fernanda Izquierdo, Birgit Ziegenhagen, Angela Martín, Claudia Mattioni, and Leonardo A. Gallo

7.1.1 Introduction

Pewen, *Araucaria araucana* (Molina) K. Koch (also known as monkey puzzle tree), is a dioecious large-seeded emergent conifer endemic to the northern region of the temperate forests of Argentina and Chile. The species integrates the Araucariaceae family, with an ancient origin in the Triassic (ca. 250 My BP), being the only one among the 19 species of the taxa which lives in temperate climate (Hill et al. 1995; Kershaw and Wagstaff 2001).

P. Marchelli (✉) · L. A. Gallo
Instituto de Investigaciones Forestales y Agropecuarias Bariloche (IFAB) INTA-CONICET, Bariloche, Argentina
e-mail: marchelli.paula@inta.gob.ar

J. Sanguinetti
Administración de Parques Nacionales, Parque Nacional Lanín,
San Martín de los Andes, Argentina

F. Izquierdo
INTA AER S. Martín de los Andes, San Martín de los Andes, Argentina

B. Ziegenhagen
Philipps-Universität Marburg, Marburg, Germany

A. Martín
Department of Genética, Universidad de Córdoba, Córdoba, Spain

C. Mattioni
Istituto di Ricerca sugli Ecosistemi Terrestri (IRET) Consiglio Nazionale delle Ricerche, Porano, TR, Italy

© Springer Nature Switzerland AG 2021
M. J. Pastorino, P. Marchelli (eds.), *Low Intensity Breeding of Native Forest Trees in Argentina*, https://doi.org/10.1007/978-3-030-56462-9_7

Araucaria araucana is currently at risk of extinction (Farjon and Page 1999), included in the Appendix I of CITES (http://www.cites.org/eng/app/appendices. shtml) and listed in the 2012 IUCN Red List of Threatened Species (http://www. iucnredlist.org) as an endangered species. Although currently protected in both countries, the species was overexploited for its timber of extraordinary value. Its high morphic coefficient (> 0.80) and the basic density of its wood (0.50–0.60) (CONAF 2013), appropriate for the industry, were the two main reasons for its exploitation. However, the threat is increased given its restricted distribution, slow growth and limited dispersal ability and mainly because of its poor regeneration and continued decline due to diverse anthropogenic causes, such as fires, seed harvest, overgrazing and exotic wild mammal invasions (Gallo et al. 2004a; Sanguinetti 2008; Premoli et al. 2013; Mundo et al. 2013; Tella et al. 2016). In addition, the species is recently going through a general decay expressed as foliar damage in branches and crowns, causing sometimes the death of the tree (Saavedra and Willhite 2017; Vélez et al. 2018).

Pewen trees can reach a height of 50 m, a diameter of 2 m and ages of up to 1000 years (Aguilera-Betti et al. 2017). Young trees develop a pyramidal form, while matures often lose lower branches, giving the tree an umbrella-like shape with the distal part of the branches upwards oriented ("Pewen" in the Mapuche language means "the one who looks at the sky") (Fig. 7.1). It is a dioecious tree; male cones are 8–12 cm long and female cones 15–20 cm in diameter containing 120–180 seeds of 4–5 cm long (Fig. 7.1) (Gut 2008). Pewen has exceptional cultural and economic relevance (Aagesen 1998) and is considered sacred by the Mapuche Pewenche people (Herrmann 2006). The edible seed is a pine nut (locally called *piñón*) and has long been a major element of the diet of the Mapuche Pewenche communities living in and around its natural distribution range (Aagesen 1998). Seeds are also used to feed livestock and constitute a family income when selling and/or exchanging them for food (Herrmann 2006), also as processed products (Cortés et al. 2019). The management of seed collection performed by the Mapuche people involves socio-cultural ancestral knowledge, but the economic market pressure for this non-timber forest product and the new assemblage of wild non-native seed feeders creates a novel and more complex socioeconomic and environmental scenario for conservation.

7.1.2   *Distribution and Ecology*

The continental drift, the change towards a drier and cooler climate, together with the elevation of the Andes mountain range and the successive glaciations, caused the retraction of the original distribution of the *Araucaria* species in southern South America (Kershaw and Wagstaff 2001). Currently, Pewen forests have a relic distribution and occur within 500–2000 m elevation range along a narrow latitudinal band of the Andes from 37.2°S to 40.2°S, in a region with high volcanic activity

Fig. 7.1 Different features of *Araucaria araucana*. (**a**) Mature forest near Lanín volcano; (**b**) female cone; (**c**) male cones with *Enicognathus ferrugineus*, the bird that disperse the seeds; (**d**) natural regeneration; (**e**) seeds and derived products exhibited in a local market. (Photos: **a–d**: J. Sanguinetti; **e**: B. Ziegenhagen)

(Fig. 7.2). Moreover, there are two disjunct locations in the Coastal Cordillera of Chile (Nahuelbuta Cordillera) (37.2°S–38.4°S) (Veblen 1982). The relatively small geographic range of the species is constrained by complex topographic, climate and vegetation gradients oriented west-east across the Andes (Kitzberger 2013). As a result, the species occurs in a range of environments varying from Mediterranean ecosystems on the western side of the Andes to the altitudinal tree-line and the semi-arid Patagonian steppe on the eastern slope of the Andes (Veblen et al. 1995).

Fire is the main disturbance in *Araucaria* forest where annual synchrony of regional widespread wildfires reflects the interannual climate variability. Years of widespread fire occurred with above-mean summer temperatures during the previous and current growing seasons concurrent with low rainfall in current spring, conditions associated with negative deviations of both "Niño" 3.4 and Pacific Decadal Oscillation indexes (i.e. "La Niña" conditions), as well as coincident phases of positive Southern Annular Mode (SAM) and "La Niña" events (Mundo et al. 2013). Temporal variation in the *Araucaria* fire history in Argentina clearly shows the combined effect of human and climate influences on fire regimes. A comparison with previous fire history studies in the *Araucaria* forests of Chile reveals substantial divergences related to differences in human activities on both sides of

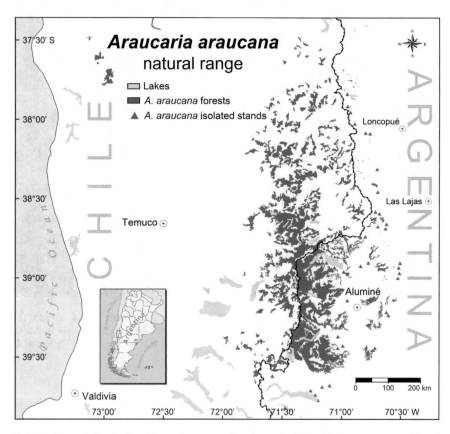

Fig. 7.2 Natural distribution range of *Araucaria araucana* in Chile and Argentina. (Figure outlined by M. Pastorino based on cartography made by INTA Bariloche Remote Sensing-GIS Laboratory)

the Andes, with fires occurring firstly in Argentina. However, the active disturbance period stopped sooner in Argentina due to an earlier implementation of protected areas (Mundo et al. 2013).

The growth of *A. araucana* is controlled by climatic signals where tree diameter increased in years with cool and wet springs and summers (Villalba et al. 1989; Mundo et al. 2012; Muñoz et al. 2014). However, the response on diameter growth due to climate variability induced by "La Niña" events and SAM conditions varied among sex and tree age (Hadad 2014; Hadad et al. 2015). On female trees, the positive conditions for growth are necessary during previous and current year (so control by "La Niña" event in the previous year). However, on male trees, only cool temperatures in the previous year and humid conditions during the current year promote diameter growth (so control by SAM conditions during growth). On the other hand, young trees showed a stronger sensitivity growth response towards climate variability (Hadad et al. 2015).

Araucaria araucana is a wind-pollinated species that shows an environmentally triggered, intermittent, moderately fluctuating and regionally synchronous seed production pattern (Sanguinetti and Kitzberger 2008). Drought conditions in December, during bud differentiation, acts as an environmental cue triggering mast events 24 months later and with a 3–7-year frequency (Sanguinetti and Kitzberger 2008; Sanguinetti 2014). This reproductive spatiotemporal pattern, enhanced by the synchrony among pollen and seed production, maximizes the wind pollination efficiency, alternatively produces the starvation and satiation on the granivory assemblage species and ultimately determines a typical seedling establishment pulse after mast years (Shepherd and Ditgen 2005, 2012; Shepherd et al. 2008; Sanguinetti and Kitzberger 2009; Díaz et al. 2012). In this way, Pewen pollen and cone production provides pulses of resources that control the distribution, reproduction and abundance of native species along the food web, including those species that disperse their seeds (Shepherd and Ditgen 2005, 2012; Sanguinetti and Kitzberger 2009; Tella et al. 2016). However, forest understory degradation and seed consumption by livestock and wild non-native animal species, together with over-collecting of seeds by humans, modified the natural seed survivorship and seedling establishment pattern. Therefore, habitat conditions for native species and their role within the ecological processes changed (Sanguinetti 2008; Sanguinetti and Kitzberger 2010; Zamorano-Elgueta et al. 2012; Shepherd and Ditgen 2013, 2016; Tella et al. 2016; Szymański et al. 2017; Milesi et al. 2017; Speziale et al. 2018).

Araucaria araucana is adapted to stressful conditions imposed by chronically low resource availability, water deficit, extreme temperatures or light limitations (Veblen et al. 1995). Their thick bark, epicormics buds that sprout after disturbance, terminal buds protected by modified leaves and long roots allow the species to resist strong winds, mudflows or wildfires. It showed a long leaf life span and an extremely slow foliage turnover that together with the self-thinning behaviour over lower branches and the ability to produce successively stem reiterations allows seedling and sapling survivorship in the shade condition for decades until a canopy gap is created.

7.1.3 Landscape Genetic Structure: Fragmentation Matters

Holocene glaciations affected South American temperate forests promoting vegetation shifts and forest retractions (e.g. Compagnucci 2011; Folguera et al. 2011), events still imprinted in the genetic diversity of long-lived species (e.g. Marchelli et al. 1998; Premoli et al. 2000; Azpilicueta et al. 2009). The glaciation in the Southern Hemisphere, which at the north of 41°S was mostly restricted to valleys (Flint and Fidalgo 1964; Rabassa and Clapperton 1990; Markgraf et al. 1996), led to speculate on forests' persistence in several small refugia. After the Last Glacial Maximum (18,000–20,000 BP), recolonization began about 14,000 years BP (Heusser et al. 1996; Moreno 1997), but the current vegetation structure was established only about 3000 years ago (Villagran 1991; Heusser et al. 1999).

The current structure and fragmentation of *Araucaria* forests is also due to the presence of humans, which in the region began some 11,000 years ago with the settlement of the continental migrations (Montané 1968). However, the real impact was more significant during the twentieth century, when growing agricultural and livestock activities led to a higher frequency and intensity of intentional forest fires. On the eastern border of the *Araucaria* native range, in Argentina, the situation is aggravated by a combination of extreme environmental conditions that increase drought stress in association with a highly intense human impact. This situation reduces natural regeneration of forests, which is the most obvious sign of Pewen forest degradation. Moreover, marginal xeric populations of *A. araucana* are characterized by a higher proportion of clonal growth, as found in other forest trees with usually less extent of clonality (e.g. Aparicio et al. 2009; Wilmking et al. 2017).

In this scenario of high habitat fragmentation, the species autoecology and related ecological processes are altered, and different genetic processes can be affected (Hartl and Clark 1989; Hanski 1998). The isolation between populations in a fragmented landscape might increase genetic differentiation, inbreeding and levels of genetic drift (Templeton et al. 2001), while gene flow can be favoured due to opening of the landscape (Robledo-Arnuncio et al. 2004). With the help of molecular markers, it is possible to evaluate the impact of habitat fragmentation on the genetic diversity and its distribution among populations.

Organelle DNA markers proved to perform best in studies on historic biogeography or phylogeography (Petit and Vendramin 2006). In Araucariaceae, chloroplast and mitochondrial DNA are paternally inherited according to cytological evidences (Kaur and Bhatnagar 1984), as well as mother-progeny comparisons with molecular markers (Marchelli et al. 2010). Therefore, both plastid genomes move with the pollen grain. With the aim of studying if the species persisted throughout glacial times in scattered and fragmented populations located towards the east of the glacial margins ("rear-edge" populations), a set of 16 populations covering the fragmented area were studied with organelle markers (Marchelli et al. 2010). Rear-edge populations are thought to be stable relics usually isolated and much older than any other population from the rest of the range (Hampe and Petit 2005). Considering the long life span of *A. araucana* with specimens that could reach more than 1000 years, relic populations could be composed of ancient genotypes.

The analysis of chloroplast and mitochondrial DNA revealed a particular result: a very low transferability of universal primers for amplifying both organelle intergenic spacer regions and chloroplast SSRs, since only 15 out of 49 markers gave a reliable amplification product. The most likely explanation for the low transferability of universal primers is the occurrence of sequence divergence between the younger taxa from which the primers were designed (mostly *Pinus* spp.) and the evolutionary old *A. araucana*. Moreover, of these 15 markers, only one showed polymorphism between populations. These low levels of genetic diversity were also reported in other members of Araucariaceae. The extreme case was that of *Wollemia nobilis*, which could represent the only living clone of an "extinct" species (Peakall et al. 2003), but also low diversity was reported for *Araucaria cunninghamii* (Scott et al. 2005), *Agathis robusta* and *Agathis borneensis* (Peakall et al. 2003), suggesting an evolutionary trend in the family. However, considerable levels of genetic diversity were detected throughout *A. araucana* range when using nuclear markers like SSRs, RAPDs or isozymes (Bekessy et al. 2002; Gallo et al. 2004a; Ruiz et al. 2007; Martín et al. 2014; see below), which could be related to a higher mutation rate of the nuclear genome compared to the chloroplast (Wolfe et al. 1987).

In spite of the low levels of polymorphism, the variation detected in the chloroplast DNA allowed the identification of five haplotypes (Fig. 7.3a). As expected for a paternally inherited plastid that moves with pollen grains, genetic differentiation was very low ($G''_{ST} = 0.267$), as in other conifer species (e.g. Vendramin et al. 1998; Schlögl et al. 2007). The low level of differentiation implies that gene flow via pollen in *A. araucana* might be extensive and therefore counterbalancing the divergence among populations (Gallo et al. 2004a).

Many of the sampled populations were located beyond the limits of the ice cap (Hollin and Schilling 1981, Fig. 7.3a) and could therefore be considered as remnants of pre-Holocene origin (i.e. before the Last Glacial Maximum that occurred *c.* 20,000 years BP). The genetic structure observed at the easternmost isolated and marginal populations (reduced allelic richness and increased genetic differentiation; Marchelli et al. 2010) is compatible with a long-lasting isolation and the effects of stochastic processes on small populations. On the other hand, fragmented populations, meaning populations currently affected by fragmentation but still large and connected with the continuous forests, hold the higher allelic richness (Marchelli et al. 2010). Multiple refugia for the species were suggested by Bekessy et al. (2002) based on variation detected with RAPDs among populations from Chile and Argentina and by Ruiz et al. (2007) among Chilean populations using isozyme markers. In addition, Martín et al. (2014) using microsatellite markers reported four gene pools among Chilean populations that clearly separated coastal from Andean populations. Thus, genetic drift due to isolation could have been counteracted by gene flow via pollen and "frozen" genetic structures due to clonal persistence. Unfortunately, a poor pollen representation of *A. araucana* which left hardly a trace (Kershaw et al. 1995) excludes genuine comparison of molecular and palaeobotanic data.

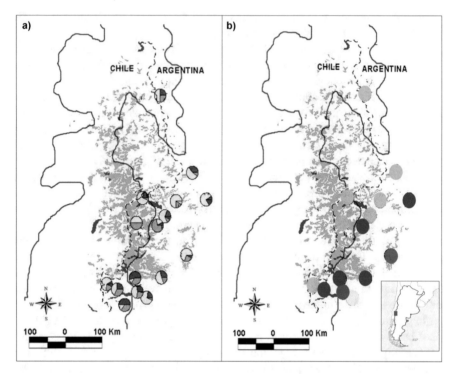

Fig. 7.3 Genetic structure and diversity studies of *Araucaria araucana* in Chile and Argentina. (**a**) The analysed populations and the frequencies of the five chloroplast haplotypes are shown as pie charts (from Marchelli et al. 2010). (**b**) Genetic clusters of nuclear microsatellites detected with BAPS. The dotted line is the international border between Chile and Argentina. The filled line is the limit of the ice cap during the Last Glacial Maximum according to Hollin and Schilling (1981)

Nuclear genetic markers are a better choice for studies on more recent evolutionary processes and particularly to evaluate genetic patterns on a landscape scale to recommend best practices for conservation and management. For this purpose, a subset of stands traditionally used by Mapuche indigenous communities was compared with unmanaged stands using isozyme genetic markers (Gallo et al. 2004a). The genetic diversity was analysed for the seed generation produced in each adult stand. This study was concentrated at a local scale in the southern natural distribution area, and samples were taken from a total of 14 populations inside the Chiuquilihuin community (moderate use of Pewen), in the Aucapan and Rucachoroi community lands (intensive use of Pewen) and in the Tromen forest (control area located inside Lanín National Park). This set of 14 populations had a West-East orientation. One of the tested hypotheses was that eastern isolated and fragmented populations produced seeds with less genetic diversity than the continuous *Araucaria* forests located at the West. We supposed that the isolation and the reduced number of adult individuals of the Eastern populations could have induced genetic drift processes and, therefore, an increase of population homozygosity. A simple sampling design was used to indirectly test this hypothesis. The isozymic genetic diversity of the set of 14 West-East oriented pairs of populations was compared. Each two compared populations were at a distance between 3 and 15 km. In all the cases, the seeds of the fragmented population located at the East had a higher genetic diversity than the seeds from the West, in some pairs even up to 1.5 times higher. The level of historical gene flow was also estimated through the number of migrants (*Ne m*) and varied between 3.04 and 27.55. This was interpreted as the result of an extensive gene flow which, through the unidirectional West-East winds characteristic of the region, moved genetic information through pollen (Gallo et al. 2004a), similarly to the movement of seeds and vegetative propagules downstream rivers (Schleuning et al. 2011). Levels of genetic diversity and differentiation were moderate and comparable with other conifers (Gallo et al. 2004a). The genetic analysis defined three clusters: one cluster consisting of populations with moderate use located in mixed *Nothofagus pumilio-Araucaria araucana* forests, a second cluster composed of eastern and severely affected populations where past and present human activities affected natural regeneration and a third cluster consisting of continuous forests under moderate use.

The development of new molecular tools for studying *A. araucaria* (e.g. SSRs, simple sequence repeats) allowed a deeper understanding of the distribution of the genetic diversity. Ten nuclear SSR markers (Marconi et al. 2011; Martín et al. 2012) were used to genotyped 254 individuals from 15 populations, located either on glaciated or non-glaciated sites from the eastern range. In general, high levels of diversity were found, and the genetic differentiation among populations was moderate and significant (Gst" = 0.349, p = 0.001). A general trend of increasing allelic richness with latitude (i.e. north to south) was detected (R = 0.71 p = 0.003). Interestingly, a significantly higher allelic richness as well as more private alleles was observed in eastern populations, i.e. those located outside the limits of the Last Glacial Maximum, which coincides with the most fragmented and disturbed populations (Table 7.1). Bayesian clustering (BAPS; Corander et al. 2008) revealed the existence of five clusters, three of them composed of a single population (Fig. 7.3b). Similarly, discriminant analysis (DAPC; Jombart et al. 2010), after running the *find. clusters* routine, encountered five clusters with the two first discriminant functions separating three groups of clusters: cluster 2, clusters 1 and 3 and clusters 4 and 5 (Fig. 7.4). Clusters 4 and 5 predominated in populations located at eastern and more fragmented populations outside the limits of the LGM (Fig. 7.4).

To deepen the knowledge on the levels of effective gene flow, a parentage analysis was performed using highly polymorphic microsatellite markers (Moreno et al. 2009, 2011; Moreno 2012). Two areas with different characteristics were chosen: the area of Tromen (39° 37′ 02″ S, 71° 20′ 23″ W) has a continuous forest towards the west and extends to more fragmented locations to the east, while the other area, named

Table 7.1 Genetic diversity and differentiation between groups of populations of *Araucaria araucana* according to their location during the Holocene glaciations

Parameter	Glaciated	Non-glaciated
A_L	5.233 (0.330)	6.1 (0.519)
r_g	**4.528 (0.560)**	**5.948 (0.480)**
A_P	**0.211 (0.033)**	**0.483 (0.312)**
H_O	0.515 (0.029)	0.478 (0.035)
Uhe	0.577 (0.028)	0.590 (0.036)
F_{IS}	0.114 (0.046)	0.189 (0.048)
G''_{ST}	0.336 (0.091)	0.387 (0.125)
N° bottlenecked populations	3	1

A_L mean number of alleles per locus, r_g allelic richness, A_P mean number of private alleles per population, H_o observed heterozygosity, UH_E unbiased expected heterozygosity, F_{IS} inbreeding coefficient, G''_{ST} Hedrick's standardized genetic differentiation, Numbers in bold are significant comparisons

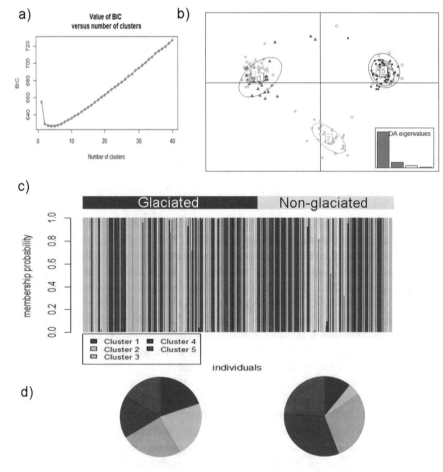

Fig. 7.4 Genetic structure in the evaluated populations as determined by Discriminant Analysis of Principal Components (DAPC), for individuals genotyped with ten genomic SSRs. (**a**) Inference of the best number of clusters according to the *find.clusters* function. (**b**) Scatterplots of DAPC of the inferred clusters showing the first two principal components of the DAPC. (**c**) STRUCTURE-like graph of the identified clusters. Each bar is an individual, and they are arranged in relation to their location during the Last Glacial Maximum. (**d**) Pie chart of the frequencies of the clusters for the two groups in (**c**)

Pulmarí (39°7′ 12″S, 71°5′ 49″W), is a continuous dense forest. A total of 79 and 30 adult trees and 535 and 300 seeds were analysed in Tromen and Pulmarí, respectively, with five microsatellites. Through this analysis of individual mother trees and their progeny (seeds), applying the TwoGener approach (Smouse et al. 2001), an average pollen dispersal distance of 1248 m was estimated in the Tromen area, while a much more reduced dispersal was detected in the denser forest of Pulmarí (71.5 m) (Moreno et al. 2009; Moreno 2012). Also, to evaluate the proportion of vegetative vs. generative reproduction, clusters of regeneration surrounding mother trees were genotyped. A higher proportion of sexual reproduction was evident since only 7% of the seedlings were identical (at least with the analysed markers) to the mother tree. Moreover, only 40% of the seedlings were assigned as progeny of the central mother tree, so the seed dispersal effects of rodents (Shepherd and Ditgen 2005, 2013) have genetic consequences, with an average distance for seedling recruitment of 5–10 m from the mother tree, related to the seed shadow and the typical spatial distribution of seeds and recruits in this species (Sanguinetti and Kitzberger 2009).

Altogether, the genetic studies suggest that *A. araucana* from Argentina has a weak genetic structure between populations, probably related to the extensive pollen flow and the transportation of the edible seeds through the original human communities during perhaps 10,000 years. Eastern and more fragile populations still harbour high levels of genetic diversity, and "frozen genetic diversity" might be maintained by clonal reproduction. However, local extinction of a species can occur with delay after habitat degradation due to the so-called relaxation time (Diamond 1972), after which small and isolated populations will eventually go extinct in a process described as "extinction debt" (Tilman et al. 1994), meaning that species are fated to extinction, but the actual event has not yet occurred due to the persistence for a time in small, isolated and degraded habitats. Extinction debt poses a challenge for conservation because the threats to biodiversity are underestimated (Hanski and Ovaskainen 2002). Particularly species with long generation times, like trees, are more likely to have an extinction debt (Kuussaari et al. 2009; Krauss et al. 2010).

7.1.4 Adaptive Genetic Variation

The analysis of genetic variation at quantitative traits allows studying adaptive processes, because their expression on the phenotype makes them potentially adaptive. The traditional provenance and progeny tests are a useful tool for this purpose. These common garden trials allow the estimation of the additive genetic variation and the heritability, i.e. the portion of the variation that is transmitted from generation to generation. The ability of a population to evolve depends on the existence of additive genetic variance (Lande and Shannon 1996).

Table 7.2 Natural populations of *Araucaria araucana* sampled to install provenance and progeny tests

Population	Latitude S	Longitude W	Altitude m asl	Ann. precip. mm
Primeros Pinos	38°52'22"	70°34'42"	1500	700
Lake Pulmarí	39°07'12"	71°05'49"	1050	1100
Moquehue	38°51'36"	71°15'22"	1290	1800
Lonco Loan	38°54'03"	70°50'30"	1565	800
Pino Hachado	38°37'53"	70°45'23"	1400	700
Caviahue	37°51'22"	71°02'08"	1675	1096
Aluminé River	39°01'56"	71°01'05"	1090	1000
Paimún	39°40'30"	71°38'30"	1050	3350
Tromen East	39°37'02"	71°20'23"	980	1700
Aucapan	39°37'22"	71°07'52"	1300	1200
Rucachoroi East	39°13'05"	71°10'00"	1250	1400

With the aim of studying adaptive responses of *A. araucana* to its environment, two nursery trials were conducted (Izquierdo 2009). In March 2001, seeds were collected from the ground in 11 natural populations of the species in Argentina (Table 7.2), representing the ample variety of environmental conditions of its range. In June of the same year, the seeds were sown in the nursery of Bariloche Experimental Station of INTA (41° 7' 23"S; 71°14' 56" W; 790 m asl). After 7 months, seedlings were transplanted to 800 cm³ pots filled with loam soil and were arranged in a randomized block design, with three repetitions (20 seedlings each; N = 660). Irrigation was provided.

Additionally to this provenance trial, a progeny test was installed with 20 families of open-pollinated seeds collected directly from the crowns of trees of a population located close to the Aluminé River, and it was proceeded in the same way with the same experimental design. In both trials, height of each seedling at the end of the second (H2) and third year (H3) was measured, and the corresponding height increment between years (I) was calculated. Also the number of branches at the third year was counted. As expected, given the low growth rate of *A. araucana* (Rechene et al. 2003), the total height and the annual increment were very low, both for provenances and progenies (H3 = 14.38 ± 2.18 cm and I = 2.27 ± 1.15 cm in average of the provenances; H3 = 13.04 ± 1.9 cm and 1.91 ± 1.0 cm in average of the Aluminé River progenies).

The progeny test was established with seeds of known mother, but unknown father (open-pollinated families). Therefore, the heritability of the analysed traits was estimated considering both half-siblings (factor 4) and full-siblings (factor 2). The estimated values both for heritability (h^2) and additive genetic coefficient of variation (CV_A) were relatively high, when comparing with other species (Table 7.3), and might be related with maternal effects because of the high proportion of carbohydrates in the seed (Cardemil and Reinero 1982). Seed weight was positively

Table 7.3 Intrapopulation variance for traits measured in seedlings of *Araucaria araucana* considering full- (factor 2) or half-siblings (factor 4)

	H2 (cm)		H3 (cm)		H Increment (cm)		N° branches	
Average	11.13 (0.337)		13.04 (0.225)		1.91 (1.094)		2.13 (1.54)	
	Factor 2	Factor 4	Factor 2	Factor 4	Factor 2	Factor 4	Factor 2	Factor 4
V_A	2.28	4.55	1.89	3.79	0.10	0.20	0.93	1.86
CV_A (%)	13.57	19.19	10.56	14.94	16.07	22.72	50.69	71.69
h^2	0.49	0.97	0.40	0.80	0.073	0.14	0.31	0.62
Q_{ST}	0.14	0.08	0.22	0.12	0.20	0.12	0.05	0.03

H2 height at the second year, *H3* height at the third year, H Increment: growth (difference between height at both growing seasons), V_A additive genetic variance, CV_A additive genetic coefficient of variation, h^2 heritability, Q_{ST} quantitative differentiation between populations. SD in brackets

correlated with height in seedlings of 2 and 3 years (Izquierdo and Gallo 2006), and therefore, maternal effects could be causing an overestimation of the heritability. The high value of the estimated CV_A (families mean CV_A = 72.7%) for number of branches suggests a potential for early selection.

Considering the h^2 value estimated with the progeny test, the quantitative differentiation (Q_{ST}) among the provenances essayed could be estimated, resulting in relatively low values for the four traits, similar to the neutral genetic differentiation (F_{ST}). Therefore, the genetic differentiation could be due to the random effects of drift and not merely to the result of divergent selection. Similarly, Bekessy et al. (2003) demonstrated a discrepancy between population differentiation at neutral markers (RAPDs) and quantitative traits (carbon isotope composition and root/mass ratio). However, the stronger differentiation was detected in the quantitative traits among populations at both sides of the Andes Mountains, reflecting evolutionary divergence in water use efficiency among these regions. This differentiation was associated with climate, suggesting that selective forces related with rainfall might have shaped present-day patterns of variation (Bekessy et al. 2003).

Once the nursery tests were finished, the seedlings were used to install a provenance trial (12 provenances) and a progeny trial (16 open-pollinated families) in the field. They were established in 2004 in Trevelin Forest Station of INTA, with N = 288 and N = 384 seedlings, respectively. They were not evaluated yet, but the saplings do not surpass 1 m of height after 16 years in the field (Fig. 7.5).

7.1.5 Restoration, Conservation and Breeding

Recently, many individuals of *A. araucana* show increasing foliar damage in branches and crowns, beginning from the base of the trunk and spreading to the top. The symptoms include necrotic rings on branches causing the death of the tree in some occasions (Saavedra and Willhite 2017). This decay is observed particularly in

Fig. 7.5 One of the saplings of the progeny test of *Araucaria araucana* after 16 years of installation in Trevelin Forest Station of INTA. (Photo: Mario Pastorino)

Fig. 7.6 Evidences of foliar damage in *Araucaria araucana* and detail of the desiccation in one branch. (Photos: J. Sanguinetti)

Chile and with less intensity in Argentina. Although the appearance of *Phytophthora* fungus was detected in almost all cases analysed, the primary cause is supposed to be another consequence of global climate change (Vélez et al. 2018; Ipinza et al. 2019) (Fig. 7.6). *Araucaria araucana* is listed as a vulnerable species under the current climate change scenario, and the loss of vitality expressed as leaf damage worries the scientific community because of the threats to its adaptation, evolutionary potential and survival. Therefore, an initiative for ex situ conservation began in 2017 in the Integrated Monitoring System of Native Forest Ecosystems (SIMEF/CONAF/FAO) of Chile, with the participation of the INFOR (Forest Institute of Chile) and the private sector (CMPC, Chile). The programme, led by Dr. Ipinza, seeks to support the genetic rescue of the species through assisted migration. Accordingly, seeds were collected with the participation and agreement of the representatives of the local indigenous communities, from 458 selected trees in the five genetic zones that make up the *Araucaria* natural distribution area in Chile (Martín et al. 2014), in order to obtain germplasm that represents the genetic variability of the species (Ipinza 2017). Seedlings were produced and micorrized to be then planted in places where the climate gives the species more possibilities. The site selection for the establishment of the ex situ conservation trial was based on an ecological niche modelling study considering the climatic conditions of the next 50 years in order to determine the main host areas for *Araucaria araucana* (Ipinza et al. 2019). The first demonstration trial of provenances and progenies was installed in 2019 in Coyhaique National Reserve, and the second and larger trial will be placed in 2020.

However, degradation and fragmentation of *A. araucana* forests was already observed before the initiation of this general decay of the species, particularly to the eastern xeric locations. The species occurs within different forest types associated with local environmental conditions. We find it living in communities with *Nothofagus pumilio* in the western humid sites and with *Nothofagus antarctica* in the whole range: humid, mesic and xeric sites. However, because in the xeric populations human impact is higher, there we find several fragmented and degraded *A. araucana* populations without *N. antarctica* understory. Due to the frequent use of *N. antarctica* as fuelwood and the occurrence of forest fires, most of the original understories of this species have been devastated carrying on important ecological and population consequences. Among the main impacts on the *A. araucana* forests, it can be mentioned that the drastic decrease in population size of rodents that inhabit this understory may negatively affect the long-distance dispersal of *A. araucana* seeds (Gallo et al. 2004a; Shepherd and Ditgen 2005, 2012, 2013). This means that most of the fragmented xeric populations of the species are highly degraded, and any restoration effort should be focussed on them.

Some restoration initiatives have been realized in Argentina after wildfires, including planting experiences. At the end of the summer of 2009, some 3000 ha were burned in the Lake Tromen area in the Lanín National Park. A previous stock of seedlings from the same provenance available in the nursery of INTA allowed about 1000 to be planted in the following June, registering a survival rate of over 65% in the next autumn. This activity continued the next year with the planting of

2000 additional seedlings. In turn, the large size of the seeds allowed a promising restoration experience realized over 400 ha of burned *A. araucana* forests. In 1967, the species was planted by dropping seeds from airplanes, estimating 49 years later that 19% of the dispersed seeds produced advanced regeneration (Yacubson 1967; Javier Sanguinetti, unpublished data). This recently evaluated experience encouraged the aerial seeding of *A. araucana* to be repeated in 2016, thus reaching burnt sites with almost no accessibility (Mazzuchelli et al. 2016).

With growth rates as low as those observed in the established trials, especially during the seedling and sapling stages, a classical improvement genetic programme for *A. araucana* is hardly thinkable. However, conservation purposes currently require planting the species, and thus, the genetic material to be used is of concern. A low-intensity breeding programme should be considered. For such a programme, it should be bear in mind that the seeds are characterized by rapidly losing their viability due to dehydration (90 to 120 days); therefore, they are classified as recalcitrant seeds, i.e. they cannot be stored for long periods (Chaves et al. 1999). Another characteristic of *A. araucana* seeds is that they present physiological dormancy, but germinate relatively easily after winter, or when applying a pre-germinative treatment of cold stratification (López et al. 1986; Benítez 2005; Muñoz 2010). Therefore, seeds are usually sown during the autumn, immediately after collection, thus emulating the natural stratification conditions experienced during winter (Gutiérrez 2019). Seed emergence at the nursery evaluated in 418 families of *A. araucana* from Andean and Coastal ranges in Chile exceeded 80% 7 months after sowing (Gutiérrez 2019). Lower values were reported for xeric populations of Argentina (between 40 and 50% depending on the pre-germinative treatment, Duplancic et al. (2015)).

Finally, it must be highlighted that the main use of *A. araucana* forests in the whole history has been food provisioning. Animals and people have been eating its pine nuts during 1000 years. Even introduced exotic animals as livestock (i.e. sheep, cows, goats, horses) as well as feral animals (i.e. red deer, wild pigs) have been fed on the species. In some forests, the tons of produced pine nuts feed the whole population of domestic and wild animals and constitute for some of them the main nourishment source of the whole system (Gallo et al. 2004b). Sometimes, the remaining pine nuts for collection by the Mapuche communities are very scarce, since they vary from year to year and are according to the different types of forests.

Consequently, a reasonable objective of a breeding programme for the species is the pine nut production. There are trees in the natural forest that abundantly produce cones and viable seeds almost every year. Their location is known by the usual collectors of pine nuts that belong to the Mapuche communities, and therefore, they should be involved in the programme. Mass-selected trees can be propagated vegetatively by grafting on rootstocks of the same species in order to install clonal seed orchards. This procedure is currently being conducted for *Araucaria angustifolia* in the Manuel Belgrano Forest Station of INTA in the northeastern extreme of Argentina (see Chap. 15). Such a breeding programme for the production of pine nuts is being developed in Spain for *Pinus pinea* (Mutke Regneri et al. 2007). Recently, several elite *P. pinea* clones and a clone mixture, selected for outstanding cone production, have been registered in Spain and Portugal for their grafted use in agroforestry systems or orchards (Guadaño et al. 2016). In this and other breeding programmes for seed nut production, the variability of seed nut production was high among individuals within populations, but low among populations (Gordo Alonso et al. 2007). Heritability values between 45 and 73% were found for seed productivity. Similarly, a large variation in the seed productivity has been registered for *A. araucana* among trees of the same population (Sanguinetti and Kitzberger 2009).

"Conservation through use" is essential to maintain the genetic diversity of Pewen's forests, as these are forests where communities of native people live. These native folks use the forest either for the rearing of their livestock, for the collection of pine nuts or for the use of deadwood to make handcrafts. The complexity of these systems is very large, and their study can only be addressed through systemic thinking integrating all the actors, factors and processes that determine their dynamics. The conservation of genetic diversity is the basis for the conservation of *Araucaria* forests and therefore depends on the correct approach to understanding the interactions within the system and with its externalities (Gallo et al. 2004b) (Fig. 7.7). Conservation through use seems to be the best way of conserving genetic diversity of the *A. araucana* ecosystems, as much as the livestock stock can be reduced to the "cultural auto-consume" rate. This reduction of the quantity of domestic animal presence needs the governmental management, for example, by employing people of the community to do restoration activities (i.e. planting).

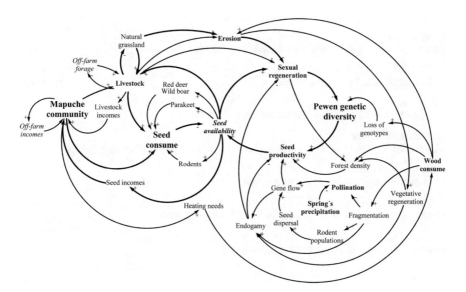

Fig. 7.7 Causal loop diagram of the conceptualization model of a traditional *Araucaria araucana* forest system where original people communities live within and from the forest. Maintaining the genetic diversity of these systems is crucial for its use in breeding programmes. (Modified from Gallo et al. 2004b)

7.2 *Salix humboldtiana*: A Very Ancient Willow and the Only Native to Argentina

Leonardo A. Gallo, Ivana Amico, Jorge Bozzi,
Marianelen Cedres Gazo, Teresa Cerrillo, Leonardo Datri, Marina Hansen,
Ilona Leyer, Hernán López, Paula Marchelli, Abel Martínez,
Juan Pablo Mikuc, Ivonne Orellana, Florencia Pomponio, Javier Puntieri,
Mariana Salgado, Susana Torales, Sergio Vincon, and Birgit Ziegenhagen

L. A. Gallo (✉)· J. Bozzi · P. Marchelli
Instituto de Investigaciones Forestales y Agropecuarias Bariloche (IFAB)
INTA-CONICET, Bariloche, Argentina
e-mail: gallo.leonardo@inta.gob.ar

I. Amico
INTA EEA Esquel, Esquel, Argentina

M. C. Gazo
Laboratorio de Biotecnología Vegetal de Viedma, Río Negro,
Argentina

T. Cerrillo
INTA EEA Delta del Paraná, Campana, Argentina

L. Datri · H. López
Universidad de Flores, Sede Comahue, Neuquen, Argentina

M. Hansen
Jardín Botánico Intercultural Bariloche, Bariloche, Argentina

I. Leyer
Hochschule Geisenheim Universität, Geisenheim, Germany

A. Martínez
INTA AER Zapala, Zapala, Argentina

J. P. Mikuc
INTA AER Chos Malal, Chos Malal, Argentina

I. Orellana · S. Vincon
Universidad Nacional de la Patagonia San Juan Bosco, Sede Esquel, Esquel, Argentina

F. Pomponio · S. Torales
INTA IRB Hurlingham, Hurlingham, Argentina

J. Puntieri · M. Salgado
IRNAD (UNRN-CONICET), San Carlos de Bariloche, Argentina

B. Ziegenhagen
Philipps-Universität Marburg, Marburg, Germany

7.2.1 An Ancient Species with a Huge Natural Distribution Area

Salix humboldtiana Willd. (Spanish names, sauce nativo, sauce amargo, sauce colorado, sauce criollo; original people names, reiwe, waik, yvira puku, wayaw), the only willow tree native to Argentina, is one of the tree species with the longest natural distribution area in the world. It spreads over more than 10,000 km covering an enormous latitudinal gradient from Central-North Mexico (23°N) to Chubut River (44°S), in the Argentine Patagonia. Consequently, it grows in a great diversity of environmental conditions, as riparian vegetation (Hauman et al. 1947; Ragonese and Rial Alberti 1958a). It occurs mostly along the Andes Mountains and adjacent regions, following watercourses in subhumid and semi-arid regions, from the sea level up to more than 3900 m asl (Flora Argentina 2014; IUCN 2020). In the subtropical and tropical climates, it is found essentially in the Yungas, in some Andean Valleys and along the Amazon River and its tributaries, in the periphery of the Amazonia towards the Andes. In Argentina, it is widely distributed, being present in almost the whole country, except for the high altitudes of the Andes and South Patagonia (Fig. 7.8).

Salix humboldtiana is a medium-size tree, growing up to 25 m in height and up to 1 m in diameter. It is a pioneer species (Parolin et al. 2002; Isebrands and Richardson 2014) and can tolerate long periods of flooding, with reports of up to 7 months with submerged roots (Parolin et al. 2002). However, it is frost sensitive, which leads to some range restrictions in the cold environments like those of high altitudes and the South Patagonian steppe.

With 20 to 25 million years of evolution, it is an ancient and phylogenetically basal species of the genus and probably the origin of many other *Salix* species (Abdollahzadeh et al. 2011; Wu et al. 2015). While most of the willows are polyploids (Abdollahzadeh et al. 2011; Lauron-Moreau et al. 2015; Wu et al. 2015, 2018), *S. humboldtiana* is a diploid. Its evolutionary origin occurred in the subtropical New World and from there migrated to the southern part of South America adapting to low temperatures (Argus 1997).

It regenerates mainly through seeds dispersed by wind or water and commonly builds sapling stands on sand or lime banks (Thomas and Leyer 2014). It has very narrow leaves (0.6–0.8 mm), not so long (6–13 cm), some of them slightly curved and sawed (Ragonese and Rial Alberti 1958a; Menendez 2012; Fig. 7.9). It is a perennial species when it occurs in warm tropical and subtropical climate conditions, but deciduous in climates with a definite cold or dry season. Male flowers have between five and eight stamens, which is also a discriminating trait in relation to most other willows, which normally have two stamens. Pollination, like in other willow species, has two vectors: insects and wind (Tamura and Kudo 2000).

The main adult trunk has a rough bark with irregular plates, and its architecture is very characteristic and quite different from that of the exotic feral willows abundant in the same ranges, since it has a very straight, monopodial (or

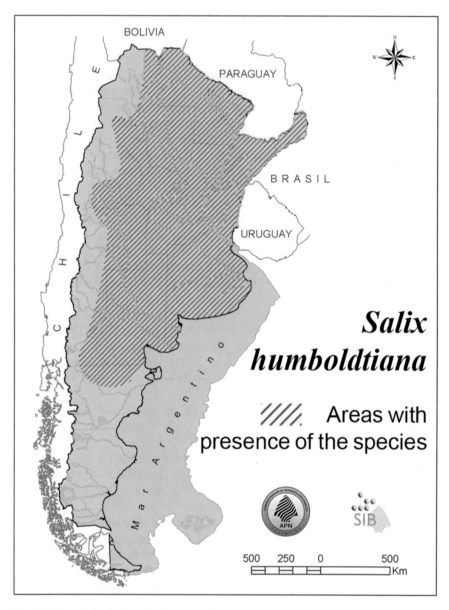

Fig. 7.8 Natural distribution of *Salix humboldtiana* in Argentina. (From APN-SIB 2020)

pseudomonopodial) trunk of up to 7 m in height (Fig. 7.9). A fastigiated variety (*S. humboldtiana* var. *fastigiata*, probably a mutation) is found in the middle part of its natural distribution area (10°–24°S) (Menendez 2012). Another variety (*S. humboldtiana* var. *martiana*) whose regular catkins have female, bisexual and male flowers occurs from North Argentina to Colombia (Rohwer and Kubitzki 1984).

Fig. 7.9 (**a**) Typical bark and leaves of *S. humboldtiana* (photo: F. De Durana). (**b**) Leaves and inflorescences of *S. humboldtiana* (female, left; male, centre) and *Salix* x *fragilis* f. *fragilis* (female, right). (Photo: L. Gallo). (**c**) Isolated group of few *S. humboldtiana* trees, at Cerro Cóndor, 400 m from the current riverbed of Chubut River, in the southernmost limit of the species, under very harsh environmental conditions. (Photo: L. Gallo). (**d**) Foreground, two *S. humboldtiana* natural individuals in San Javier population (Province of Río Negro; their trunk shape stands out in comparison to the introduced willows in the background). (Photo: L. Gallo)

Salix humboldtiana has been used since at least 10,000 years ago by original folks, among others, to make tool handles and arrows during the early and mid-Holocene (Rodríguez 1999, 2000) in the North of the country and during the Pleistocene-Holocene transition in the Tandilia Range, Province of Buenos Aires (Brea et al. 2014). Its wood is reddish and has a basic density (0.35–0.50 g/cm^3) higher than most of all other willows introduced in Argentina (Tortorelli 2009; Cerrillo et al. 2016). The settlers prefer its wood, and, therefore, even in its southernmost range, where few natural specimens of this species remain standing, trees are being cut clandestinely. Also in the Paraná River Delta, its wood is considered of a better quality compared to other introduced members of the Salicaceae family, and it has been used as a hard and durable wood for very special uses in that humid environment (Luis Hansen, pers. comm.). It is used as fuel and for construction and manufacture of barrels, hoops, poles, drawers, joinery, tool handles, packaging, furniture and openings. Thin and flexible branches were and are still used to make wicker baskets and furniture. The bark contains salicin, phenols and oxalates that have febrifuge, analgesic, sedative, tonic, astringent and antispasmodic properties for medicinal and veterinary use. It is also a honey plant and an ornamental tree used in gardens, urban green spaces and roads (Larroulet et al. 2011; Rosso and Scarpa 2019).

7.2.2   The Main Threats to Its Conservation

Although *Salix humboldtiana* has a continental distribution and in Argentina crosses various ecosystems, in the marginal populations of the Patagonian steppe, its conservation acquires a high value. In this sense, it is perhaps among the most threatened tree species in Patagonia. A great loss of its genetic diversity in its southernmost range can be presumed to have occurred in the last 250 years. The extinction of several populations has been confirmed when compared to historical registers (Falkner 1774; Moreno 1876; Entraigas 1960), and in several river shores, just very few isolated individuals constitute the entire current presence of the species. In turn, in most of the rivers of Argentina, the natural distribution of *S. humboldtiana* is highly fragmented. One of the main causes of the critical situation of this species in its southern range is the massive use of its wood during the European colonization of the Negro River valley, close to the northern border of Patagonia. There are mentions of thousands of trees cut down per week (Entraigas 1960; Gómez Otero and Bellelli 2006) to build the first villages (houses, fortifications, churches, fences and even boats). Remains of willow beams can still be seen in very old houses of Carmen de Patagones, built at the end of the eighteenth and the beginning of the nineteenth centuries.

More recently, the introduction of highly vigorous and invasive exotic *Salix* clones has affected the remaining natural stands of *S. humboldtiana*. The higher vegetative propagation capacity of these clones (Thomas et al. 2012; Lewerentz et al. 2019) contributed to the reduction of the native willow's habitat. Additionally, interspecific hybridization occurs and has been reported in systems dominated by the exotic willows. The low clonal capacity of *S. humboldtiana* compared with Eurasian willows and poplars made this species vulnerable to competitive exclusion (Thomas et al. 2012; Thomas and Leyer 2014). Already Hauman et al. (1947) reported the scarcity of *S. humboldtiana* along the Patagonian river banks in the mid-twentieth century. Alluvial floodplains, low-lying surfaces composed of fluvial deposits adjacent to river channels (Balian and Naiman 2005), are typical *Salix humboldtiana* habitats in most of the rivers of Argentina. However, nowadays they are covered mainly by exotic species.

In Limay River (a main affluent of the Negro River), for example, the synchronized colonization of *Populus nigra* cv. *italica* and *Salix fragilis* f. *fragilis* after floods higher than 1290 m^3/s installed a new emerging interspecific competition scenario with the native willow in the channels, strips and island plains. This unprecedented event affected the native *S. humboldtiana* populations in advanced rather than early successional stages, due to the massive invasion of *P. nigra* cv. *italica*, occupying the sand banks that are necessary for the native willow's sexual regeneration. Energy dams in some of the Patagonian rivers avoid the consolidation of sand and lime banks that are essential for the establishment of *S. humboldtiana* seedlings. Similarly, in the river basins of Futaleufú River and Chubut River, a high density of interspecific exotic hybrids was found (Orellana et al. 2016), and mechanical and chemical methods to control the exotic willows' invasion were proposed (Amico and Orellana 2014).

Salix x *fragilis* f. *fragilis* (previously called *S.* x *rubens*) is the natural hybrid between *Salix alba* L. and *Salix euxina* I. (ex *S. fragilis*) (Belyaeva 2009) and has an extraordinary vegetative propagation capacity. A study with molecular markers determined that in the Patagonian rivers, most of the willow trees (93%) belong to the same and only female clone of *Salix* x *fragilis* f. *fragilis* (Budde et al. 2011). In a greenhouse experiment, an inferior vegetative propagation capacity of *S. humboldtiana* was observed when compared with exotic introduced clones (*S.* x *fragilis* f. *fragilis*, *S. babylonica* hybrid) and *Populus* spp., under different humidity conditions (Thomas et al. 2012; Fig. 7.10).

In Patagonia, hydro-geomorphological processes regulate the plant succession on floodplains, establishing a zoning associated with river geoforms. They are frequently flooded and normally covered by plant communities that change through time. Pioneer vegetation is highly adapted to extremely variable river flows and morphology (Gray et al. 2006) and can trap and stabilize sediments that influence their own distribution patterns in the river channel. A closer and more complex interplay between riparian trees, woody debris, flood pulses, sedimentation, and island development results in new plant systems that grow by sexual and vegetative reproduction (Datri et al. 2016a, 2017).

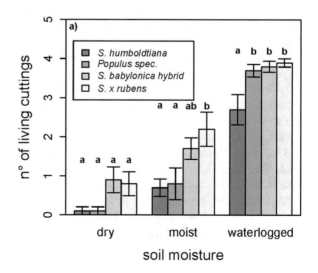

Fig. 7.10 Comparative vegetative propagation capacity of *S. humboldtiana* cuttings vs. exotic willows and poplars along a humidity gradient in a greenhouse experiment. (From Thomas et al. 2012). (*S.* x *rubens* is at present named *S.* x *fragilis* f. *fragilis*)

Fig. 7.11 Invasion of Patagonian river shores by clones of exotic willows, in Middle Valley, Negro River. (Photo: J. Bozzi)

In the arid Patagonian environment, the invasion of Eurasian willows and poplars was limited to the riverbed and areas below the 1.5 m flood level. Above this level, a greater diversity of exotic and native species occurred. Beyond the flooding area, there was a natural limit to the invasion of riparian species and an overall reduction in diversity. In terms of conservation, the invaded areas posed a threat to *S. humboldtiana* (Datri et al. 2016b, 2017) (Fig. 7.11).

As said above, the Chubut River is home to the most austral populations of *S. humboldtiana*. Marginal populations are valuable for biological conservation because they may contain exclusive genetic variants (Lowe et al. 2004), and in this sense, Chubut River populations are of special concern. A field survey of 300 km along the Chubut River was carried out registering, at each *Salix humboldtiana* population, frequencies of individuals, sex ratio, diametric frequencies, height and number of stems per individual. The species is absent in the upper Chubut River where the riparian areas were invaded by *Salix* x *fragilis* f. *fragilis* hybrids and some isolated individuals of *Salix* x *fragilis* f. *vitellina* (Belayaeva et al. 2018). The first native willow individuals occur at about 150 km southeast of the river origin. Along the following 50 km of the river, only 23 *S. humboldtiana* "populations" or, better said, sites with a handful of individuals were found. In total, along 300 km of river shores, only 200 adult trees were registered, at the rate of 1 to 40 individuals per site (80% of the sites have less than 15 individuals).This agrees with previous estimations that assumed no more than 1000 adult individuals in the total length of the river (810 km) (Gallo et al. 2018). Most of these populations (70%) have a balanced sex ratio. The size of the individuals ranged from 5 to 100 cm DBH with an average of 38.7 cm. The thicker (older) individuals were usually found at about 500 m out of the current riverbed, in the floodplain. In these sites, the soil is more or less consolidated, and the vegetation is shrubby and typical of the Patagonian steppe. The average annual diameter growth measured in wood cores at breast height of 15 prospected adult trees from different sites and rivers was 9 mm, reaching in some sites and years up to 22 mm. The average height was 6 m and the tallest individuals reached 13.6 m. The harsh environmental conditions, with constant strong winds, produce serious damages on the willow shoots. The bark of the adult trees, especially the oldest ones, had clear signs of the grinding effect of the sand moved by the wind, making irregular smooth plates. Most trees lost a part of their canopy, and had numerous dead branches, probably due to the drought stress experienced during the summer. The average proportion of living canopy in these old trees was around 45%. Most trees had a single main stem, and in some cases, large old trees were fully or partially prostrated with some lateral branches taking a dominant role as main stems (Orellana et al. 2016). Natural regeneration was found only in 3 out of the 23 monitored sites, with less than 100 juveniles, severely affected by livestock herbivory and always restricted to the edge of the current riverbed. Putative hybrids between *S. humboldtiana* and *S.* x *fragilis* f. *fragilis* were observed also in three sites.

7.2.3 Landscape Genetic Structure

The huge invasion of all river shores by exotic willow clones is important for the hybridization process given the high demographic imbalance with hundred thousands of exotic *Salix* trees and just few surviving native willows. The probability that an insect and/or the wind carrying native willow pollen fecundates a native female tree is very small. On the contrary, the probability of fecundating the female exotic willow clone is very high. Therefore, in most of the rivers, when phenological windows overlap, hybrid seeds are produced.

Salix humboldtiana hybridizes specially with the introduced *Salix* species of the section *Salix* (Wu et al. 2015) with a remarkable introgression and consequently genetic dilution of its genetic information. In order to identify hybridization processes among the invasive willow taxa (*S. alba, S. fragilis, S. babylonica, S. matsudana, S. viminalis*) and *S. humboldtiana*, a molecular toolbox for taxa discrimination was created. It combines three maternally inherited chloroplast microsatellite with ten biparentally inherited nuclear microsatellite markers. This toolbox is suitable to unambiguously distinguish first-generation *S. humboldtiana* hybrids. The taxa can be discerned by chloroplast microsatellite markers, and in addition, *S. humboldtiana* presented a private allele at one of the nuclear microsatellite loci (Bozzi et al. 2012). The analysis of 1300 individuals from five Patagonian rivers allowed detecting 86 *Salix humboldtiana* hybrids: 76% were juvenile and only 24% were adults. In most cases, *S. humboldtiana* was the pollen donor (Bozzi et al. 2012), which is expected, since the most distributed exotic willow clone in the study area is female (Budde et al. 2011).

Additionally, a molecular pair-comparison was made in 15 sites of the Negro River, where adult trees (> 20 years) and groups of juveniles (< 5 years) grow together. It was found that the number of populations with hybrids increased twice in the last 20 years (50% of the juvenile groups analysed had hybrids) and that the number of hybrid individuals increased three times in the last 20 years (16% of the regeneration) (Gallo et al. 2016, 2018). The higher proportion of hybrids at the juvenile stage could indicate the intensification of the hybridization process. Processes like genetic assimilation, demographic swamping and loss of genetic integrity are some of the possible outcomes (Thomas and Leyer 2014).

To assess the genetic diversity of *S. humboldtiana* and to unravel the population genetic structure along Patagonian rivers, samples were collected at four locations, most of them along the Negro River and just a few in the confluence of its main tributaries, the rivers Limay and Neuquén. By genotyping at five nuclear microsatellites (Bozzi et al. 2015), a lower genetic diversity was found (Ho = 0.568; He = 0.555) as compared with other riparian species (e.g. Lin et al. 2009; Mosner et al. 2012). A trend of increasing genetic diversity was observed downstream along the Negro River, which is in line with some other studies of riparian species indicating unidirectional gene flow mediated by water dispersal (Huang et al. 2015; Schleuning et al. 2011). However, no structuring was detected among the 34 analysed sites.

The genetic diversity of the main North Patagonian rivers and a subtropical population was also compared with some of the same markers used in the previous analysis of the Negro River genetic diversity (Pomponio et al. 2018). Samples of leaves from typical trees in Victoria (Province of Entre Ríos, 32°37′S) and on the banks of the rivers Neuquén (38°30′S), Negro (39°S) and Chubut (42°30′S) in North Patagonia were collected. Samples of introduced species of *Salix* inhabiting the same region were also included to estimate the existence of hybridization. Preliminary molecular studies were carried out from a total of 178 samples belonging to four populations using five microsatellite markers: Shum_49, Shum_71 (Bozzi et al. 2015), gSIMCO24, ORPM_446 and PMGC-223 (http://www.ornl.gov/sci/ipgc/ssr_resource.htm); the latter was used as taxa diagnostic for hybrid estimation (Bozzi et al. 2012).

The sampled trees were selected phenotypically according to leaves, buds, bark, architecture and phenological traits. The exotic species that inhabit the natural distribution of *S. humboldtiana*, such as *S. alba, S. euxina (ex S. fragilis) and S. babylonica* in North Patagonia and *S. nigra, S. alba* and *S. babylonica* in Entre Ríos (NE of Argentina) were analysed with the same SSRs markers as a control. Only two putative hybrids were found, indicating that phenotypical selection of pure individuals can be done with minimum bias. When both analysed regions were compared, a notable difference was found in all genetic diversity parameters. Patagonian populations showed a low genetic diversity over loci (Ho = 0.180; He = 0.236), whereas the subtropical population at Victoria (Province of Entre Ríos) presented moderate values (Ho = 0.515; He = 0.612) (Table 7.4). The

Table 7.4 Preliminary genetic diversity results in the southern natural range of *Salix humboldtiana* in comparison to a subtropical population (Victoria). (From Pomponio et al. 2018)

Population	N	Na	Ho	He
Victoria (32° 37′S)	23	17	0.515	0.612
Neuquén (38° 30′S)	50	12	0.198	0.316
Negro (39° 00′S)	34	17	0.201	0.220
Chubut (42° 30′S)	71	13	0.141	0.174

N number of samples, *Na* number of alleles, *Ho* observed heterozygosity, *He* expected heterozygosity

southern limit of the natural distribution of the native species (North Patagonia region as a whole) showed therefore an incipient and expected cline in the genetic diversity from north to south following the ancient colonization route of the species (Argus 1997; Wu et al. 2015).

7.2.4 First Steps Towards a Breeding programme

Commercial willow plantations in Argentina are based on exotic species and mainly located in the Paraná River Delta, northeast of the country, with a cultivated area of 68,862 ha (Borodowski 2017). Since the 1950s, sequential genetic improvement programmes of willow have been developed to obtain a wider availability of productive clones for cultivation (Ragonese and Rial Alberti 1958b; Arreghini and Cerrillo 1996). These strategies have almost exclusively considered non-native germplasm, with only a few cases of controlled crossings using individuals of *Salix* x *argentinensis*, which is a natural hybrid between *S. babylonica* and *S. humboldtiana*. However, the results were not outstanding for the pulp industry, which is the main consumer of willow wood in this region, because of its reddish colour. Notwithstanding, this natural hybrid was cultivated for decades in the Paraná River Delta due to its resistance to some important diseases (Ragonese and Rial Alberti 1958a, b).

In the Paraná River Delta, natural populations of *Salix humboldtiana* were widely distributed along streams and rivers. However, due to the easy hybridization with commercial clones containing non-native germplasm, the native pure species is gradually disappearing. Considering the genetic and economic value of indigenous germplasm, a new research proposal began in 2013 aiming at conserving and studying the variability of native populations in the region and including them into the *Salix* sp. genetic improvement programme of INTA (Instituto Nacional de Tecnología Agropecuaria) (Cerrillo et al. 2019).

In the frame of this new research proposal, native material was collected at two locations of the Paraná River Delta: near the cities of Victoria and Diamante (Province of Entre Ríos, 34°29′ S) and the cities of Campana and Baradero (Province of Buenos Aires, 33° 37′S). As part of the operative methodology, 0.50-m-long cuttings, taken from branches of 52 trees, were planted in rows in a field clone bank at Paraná Delta Experimental Station of INTA (34° 10′ 28″ S; 58° 51′ 45″ W; 0 m asl), with ten plants per genotype. After the first year, 89% of the genotypes had survived, with significant differences in survival rate between them (from 0% to 100% surviving plants/genotype). The average total height of the dominant stem in the first year was 2.90 m, with a mean of five stems per plant. A low incidence of the main leaf diseases (rusts caused by *Melampsora* spp. and anthracnose by *Marssonina* spp.) was registered using a scale of six grades (no differences in susceptibility were observed between genotypes) (Cerrillo et al. 2018). Recently (2019), a field test was established in Paraná Delta Experimental Station with a selection of the most promising 16 genotypes of that collection (Teresa Cerrillo, pers. comm.).

In addition, some variation studies were performed with Patagonian populations. Variation of frost susceptibility among provenances was studied in a clonal bank installed in Bariloche Experimental Station of INTA (41° 7′23″S; 71°14′ 56″ W; 790 m asl) with 39 genotypes of the Negro River provenance and 33 genotypes of the Chubut River provenance (Leonardo Gallo, data not published). Phenological differences were preliminary observed between those two provenances. Chubut River flows from the Andes Range to the Atlantic Ocean at about 43° S. Its mean annual temperature is lower than that of Negro River, which flows in the same direction but about 4 latitudinal degrees northwards. At the clonal bank, spring sprouts developed 3 weeks later in Chubut River clones than in Negro River clones, whereas autumn leaf drop took place up to 1 month earlier in the Chubut River clones. This means that Chubut River clones had an almost 2-month shorter growing season. The genetic signal from the home environment is maintained in Bariloche, that is, at an intermediate latitude between the two provenances. This has consequences in frost resistance. The trial included 5 ramets per clone (N = 360). Frost damage was described with six categories considering the necrosis in the main shoot. Category 1, no damage at all; 2, up to 10% of the main shoot necrotic; 3, 10 to 30%; 4, 30 to 60%; 5, more than 60% necrotic but plant still alive; and 6, plant dead. The average value of all Negro River clones was 2.33, whereas that of Chubut River clones was 1.30. The clones of the two provenances were compared statistically with a T test for paired comparisons. Significant differences were found between the two provenances in the average early frost damage (Fig. 7.12).

A study of the architecture of the species was initiated also with the Patagonian provenances (Salgado et al. 2018, 2019). In the clonal bank of Bariloche, the length, diameter and number of nodes of annual shoots of 15 clones of Negro and Chubut provenances were recorded every 10 days in the 2017/2018 season. Maximum and average growth during the growing season were also calculated. The moment at which the maximum growth rate occurred was more variable in clones of Chubut than in those of Negro, being the date for this trait in both provenances between December 17 and January 7. No differences between provenances were found for the number of nodes and the maximum and average growth rates. However, shoots of Negro provenance were longer and thicker, which could be associated with the longer duration of their growth period.

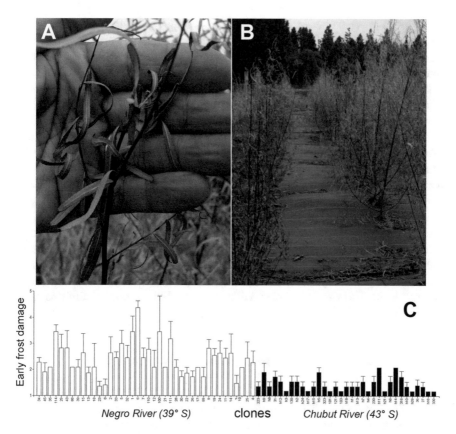

Fig. 7.12 Early frosts damage in Negro River and Chubut River clones. (**a**) Severe early frost damage on the leaves and shoots of one of the Negro River clones. (**b**) Frost damage on the group of Negro River clones (left) and almost no damage on Chubut River clones (right). (Photos: L. Gallo). (**c**) Mean frost damage per clone and their standard errors for the Chubut and Negro provenances

The ratios between the annual shoots of the main axis and the shoots of the branches for length, diameter and number of nodes were studied in the field on natural growing trees. These three ratios were greater than 1 in both analysed provenances. The length-diameter and number of nodes-diameter ratios of the annual shoots differed between the positions of the shoots, being in both provenances larger in the branch than in the main axis. These evaluations indicate clear differentiation levels between trunk and branches in terms of annual shoot morphology.

The differences observed in phenology and tested in frost resistance between provenances and clones are indicative of the possibility of population and individual selection in future breeding programmes. Even in Negro provenance, which has a longer growing season, some clones resisted early frosts similarly to those of Chubut provenance. Such traits of important adaptability and productive meaning (probably of high heritability), combined with the fact that *S. humboldtiana* is able to produce seeds just 3–4 years after rooting in the North Patagonian conditions, facilitate the development of a breeding programme for the southernmost area of the species range. On the other hand, *Salix* spp. breeding programmes are mainly based on interspecific hybridization to maximize genotypic variance by combining the outstanding features of the species involved. Thus, some traits of *S. humboldtiana* are interesting to consider for the general genetic programme of the genus. The higher density and pale reddish colour of its wood stand out for use as solid wood, such as in carpentry and openings. Its resistance against several diseases and plagues must also be mentioned. Besides the above cited resistance to *Melampsora* spp. and *Marssonina* spp. in Paraná River Delta, in Patagonia *S. humboldtiana* is rarely attacked by the two most important plagues of willows in the region: sawfly (*Nematus desantisi*) and the giant willow aphid (*Tuberolachnus salignus*) (Sopow et al. 2017). In the first case, the wasp was firstly reported in Argentina in 1980 in the Chubut River on exotic willows, and from there it has spread fast and extensively throughout the whole country (Dapoto and Giganti 1994). However, so far only a very light attack of this very aggressive defoliator has been observed on the native willow. Similarly, in the case of the giant aphid, the exotic willows show very strong attacks. However, the native willow has very little or no attack of this aphid in the wild.

7.2.4.1 Launching of a Participatory Genetic Rescue programme

As we mentioned before, *S. humboldtiana* is severely threatened, especially in Patagonia (Bozzi et al. 2014), where the extinction of some populations is evident when comparing its current absence in places where historical references mention the presence of the species (Moreno 1876) and in some cases of thousands of trees (Entraigas 1960). A rescue programme was consequently initiated in 2018, taking into account some genetic principles and convening the local community to be involved in this incipient long-lasting process. This was called *Participatory Genetic Rescue Programme of Sauce Criollo in Patagonia* (Gallo et al. 2018, Fig. 7.13).

This programme was initiated by prospecting more than 2500 km of riverbanks along seven different rivers of North Patagonia: Chubut, Limay, Collón Cura, Negro, Futaleufú, Neuquén and Agrio. In these explorations, putatively pure *S. humboldtiana* individuals were identified and sampled. Branches were collected and the sampled genotypes were cloned, most of them at the Bariloche Experimental Station of INTA. Then a set of clones were distributed to four different propagation nurseries in Patagonia, according to their location and corresponding area of influence, for restoration purposes: Viedma Nursery from the Province of Río Negro, Mariano Moreno Nursery from the Province of Neuquén, Cipolletti Municipal Nursery and the nursery of Trevelin Forest Station (INTA). Meanwhile, at the nursery of Bariloche Experimental Station, a clonal bank with all clones (more than 300) from all rivers was installed aiming at the ex situ conservation of the remaining genetic diversity of the species in Patagonia. In all these nurseries, cuttings were rooted in order to produce seedlings for restoration purposes.

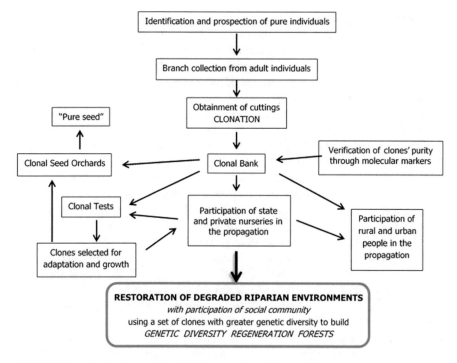

Fig. 7.13 Functional scheme of the Participatory Genetic Rescue Programme of *Salix humboldtiana* in Patagonia. (From Gallo et al. 2018)

According to the degree and type of threat that a species can have in nature, different alternatives of rescue have been proposed. Sometimes the main long-term threats can be attributed to genetic or evolutionary processes and mechanisms; other times, the main cause is related to demographic factors (Carlson et al. 2014). In the case of the Patagonian populations of the native willow, due to its very small, fragmented and isolated "populations", the species might have a reduced ability to evolve and an elevated extinction risk. Therefore, the three different proposed types of rescue, demographic, genetic and evolutionary, fit very well.

Every genetic rescue programme finishes when the recovered genetic diversity returns to nature (Frankham 2015), in our case to the rivers. Our aim is to install small riparian plantations that should act as "regeneration nuclei" and to conserve the few natural regeneration sites found in some of the rivers. Since *S. humboldtiana* propagates essentially by seeds on lime and sand banks, these nuclei will act in the future as genetic diversity regeneration stands. From them, a passive restoration process could be induced as has been already proposed for other Salicaceae (Zhang et al. 2019).

Because of the strong degradation of the native riparian forests, and the few remaining individuals of *S. humboldtiana* in Patagonia, there is scarce production of *S. humboldtiana* seeds in nature. Even more, the seeds produced are most likely hybrid between *S. humboldtiana* and *Salix* x *fragilis* f. *fragilis*. Therefore, just by putting together some 30 to 50 different genotypes (clones) in nuclei on the shores of the rivers, we will manage to obtain seeds from the native willow with a higher degree of purity than the current natural production. Although the exotic willow clones have invaded most of the Patagonian rivers, with the conformation of these nuclei, we could obtain a good amount of seeds to give the native willow a chance to recover part of the river shores.

So far, restoration activities have started at five sites of five Patagonian rivers: (Chos Malal, 37°23′49″ S, 70°16′15″ W; Plottier, 38°58′42″ S, 68°13′17″ W; Isla Jordan, 38°59′ 28″S, 67°59′22″W; Pichi Leufu, 41°08′37″S, 70°51′13″ W; Piedra Parada, 42°39′33″ S, 70° 05′43″ W). In each of them, plantations were carried out with the collaboration of the local community and different institutions (INTA, University of Flores, Province of Neuquén, Province of Río Negro, Province of Chubut, Intercultural Botanic Garden Bariloche, Botanic Garden Plottier, Gualjaina Village, Plottier City, Cipolletti City, Conesa City, Viedma City, Bariloche City, rural schools and private nurseries, among others).

Taking into account the first genetic diversity studies with molecular markers (nSSR) in some Patagonian rivers (Bozzi et al. 2012) and the recent report on genetic structuration of the species in Central Mexico (cpDNA, nSSR) (Hernández-Leal et al. 2019), we decided to consider all the individuals coming from the same river as a "population". This is because of the unidirectional gene flow that tends to homogenize the genetic information within the same river and that promotes genetic differentiation among rivers and hydrological basins (Hernández-Leal et al. 2019). The genetic material that is being used for restoration purposes in one river is there-fore a sampling of at least 30 to 50 clones coming from the same river or basin.

Since the vegetative propagation ability of *S. humboldtiana* is lower than that of other *Salix* species (Thomas et al. 2012), it was necessary to search for other pos-sible propagation methods, such as in vitro culture of plant tissues. The use of this technique in forest species has facilitated the cloning of selected phenotypes. It allows large-scale multiplication of clones in a short time, at any time of the year, with a reduced amount of tissue from very old dying trees, in a limited space, thus constituting a very important tool in conservation and germplasm exchange (Sharry et al. 2011).

Cloning can be done by somatic embryogenesis and adventitia organogenesis, two morphogenic processes widely used in in vitro culture of plant species. Preliminary results of an in vitro vegetative propagation assay by means of micro cuttings showed a rooting proportion of 55% of the explants (Marianelen Cedres, data not published). After 10 days, the material could grow with culture media with and without additional hormones. Differences were found according to the thick-ness of the micro cuttings (> 9 mm; between 9 and 5 mm, < 5 mm), thicker ones presenting more contamination.

In order to produce seeds of absolute specific purity, which can hardly be found in nature, the installation of clonal seed orchards is taken into account for advanced stages of the Programme. One orchard of 41 clones (with 8 replicates each) from the Chubut River has been already installed in Trevelin Forest Station in an isolated site, thus preventing contamination with exotic willows. The clonal bank at Bariloche Experimental Station with more than 300 clones has been started to be managed also to produce pure seeds. Two more seed orchards are planned at provincial nurseries too. The seeds of these orchards will allow producing seedlings of pure *S. humboldtiana* germplasm with new genotypes, thus enhancing the genetic diversity of future stands.

References

Aagesen DL (1998) Indigenous resource rights and conservation of the monkey-puzzle tree (*Araucaria araucana*, Araucariaceae): a case study from southern Chile. Econ Bot 52:146–160

Abdollahzadeh A, Kazempour S, Maassoumi AA (2011) Molecular phylogeny of the genus *Salix* (Salicaceae) with an emphasize to its species in Iran. Iran J Bot 17:244–253

Aguilera-Betti I, Muñoz AA, Stahle D, Figueroa G, Duarte F, González-Reyes A et al (2017) The first millennium-age *Araucaria Araucana* in Patagonia. Tree-Ring Res 73(4):53–56

Amico I, Orellana I (2014) Aportes al manejo de las invasiones de sauces en el Valle 16 de Octubre. Cartilla Técnica INTA: 107–114. https://inta.gob.ar/sites/default/files/script-tmp-inta_medio-ambiente24_manejo_y_control_de_sauces.pdf

Aparicio A, Pastorino M, Martinez-Meier A, Gallo L (2009) Propagación vegetativa del ciprés de la cordillera, una especie vulnerable del bosque subantártico de Sudamérica. Bosque 30:18–26

APN-SIB (2020) Administración de Parques Nacionales. Sistema de Información de Biodiversidad. sib.gob.ar. https://doi.org/10.4067/S0717-92002009000100004

Argus GW (1997) Infrageneric classification of *Salix* (Salicaceae) in the New World. Syst Bot Monogr 52:1

Arreghini R, Cerrillo T (1996) Willows in Argentina. 20th Session of the Poplar International Commission, Budapest, October 1–4 1996, pp 715–768

Azpilicueta MM, Marchelli P, Gallo LA (2009) The effects of quaternary glaciations in Patagonia as evidenced by chloroplast DNA phylogeography of southern beech *Nothofagus obliqua*. Tree Genet Genomes 5:561–571

Balian E, Naiman R (2005) Abundance and production of riparian trees in the lowland floodplain of the Queets River, Washington. Ecosystems 8:841–861

Bekessy SA, Allnutt TR, Premoli AC, Lara A, Ennos RA, Burgman MA et al (2002) Genetic variation in the monkey puzzle tree (*Araucaria araucana* (Molina) K.Koch), detected using RAPDs. Heredity 88:243–249

Bekessy SA, Ennos RA, Burgman MA, Newton AC, Ades PK (2003) Neutral DNA markers fail to detect genetic divergence in an ecologically important trait. Biol Conserv 110:267–275

Belyaeva I (2009) Nomenclature of *Salix fragilis* and a new species. Taxon 58:1344–1348

Belyaeva I, Epantchintseva O, Govaerts R, McGinn K, Hunnex J, Kuzovkina Y (2018) The application of scientific names to plants in cultivation: *Salix vitellina* L. and related taxa (Salicaceae). Skvortsovia 4:42–70

Benítez C (2005) Viabilidad de las semillas y crecimiento inicial de plántulas de *Araucaria araucana* (Mol.). K. Koch de la Cordillera de Nahuelbuta en la IX región de Chile. Tesis Fac. Ciencias Agropecuarias y Forestales, Universidad Católica de Temuco

Borodowski E (2017) Situación actual del cultivo y uso de las Salicáceas en Argentina. IV Congreso Internacional de Salicáceas en Argentina. La Plata, March 19–21, 2014

Bozzi J, Leyer I, Mengel C, Marchelli P, Ziegenhagen B, Thomas LK, Gallo LA (2012) Assessment of hybridization and introgression between the native *Salix humboldtiana* and invasive *Salix* species at the Rio Negro, Patagonia. Verh Ges Ökol 42. S.47

Bozzi J, Marchelli P, Gallo LA (2014) Sauce criollo: una especie nativa amenazada en Patagonia. Revista Presencia INTA 62:29–33

Bozzi J, Liepelt S, Ohneiser S, Gallo LA, Marchelli P, Leyer I, Ziegenhagen B, Mengel C (2015) Characterization of 23 polymorphic SSR markers in *Salix humboldtiana* (Salicaceae) using NGS and cross-amplification. Appl Plant Sci 3:1400120

Brea M, Mazzanti D, Martínez G (2014) Selección y uso de los recursos madereros en cazadores-recolectores de la transición Pleistoceno-Holoceno y Holoceno medio, sierras de Tandilia oriental. Argentina Rev Mus Argentino Cienc Nat 16:129–141

Budde KB, Gallo LA, Marchelli P, Mosner E, Liepelt S, Ziegenhagen B, Leyer I (2011) Wide spread invasion without sexual reproduction? A case study on European willows in Patagonia, Argentina. Biol Invasions 13:45–54

Cardemil L, Reinero A (1982) Changes of *Araucaria araucana* seed reserves during germination and early seedling growth. Can J Bot 60:1629–1638

Carlson SM, Cunningham CJ, Westley PAH (2014) Evolutionary rescue in a changing world. Trends in Ecology & Evolution Vol. 29 No. 9: 521–530.

Cerrillo T, Iribarren R, Cobas AC, Monteoliva S (2016) Evaluación xilológica de familias mejoradas de sauces con destino industrial maderero. Rev Fac Agron La Plata 115:99–106

Cerrillo T, Torales S, Pomponio F, Loval S, Dieta V (2018) Collection and characterization of the native species *Salix humboldtiana* Willd in the Delta of Paraná, towards their conservation and improvement. VII Simposio Internacional IUFRO del Álamo, Buenos Aires, Oct 28–Nov 4, 2018

Cerrillo T, Loval S, Casaubón E, Thomas E, Grande J, Monteoliva S (2019) Willow breeding for diversified and sustainable applications in Argentina. XXV Congreso Mundial IUFRO, Curitiba, Sept 29–Oct 10, 2019

Chaves A, Mugridge A, Fassola H, Alegranza D, Fernandez R (1999) Conservación refrigerada de semillas de *Araucaria angustifolia* (Bert.) O. Kuntze. Bosque 20:117–124

Compagnucci RH (2011) Atmospheric circulation over Patagonia from the Jurassic to present: a review through proxy data and climatic modelling scenarios. Biol J Linn Soc 103:229–249

CONAF (2013) Funciones alométricas para la determinación de existencias de carbono forestal para la especie *Araucaria araucana* (Molina) K. Koch. 50pp. Unidad de Cambio Climático. Gerencia Forestal Corporación Nacional Forestal. Programa Bosques PROCARBONO. Universidad Austral de Chile. Revisión Técnica: Angelo Sartori, Fabián Milla, Francoise Pincheira, Cristian Perez

Corander J, Marttinen P, Siren J, Tang J (2008) Enhanced Bayesian modelling in BAPS software for learning genetic structures of populations. BMC Bioinf 9:539

Cortés J, Ugalde I, Caviedes J, Ibarra JT (2019) Semillas de montaña: recolección, usos y comercialización del piñón de la araucaria (*Araucaria araucana*) por comunidades Mapuche-Pewenche del sur de los Andes. Pirineos 174:48. https://doi.org/10.3989/pirineos.2019.174008

Dapoto G, Giganti H (1994) Bioecología de *Nematus desantisi* Smith (Hymenoptera: Tenthredinidae: Nematinae) en las provincias de Río Negro y Neuquén (Argentina). Bosque 15:27–32

Datri LA, Faggi AM, Gallo LA, Carmona F (2016a) Half a century of changes in the riverine landscape of Limay River: the origin of a riparian neoecosystem in Patagonia (Argentina). Biol Invasions 18:1713–1722

Datri LA, Faggi AM, Gallo LA (2016b) Entre el orden y el caos: invasiones con dinámicas no lineales de sauces y álamos en el norte de la Patagonia. Rev Asoc Argent Ecol Paisajes 6:12–22

Datri LA, Maddio R, Faggi AM, Gallo LA (2017) A dendrogeomorphological study of the local effect of climate change. Eur Sci J 13:176–192

Diamond JM (1972) Biogeographic kinetics: estimation of relaxation times for Avifaunas of Southwest Pacific Islands. Proc Natl Acad Sci 69:3199–3203

Díaz S, Kitzberger T, Peris S (2012) Food resources and reproductive output of the Austral Parakeet (*Enicognathus ferrugineus*) in forests of northern Patagonia. Emu 112:234–243

Duplancic M, Martínez Carretero E, Cavagnaro B et al (2015) Factores que inciden en la germinación de *Araucaria araucana* (Araucariaceae) del bosque xérico. Rev FCA UNCUYO 47:71–82

Entraigas RA (1960) El fuerte del Río Negro. Librería Don Bosco, Buenos Aires, 308 pp.

Falkner Th (1774) A description of Patagonia and the adjoining parts of South America. The Associates of the John Carter Brown Library. 144 pp.

Farjon A, Page CN (1999) Conifers. Status survey and conservation action plan. IUCN/SSC Conifer Specialist Group. IUCN, Gland/Cambridge

Flint RF, Fidalgo F (1964) Glacial geology of the east flank of the Argentine Andes between latitude 39°10'S and latitude 41° 20'S. Geol Soc Am Bull 75:335–352

Flora Argentina (2014). www.floraargentina.edu.ar. Instituto Darwinion

Folguera A, Orts D, Spagnuolo M, Vera ER, Litvak V, Sagripanti L et al (2011) A review of late cretaceous to quaternary palaeogeography of the southern Andes. Biol J Linn Soc 103:250–268

Frankham R (2015) Genetic rescue of small-inbred populations: meta-analysis reveals large and consistent benefits of gene flow. Mol Ecol 24:2610–2618

Gallo L, Izquierdo F, Sanguinetti LJ (2004a) *Araucaria araucana* forest genetic resources in Argentina. In: Vinceti B, Amaral W, Meilleur M (eds) Challenges in managing forest genetic resources for livelihoods: examples from Argentina and Brazil. IPGRI, Rome, pp 105–131

Gallo LA, Letourneau F, Vinceti B (2004b) A modelling case study: options for Forest. Genetic resources management in *Araucaria araucana* ecosystems. In: Vinceti B, Amaral W, Meilleur B (eds) Challenges in managing forest genetic resources for livelihoods. Examples from Argentina and Brazil. International Plant Genetic Resouces Institute, Rome, pp 187–210

Gallo LA, Martinez A, Bozzi J, Amico I, Hansen M (2016) Hacia el rescate genético del sauce criollo (*Salix humboldtiana*). Programa para su conservación y la restauración de ecosistemas ribereños patagónicos. Revista Presencia INTA 66:13–17

Gallo LA, Amico I, Martinez A, Hansen M (2018) Participatory genetic rescue of severely menaced populations of *Salix humboldtiana* Willd. In: Patagonia. VII Simposio Internacional IUFRO del Álamo, Buenos Aires, Oct 28–Nov 4, 2018

Gómez Otero J, Bellelli C (2006) Arqueología de la provincia del Chubut. In: Bandieri S (ed) Patagonia Total. Primera Parte: Historias de la Patagonia. Sociedades y espacios en el tiempo. Barcelbaires Ediciones y Alfa Centro Literario, Barcelona, pp 27–51

Gordo Alonso J, Mutke Regneri S, Gil Sánches L (2007) Ausencia de diferenciación ecotípica entre rodales selectos de pino piñonero en la cuenca del Duero. Investig Agrar Sist y Recur For 16:253–261

Gray D, Scarsbrook M, Harding J (2006) Spatial biodiversity patterns in a large New Zealand braided river. N Z J Mar Fresh 40:631–642

Guadaño C, Iglesias S, Leon D et al (2016) Establecimiento de plantaciones clonales de *Pinus pinea* para la producción de piñón mediterráneo. Monogr INIA Ser For 28:79

Gut B (2008) Trees in Patagonia. Birkhäuser, Basel, 283pp

Gutiérrez B (2019) Análisis de la emergencia de plántulas durante la viverización de una colección de semillas de 418 familias de *Araucaria araucana*. Cienc e Investig For 25:21–38

Hadad MA (2014) Efecto del clima en los anillos de crecimiento de *Araucaria araucana* en el norte de la Patagonia Argentina. Ecosistemas 23:109–111

Hadad MA, Roig Juñent FA, Boninsegna JA, Patón D (2015) Age effects on the climatic signal in *Araucaria araucana* from xeric sites in Patagonia, Argentina. Plant Ecol Divers 8:343–351

Hampe A, Petit RJ (2005) Conserving biodiversity under climate change: the rear edge matters. Ecol Lett 8:461–467

Hanski I (1998) Metapopulation dynamics. Nature 396:41–49

Hanski I, Ovaskainen O (2002) Extinction debt at extinction threshold. Conserv Biol 16:666–673

Hartl DL, Clark AG (1989) Principles of population genetics. Sinauer Associates, Sunderland

Hauman L, Burkhart A, Parodi LR, Cabrera AL (1947) La vegetación de la Argentina. Sociedad Argentina de Estudios Geográficos, Buenos Aires

Hernández-Leal MS, Suarez-Atilanos M, Piñero D, Gonzalez-Rodriguez A (2019) Regional patterns of genetic structure and environmental differentiation in willow populations (*Salix humboldtiana* Willd.) from Central Mexico. Ecol Evol 00:1–16

Herrmann TM (2006) Indigenous knowledge and management of *Araucaria araucana* forest in the Chilean Andes: implications for native forest conservation. Biodivers Conserv 15:647–662

Heusser CJ, Lowell TV, Heusser LE et al (1996) Full-glacial – late-glacial paleoclimate of the southern Andes: evidence from pollen, beetle and glacial records. J Quat Sci 11:173–184

Heusser CJ, Heusser LE, Lowell TV (1999) Paleoecology of the southern chilean lake district-Isla Grande de Chilo, during middle-late Llanquihue glaciation and deglaciation. Geogr Ann 81A :231–284

Hill RS, Enright NJ, Hill RS (1995) Conifer origin, evolution and diversification in the southern hemisphere. In: Ecology of the southern conifers. Melbourne University Press, Melbourne, pp 10–29

Hollin JT, Schilling DH (1981) Late Wisconsin-Weichselian mountain glaciers and small ice caps. In: Denton GH, Hughes TJ (eds) The last great ice sheets. Willey, New York/Chichester/Brisbane/Toronto, pp 179–206

Huang CL, Chang CT, Huang BH, Chung JD, Chen JH, Chiang YC, Hwang SY (2015) Genetic relationships and ecological divergence in *Salix* species and populations in Taiwan. Tree Genet Genomes 11:39

Ipinza R (2017) Migración Asistida. Una opción para la Conservación de la Araucaria. In: Reunión Internacional Daño Foliar de *Araucaria araucana*. Villarrica, Chile

Ipinza R, Gutiérrez B, Muller-Using S et al (2019) La migración asistida de la *Araucaria araucana*. Plan operacional. Cienc e Investig For INFOR 25:75–88

Isebrands JG, Richardson J (2014) Poplars and willows: trees for society and the environment. Ed. Isebrands JG, Richardson J. CABI/ FAO, 634 pp.

IUCN (2020) The IUCN red list of threatened species. Version 2020–1. https://www.iucnredlist.org

Izquierdo F (2009) Análisis de la diversidad y diferenciación genética del pehuén (*Araucaria araucana*). MSc thesis, Fac. de Agronomía, Universidad de Buenos Aires

Izquierdo F, Gallo LA (2006) Variación de la altura promedio en plantines de poblaciones argentinas de *Araucaria araucana* en relación al peso de sus semillas. In: IUFROLAT, segundo congreso Latinoamericano IUFRO. La serena, Chile

Jombart T, Devillard S, Balloux F (2010) Discriminant analysis of principal components: a new method for the analysis of genetically structured populations. BMC Genet 11:94

Kaur D, Bhatnagar S (1984) Fertilization and formation of neocytoplasm in *Agathis robusta*. Phytomorphology 34:56–60

Kershaw P, Wagstaff B (2001) The southern conifer family Araucariaceae: history, status, and value for Paleoenvironmental reconstruction. Annu Rev Ecol Syst 32:397–414

Kershaw AP, McGlone M, Enright NJ, Hill RS (1995) The quaternary history of the southern conifers. In: Ecology of the southern conifers. Melbourne University Press, Melbourne, pp 30–63

Kitzberger T (2013) Ecotones as complex arenas of disturbance, climate, and human impacts: the trans-Andean forest-steppe ecotone of northern Patagonia. In: Ecotones between forest and grassland. Springer, New York, pp 59–88

Krauss J, Bommarco R, Guardiola M et al (2010) Habitat fragmentation causes immediate and time-delayed biodiversity loss at different trophic levels. Ecol Lett 13:597–605

Kuussaari M, Bommarco R, Heikkinen RK et al (2009) Extinction debt: a challenge for biodiversity conservation. Trends Ecol Evol 24:564–571

Lande R, Shannon S (1996) The role of genetic variation in adaptation and population persistence in a changing environment. Evolution 50:434–437

Larroulet A, Résico C, Arbeletche G, Benmuyal L, Bejar W (2011) Usos no madereros de *Salix humboldtiana*. III Congreso Internacional de Salicáceas en Argentina. Neuquén, March 14–19, 2011

Lauron-Moreau A, Pitre FE, Argus GW, Labrecque M, Brouillet L (2015) Phylogenetic relationships of American willows (*Salix* L., Salicaceae). PLoS One 10:e0121965

Lewerentz A, Egger G, Householder JE, Reid B, Braun AC, Garófano-Gómez V (2019) Functional assessment of invasive *Salix fragilis* L. in north-western Patagonian flood plains: a comparative approach. Acta Oecol 95:36–44

Lin J, Gibbs JP, Smart LB (2009) Population genetic structure of native versus naturalized sympatric shrub willows (*Salix*; Salicaceae). Am J Bot 96:771–785

López H, Jiménez G, Reyes B (1986) Algunos antecedentes sobre cosecha, procesamiento y viverización de varias especies nativas. Revista Chile forestal. Documento técnico N°5

Lowe A, Harris S, Ashton P (2004) Ecological genetics: design, analysis, and application. Wiley-Blackwell, 344 pp.

Marchelli P, Gallo L, Scholz F, Ziegenhagen B (1998) Chloroplast DNA markers reveal a geographical divide across Argentinean southern beech *Nothofagus nervosa* (Phil.) dim. Et mil. distribution area. Theor Appl Genet 97:642–646

Marchelli P, Baier C, Mengel C, Ziegenhagen B, Gallo LA (2010) Biogeographic history of the threatened species *Araucaria araucana* (Molina) K. Koch and implications for conservation: a case study with organelle DNA markers. Conserv Genet 11:951–963

Marconi G, Martín MA, Cherubini M et al (2011) Primer note: microsatellite-AFLP development for *Araucaria araucana* (Mol.) K. Koch, an endangered conifer of Chilean and Argentinean native forests. Silivae Genet 60:285–288

Markgraf V, Romero EJ, Villagran C, Veblen TT, Hill RS, Read J (1996) History and paleoecology of south American *Nothofagus* forests. In: The ecology and biogeography of *Nothofagus* forests. Yale University Press, New Haven/London, pp 354–386

Martín MA, Mattioni C, Lusini I, Drake F, Cherubini M, Herrera MA et al (2012) Microsatellite development for the relictual conifer *Araucaria araucana* (Araucariaceae) using next-generation sequencing. Am J Bot 99:e213–e215

Martín MA, Mattioni C, Lusini I, Molina JR, Cherubini M, Drake F et al (2014) New insights into the genetic structure of *Araucaria araucana* forests based on molecular and historic evidences. Tree Genet Genomes 10:839–851

Mazzuchelli M, Sanguinetti J, Catalán M, González Musa R, Ceballos S, Szychowski A (2016) Ensayo de siembra aérea de *Araucaria araucana* con helicóptero en el área Moquehue, Provincia del Neuquén. In: V Jornadas Forestal Patagónicas. Esquel, Argentina

Menendez E (2012) Revisión del género *Salix* (Salicaceae) en la Provincia de Mendoza. Argentina Rev Fac Cs A UN Cuyo 44:157–192

Milesi FA, Guichón ML, Monteverde MJ, Piudo L, Sanguinetti J (2017) Ecological consequences of an unusual simultaneous masting of *Araucaria araucana* and *Chusquea culeou* in North-West Patagonia, Argentina. Austral Ecol 42:711–722

Montané J (1968) Paleo-indian remains from Laguna de Tagua Tagua, Central Chile. Science 161:1137–1138

Moreno FP (1876) Viaje a la Patagonia Austral. Sociedad de Abogados Editores. Administrador: Aldo de Rosso. Sarmiento 1411, Buenos Aires. 240pp

Moreno P (1997) Vegetation and climate near Lago Llanquihue in the Chilean Lake district between 20200 and 9500 14 C yr BP. J Quat Sci 12:485–500

Moreno AC (2012) Estudio del flujo génico mediado por polen en poblaciones fragmentadas de *Araucaria araucana*. Doctoral thesis, Universidad Nacional del Comahue

Moreno AC, Marchelli P, Gallo LA (2009) Distancia de polinización en dos especies anemófilas patagónicas. In: Congreso Forestal Mundial. Buenos Aires, Argentina

Moreno AC, Marchelli P, Vendramin GG, Gallo LA (2011) Cross transferability of SSRs to five species of araucariaceae: a useful tool for population genetic studies in araucaria araucana. For Syst 20:303–314

Mosner E, Liepelt S, Ziegenhagen B, Leyer I (2012) Floodplain willows in fragmented river landscapes: understanding spatio-temporal genetic patterns as a basis for restoration plantings. Biol Conserv 153:211–218

Mundo IA, Juñent FAR, Villalba R, Kitzberger T, Barrera MD (2012) *Araucaria araucana* tree-ring chronologies in Argentina: spatial growth variations and climate influences. Trees Struct Funct 26:443–458

Mundo IA, Kitzberger T, Roig Juñent FA, Villalba R, Barrera MD (2013) Fire history in the *Araucaria araucana* forests of Argentina: human and climate influences. Int J Wildl Fire 22:194

Muñoz F (2010) Evaluación del almacenamiento, germinación de semillas y producción de plantas de *Araucaria araucana* (Mol.) K. Koch, procedentes de la comuna de Lonquimay, IX Región. Universidad de Chile

Muñoz AA, Barichivich J, Christie DA, Dorigo W, Sauchyn D, González-Reyes A et al (2014) Patterns and drivers of *Araucaria araucana* forest growth along a biophysical gradient in the northern Patagonian Andes: linking tree rings with satellite observations of soil moisture. Aust Ecol 39:158–169

Mutke Regneri S, Iglesias Sauce S, Gil Sánchez L (2007) Selection of Mediterranean stone pine clones for cone production. For Syst 16:39–51

Orellana IA, Amico I, Fasanella M, Pildain MB, Premoli A, Bonansea TB (2016) Evaluación y propuesta de manejo de la invasión de sauces en el noroeste de la Provincia del Chubut. In: Gingins M, Álvarez G, Llavallol CI (eds) Investigación Forestal 2011–2015: Los Proyectos de Investigación Aplicada. Min Agri. – UCAR, Buenos Aires, pp 75–77. https://es.scribd.com/doc/299077824/Investigacion-Forestal-2011-2015-Los-Proyectos-de-Investigacion-Aplicada

Parolin P, Oliveira AC, Piedade MTF, Wittmann F, Junk WJ (2002) Pioneer trees in Amazonian floodplains: three key species form monospecific stands in different habitats. Fol Geobot 37:225–238

Peakall R, Ebert D, Scott LJ, Meagher PF, Offord CA (2003) Comparative genetic study confirms exceptionally low genetic variation in the ancient and endangered relictual conifer, *Wollemia nobilis* (Araucariaceae). Mol Ecol 12:2331–2343

Petit RJ, Vendramin GG (2006) Plant phylogeography based on organelle genes: an introduction. In: Weiss S, Ferrand N (eds) Phylogeography of southern European refugia. Springer, Dordrecht, p 93

Pomponio F, Cerrillo T, Gallo LA, López M, Torales S (2018) Populations genetic diversity analysis as support for Native Willow Rescue in Argentina. VII Simposio Internacional IUFRO del Álamo, Buenos Aires, Oct 28–Nov 4, 2018

Premoli AC, Kitzberger T, Veblen TT (2000) Isozyme variation and recent biogeographical history of the long-lived conifer *Fitzroya cupressoides*. J Biogeogr 27:251–260

Premoli A, Quiroga P, Gardner M (2013) *Araucaria araucana*. The IUCN red list of threatened species

Rabassa J, Clapperton CM (1990) Quaternary glaciations in the southern Andes. Quat Sci Rev 9:153–174

Ragonese AF, Rial Alberti F (1958a) Sauces híbridos originados naturalmente en la Rep. Argentina Rev Inv Agric 12: 11–153, 17f, 22l

Ragonese AF, Rial Alberti F (1958b) Mejoramiento de sauces en la República Argentina. Rev Inv Agric 12:225–246

Rechene C, Bava JO, Mujica R (2003) Los bosques de *Araucaria araucana* en Chile y Argentina. Programa de Apoyo Ecológico (TOEB), GTZ, Agencia de Cooperación Alemana, Informe TWF-V/40s, Eschborn, Alemania

Robledo-Arnuncio JJ, Alia R, Gil L (2004) Increased selfing and correlated paternity in a small population of a predominantly outcrossing conifer, *Pinus sylvestris*. Mol Ecol 13:2567–2577

Rodriguez MF (1999) Arqueobotánica de Quebrada Seca 3 (Puna Meridional Argentina): Especies vegetales utilizadas en la confección de artefactos durante el aracaico. Relaciones de la Sociedad Argentina de Antropología XXIV, Buenos Aires

Rodríguez MF (2000) Woody plant species used during the Archaic period in the southern argentine Puna. Archaeobotany of Quebrada Seca 3. J Archaeol Sci 27:341–361

Rohwer JG, Kubitzki K (1984) *Salix martiana*, a regularly hermaphrodite willow. Plant Syst Evol 144:99–101

Rosso C, Scarpa G (2019) Etnobotánica médica moqoit y su comparación con grupos criollos del Chaco argentino. Bol Soc Argent Bot 54:637–662

Ruiz E, González F, Torres-Diaz C, Fuentes G, Mardones M, Stuessy T et al (2007) Genetic diversity and differentiation within and among Chilean populations of *Araucaria araucana* (Araucariaceae) based on allozyme variability. Taxon 56:1221–1228

Saavedra A, Willhite E (2017) Observaciones y recomendaciones relacionadas con la muerte de ramas y follaje (daño foliar de la Araucaria) en *Araucaria araucana* en los parques nacionales del Sur. Centro de Chile, Washington DC

Salgado M, Puntieri JG, Gallo LA (2018) Architectural development and phenotypic plasticity in *Salix humboldtiana* Willd. from Patagonia. VII Simposio Internacional IUFRO del Álamo, Buenos Aires, Oct 28–Nov 4, 2018

Salgado M, Puntieri JG, Gallo LA (2019) Variabilidad arquitectural de *Salix humboldtiana* Willd. en el borde sur de su distribución. XXXVII Jornadas Argentinas de Botánica. Tucumán, Sept 9–13, 2019

Sanguinetti J (2008) Producción y predación de semillas, efectos de corto y largo plazo sobre el reclutamiento de plántulas. Caso de Estudio: *Araucaria araucana*. Doctoral thesis, Universidad Nacional del Comahue, Bariloche

Sanguinetti J (2014) Producción de semillas de *Araucaria araucana* (Molina) K. Koch durante 15 años en diferentes poblaciones del Parque Nacional Lanín (Neuquén-Argentina). Ecol Austral 24:165–175

Sanguinetti J, Kitzberger T (2008) Patterns and mechanisms of masting in the large-seeded southern hemisphere conifer *Araucaria araucana*. Austral Ecol 33:78–87

Sanguinetti J, Kitzberger T (2009) Efectos de la producción de semillas y la heterogeneidad vegetal sobre la supervivencia de semillas y el patrón espacio-temporal de establecimiento de plántulas en *Araucaria araucana*. Rev Chil Hist Nat 82:319–335

Sanguinetti J, Kitzberger T (2010) Factors controlling seed predation by rodents and non-native Sus scrofa in *Araucaria araucana* forests: potential effects on seedling establishment. Biol Invasions 12:689–706

Schleuning M, Becker T, Vadillo GP, Hahn T, Matthies D, Durka W (2011) River dynamics shape clonal diversity and genetic structure of an Amazonian understorey herb. J Ecol 99:373–382

Schlögl PS, de Souza AP, Nodari RO (2007) PCR-RFLP analysis of non-coding regions of cpDNA in *Araucaria angustifolia* (Bert.) O. Kuntze. Genet Mol Biol 30:423–427

Scott LJ, Sheperd MJ, Nikles DG, Henry RJ (2005) Low efficiency of pseudotestcross mapping design was consistent with limited genetic diversity and low heterozygosity in hoop pine (*Araucaria cunninghamii,* Araucariaceae). Tree Genet Genomes 1:124–134

Sharry S, Adema M, Basiglio Cordal MA, Villarreal B, Nikoloff N, Briones V, Abedini W (2011) Propagation and conservation of native Forest genetic resources of medicinal use by means of in vitro and ex vitro techniques. Nat Prod Commun 6:985–988

Shepherd J, Ditgen R (2005) Human use and small mammals communities of *Araucaria* forests in Neuquén, Argentina. Mastozoología Neotrop 12(217):226

Shepherd J, Ditgen R (2012) Predation by *Rattus norvegicus* on a native small mammal in an *Araucaria araucana* forest of Neuquén, Argentina. Rev Chil Hist Nat 85:155–159

Shepherd JD, Ditgen RS (2013) Rodent handling of *Araucaria araucana* seeds. Aust Ecol 38:23–32

Shepherd JD, Ditgen RS (2016) Small mammals and microhabitats in *Araucaria* forests of Neuquén, Argentina. Mastozoología Neotrop 23:467–482

Shepherd JD, Ditgen RS, Sanguinetti J (2008) *Araucaria araucana* and the austral parakeet: Predispersal seed predation on a masting species. Rev Chil Hist Nat 81:395–401

Smouse PE, Dyer RJ, Westfall RD, Sork VL (2001) Two-generation analysis of pollen flow across a landscape. I. Male gamete heterogeneity among females. Evolution (NY) 55:260–271

Sopow SL, Jones T, McIvor I, McLean JA, Pawson SM (2017) Potential impacts of *Tuberolachnus salignus* (giant willow aphid) in New Zealand and options for control. Agric For Entomol 19:225–234

Speziale KL, Lambertucci SA, Gleiser G, Tella JL, Hiraldo F, Aizen MA (2018) An overlooked plant-parakeet mutualism counteracts human overharvesting on an endangered tree. R Soc Open Sci 5:171456

Szymañski C, Fontana G, Sanguinetti J (2017) Natural and anthropogenic influences on coarse woody debris stocks in *Nothofagus - Araucaria* forests of northern Patagonia, Argentina. Austral Ecol 42:48–60

Tamura S, Kudo G (2000) Wind pollination and insect pollination of two temperate willow species, *Salix miyabeana* and *Salix sachalinensis*. Plant Ecol 147:185–192

Tella JL, Lambertucci SA, Speziale KL, Hiraldo F (2016) Large-scale impacts of multiple co-occurring invaders on monkey puzzle forest regeneration, native seed predators and their ecological interactions. Glob Ecol Conserv 6:1–15

Templeton AR, Robertson RJ, Brisson J, Strasburg J (2001) Disrupting evolutionary processes: the effect of habitat fragmentation on collared lizards in the Missouri Ozarks. Proc Natl Acad Sci USA 98:5426–5432

Thomas L, Leyer I (2014) Age structure, growth performance and composition of native and invasive Salicaceae in Patagonia. Plant Ecol 215:1047–1056

Thomas L, Tölle L, Ziegenhagen B, Leyer I (2012) Are vegetative reproduction capacities the cause of wide spread invasion of Eurasian Salicaceae in Patagonian river landscapes? PLoS One 7:e50652

Tilman D, May RM, Lehman CL, Nowak MA (1994) Habitat destruction and the extinction debt. Nature 371:65–66

Tortorelli L (2009) Maderas y bosques argentinos. Orientación Gráfica Editora, Buenos Aires

Veblen TT (1982) Regeneration patterns in *Araucaria araucana* forests in Chile. J Biogeogr 9:11

Veblen TT, Burns BR, Kitzberger T, Lara A, Villalba R, Enright NJ, Hill RS (1995) The ecology of the conifers of southern south america. In: Ecology of the southern conifers. Melbourne University Press, Melbourne, pp 120–155

Vélez M, Salgado Salomón M, Marfetan A, Tirante SI, Mattes Fernández H, Avila M, et al (2018) Caracterización de la desecación del dosel y sanidad de *Araucaria araucana* en Argentina. Technical report, 32pp

Vendramin GG, Anzidei M, Madaghiele A, Bucci G (1998) Distribution of genetic diversity in *Pinus pinaster* Ait. As revealed by chloroplast microsatellites. Theor Appl Genet 97:456–463

Villagran C (1991) Historia de los bosques templados del sur de Chile durante el Tardiglacial y Postglacial. Rev Chil Hist Nat 64:447–460

Villalba R, Boninsegna J, Cobos D (1989) A tree-ring reconstruction of summer temperature between AD. 1500 and 1974 in western Argentina. In: Extended abstracts, third international conference on southern hemisphere meteorology & oceanography. American Meteorological Society, Buenos Aires, pp 196–197

Wilmking M, Buras A, Eusemann P, Schnittler M, Trouillier M, Würt D et al (2017) High frequency growth variability of white spruce clones does not differ from non-clonal trees at Alaskan treelines. Dendrochronologia 44:187–192

Wolfe KH, Li WH, Sharp PM (1987) Rates of nucleotide substitution vary greatly among plant mitochondrial, chloroplast, and nuclear DNAs. Proc Natl Acad Sci USA 84:9054–9058

Wu J, Nyman T, Wang D-C, Argus GW, Yang Y-P, Chen J-H (2015) Phylogeny of *Salix* subgenus *Salix* s.l. (Salicaceae): delimitation, biogeography, and reticulate evolution. Evol Biol 15:31

Wu Q, Liang X, Dail X, Chen Y, Yin T (2018) Molecular discrimination and ploidy level determination for elite willow cultivars. Tree Genet Genomes 14:65

Yacubson N (1967) Nota preliminar sobre siembra aérea en la cordillera Patagónica. Notas Silvícolas Adm Nac Bosques Dir Investig For Buenos Aires, Argentina 23, pp 1–3

Zamorano-Elgueta C, Cayuela L, González-Espinosa M, Lara A, Parra-Vázquez MR (2012) Impacts of cattle on the south American temperate forests: challenges for the conservation of the endangered monkey puzzle tree (*Araucaria araucana*) in Chile. Biol Conserv 152:110–118

Zhang P, Deng X, Long A, Xu H, Ye M, Li J (2019) Patterns and regeneration of *Populus euphratica* under different surface soil salinity conditions. Sci Rep 9:9123

Part II
Subtropical Dry Forests

Chapter 8
Subtropical Dry Forests: The Main Forest Ecoregion of Argentina

Aníbal Verga and Diego López Lauenstein

8.1 Main Physiographic and Physiognomic Features of the Argentine Chaco

According to the Royal Spanish Academy (https://dle.rae.es/chaco), the word "chaco" is derived from the Quechua word "chacu." "Chacu" refers to a type of hunting historically done by indigenous communities in South America in which hunters would circle around the targeted animal before closing in to kill it. This etymology reflects the cultural rather than geographical meaning of the term, highlighting that the Chaco is not merely a territory, but the vital space of an ancestral human group.

The Great American Chaco is a sparsely populated, lowland natural region of the Río de la Plata basin, of approximately 850,000 km² divided among eastern Bolivia, western Paraguay, northern Argentina, and a small portion of southern Brazil. It extends into both tropical and temperate zones and is one of the major wooded grasslands in South America (Pacheco Balanza et al. 2012; Spensley et al. 2013). The Argentine portion occupies approximately 60% of the total surface of the Great American Chaco, and for the country, it represents the largest native forest area. The rainfall regime is monsoonal, with concentrated rains during the summer (between October and March). The entire region is between the mean annual precipitation contour lines of 1200 and 300 mm (Fig. 8.1). In general, the gradient goes from the

A. Verga (✉)
Agencia de Extensión Rural INTA, La Rioja, Argentina

Instituto de Fisiología y Recursos Genéticos Vegetales, Unidad de Estudios Agropecuarios (IFRGV, UDEA) INTA-CONICET, Córdoba, Argentina
e-mail: verga.anibal@inta.gob.ar

D. López Lauenstein
Instituto de Fisiología y Recursos Genéticos Vegetales, Unidad de Estudios Agropecuarios (IFRGV, UDEA) INTA-CONICET, Córdoba, Argentina

© Springer Nature Switzerland AG 2021
M. J. Pastorino, P. Marchelli (eds.), *Low Intensity Breeding of Native Forest Trees in Argentina*, https://doi.org/10.1007/978-3-030-56462-9_8

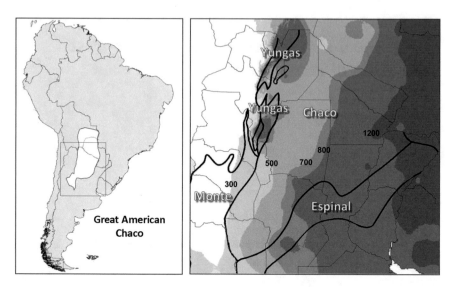

Fig. 8.1 Great American Chaco in South America and the Argentinean sector considered in this Chapter (**a**) (Spensley et al. 2013). Argentine Chaco and its neighbor phytogeographic regions (Cabrera 1976) over a map of mean annual precipitation contour plots (Hijmans et al. 2005) (**b**)

most humid sites in the east, to the driest ones in the west and is the main factor in defining the subregions and productive activities, also marking a steep climatic gradient (Spensley et al. 2013).

The humid Chaco subregion is the area between 1200 and 700–800 mm of annual precipitation, the semi-arid or subhumid Chaco is that between 700–800 mm and 500 mm, while the arid Chaco subregion lies between 500 and 300 mm. On the other hand, the Chaco Serrano region includes the Chaco vegetation that ascends the slopes of the Pampas Sierras located on the extreme west and southwest (Cabrera 1976).

Except for the Pampas Sierras, the Chaco is an enormous plain with a very low slope, without relief, which results in the existence of very few rivers, with erratic behavior, modeling soils and microreliefs far beyond their current courses (Naumann 2006). In this enormous plain there are only four important rivers: Pilcomayo, Bermejo, and Salado, which are part of the Río de la Plata basin, and Dulce, which empties into the huge but shallow salt lake called Mar Chiquita (the end of this endorheic basin). Although the Paraná and Paraguay rivers delimit the humid Chaco region to the east, their vegetation corresponds to gallery forests that descend from the Alto Paraná Rainforest or to large wetlands. These riverside forests mix elements from both phytogeographic regions.

In the northwest, the semi-arid Chaco subregion borders the Yungas Rainforest, one of the most diverse ecosystems in the world that ranges from northern Argentina to Venezuela on the humid and cloudy slopes of the Andes. In this area, rainfall on the Chaco plain begins to increase toward the west, inverting the gradient. The arid Chaco is located to the southwest, which limits to the west with the semi-desert region of Monte, with rainfall less than 300 mm. On the other hand, the semi-arid Chaco and the humid Chaco border the Espinal region to the south, which is a transition area with the Pampas region (Cabrera 1976; Naumann 2006).

From a physiognomic point of view, the Chaco region is characterized by the presence of an open semi-deciduous xerophytic forest, with dominance of "quebrachos" (*Schinopsis* spp. and *Aspidosperma quebracho-blanco*), interspersed with large bushy areas caused by fire, clearing, and subsequent abandonment (where species of the *Acacia* genus dominate), and "low" zones (i.e., old river beds, microrelief) with more closed forests dominated by "algarrobos" (*Prosopis* spp.) or *Enterolobium contortisiliquum* (mainly to the north). In floodable areas *Prosopis ruscifolia* dominates, and on the humid Chaco palm groves can be found, with grasslands and algarrobos as secondary woody species (Cabrera 1976; Brown and Pacheco 2006; Oyarzabal et al. 2018). On the other hand, the Chaco is a region deeply modified by human activity, mainly by selective logging, overgrazing, fire, and deforestation.

The Chaco soils are, in general, heavy and with hydromorphism problems to the east, while the central strip, between the isohyets of 700 and 800 mm, presents developed soils, some with good agricultural quality (locally known as the "agricultural dome" or "Chaco dome"). To the west, in the semi-arid and arid Chaco, soils are less developed, in some areas with excessive drainage or salinity problems, typical of arid and semi-arid regions (Giménez and Moglia 2003a, b; Panigatti 2010).

The forests of the Chaco are currently in a state of great vulnerability to climate change. Considering a moderate scenario, such as A2 according to the report of the Intergovernmental Panel on Climate Change (IPCC 2014), a warming is projected for the entire region with respect to the period 1960–2010 that ranges from 0 to 1 °C in the near future (year 2050) and up to 2.5–3.5 °C in the distant future (year 2100) in the north of the region. Regarding rainfall, no major changes in the average annual precipitation are foreseeing. However, the greatest climatic risk would come from the variability of rainfall magnified by the change in land use and the high probability of extreme drought events (MAyDS 2015).

The four Chaco subregions are characterized by the dominance of some of the quebracho forest species, emblematic both from a floristic and a cultural point of view. The humid Chaco is characterized by the presence of *Schinopsis balansae*, the semi-arid Chaco by *Schinopsis lorentzii* and the Chaco serrano by *Schinopsis haenkeana*. In the arid Chaco, these three species of red quebracho disappear and only the white quebracho *Aspidosperma quebracho-blanco* becomes dominant, whose distribution covers the entire Chaco region (Morello and Adámoli 1974; Cabrera 1976). In the strip that separates the humid from the semi-arid Chaco (700–800 mm of average annual precipitation, agricultural quality soils), a particular forest develops, known as "of the three quebrachos" (*S. balansae, S. lorentzii,* and *A. quebracho-blanco*). Also in this strip grows *Schinopsis heterophylla*, a putative hybrid between *S. balansae* and *S. lorentzii* (Muñoz 2000).

Along with quebrachos, algarrobos (*Prosopis* spp.) complete the most emblematic native forest species in the region. According to the traditional taxonomy of the genus, both the humid and semi-arid Chaco are characterized by *Prosopis alba* and

Prosopis nigra, while *Prosopis chilensis* and *Prosopis flexuosa* appear in the arid Chaco (Galera 2000). However (see Chap. 9), genetic, taxonomic, and ecophysiological studies identify important differences between populations of *P. alba* distributed in the semi-arid and humid Chaco (Verga et al. 2009; Verga 2014). On the other hand, a fifth species of algarrobo stands out in the north of the humid Chaco subregion: *Prosopis hassleri*. In Argentina it occupies only a strip of no more than 50 km wide south of the Pilcomayo River (which forms the border with Paraguay), ranging from the Paraguay River to the limit with the semi-arid Chaco. This is the southern end of the distribution area of this species, whose center is located in the Paraguayan humid Chaco (Burkart 1976; Kees et al. 2011). Interspecific hybrids also occur between the different *Prosopis* species, giving rise to hybrid swarms in the contact areas, the magnitude and extent of which would be influenced by the degree of natural or anthropogenic disturbance of the environment they occupy (Saidman et al. 2000; Mottura et al. 2005; Verga and Gregorius 2007).

In a dendrological study, Giménez and Moglia (2003a, b) identified 83 native forest tree species (54 of them with at least 20 m of mean height) in 17 sampling areas of the Chaco. In turn, the National Institute of Industrial Technology of Argentina (INTI) highlighted 29 of them as of industrial interest. Among the most important species due to their wood quality, chemical extracts, natural range, ecological role, silviculture potential, and even cultural value, the following must be mentioned: the red quebrachos (*Schinopsis* spp.), *Prosopis kuntzei*, *Gonopterodendron sarmientoi*, *Caesalpinia paraguariensis*, *Handroanthus heptaphyllus*, algarrobos (*Prosopis* spp.), *Ziziphus mistol*, *Aspidosperma quebracho-blanco*, and *Enterolobium contortisiliquum*.

8.2   Chaco and Deforestation: Two Words That Have Become Synonymous in the Last Two Decades

The Chaco is one of the ecosystems with the highest deforestation rates of the world (Hansen et al. 2013; Volante and Seghezzo 2018). The humid and dry Chaco subregions of Argentina have a great biological and productive diversity, cultural richness, as well as high levels of poverty and inequality (Paolasso et al. 2012). It has experienced drastic changes in land use in recent times, leading to a dramatic deforestation throughout its entire extension. Likewise, during the first half of the twentieth century, the forests of *S. balansae* were devastated for tannin and sleeper production. From the end of the twentieth century to the present day, in particular, the semi-arid Chaco has suffered an intense deforestation in order to extend livestock farming and agriculture.

The social actors that occupy this region include large-scale capitalized farmers, small-scale farmers ("puesteros'), indigenous communities and Mennonite settlers, among others (Baldi et al. 2015). These actors interact with the environment in multiple ways by capturing ecosystem services and modifying the environment (e.g., deforestation for industrial agriculture, thinning, and land clearings for livestock farming, selective logging, hunting and gathering). The rapid conversion of native forests to annual crops and pastures makes the Chaco a global deforestation hotspot (Hansen et al. 2013; Vallejos et al. 2015). This transformation brought an increase in the production of commodities (i.e., crops, meat) for the foreign markets, which are relevant for the national economy. However, it is also compromising the supply of multiple ecosystem services (Grau et al. 2005; Paruelo et al. 2011; Mastrangelo and Laterra 2015), creating social inequalities (Sacchi and Gasparri 2015; Laterra et al. 2019) and increasing social conflicts (REDAF 2012; Cáceres et al. 2015; Aguiar et al. 2017). Despite some legislation efforts (e.g., National Law N° 26,331), the territorial configuration responds more to an economic rationality, than to ecological, social and cultural criteria (Aguiar et al. 2017).

The vast majority of the deforested areas until 2018 (Fig. 8.2) has been transformed into rainfed agricultural lands, and to a lesser extent are being used for livestock or mixed production, or irrigated agriculture (Caballero et al. 2014; Vallejos et al. 2015). In addition, the remaining forest is affected by: (i) selective logging (historical and current), (ii) the increase in grazing load due to the displacement of livestock activity as a secondary effect of the expansion of the agricultural frontier, (iii) the use of fire by livestock producers, and (iv) the progress of urbanization, to a lesser extent.

From the end of the 1990s, with the introduction of soybean crop in the region, there was a general raise of land clearing with increasing intensity around the year 2000 (Fig. 8.2), especially toward drier areas (Fig. 8.3a, b). Medium or large landowners and lesser corporations (some of them supported by financial capitals) have mainly carried out this process (Baldi et al. 2019). Prior to that decade, in the middle

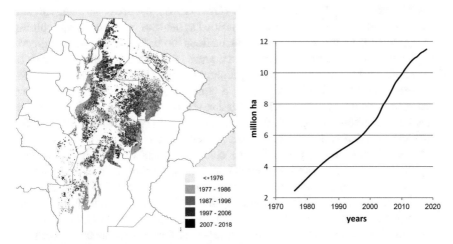

Fig. 8.2 Geographical distribution of land clearings in the last five decades until 2018 in the Argentine semi-arid Chaco (**a**). Accumulated cleared area in millions of ha in the same region and period (**b**). (Source: Deforestation monitoring project in the dry Chaco. REDAF, FAUBA, LART, INTA 2020. http://www.monitoreodesmonte.com.ar)

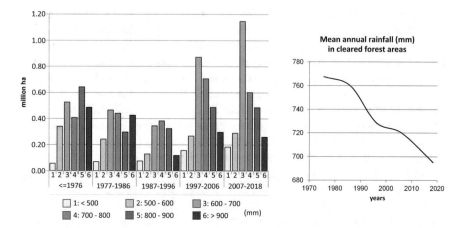

Fig. 8.3 Distribution of absolute frequencies in millions of ha of cleared areas in the semi-arid Chaco for the last five decades according to the annual mean precipitation range corresponding to each clearing site (**a**). Advance of clearing toward more arid areas (**b**). The graph indicates the average annual precipitation of the cleared areas until 2018. Own elaboration based on climatic information (Hijmans et al. 2005) and the vector layer of cleared areas (REDAF, FAUBA, LART, INTA 2020)

of the twentieth century, the clearing responded to the settlement of farmers. They received agricultural plots mostly in areas with rainfall greater than 800 mm and they mainly engaged in rainfed production of cotton (in the Province of Chaco) and beans (in the Province of Salta), using family labor (Galafassil 2005), or indigenous labor in larger cases (Iñigo Carrera 2013). Additionally, some clearing in more arid areas corresponded to livestock farming and irrigated agriculture.

According to an OEA report of 2009 (OEA, 2009), the areas between 750 and 900 mm of precipitation were already classified as being of high risk for agriculture, taking into account the strong annual fluctuations in rainfall, with alternating periods of high and low humidity. For cleared areas below 900 mm there is a risk of "reversion", that is, abandonment of agricultural activity that gives rise to a secondary forest called "fachinal", made up of a closed scrub, almost without grass and practically worthless for livestock production (Cozzo 1995). The OEA report analyzes land clearings during the 1996–2004 period and concludes that 30 % of the land cleared in that period presents a risk of "reversion".

As mentioned, due to global climate change, a slight increase in the average annual rainfall in the region is observed but at the same time a significant increase in the occurrence of extreme drought events and heavy rains (IPCC 2014; MAyDS 2015). In this context, we can consider with a high degree of probability that the clearings carried out below 700 mm in this region, in the medium term, will be unfeasible for agricultural activity. This effect would be a consequence not only of the fluctuations of the rains, but also of the deterioration of the soil that the dominant agricultural model in the region entails (Vallejos et al. 2017). It is proven that the soils subjected to this type of exploitation suffer significant losses of organic matter, accompanied by a significant impoverishment of their microbial flora. As a result, a severe process of soil compaction is triggered (Albanesi et al. 2001; Baridón and Casas 2014; Rojas et al. 2016). These combined factors, not only decrease fertility but also infiltration capacity and water retention of soils (Magliano et al. 2017). Therefore, small decreases in rainfall lead to increasing yield losses and, in dry years, to the loss of the harvest. On the other hand, the decrease in the infiltration capacity of the soil increases the runoff. Given that the rains in the summer period are usually torrential in this region, in slightly undulating areas such as those in the northeast of the Province of Santiago del Estero and southwest of the Province of Chaco, water erosion processes of increasing magnitude are triggered.

The area between 600 and 700 mm of annual rain that we consider at high risk of "reversion" includes 3.3 million ha cleared, which is equivalent to 29% of the total cleared surface in the entire semi-arid Chaco (Fig. 8.4a). Note that of the 3.3 million ha cleared, more than 1.5 million corresponds to areas with less than 650 mm of average annual precipitation (Fig. 8.4b). The cleared areas to the southwest, below 600 mm, are mostly dedicated to irrigated agriculture, which means other problems that contribute to this general picture of current and potential desertification of the region. The agricultural frontier shift in the border between the provinces of Santiago del Estero and Chaco, comparing the years 1984 and 2016 is clearly seen (Fig. 8.5).

It is noteworthy that the deforestation process responds not only to production prospects but also to the logic of the real estate business and to international commodity prices. Both factors fluctuate according to the markets and these fluctuations imply determining effects on business decision-making for advance toward higher risk areas. These decisions are independent of the sustainability of the activity

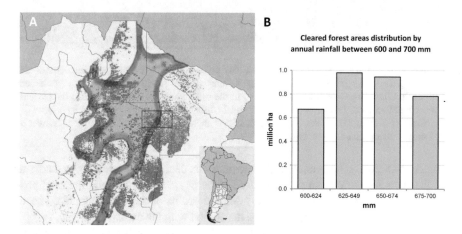

Fig. 8.4 Clearings in the semi-arid Chaco (green) (**a**) Red border zone, gray background: area between 600 and 700 mm of annual rainfall considered to be at high risk of "reversion." A red rectangular area is indicated in the center of the map, which is detailed in Fig. 8.5. (**b**) Distribution of the cleared areas between 600 and 700 mm of average annual precipitation. Own elaboration from climatic information (Hijmans et al. 2005) and the vector layer of cleared areas (REDAF, FAUBA, LART, INTA. 2020)

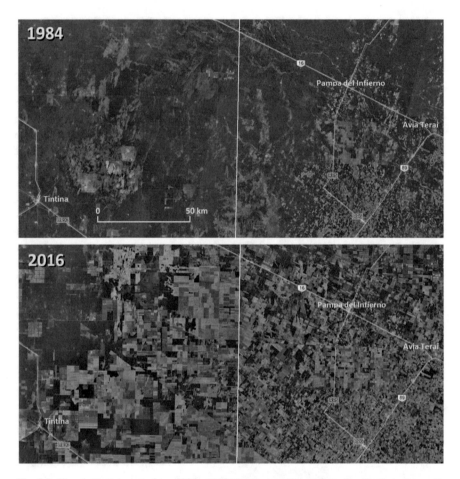

Fig. 8.5 Google Earth images from 1984 and 2016 corresponding to the red rectangle in the center of the map in Fig. 8.4. Already in 1984 some deforested areas can be seen to the east (settlement period). In 2016, large clearings are observed in the west, mostly dedicated to industrial agricultural production with "reversion" risk. The vertical white line corresponds to the limit between the provinces of Santiago del Estero and Chaco, which approximately coincides with the isohyet of 700 mm of average annual precipitation

(Baldi et al. 2019). Instead, its implementation is decided according to an estimate of the costs for the clearing (eventually for the payment of fines and the final productive abandonment of the property) and the profits during its productive phase, thus guaranteeing an adequate earning in relation with the capital invested.

Irrigation areas also show environmental deterioration problems. Due to the lack of adequate drainage infrastructure and the low quality of the irrigation water, there is a widespread and growing salinization process. This has led to the abandonment of agricultural activities, in some areas almost completely (Taleisnik et al. 2008), with consequent social and economic problems.

To get a complete environmental picture, due to the drastic anthropogenic changes, an increase of the water table level above 800 mm of precipitation is expected. This is a consequence of the low water consumption of the massive monoculture of soybeans compared to the original forest and even to the traditional "farms," which rotated crops in combination with livestock. Many of the soils where agricultural activity is located in the "Chaco dome" have a subsurface layer of salts that rises along with the water table. This process, although slow, in the long term has devastating effects when begins to appear on the surface (Jobbágy et al. 2008).

In summary, the Argentine semi-arid Chaco is in emergency from an environmental and productive point of view. It can be said that the current production model is not viable in the medium term. This affirmation is based not only on the environmental and productive problems presented but also on the enormous social and cultural cost that has accompanied this process initiated some three decades ago and whose analysis exceeds the scope of this book. If the State does not perform land-use planning and if society as a whole does not adequately value the natural resources of the Chaco, the future of its forests will be seriously compromised (Naumann 2006).

8.3 The Role of Trees in the Primary Production of the Chaco

Forest plantations and industry in Argentina concentrate in the so-called Mesopotamia Region (provinces of Entre Ríos, Corrientes and Misiones), where 90% of the Argentine productive plantations are located. In turn, 95% of the existing forest plantations throughout the country correspond to introduced species: pines, eucalypts, willows, and poplars (Schlichter 2012; Marcó et al. 2016).

Although the Chaco is an eminently forest region, all the social actors linked to primary productive activities (i.e., settlers, native communities, small livestock producers, agricultural corporations, real estate investors, and agricultural investment funds) do not have a forest culture. In other words, they do not consider forest plantation or the sustainable management of native forest under silvicultural norms, as an important regional productive alternative. The original communities, currently confined to the least productive areas, mostly come from hunting and gathering culture associated with the native forest, which today they observe on the way to extinction. The former settlers and their descendants are farmers who carry out their activity exclusively on cleared areas. The forestry activities of small livestock producers, and to some extent native communities, are restricted to the extraction of wood for fence posts, firewood, and charcoal production (Fig. 8.6). They eventually sell the standing timber to small entrepreneurs with little capitalization, who form logging and transport crews. In this way, raw material is provided mainly to regional sawmills (Fig. 8.6). This extractive activity is almost exclusively restricted to the quebrachos, to algarrobos, and, to a much lesser extent, to other minor species (Kees and Michela 2016).

Fig. 8.6 Argentine half orange or beehive brick kiln, typical in Chaco for the production of charcoal (photo: Clelia Gomez). *Prosopis alba* trunks, harvested in the natural forest and on the way to a regional sawmill (photo: A. Verga)

8.4 Agrobusiness: The New Productive Paradigm in the Chaco and Its Social Consequences

In the last two decades, three new social actors entered the region: (i) medium and large agricultural entrepreneurs, (ii) real estate investors, and (iii) agricultural investment funds. The vast majority of agricultural entrepreneurs came from the Pampas plain (provinces of Córdoba, Santa Fe, and Buenos Aires) (González and Román 2009), and they transferred the region's own production model to the Chaco, aiming to agriculturize the forest. In this way, they extracted any trace of woody species, implemented no-till farming, applied enormous quantities of agrochemicals and fertilizers, and impose the "industrial" monoculture of soybean, eventually accompanied by some rotation of corn, sunflower, sorghum, wheat, or cotton (Paruelo et al. 2005; Aguiar et al. 2016). Both these agricultural entrepreneurs and real estate investors engaged in the purchase and clearing of land, the majority of which have already been occupied by native communities and settlers during hundreds of years (González and Román 2009; Aguiar et al. 2016). This gave rise to an increasing social conflict that in many cases led to armed confrontations with loss of life (Seghezzo et al. 2011). The direct effect was the expulsion of a good part of the original rural population from vast regions (Román and González 2012), whose destination has been mostly the poverty belts of urban centers or, to a much lesser extent, their transformation into rural laborers.

Much of the land cleared by real estate investors and partly by agricultural entrepreneurs is leased to agricultural investment funds. These investment funds produce exclusively on leased land, hiring technical teams, and contracting companies for agricultural machinery and aerial spraying (also emerging players in the region). Small, medium, and even large agricultural entrepreneurs, who produce on their land and therefore have an interest in achieving more sustainable agriculture, are permeable to incorporate crop rotation and possible changes in production logic. In contrast, investment funds make decisions exclusively to maximize the return on invested capital, with a very short-term vision. Their interests are independent of the sustainability of the activity in the medium and long term, as mentioned above (González and Román 2009).

In view of the foregoing, forestry activity in the region never played a prominent role from a productive point of view, despite the fact that the region is itself a forest. Productive forest activity in the region mainly involves the exploitation of valuable species with a relative level of control regarding sustainability. The list of native tree species currently exploited in the Chaco for wood production are quebrachos (370,000 m³), algarrobos (with the highest volume of native wood for sawing marketed in the country: 115,000 m³), *Aspidosperma quebracho-blanco* (60,000 m³), *Prosopis kuntzei* (15,000 m³), *Handroanthus* spp. (4700 m³), *Gonopterodendron sarmientoi* (3700 m³) and *Caesalpinia paraguariensis* (3200 m³). The volumes indicated correspond to those marketed for all items according to official forest statistics for 2016 (MAyDS 2018). It is noteworthy that forestry activity in the Chaco has a very high informality rate, so these volumes widely underestimate the actual extraction of wood in the region. Nor are the wood extractions carried out for "own" use reflected here, which are generally utilized for house building and rural fences. The commercialization of the rest of the native forest species of Chaco does not present significant volumes. Between 2010 and 2014, the Chaco provided 94% of the native wood exploited throughout the country (MAyDS 2016).

8.5 The Need for a New Tree-Centered Production Paradigm

Profound changes are expected in the medium and long term in the Chaco, triggered by the increasing environmental deterioration and the exhaustion of a good part of the productive systems that dominate today. In this context, it is necessary to advance in a transition aimed at reconversion of the productive system, not only at the farm level, but also at the landscape and regional levels. It is essential to stop the course of deterioration and recover productive capacity. In this process of productive and environmental reconversion, native forest tree species must play a relevant role, particularly those with high productive potential (regarding timber and non-timber products) and/or fulfill a key ecological function, including resilience to climate change and tolerance of soil degradation conditions.

Native forest tree species can play a key role in the Chaco region through:

- Afforestation oriented to the production of high-value wood for use in the local forestry industry, as a replacement for wood from the native forest
- Silvopastoral and agroforestry production systems
- Programs to enrich the degraded native forest
- Restoration programs of degraded areas destined for the generation of environmental and productive services

These production systems are applicable at the farm and landscape levels, coming into consideration the spatial planning, the diversification of production and the environmental impact (e.g., water and nutrient cycles, salinity) (Paruelo et al. 2005).

At the farm level, long-term rotation between agriculture, livestock, and forestry can be considered, maintaining productivity and soil health, and supporting productive diversification, heterogeneity of the territory and job positions. At the regional level, territorial planning, the establishment of biological corridors, the creation of protected areas (and the strengthening of existing ones), and the development of forest industry clusters are desirable. Thus, native tree species are basic bricks of this building, and genetic improvement is a necessary procedure to build with the best bricks.

To end this chapter, we present a group of species not treated in the rest of the book that we consider key to put in practice the reconversion of the productive systems in the Chaco. They are useful for timber and non-timber products, and at the same time, several of them are structuring of fundamental ecosystems in the region. However, these species are not yet the object of breeding programs. Their actual distribution areas are not known in detail, and they do not count with basic and sufficient biological, genetic, and ecophysiological studies for the management of their genetic resources. There are not any natural or planted seed-producing areas of these species. Most of them have no specific protocols for DNA extraction or PCR amplification nor have specific molecular markers been isolated.

8.5.1   Red Quebrachos: Schinopsis spp. (Anacardiaceae)

Schinopsis is a small genus that includes only seven species distributed in the tropical and subtropical dry forests of South America (Mogni et al. 2017). The red quebrachos cited for the Argentine Chaco (*S. balansae*, *S. lorentzii*, *S. heterophylla*, and *S. haenkeana*) are the dominant species of the humid, semi-arid, and Serrano Chaco (Fig. 8.7). Their wood of extreme density (> 1 kg/dm^3), hardness and durability characterizes them. They have a high content of tannins that in the heartwood can reach 42% in *S. balansae*. This species was declared the "National Forest Tree" of Argentina (decree 15190/56).

Quebrachos are trees that reach 20–25 m in height with diameters that can exceed 1.5 m. The crown has a characteristic inverted cone shape that allows it to be easily distinguished. They have alternate, simple, or imparipinnately compound leaves; lateral or terminal paniculate inflorescences; staminate or pistillate flowers with vestigial ovaries or stamens depending on the case; samaroid fruits with a woody endocarp and a persistent calyx; seeds with membranous testa (Muñoz 2000). The genus is polygamous-dioecious, with hermaphrodite and unisexual flowers on the same or different trees (Muñoz 2000). However, Barberis et al. (2012) in a very detailed review of *S. balansae* state that this species is dioecious.

From the end of the nineteenth century to the mid-twentieth century, the largest exploitation in history of red quebrachos occurred. Trees were felled for the production of railway sleepers and fence posts, and, to a much greater extent, for the extraction of tannins. A single company of English origin (the sadly famous "La Forestal") is estimated to have exploited around 2 million ha of forests during this

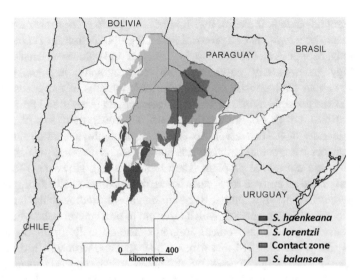

Fig. 8.7 Geographical distribution of red quebrachos in the Argentine Chaco. The dark green area corresponds to the contact zone between *S. lorentzii* and *S. balansae*, with the presence of *S. heterophylla* (putative hybrid between both previous species (Oyarzabal et al. 2018)

period, extracting red quebracho without following any silvicultural sustainability criteria. In the early 1900s, it had founded 40 villages, built 400 km of railroad tracks, and even minted its own currency. Tannin was extracted in some 30 factories distributed throughout the north of the Province of Santa Fe. It became the first exporter of the world of this product used in leather tannery. The exploitation of the red quebracho affected practically its entire distribution area (Barberis et al. 2012; Bitlloch and Sormani 2012).

Currently only two tannin factories remain. There are also restrictions for the exploitation of the red quebrachos: only individuals with diameters greater than 45 can be harvested (Provincial Law of Chaco 2,386, Regulatory Decree 1,195/80). All tannin production is still based on the exploitation of the native forest; there are no commercial plantations of these species. Between 2008 and 2014, an average of 60,000 tons of tannin was produced annually, from an annual average consumption of 250,000 m^3 of quebracho logs (MAyDS 2016). According to the estimated growth of these species in the natural forest (about 2 m^3/ha per year) it can be estimated that to supply the current tannin industry the exploitation of about 125,000 ha of forest are necessary. The exploitation of these species for the production of firewood, charcoal, sleepers, posts, and rural buildings should be additionally considered. The official statistics indicate some 370,000 m^3 of red quebracho sold in Argentina in 2016 (including *S. balansae* and *S. lorentzii*) for all destinations (MAyDS 2018).

Despite its environmental, productive and cultural importance, there are no publications analyzing the distribution of genetic variation, the mating system and the relationships between genetic structures and ecophysiology in red quebracho species. These are key issues for genetic conservation and improvement. In this sense, an important advance is the work by Mogni et al. (2016) on optimization of DNA extraction and development of a protocol for PCR amplification for *Schinopsis*. In a recent study, Mogni et al. (2018), based on nuclear and chloroplast markers, foliar morphology and anatomy, palynology, and biogeographic features, postulated hypotheses about phylogenetic and biogeographic patterns of the genus. There is evidence of the possible hybrid origin of *S. heterophylla* (Meyer and Barkley 1973; Hunziker 1998), but this is not confirmed yet. Moreover, there is still discussion about the identity of *S. haenkeana*. According to morphological studies, no significant differences were found between this species and *S. lorentzii*, which could be attributed to the altitudinal gradient (Flores et al. 2013). However, Mogni et al. (2017) designated an epitype for *S. haenkeana*, thus taxonomically distinguishing it from *S. lorentzii*. Notwithstanding, there are no genetic studies analyzing the relationship between the two species, which turns out to be a central issue when considering conservation and improvement programs. The difficulty in the taxonomy of the genus (Mogni et al. 2014, describes a new species, with which there would already be eight taxonomic species, but not necessarily biological sepecies as evolutionary units), the description of possible interspecific hybrids and the existence of related sympatric species, difficult to differentiate morphologically, support the hypothesis that the genus could constitute a complex of species among which there are significant genetic flows, as in the case of algarrobo trees (Prosopis sp.). In this context, a favorable contribution to this hypothesis is the fact that *S. balansae* and *S. haenkeana* (La Porte, 1962; Cáceres and Ávila 1987) are diploid 2n = 28. It remains to know the ploidy of the rest of the species that could be eventually part of a possible homogametic complex.

The contact area between *S. balansae* and *S. lorentzii* coincides with the highest density of clearings for agricultural use, where there are practically no natural protected areas. It is very likely that this area has an important conservation value for quebrachos and the associated ecosystems, considering the existence of both species and the possible interspecific hybrid *S. heterophylla* (the three quebrachos forest). A study on the conservation status of the vegetation carried out by Torrella et al. (2011) on remaining forest fragments of between 5 and 1000 ha in this region stated the feasibility of constituting protected areas and biological corridors. From the comparison of these forests with those from two existing protected areas (Copo National Park in the semi-arid Chaco and El Bagual Reserve in the humid Chaco), the authors concluded that the state of conservation of the remaining fragments in the central subhumid Chaco is "surprisingly good" despite the high degree of fragmentation. They also mentioned that from the stump count in the sampling plots, it seems that the selective extraction was not of great intensity.

Schinopsis balansae stands out against the rest of the quebrachos regarding the needs of study, organization, and management of their genetic resources for regional development. As we mentioned, the logging of this species in the natural forest is mainly intended for the extraction of tannin. Thus, there would be three main purposes for base propagation materials for the plantation of this species: (1)

production of solid wood to supply the market for sleepers, rural constructions, posts, and firewood; (2) enrichment and restoration of the remaining native forest; and (3) tannin production. Studies on the genetic variation patterns of the species, its genetic relationship with *S. lorentzii* in the contact area, methodologies for the identification of hybrids and an ecophysiological background are needed to advance toward the identification and creation of base materials. The identification of natural seed-producing areas should be the first action for a conventional low-intensity breeding program.

The approach toward tannin production changes the paradigm of forest production for timber purposes. The dimensions of the material to be produced are no longer important; diameter and height disappear from the decision-making process. The production of woody biomass in a rotation as short as possible and the tannin content per unit of biomass become the main yield components in such an eventual system. These particularities require the development of a specific forest genetic improvement program. In this sense, planting at high densities selected material with the following characteristics appears as a possible alternative production system:

- High density of heartwood
- Earliness of wood duraminization
- High tannin content in heartwood at early ages
- High initial growth
- High biomass production at equal basal area
- Low capacity of intraspecific competition (to enhance the duraminization process and increase the heartwood/sapwood ratio)

This intensive productive system (i.e., fertilization, soil tillage, weeding, diseases, and pest control) leads genetic improvement and silviculture to very particular approaches, including vegetative propagation and clonal forestry. According to Giménez et al. (2000), from the moment heartwood begins to form, for each new active growth ring, the heartwood also increases by one ring. Thus, the planting density must be such that the plants compete at the moment of starting the duraminization, in such a way that their growth decreases from that point, which implies an increase in the heartwood/sapwood ratio.

Based on data from field trials of the tannin production company Unitán, located in the provinces of Formosa and Chaco, Verga built growth curves in diameter, volume, and basal area and developed a high-density production model (4465 trees/ha, no thinning) for *S. balansae* (not published). According to this model and to the average content of tannin per ton of heartwood (38%) and of heartwood per ton of logs, an annual harvest of 2315 ha would be sufficient to supply the mean annual consumption of the two existing tannin industries. The rotation that maximizes tannin production per ha under this model would be 20 years, therefore, a total of 46,302 ha of crop would be necessary, which is approximately a third of the surface exploited today.

8.5.2 Aspidosperma Quebracho-Blanco (Apocynaceae)

This is an evergreen sclerophyllous tree species, 6–20 m in height and up to 1 m in diameter. Its trunk is straight; the bark is grayish brown with deep longitudinal grooves (Demaio et al. 2002). The species has two biological forms: an erect form with an ovoid crown and a pendulum form with a crown of decumbent branches (Giménez and Moglia 2003a, b). The fruit is a woody, ovoidal rounded or elongated capsule, 7–11 cm long by 4–6 cm wide and 1–2 cm thick, grayish green. Seeds are suborbicular, with a diameter between 5 and 7 cm including a membranous wing (Orfila 1995), which facilities their dispersion by the wind. Its wood is hard and heavy (0.85 kg/dm³), white-yellowish to ocher yellow, with soft grain and no difference between sapwood and heartwood. The wood is predominantly used for the elaboration of traditional products (charcoal, firewood, and sleepers but also sawn wood); it has problems in some applications due to instability and a certain tendency to suffer contractions, swellings, and warping (Giménez and Moglia 2003a, b). It is extremely slow growing, finding trees in the forests that are 40 cm in diameter and over 200-year old (Chaca and Saravia 2014).

Aspidosperma quebracho-blanco is the Chaco forest tree species with the widest geographical distribution (Fig. 8.8), due to which it has aroused interest as a carbon sink (Gaillard de Benitez et al. 2002; Barrionuevo et al. 2013). It even ranges to the Espinal and the Monte (Barchuk and Valiente-Banuet 2006) and is also distributed in the southeast of Bolivia, western Paraguay, and western Uruguay. In Argentina, it can be found in some sectors of Mesopotamia with annual rainfall of more than 1100 mm and in the Monte with less than 300 mm per year. It also occupies an enormous diversity of soils. It is not an "azonal" species but a "climax" one, since

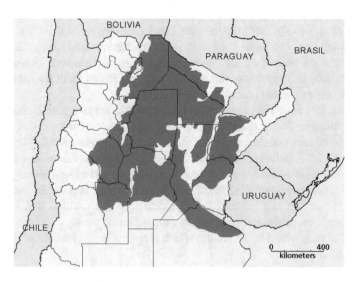

Fig. 8.8 Schematic Argentine range of *Aspidosperma quebracho-blanco* (Oyarzabal et al. 2018)

it shares the dominant layer with the red quebrachos. In the arid Chaco, it is the only dominant species. Here it has been verified that for its installation it requires the facilitation of "nurse" plants, generally shrubs of the genus *Larrea* (Barchuk and Díaz 2000).

Barchuk and Valiente-Banuet (2006) found significant differences in the specific leaf area and in the leaf exposure angle among three populations of this species located in humid, semi-arid, and arid Chaco. They demonstrated that populations in more arid areas have a distribution of leaf angles closer to 90° (this increased the frequency of individuals with pendulous branches) and a lower specific leaf area, both factors that decrease water consumption. Likewise, Moglia and López (2001) found that the xylem of this species in the semi-arid Chaco works "as an interconnected network," where the vessels participate, guaranteeing a high efficiency in driving, as well as the supporting elements form a subsidiary system that ensures water conduction under conditions of high drought stress.

Regarding the genetic pattern of the species, Torres Basso (2014) evaluated the distribution of its genetic diversity in the arid Chaco through the use of RAPDs (Random Amplified Polymorphic DNA) molecular markers. He found an important differentiation between relatively close populations (relative to its huge range), significant correlation between geographical and genetic distances, and a certain grouping of populations with respect to their genetic composition based on the hydric conditions of the studied sites. Recently, Messias et al. (2020) have developed 16 SSR markers for *Aspidosperma pyrifolium*, and transferred to 11 other species of the genus, including *A. quebracho-blanco*, which is undoubtedly a significant contribution to the study of its genetic resources.

8.5.3   *Enterolobium contortisiliquum (Fabaceae)*

It is a 10–20 m high tree (it can reach 30 m), with an extended crown and deciduous light green foliage, with bipinnate compound leaves (Slanis 2018). The trunk is cylindrical, slightly tortuous, with a basal diameter of up to 1.5 m, with gray, smooth bark. It has a wide range of distribution, across Brazil, Argentina, Uruguay, and Paraguay. In Argentina it can be found in the piedmont of the Yungas, in the Alto Paraná Rainforest, in the Chaco and in the Espinal (Fig. 8.9).

Its wood is light (specific weight: 0.39 kg/dm^3) and resistant to water thanks to its resin, which made the Wichí aborigines to used it to make canoes by hollowing out their trunks. The heartwood is brown-pinkish in color, with yellowish sapwood (Giménez and Moglia 2003a, b). It is used for openings, outdoor furniture and naval carpentry. It is suitable for peeling and plywood manufacturing. As a tree it is appreciated for its ornamental value and for the shade of its wide crown, which makes it suitable for large green spaces. It is a fast growing species (up to 4 cm in diameter and 2.5 m in height per year; Giménez and Moglia 2003a, b), and it has been suggested for active restoration of degraded forests, especially in riparian areas

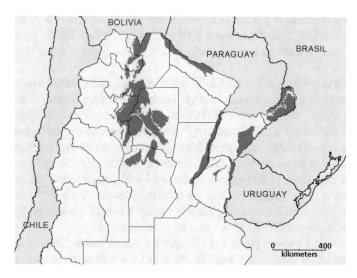

Fig. 8.9 Schematic Argentine range of *Enterolobium contortisiliquum* (Morello et al. 2012)

(Santos Júnior et al. 2004; Ferreira da Silva et al. 2011; Jesus et al. 2019; Bevilacqua Marcuzzo et al. 2020).

The fruit is an indehiscent, rounded, laterally flattened black pod of about 4–7 cm in length (Slanis 2018). Its ingestion is harmful to cattle, causing abortion and severe injuries in the skin (Bonel-Raposo et al. 2008; Costa et al. 2009). The seeds show physical dormancy, and the mechanical and chemical scarification methods are the most efficient to break dormancy and thus obtain a germination power close to 100% (Lozano et al. 2016). There are studies on germination and seedling production (de Lima et al. 1997; Malavasi and de Matos 2004; Parra de Araújo and de Paiva 2011) and works on specific mycorrhizae and rhizobia associated with the species (de Souza Moreira et al. 2015; Abreu et al. 2018).

Some genomics resources were already developed for this species, and there are antecedents of studies on genetic variation patterns, the vast majority carried out in Brazil. Nine SSR microsatellite markers have been transferred and characterized (Moreira et al. 2012; Niella et al. 2017), previously developed in *Enterolobium cyclocarpum* (Peters et al. 2008). Population studies have also been carried out using ISSR fragments (de Abreu et al. 2015) and RAPDs (da Cruz et al. 2008) in order to select areas potentially important for the conservation of the species in Brazil. In Argentina, the only antecedent of studies with molecular markers on this species is the verification of the SSR transfer (Niella et al. 2017). Basic genetic parameters were calculated from open-pollinated progeny tests in Brazil (Zaffani Sant'Ana et al. 2013).

8.5.4 *Caesalpinia Paraguariensis (Fabaceae)*

This is a tree species native to Argentina, Bolivia, Brazil, Paraguay, and Uruguay. In Argentina it is distributed in very contrasting environments (Fig. 8.10), from the Monte region with average annual rainfall of less than 300 mm to the Espinal and Yungas, where it is found up to 1200 m asl with more than 1200 mm. In the Chaco it is mainly distributed in the semi-arid region, and it is also found in the humid Chaco but more sparsely (Giménez et al. 2017).

It is a tree up to 18 m tall, generally with a short bole (up to 2 m high) and diameter up to 1 m, slow growing (growth rings between 1 and 2 mm wide). The bark is smooth, very thin, greenish brown in color, dehiscent into irregular plates with rounded edges that reveal the greenish gray new bark. Wood is heavy (1.2 kg/dm3) and due to its hardness, color, and physical-mechanical characteristics, it is considered a good substitute for ebony (Giménez et al. 2017). There are studies on the characteristics of its wood, growth, and botanical description (Giménez and Moglia 2003a, b; Gasson et al. 2009; Giménez et al. 2017).

The seeds and fruits have been characterized (Giamminola and De Viana 2013) and pre-germinative treatments have been tested (Ortega Baes et al. 2020; de Noir et al. 2004). Its foliage is consumed as forage (Aronson and Toledo 1992; Camardelli and Pérez de Bianchi 2003; García et al. 2017). Ethnobotanical and pharmacological studies have been conducted about the antibacterial properties of extracts of its bark (Scarpa 2007; Sgariglia et al. 2010).

It is known to be a 2n = 28 diploid species with the presence of tetraploid cells (Cangiano and Bernardello 2005), but there are no basic works for the management of its genetic resources.

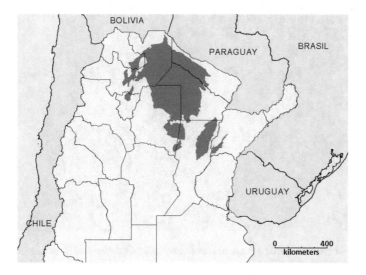

Fig. 8.10 Schematic Argentine range of *Caesalpinia paraguariensis* (Oyarzabal et al. 2018)

8.5.5 *Handroanthus Heptaphyllus (Bignoniaceae)*

In Argentina there are two species of the genus *Handroanthus* (Lozano and Zapater Cano 2008), *H. impetiginosus* that is distributed in the Pedemontana Rainforest of the Yungas and will be presented in Chap. 16 and *H. heptaphyllus*, mainly distributed in the humid Chaco (Fig. 8.11). Both are highly appreciated for their ornamental value and the quality of their wood. They are heliophilous species, but they tolerate shade in the juvenile stage.

Handroanthus heptaphyllus can be planted in the open, in pure or mixed stands (associated with pioneer species), and it can also be used in forest enrichment (Carvalho 1994; Montagnini et al. 2006). Seeds stored at room temperature without any treatment quickly lose their viability, but with a moisture content of 5–7% in closed containers at low temperatures, maintained the viability between 60% and 90% (Eibl et al. 2010). Its wood is very heavy (1.01 kg dm³), highly resistant, the sapwood color is white-yellowish and the heartwood is greenish brown. It is suitable for general carpentry, bodywork, frames, floors and widely used in outdoor constructions due to its great durability and stability (Giménez and Moglia 2003a, b).

Several studies have been carried out in Brazil on the genetic variability of *H. heptaphyllus*, on the effects of climate change on its future distribution and on the possibilities of effective conservation of its natural populations (Scarante et al. 2017). In addition, microsatellites were used to understand its mating system and the implications for seed collection with genetic conservation and breeding purposes (Mori et al. 2015).

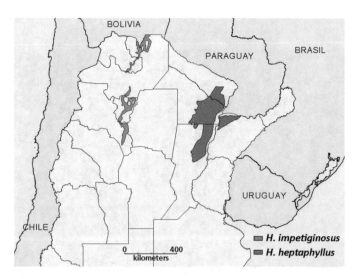

Fig. 8.11 Schematic Argentine range of *Handroanthus heptaphyllus* and *Handroanthus impetiginosus* (Oyarzabal et al. 2018)

8.5.6 Ziziphus mistol (Rhamnaceae)

This is a small tree present mainly in the semi-arid and arid Chaco (Fig. 8.12). It has a short and tortuous trunk. Although of multiple use, the main interest lies in its flowers and fruits, since it is a melliferous and wild fruit species, adapted to arid and saline areas (Cerino et al. 2015). The fruit is a glabrous, spherical drupe, about 15 mm in diameter, reddish brown at maturity. It can be consumed fresh or in different preparations (i.e., "arrope," liquor, "bolanchao"). Roasting and grinding this fruit produces a coffee substitute called mistol coffee, currently used in diets for its nutritional values (Colares and Arambarri 2008). Cardozo et al. (2011) studied the nutritional qualities and antioxidant and anti-inflammatory effects of the fruit.

The wood is hard and heavy (0.95 kg/dm³), with yellowish sapwood and reddish brown heartwood, although scarce. Its texture is fine and homogeneous and has a soft grain. It is used in carpentry and is suitable for turning works, but mostly used in charcoal production (Giménez and Moglia 2003a, b).

According to a study of its floral biology (Cerino et al. 2015), Z. mistol is an allogamous species, with entomophilic pollination, self-incompatible. A high inbreeding depression of progenies obtained by self-fertilization through artificial pollination has been reported (Cerino et al. 2015). There are some studies on germination (Araoz and Del Longo 2006). Only a single exploratory study on genetic variability was found in the literature (Tomas et al. 2017) performed through ISSR in a single population.

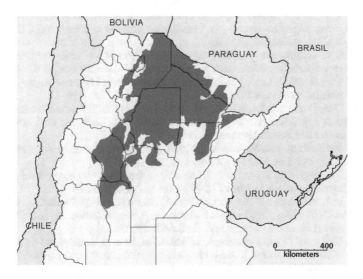

Fig. 8.12 Schematic Argentine range of *Ziziphus mistol* (Oyarzabal et al. 2018)

References

Abreu GM, Schiavo JA, Abreu PM, Bobadilha GS, Rosset JS (2018) Crescimento inicial e absorção de fósforo e nitrogênio de *Enterolobium contortisiliquum* inoculada com fungos micorrízicos arbusculares. Rev de Ciências Agrárias 41:156–164

Aguiar S, Texeira M, Paruelo J, Román M (2016) Conflictos por la tenencia de la tierra en la provincia de Santiago del Estero y su relación con los cambios en el uso de la tierra. In: Román ME, González MC (eds) Transformaciones agrarias argentinas durante las últimas décadas. Una visión desde Santiago del Estero y Buenos Aires. Editorial Facultad de Agronomía, Buenos Aires, pp 199–225

Aguiar S, Camba Sans G, Paruelo JM (2017) Instrumentos económicos basados en mercados para la conservación de la biodiversidad y los servicios ecosistémicos en Latinoamérica: ¿panacea o rueda cuadrada? Ecol Austral 27:146–161

Albanesi A, Anriquez A, Polo Sánchez A (2001) Efectos de la agricultura convencional en algunas formas del N en una toposecuencia de la Región Chaqueña, Argentina. Agri 18:3–11

Araoz SD, Del Longo OT (2006) Tratamientos pregerminativos para romper la dormición física impuesta por el endocarpo en *Ziziphus mistol* Grisebach. Quebracho 13:56–65

Aronson J, Toledo CS (1992) *Caesalpinia paraguariensis* (Fabaceae): forage tree for all seasons. Econ Bot 46:121–132

Baldi G, Houspanossian J, Murray F, Rosales A, Rueda C, Jobbágy E (2015) Cultivating the dry forests of South America: diversity of land users and imprints on ecosystem functioning. J Arid Environ 123:47–59

Baldi G, Murray F, Jobbagy E (2019) Capítulo D.4. La heterogeneidad de estrategias productivas agrícolas en sistemas semiáridos de Sudamérica. In: El lugar de la naturaleza en la toma de decisiones: servicios ecosistémicos y ordenamiento territorial rural / José M. Paruelo ... [et al.]; editado por Paruelo JM, Laterra P. – 1a ed. – Ciudad Autónoma de Buenos Aires: Fundación CICCUS

Barberis I, Mogni V, Oakley L, Alzugaray C, Vesprini J, Prado D (2012) Biología de especies australes: *Schinopsis balansae* Engl. (Anacardiaceae). Kurtziana 37:59–86

Barchuk AH, Díaz MP (2000) Vigor de crecimiento y supervivencia de plantaciones de *Aspidosperma quebracho-blanco* y de *Prosopis chilensis* en el Chaco árido. Quebracho – Revista de Ciencias Forestales 8:17–29

Barchuk AH, Valiente-Banuet A (2006) Comparative analysis of leaf angle and sclerophylly of *Aspidosperma quebracho-blanco* on a water deficit gradient. Austral Ecol 31:882–891

Baridón JE, Casas R (2014) Quality indicators in subtropical soils of Formosa, Argentina: changes for agriculturization process. ISWCR 2:13–24

Barrionuevo SA, Pan E, Medina JC, Taboada R, Ledesma R (2013) La contribución ambiental de rodales de *Aspidosperma quebracho-blanco* Schltdl. en la fijación de CO^2: bases para una gestión sustentable. Foresta Veracruzana 15:31–36

Bevilacqua Marcuzzo S, Viera M, Salin M (2020) Regeneration under the canopies of native species in a restoration area. Floresta e Ambiente 27:e20170521

Bitlloch E, Sormani H (2012) Formación de un sistema productivo: los enclaves forestales de la región chaqueño-misionera (Siglos XIX-XX). Rev Indias 72:551–580

Bonel-Raposo J, Riet-Correa F, Normanton Guim T, Duarte Schuch D, Boreli Grecco F, Gevehr Fernandes C (2008) Intoxicação aguda e abortos em cobaias pelas favas de *Enterolobium contortisiliquum* (Leg. Mimosoideae). Pesq Vet Bras 28:593–596

Brown AD, Pacheco S (2006) Propuesta de actualización del mapa ecorregional de la Argentina. In: Brown AD, Martínez Ortíz U, Acerbi M, Corcuera J (eds) La Situación Ambiental Argentina 2005. Fundación Vida Silvestre Argentina, Buenos Aires, pp 28–31

Burkart A (1976) A monograph of the genus *Prosopis* (Leguminosae subfam. Mimosoidae). J Arn Arb 57:219–525

Caballero J, Palacios F, Arébalos F, Rodas O, Yanosky A (2014) Cambio de uso de la tierra en el Gran Chaco Americano en el año 2013. Paraquaria Nat 2:21–28

Cabrera AL (1976) Regiones Fitogeográficas Argentinas. Enciclopedia Argentina de agricultura y jardinería, Tomo 2, fasc. 1. ACME, Buenos Aires

Cáceres M, Ávila TP (1987) Cromosomas mitóticos de *Schinopsis haenkeana* y *Loxopterigium grisebachii* (Anacardiaceae). Kurtziana 19:47–51

Cáceres D, Tapella E, Quétier F, Díaz S (2015) The social value of biodiversity and ecosystem services from the perspectives of different social actors. Ecol Soc 20:62

Camardelli MC, Pérez de Bianchi SM (2003) Characterization of feeding plants consumed by goats in a semi-arid woodland area of Salta, Argentina. Revista Agronómica del Noroeste Argentino 31:49–60

Cangiano MA, Bernardello G (2005) Karyotype analysis in Argentinean species of *Caesalpinia* (Leguminosae). Caryologia 58:262–268

Cardozo ML, Ordoñez RM, Alberto MR, Zampini IC, Isla MI (2011) Antioxidant and antiinflammatory activity characterization and genotoxicity evaluation of *Ziziphus mistol* ripe berries, exotic Argentinean fruit. Food Res Int 44:2063–2071

Carvalho PE (1994) Espécies florestais brasileiras: recomendações silvilculturais, potencialidades e uso da madeira. EMBRAPA/CNPF, Brasília, 640 p

Cerino MC, Richard GA, Torreta JP, Gutiérrez HF, Pensiero JF (2015) Reproductive biology of *Ziziphus mistol* Griseb. (Rhamnaceae), a wild fruit tree of saline environments. Flora 211:18–25

Chaca R, Saravia P (2014) Determinación del crecimiento de la cacha (*Aspidosperma quebracho blanco*) mediante el método de dendrocronología. Universidad Gabriel René Moreno. (Documento científico N° 5). Santa Cruz, Bolivia, 123 p

Colares MN, Arambarri AM (2008) *Ziziphus mistol* (Rhamnaceae): Morfo-anatomía y Arquitectura Foliar. Lat Am J Pharm 27:568–577

Costa RL, Marini A, Tanaka D, Berndt A, de Andrade FM (2009) Um caso de intoxicaçao de bovinos por *Enterolobium contortisiliquum* (timboril) no Brasil. Arch Zootec 58:313–316

Cozzo D (1995) Interpretación forestal del sistema fachinal de la Argentina y faxinal del Brasil. Quebracho:5–12

da Cruz Santana G, Mann RS, Ferreira RA, Bomfim Gois I, dos Santos Oliveira A, Boari AJ, Alvares Carvalho SV (2008) Diversidade genética de *Enterolobium contortisiliquum* (Vell.) Morong. no Baixo Rio São Francisco, por meio de marcadores RAPD. Rev. Árvore 32

de Abreu MP, Malveira Brandão M, Hayashida de Araujo N, Alves de Oliveira D, Fernandes WG (2015) Genetic diversity and structure of the tree *Enterolobium contortisiliquum* (Fabaceae) associated with remnants of a seasonally dry tropical forest. Flora 210:40–46

de Lima CM, Borghetti F, de Sousa MV (1997) Temperature and germination of the Leguminosae *Enterolobium contortisiliquum*. Rev Bras Fisiol Veg 9:97–102

de Noir AF, Gulotta de Maguna M, Abdala R (2004) How to improve germination in *Caesalpinia paraguariensis* Burk. Seed Sci Technol 32:235–238

de Souza Moreira FM, Ferreira PA, Vilela LA, Carneiro MA (2015) Symbioses of plants with rhizobia and Mycorrhizal Fungi in heavy metal-contaminated tropical soils. In: Sheramети I, Varma A (eds) Heavy metal contamination of soils. Soil biology. Springer, Cham

Demaio P, Karlin UO, Medina M (2002) Árboles nativos del centro de Argentina. L.O.L.A (Literature of Latin América). Editorial: Colin Sharp. Buenos Aires, 210 pp.

Eibl B, González C, Mattes L (2010) Manejo de frutos y semillas, producción de plantines y establecimiento a campo de especies nativas *Handroanthus heptaphyllus* (Vell.) Mattos. (Lapacho negro). Yvyrareta 17:51–52

Ferreira da Silva R, Lupatini M, Antoniolli ZI, Leal LT, Moro Junior CA (2011) Comportamento de *Peltophorum dubium* (Spreng.) Taub., *Parapiptadenia rigida* (Benth.) Brenan e *Enterolobium contortisiliquum* (Vell.) Morong cultivadas em solo contaminado com cobre. Ciênc. Florest 21

Flores C, Zapater M, Sühring S (2013) Identidad taxonómica de *Schinopsis lorentzii* y *Schinopsis marginata* (Anacardiaceae). Darwin 1:25–38

Gaillard de Benitez C, Pece M, Juárez de Galíndez M, Maldonado A, Acosta VH, Gómez A (2002) Biomasa aérea de ejemplares de quebracho blanco (*Aspidosperma quebracho-blanco*) en dos localidades del Parque Chaqueño Seco. Quebracho 9:115–127

Galafassil G (2005) Rebelión en el campo. Las Ligas Agrarias de la Región Chaqueña y la discusión del modelo dominante de desarrollo rural (1970–1976). In: Lázzaro, Silvia y Guido Galafassi (comp.), Sujetos, políticas y representaciones del mundo rural. Argentina. Buenos Aires, Siglo XXI, pp 1930–1976. ISBN: 987-1013-39-6

Galera F (2000) Los algarrobos: Las especies del género *Prosopis* (algarrobos) de América Latina con especial énfasis en aquellas de interés económico. Graziani Gráfica, Córdoba, 269 pp

García EM, Cherry N, Lambert BD, Muir JP, Nazareno MA, Arroquy JI (2017) Exploring the biological activity of coo+densed tannins and nutritional value of tree and shrub leaves from native species of the Argentinean dry Chaco. J Sci Food Agric 97:5021–5027

Gasson P, Warner K, Lewis G (2009) Wood Anatomy of *Caesalpinia* S.S., *Coulteria, Erythrostemon, Guilandina, Libidibia, Mezoneuron, Poincianella, Pomaria* and *Tara* (Leguminosae, Caesalpinioideae, Caesalpinieae). IAWA J 30:247–276

Giamminola EM, De Viana ML (2013) Caracterización morfológica de frutos y semillas de dos accesiones de *Prosopis nigra* (Griseb.) Hieron. y *Caesalpinia paragueriensis* (D. Parodi) Burkart., conservadas en el Banco de Germoplasma de Especies Nativas de la Universidad Nacional de Salta, Argentina. Lhawet 2:23–29

Giménez A, Moglia J (2003a) Árboles del Chaco Argentino. Guía para el reconocimiento dendrológico. Editorial El Liberal, Santiago del Estero, 307pp

Giménez AM, Moglia JG (2003b) Arboles del Chaco Argentino. Guía para el reconocimiento dendrológico. Editorial El Liberal, Santiago del Estero, 307pp

Giménez AM, Ríos NA, Moglia J (2000) Relación albura-duramen en tres especies arbóreas de la Región Chaqueña seca. Quebracho 8:56–63

Giménez AM, Bolzon Muniz G, Moglia JV, Nigosky S (2017) Ecoanatomia del ébano Sudamericano: "guayacán" (*Libidibia Paraguariensis*, Fabaceae). Bol Soc Argent Bot 52:45–54

González MC, Román M (2009) Expansión agrícola en áreas extrapampeanas de la Argentina. Una mirada desde los actores sociales Cuadernos Des Rural 6:99–120

Grau HR, Gasparri NI, Aide TM (2005) Agriculture expansion and deforestation in seasonally dry forests of north-west Argentina. Environ Conserv 32:140–148

Hansen MC, Potapov PV, Moore R, Hancher M, Turubanova S, Tyukavina A, Thau D et al (2013) High-resolution global maps of 21st-century forest cover change. Science 342:850–853

Hijmans RJ, Cameron SE, Parra JL, Jones PG, Jarvis A (2005) Very high resolution interpolated climate surfaces for global land areas. Int J Climatol 25:1965–1978

Hunziker A (1998) Los nombres científicos correctos de los "quebrachos colorados" (*Schinopsis,* Anacardiaceae) del centro y noroeste de Argentina. Kurtziana 26:55–64

Iñigo Carrera V (2013) Trabajadores indígenas en el Chaco argentino: Algunos sentidos estigmatizadores. Antipod Rev Antropol Arqueol 7:229–251

IPCC (2014) Cambio climático 2014: Informe de síntesis. Contribución de los Grupos de trabajo I, II y III al Quinto Informe de Evaluación del Grupo Intergubernamental de Expertos sobre el Cambio Climático [Equipo principal de redacción, R.K. Pachauri y L.A. Meyer (eds.)]. IPCC, Ginebra, Suiza, 157pp

Jesus JB, Ferreira RA, Gama DC (2019) Behavior of the emergence of native forest seedlings for the purpose of recovery of the riparian forest. Floresta 49:57–98

Jobbágy EG, Nosetto MD, Santoni CS, Baldi G (2008) El desafío ecohidrológico de las transiciones entre sistemas leñosos y herbáceos en la llanura Chaco-Pampeana. Ecol Austral 18:305–322

Kees S, Michela J (2016) Aspectos de la producción primaria y el mercadeo del algarrobo en Chaco, Argentina. EEA-INTA Sáenz Peña, Presidencia Roque Sáenz Peña, Argentina

Kees S, Gómez C, Vera M, Cardozo F, López Lauenstein D, Mutton F, Gon V, Verga A (2011) Predicción del área de distribución natural de *Prosopis hassleri* en la provincia de Formosa – Argentina. 1° Congreso Forestal del Chaco Sudamericano

La Porte J (1962) Observaciones sobre germinación y cromosomas de algunas especies de "*Schinopsis*" (Anacardiáceas) y de Esterculiáceas cultivadas. Revista Argent Agron 29:29–41

Laterra P, Nahuelhual L, Vallejos M, Berrouet L, Arroyo Pérez E, Enrico L, Jiménez-Sierra C, Mejía K, Meli P, Rincón-Ruíz A, Salas D, Špiri J, Villegas JC, Villegas-Palacio C (2019) Linking inequalities and ecosystem services in Latin America. Ecosyst Serv 36:100875

Lozano LC, Zapater Cano MA (2008) Delimitación y estatus *de Handroanthus heptaphyllus* y *H. Impetiginosus*. (Bignoniaceae, Tecomeae). Darwin 46:304–317

Lozano EC, Zapater MA, Mamani C, Flores CB, Gil MN, Sühring SS (2016) Efecto de pretratamientos en semillas de *Enterolobium contortisiliquum* (Fabaceae) de la selva pedemontana argentina. Bol Soc Argent Bot 51:79–87

Magliano P, Fernández R, Florio E, Murray F, Jobbágy E (2017) Soil physical changes after conversion of woodlands to pastures in Dry Chaco rangelands (Argentina). Rangel Ecol Manag 70:225–229

Malavasi UC, de Matos MM (2004) Dormancy breaking and germination of *Enterolobium contortisiliquum* (Vell.) Morong seed. Braz Arch Biol Technol 47:851–854

Marcó M, Gallo L, Verga A (2016) In Domesticación y mejoramiento de especies forestales (Marcó MA, Llavallol CE, eds). Unidad para el Cambio Rural. Ministerio de Agroindustrias. Presidencia de la Nación Argentina, Buenos Aires

Mastrangelo ME, Laterra P (2015) From biophysical to social-ecological trade-offs: integrating biodiversity conservation and agricultural production in the Argentine Dry Chaco. Ecol Soc 20:20

MAyDS - Ministerio de Ambiente y Desarrollo Sustentable. Presidencia de la Nación (2015) Tercera comunicación nacional de la República Argentina a la convención marco de las Naciones Unidas sobre el cambio climático. Ministerio de Ambiente y Desarrollo Sustentable de la Nación

MAyDS – Ministerio de Ambiente y Desarrollo Sustentable. Presidencia de la Nación (2016) Series Estadísticas Forestales 2008–2014, 82pp

MAyDS – Ministerio de Ambiente y Desarrollo Sustentable. Presidencia de la Nación (2018) Anuario de Estadística Forestal – Especies nativas 2016,

Messias PA, Alves FM, Pinheiro F, Pereira de Souza A, Koch I (2020) Development and transferability of microsatellite markers for a complex of *Aspidosperma* Mart. & Zucc. (Apocynaceae) species from South American Seasonally Dry Tropical Forests. Braz J Bot 43:139–145

Meyer T, Barkley F (1973) Revisión del género *Schinopsis* (Anacardiaceae). Lilloa 33:207–257

Moglia JG, López CR (2001) Estrategia adaptativa del leño *Aspidosperma quebracho blanco*. Madera y Bosques 7:13–25

Mogni V, Oakley L, Vera Jiménez M, Prado D (2014) A new tree species of *Schinopsis* (Anacardiaceae) from Paraguay and Bolivia. Phytotaxa 175:141–147

Mogni V, Kahan M, Paganucci de Queiroz L, Vesprini J, Ortiz J, Prado D (2016) Optimization of DNA extraction and PCR protocols for phylogenetic analysis in *Schinopsis* spp. and related Anacardiaceae. Springerplus 5:1–7

Mogni V, Prado D, Oakley L (2017) Notas nomenclaturales en el género *Schinopsis* (Anacardiaceae). Bol Soc Argent Bot 52:185–191

Mogni V, Prado D, Oakley L (2018) Consideraciones taxonómicas y biogeográficas del género *Schinopsis* (Anacardiaceae), con énfasis en las especies nativas del Paraguay. III Jornadas Paraguayas de Botánica. 30, 31 de julio y 1, 2, 3 de agosto de 2018. Universidad Nacional de Asunción. Facultad de Ciencias Exactas y Naturales, San Lorenzo – Paraguay

Montagnini F, Eibl B, Fernandez R (2006) Rehabilitation of degraded lands in Misiones Argentina. Bois Fors Trop 288:51–65

Moreira PA, Sousa SAS, Oliveira FA, Araújo NH, Fernandes GW, Oliveira DA (2012) Characterization of nine transferred SSR markers in the tropical tree species *Enterolobium contortisiliquum* (Fabaceae). Genet Mol Res 11:3729–3734

Morello J, Adámoli J (1974) Las grandes unidades de vegetación y ambiente del Chaco argentino. Segunda parte: Vegetación y Ambiente de la Provincia del Chaco. Serie Fitogeográfica N° 13. INTA, Buenos Aires, 130 pp

Morello J, Matteucci S, Rodríguez A (2012) Ecorregiones y complejos ecosistémicos argentinos. Primera edición – Buenos Aires. Orientación Gráfica Editora, 752 pp

Mori NT, Mori ES, Tambarussi EV, Moraes MLT, Sebbenn AM (2015) Sistema de cruzamento em populações de *Handroanthus heptaphyllus* (Veli.) Mattos e suas implicações para a coleta de sementes para fins de conservação e melhoramento genético. Sci For 43:675–681

Mottura M, Finkeldey R, Verga A, Gailing O (2005) Development and characterization of microsatellite markers for *Prosopis chilensis* and *Prosopis flexuosa* and cross-species amplification. Mol Ecol Notes 5:487–489

Muñoz JD (2000) Anacardiaceae. In: Hunziker AT (ed.). Flora Fanerogámica Argentina 65:1–28

Naumann M (2006) Atlas Del Gran Chaco Americano Ed. Sociedad Alemana de Cooperación Técnica (GTZ)

Niella FO, Rocha SP, Ojeda P, Petruszynski GA, Zapata PD, Otegui MB (2017) Amplificación de marcadores microsatélites en *Peltophorum dubium* (spreng.) Taub (caña fistola) y *Enterolobium contortisiliquum* (vell.) (timbo), utilizando cebadores de la especie *Koompassia malaccensis* (benth.) y *Enterolobium cyclocarpum* (Jacq.) Griseb. Yvyrareta 25:60–65

OEA. Departamento de Desarrollo Sostenible. Secretaría General (2009) Evaluación regional del impacto en la sostenibilidad de la cadena productiva de la soja: Argentina – Paraguay – Uruguay. 310 páginas. OEA documentos oficiales; OEA/Ser.D/XXIII.7. ISBN: 978-0-8270-5510-0

Orfila EN (1995) Frutos, Semillas y plántulas de la flora leñosa Argentina. Ediciones Sur. La Plata, Argentina, 155 pp

Ortega Baes P, De Viana ML, Larenas G, Saravia M (2020) Germinación de semillas de *Caesalpinia paraguariensis* (Fabaceae): agentes escarificadores y efecto del ganado. Rev Biol Trop 49:301–304

Oyarzabal M, Clavijo J, Oakley L, Biganzoli F, Tognetti P, Barberis I, Maturo HM, Aragón R, Campanello PI, Prado D, Oesterheld M, León RJC (2018) Unidades de vegetación de la Argentina. Ecol Austral 28:40–63

Pacheco Balanza D, Ontiveros M, Espinoza D, Orellana R, Suarez G, Ortuño C (2012) Cambio climático, Sequía y Seguridad Alimentaria en el Chaco boliviano. Reflexiones y Propuestas desde la Cordillera. Fundación Cordillera – Universidad de la Cordillera, La Paz

Panigatti JL (2010) Argentina 200 años, 200 suelos. Ed. INTA Buenos Aires, 345 pp

Paolasso P, Krapovickas J, Gasparri NI (2012) Deforestación, expansión agropecuaria y dinámica demográfica en el Chaco Seco argentino durante la década de los noventa. Lat Am Res Rev 47:35–63

Parra de Araújo A, de Paiva SS (2011) Germinação e produção de mudas de tamboril (*Enterolobium Contortisiliquum* (Vell.) Morong) em diferentes substratos. Árvore 35:581–588

Paruelo J, Guerschman J, Verón S (2005) Expansión agrícola y cambios en el uso del suelo. Ciencia Hoy 15:14–23

Paruelo JM, Verón SR, Volante JN, Seghezzo L, Vallejos M, Aguiar S, Amdan L, Baldassini P, Ciuffolif L, Huykman N, Davanzo B, González E, Landesmann J, Picardi D (2011) Conceptual and methodological elements for cumulative environmental effects assessment (CEEA) in subtropical forests. The case of eastern Salta. Argentina. Ecol Austral 21:163–178

Peters MB, Hagen C, Trapnell DW, Hamrick JL, Rocha O, Smouse PE, Glenn TC (2008) Isolation and characterization of microsatellite loci in the Guanacaste tree, *Enterolobium cyclocarpum*. Mol Ecol Res 8:129–131

REDAF (2012). Monitoreo de deforestación de los bosques nativos en la región chaqueña argentina – Informe N° 1: Ley de Bosques, análisis de deforestación y situación del bosque chaqueño en la provincia de Salta. http:// redaf.org.ar/wp-content/uploads/2012/12/REDAF_informedeforestacion_n1_casoSALTA.dic2012.pdf

REDAF, FAUBA, LART, INTA (2020) Proyecto de monitoreo del desmonte en el Chaco seco. http://www.monitoreodesmonte.com.ar

Rojas JM, Prause J, Sanzano G, Arce O, Sánchez MC (2016) Soil quality indicators selection by mixed models and multivariate techniques in deforested areas for agricultural use in NW of Chaco. Argentina Soil Till Res 155:250–262

Román ME and González C (2012) Juventud y migración: vivencias, percepciones, ilusiones. Un estudio en NOA y NEA. ONU mujeres. Ministerio de Agricultura, Ganadería y Pesca

Sacchi LV, Gasparri N (2015) Impacts of the deforestation driven by agribusiness on urban population and economic activity in the Dry Chaco of Argentina. J Land Use Sci 11:523–537

Saidman BO, Bessega CF, Ferreyra LI, Julio N, Vilardi JC (2000) The genus *Prosopis* (Leguminosae). Bol Soc Argent Bot 35:315–324

Santos Júnior NA, Alvarenga Botelho S, Davide AC (2004) Study of germination and survival of three species in direct seeding system, aiming riparian forest restoration. Cerne 10:103–117

Scarante AG, Matos MF, Soares MT, de Aguiar AV, Wrege MS (2017) Distribution of *Handroanthus heptaphyllus* in Brazil and future projections according to global climate change. Rev Geama 3:201–207

Scarpa GF (2007) Etnobotánica de los Criollos del oeste de Formosa: Conocimiento tradicional, valoración y manejo de las plantas forrajeras. Kurtziana 33:153–174

Schlichter T (2012) Aportes a una política forestal en Argentina: el sector forestal y el desarrollo económico, ambiental y social del país. 1a ed.; Ministerio de Agricultura, Ganadería y Pesca. MAGyP. Unidad para el Cambio Rural, UCAR. Buenos Aires, Argentina, 92 p

Seghezzo L, Volante J, Paruelo J, Somma D, Buliubasich E, Rodríguez H, Gagnon S, Hufty M (2011) Native forests and agriculture in Salta (Argentina) conflicting visions of development. J Environ Dev 20:251–277

Sgariglia MA, Soberón JR, Sampietro DA, Quiroga EN, Vattuone MA (2010) Isolation of antibacterial components from infusion of *Caesalpinia paraguariensis* bark. A bio-guided phytochemical study. Food Chem 126:395–404

Slanis AC (2018) *Enterolobium contortisiliquum*: Oreja de negro, pacará, timbó. Universo tucumano nro 7. In: Scrocchi GJ, Szumik C (eds) Unidad Ejecutora Lillo. CONICET. Funadación Lillo, Septiembre de, 10 pp

Spensley J, Sabelli A, Buenfil J (2013) Estudio de vulnerabilidad e impacto del cambio climático en el gran Chaco americano. Ed. Programa de las Nacionales Unidas para el medio ambiente – PNUMA

Taleisnik E, Grunberg K, Santa María G (2008) La salinización de suelos en la Argentina: su impacto en la producción agropecuaria. EDUCC (Editorial Universidad Católica de Córdoba), Córdoba

Tomas PA, Zietz R, Cerino MC (2017) Análisis de diversidad genética en *Ziziphus mistol* griseb. mediante marcadores moleculares ISSR. Revista FAVE – Ciencias Agrarias 16:153–162

Torrella S, Oakley L, Ginzburg L, Adámoli J, Galetto L (2011) Estructura, composición y estado de conservación de la comunidad de plantas leñosas del bosque de tres quebrachos en el Chaco Subhúmedo Central. Ecol Austral 21:179–188

Torres Basso MB (2014) Estudio de la diversidad genética poblacional de *Aspidosperma quebracho blanco* Schltdl. en Chaco árido. Tesis Doctoral. Universidad Nacional de Córdoba, 173pp

Vallejos M, Volante JN, Mosciaro MJ, Vale LM, Bustamante ML, Paruelo JM (2015) Transformation dynamics of the natural cover in the Dry Chaco ecoregion: a plot level geo-database from 1976 to 2012. J Arid Environ 123:3–11

Vallejos M, Mastrángelo M, Paruelo J (2017) Chaco y deforestación ¿Se puede producir y conservar? Aves Argentinas 50:21–25

Verga A (2014) Rodales semilleros de *Prosopis* a partir del bosque nativo. Quebracho 19:125–138

Verga A, Gregorius H (2007) Comparing morphological with genetic distances between populations: a new method and its application to the *Prosopis chilensis – P. flexuosa* Complex. Silvae Genet 56:45–51

Verga A, López Lauenstein D, López C, Navall M, Joseau J, Gómez C, Royo O, Degano W, Marcó M (2009) Caracterización morfológica de los Algarrobos (*Prosopis* sp.) en las regiones fitogeográficas Chaqueña y Espinal Norte de Argentina. Quebracho 17:31–40

Volante JN, Seghezzo L (2018) Can't see the forest for the trees: can declining deforestation trends in the Argentinean Chaco Region be ascribed to efficient law enforcement? Ecol Econ 146:408–413

Zaffani Sant'Ana V, Menezes Freitas ML, Teixeira de Moraes ML, Zanata M, Scatena Zanatto AC, de Moraes MA, Sebbenn AM (2013) Parâmetros genéticos em progênies de polinização aberta de *Enterolobium contortisiliquum* (Vell.) Morong em Luiz Antonio, SP, Brasil. Hoehnea 40:515–520

Chapter 9
Genetic Variation Patterns of "Algarrobos" from the "Great American Chaco" (*Prosopis alba*, *P. nigra*, *P. hassleri*, *P. fiebrigii*, *P. ruscifolia*, *P. chilensis*, and *P. flexuosa*)

Carmen Vega, Dana Aguilar, Cecilia Bessega, Ingrid Teich, María Cristina Acosta, Andrea Cosacov, Mauricio Ewens, Juan Vilardi, Alicia N. Sérsic, and Aníbal Verga

The genus *Prosopis* comprises approximately 44 species of trees and shrubs distributed in arid, semi-arid, and subhumid zones of America, Africa, and West Asia (Burkart 1976). Although some species are native to North Africa, Asia, and North America, the "Great American Chaco" constitutes its main center of diversity. The greatest number of endemism and species is distributed in virtually all environments of this large phytogeographic region (Burkart 1976; Ferreyra et al. 2010). Burkart (1976) defined the taxonomic delimitation of the genus and proposed an intragenus classification in five sections (*Prosopis*, *Anonichium*, *Strombocarpa*, *Monilicarpa*, and *Algarobia*) based on floral and vegetative characters. *Prosopis alba*, *P. nigra*,

C. Vega (✉) · I. Teich
Instituto de Fisiología y Recursos Genéticos Vegetales, Unidad de Estudios Agropecuarios (IFRGV, UDEA) INTA-CONICET, Córdoba, Argentina
e-mail: vega.carmen@inta.gob.ar

D. Aguilar · M. C. Acosta · A. Cosacov · A. N. Sérsic
Lab. Ecología Evolutiva – Biología Floral, IMBIV (UNC-CONICET), Córdoba, Argentina

C. Bessega · J. Vilardi
Lab. Genética, Depto. Ecología, Genética y Evolución, IEGEBA (CONICET- UBA), Buenos Aires, Argentina

M. Ewens
Estación Experimental Fernández-UCSE (Prov. Santiago del Estero- Universidad Católica Santiago del Estero), Santiago del Estero, Argentina

A. Verga
Instituto de Fisiología y Recursos Genéticos Vegetales, Unidad de Estudios Agropecuarios (IFRGV, UDEA) INTA-CONICET, Córdoba, Argentina

Agencia de Extensión Rural INTA, La Rioja, Argentina

© Springer Nature Switzerland AG 2021
M. J. Pastorino, P. Marchelli (eds.), *Low Intensity Breeding of Native Forest Trees in Argentina*, https://doi.org/10.1007/978-3-030-56462-9_9

P. hassleri, *P. chilensis*, and *P. flexuosa* belong to the Algarobia section, which contains most economically important species (Burkart 1976).

The identification of management and conservation units of "algarrobo" (the Spanish word for carob) is hampered by characteristics of the genus *Prosopis*, which is shown as a continuum from the morphological and adaptive point of view (Verga 1995). Furthermore, its taxonomic species have the ability to exchange genetic information, generating high diversity in multiple points of contact and giving rise to hybrid swarms (Hunziker 1986; Saidman et al. 2000). These areas are often active sites of evolutionary changes in which hybridization and introgression may lead to increases in intraspecific genetic diversity, transfer of genetic adaptations, and even the emergence of new ecotypes or species (Petit et al. 1999). Moreover, the existence of interspecific hybridization is accepted as the major mechanism for generating evolutionary novelties in the plant kingdom (Rieseberg and Gerber 1995; Rieseberg 1997; Arnold 1997; Rieseberg and Carney 1998; Barton 2001).

Problems of species delimitation in the genus *Prosopis* occur mainly in the section Algarobia, in which it is very common the appearance of intermediate forms originated presumably by interspecific hybridization in areas where different species come into contact. Numerous authors have presented morphological, cytological, and molecular evidences of natural hybridization between several species of the Algarobia section that grows in the Chaco Domain (Burghardt and Palacios 1981; Palacios and Bravo 1981; Hunziker 1986; Verga 1995; Vázquez-Garcidueñas et al. 2003; Vega and Hernandez 2005; Ferreyra et al. 2013). These authors have suggested that the absence of strong interspecific reproductive barriers and the fact that some species are sympatric facilitate hybridization and introgression and contribute to the great morphological and genetic variation observed in natural populations.

In *Prosopis*, the existence of hybrids and introgressed individuals is very common. In this chapter we will present the results that have been obtained in wild pure populations and in zones of hybrid swarms of *Prosopis* together with those obtained from installed progeny trials. This knowledge will help to understand the functioning of this complex of species and will serve as the basis for the design of intervention programs for the dynamic conservation of genetic resources, improving the evolutionary capacity of the complex and facilitating adaptation to changing environments.

9.1 Morphological Analyses as a First Approach to the Study of the Algarrobos' Genetic Resources in the Chaco Region

Describing patterns of morphological variability among taxa and among individuals within taxa, and unraveling the processes leading to such patterns, is one of the fundamental questions in biology. In the Chaco region, although some *Prosopis* species are easily recognized, there are great morphological leaf variability within species (Verga et al. 2014) and high morphological similarities among certain species (Burkart 1976). Moreover, interspecific hybridization creates intermediate phenotypes (Burkart 1976; Palacios and Bravo 1981; Hunziker 1986; Saidman 1990; Verga 1995; Teich et al. 2015), hampering their taxonomical classification. It has even been proposed that the complex of *Prosopis* species in the Great American Chaco may conform a single evolutionary unit (syngameon) (Verga et al. 2014) in which their ability to exchange genetic information may act as a mechanism that increases their variability and evolutionary potential, generating high diversity in contact areas. In this context, taxonomic methods based on the subjective observation of morphological traits, such as systematic keys (Burkart 1976), may not be enough to classify individuals. On the other hand, numerical taxonomy, based on leaf and fruit traits, has proved to be very useful to obtain groups of *Prosopis* individuals in a much greater degree of detail than that achieved through classical systematics, even among species with similar genetic characteristics (Verga 1995; Joseau and Verga 2005; Verga and Gregorius 2007).

Traditionally, *Prosopis* species in the Chaco region are classified in two large groups: the White algarrobos or "algarrobos blancos" (WA) and the Black algarrobos or "algarrobos negros" (BA). Though this classification is not taxonomical, these two groups are easily differentiated by the characteristics of the fruits, being the two main differences that WA fruits are yellowish and thinner, with less mesocarp, than BA fruits, which are always dark. At a macrogeographic scale, species within each group are not sympatric; they replace each other establishing contact areas. On the contrary, black and white algarrobos coexist, although at the microgeographic scale, they occupy different niches (Vega et al. 2020).

Figure 9.1 shows the central areas of distribution of the pure species of white algarrobos in Argentina. As a whole, they occupy the totality of the Chaco region, always depending on the local environmental conditions that determine their presence or absence on a more detailed scale. The distribution of *P. chilensis* appears fragmented because its distribution depends to a great extent on the existence of water in the subsoil, so this species in the Chaco is always associated with the foothills of the Sierras Pampeanas. In the central part of the arid Chaco, *P. flexuosa* (black algarrobo) and *Aspidosperma quebracho-blanco* (the white quebracho) are dominant, while contact areas between *P. flexuosa* and *P. chilensis* appear on the piedmont. In the rest of the Chaco, the white algarrobos are in contact with *P. nigra*. To the north of the Chaco, *P. ruscifolia* ("vinal") appears completing this complex of species (Verga et al. 2009).

Many genetic, ecological, and evolutionary studies of *Prosopis* species in the Chaco have used a set of leaf and fruit traits to characterize and identify groups (Burkart 1976; Pasiecznik et al. 2001; Bessega et al. 2009). Regarding leaf traits, the most frequently used are petiole length (PEL), number of pairs of pinnae (NPI), pinna length (PIL), number of pairs of leaflets per pinna (NLP), leaflet length (LEL), leaflet width (LEW), leaflet area (LEA), leaflet apex (LAPX), leaflet length/leaflet width (LEL/LEW), and leaflet apex/total area (LAPX/LEA) (Fig. 9.2). Fruit morphometric characters usually include fruit length (FrL), fruit width (FrW), and fruit thickness (FrTh) (Fig. 9.2). *Prosopis* fruits are also characterized according to their

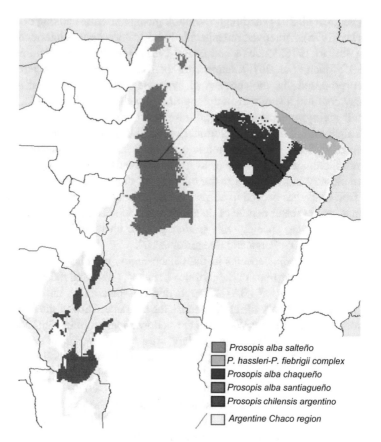

Fig. 9.1 Distribution of the "core" areas of the white algarrobos species in the Argentine Chaco, modeled by "Bioclim"

shape (FSh), the fruit edge shape (FESh), and the color (FrC) (Burkart 1976; Palacios and Brizuela 2005). The classes of the last three variables (FSh, FESh, and FrC) represent a continuum in shape (from straight to curly) and in color (from yellow to black) and are therefore usually considered ordinal (Joseau et al. 2013). The software HOJA 3.4 (Verga 2015) has facilitated measuring these traits and is widely used in *Prosopis* studies (Ferreyra et al. 2013; Joseau et al. 2013). It is well-known that arrangement, size, shape, and anatomy of leaves and fruits differ greatly in plants growing in different environments. These often easy-to-measure traits, such as leaf shape, size, or thickness, are associated to plant physiological responses to their immediate environment (e.g., through photosynthetic rates), which in turn affect performance, growth, and survival and are often regarded as "functional traits" (Geber and Griffen 2003; Reich et al. 2003).

The ways in which functional traits indirectly affect fitness vary with the environmental context, and environmentally linked differences in leaf size and shape are often thought to reflect a legacy of adaptive evolution in plants. Broad-scale correlations between traits and environments could arise through a combination of

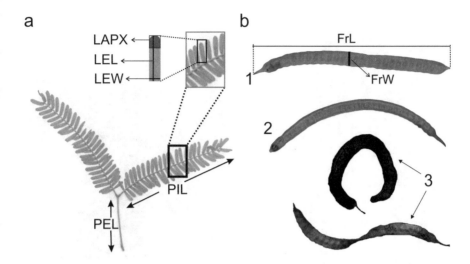

Fig. 9.2 Leaf (**a**) and fruit (**b**) characters frequently used in ecological, genetic, and evolutionary studies of *Prosopis*. PEL, petiole length; PIL, pinna length; LEL, leaflet length; LEW, leaflet width; LAPX, leaflet apex; FrL, fruit length; FrW, fruit width. Numbers 1–3 indicate classes of fruit shape (from straight to curly) (from Vega et al. 2020)

adaptive evolution in situ, differences in colonization success, and phylogenetic constraints. However, studies that control for phylogeny or focus on single lineages often (but not always) find that morphological differences predominantly reflect adaptive rather than non-adaptive divergence (Carlson et al. 2016 and references therein). The study of local adaptation in plants is of fundamental importance in evolutionary, population, conservation, and global-change biology. For example, understanding the association of quantitative traits to particular environmental conditions is key to evaluate the probability of success of selected phenotypes to grow in different regions.

In *P. alba,* the joint analysis of morphological characters and neutral genetic markers, revealed that the patterns of morphological differentiation respond to differential selection pressures (Bessega et al. 2015). *Prosopis alba* is one of the primary components of the dry forests in the Gran Chaco Region with a large geographical distribution, being found in areas ranging from 500 to 1200 mm of annual precipitation, with extreme temperatures between 48 °C absolute maximum and −10 °C absolute minimum. Throughout its distribution, at least three morphological types of *P. alba* have been described that can be differentiated by their foliar traits and that occupy areas with significantly different bioclimatic environments: *P. alba* "chaqueño," *P. alba* "santiagueño," and *P. alba* "salteño" (Verga et al. 2009; Verga et al. 2014). Interestingly, heritability of these foliar characters in *P. alba* ranges from 0.11 to 0.89, with total leaf area and the relation between the length and width of folioles (LEL/LEW) showing the highest heritability (Teich et al. in prep). Bessega et al. (2009) also reported high heritability for LEL/LEW.

Within the group of White algarrobos, the complex of *P. alba*, *P. hassleri*, *P. chilensis*, and *P. fiebrigii* shows the largest distribution in the Great American

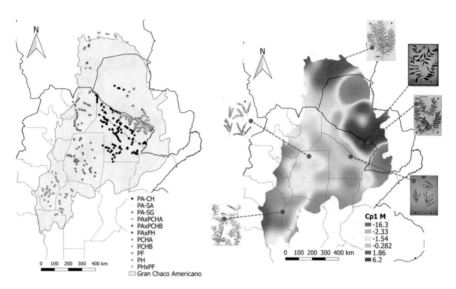

Fig. 9.3 Spatial distribution of sampled trees in the Chaco Region (left), different colors correspond to morphological groups: PA-CH, *P. alba* "chaqueño"; PA-SA, *P. alba* "salteño"; PA-SG, *P. alba* "santiagueño"; PCHB, *P. chilensis* "boliviano"; PCHA, *P. chilensis* "argentino"; PF, *P. fiebrigii*; PH, *P. hassleri*; and hybrids. Right: map resulting from the spatial interpolation of the first principal component of a Principal Components Analysis of 11 morphological traits in 927 white algarrobos trees (from Teich et al. 2017)

Chaco. The spatial pattern of leaf morphological variation (studied in 30 to 200 individuals within each morphological group) throughout its distribution range and its association with climatic characteristics shows that as temperatures and rainfall decrease to the south and to the west, leaves are smaller, with more pinnae and leaflets. However, in the south, where the *P. chilensis* "argentino" morphotype predominates, no such association is observed (Teich et al. 2017). These associations between phenotypic variation and environment may reflect evolutionary processes of local adaptation, suggesting that this set of species/morphotypes would respond as an evolutionary complex to environmental variations, in which at a macrogeographic scale each morphotype occupies a portion of the environmental gradient of the Great American Chaco. Hotspots of morphological diversity were found in the contact areas among morphotypes/species (Fig. 9.3).

9.2 Variability and Genetic Differentiation Through Isozyme Analysis

In population genetics, the study of causes and effects of genetic variation within and among populations is essential, and until early 2000s, isoenzymes have been the most widely used biochemical markers for this purpose. Although they have already been largely replaced by more informative molecular markers (such as direct DNA sequencing, single-nucleotide polymorphisms and microsatellites), they are still among the fastest and cheapest marker systems and are an excellent choice to identify high levels of genetic differentiation, e.g., incipient speciation. These initial studies provided the first clues about the genetic relationships between the main species of the Algarobia section, of great environmental and productive importance for the Chaco region.

Palacios and Bravo (1981) performed morphological and chromatographic studies on some species of *Prosopis* (*P. alba*, *P. fiebrigii*, *P. hassleri*, *P. nigra*, *P. ruscifolia*, *P. vinalillo*) and natural hybrids of the Algarobia section. They postulated that from a biological point of view, each taxonomic species would be equivalent to a semi-species, and the sympatric semi-species community would constitute a syngameon (sensu Grant 1981). Isozyme studies in these and other species of the Algarobia section showed high variability and low genetic differentiation (Saidman 1985, 1986, 1988, 1993; Saidman and Vilardi 1987, 1993; Verga 1995; Saidman et al. 1998; Joseau et al. 2013) which would support the hypothesis proposed by Palacios and Bravo (1981).

As a reference we can mention the study carried out by Saidman and Vilardi (1987) where they analyzed the genetic similarities between seven *Prosopis* species (*P. alba*, *P. nigra*, *P. alpataco*, *P. caldenia*, *P. flexuosa*, *P. ruscifolia*, and *P. hassleri*). The study was based on data from 25 enzyme loci encoding seven enzymes (Saidman 1985, 1986; Saidman and Naranjo 1982). Only six out of 25 loci showed alleles that characterized different groups of species. *Prosopis caldenia* is the most identifiable species based on the pattern retrieved in two loci, and *P. ruscifolia* can be identified by the band corresponding to a locus. The rest of the species could not be identified because their zymograms were very similar. However, groups of species were retrieved such as *P. alpataco-P. flexuosa* or *P. alba-P. hassleri*. *Prosopis nigra* was similar to the *P. alpataco-P. flexuosa* group for one of the locus but differed from them for another locus. The genetic distances between the studied species were very low, with the exception of *P. caldenia*, which was the most easily identifiable species through the isoenzyme technique, probably because it is the only species in the Algarobia section that presents an isolation mechanism.

Although there are clear morphological differences that allowed Burkart (1976) to consider the taxa analyzed here as taxonomically different, discrepancies have been observed between morphological (through classical taxonomy) and molecular data. Saidman and Vilardi (1987) concluded that this disagreement can be explained by the high genetic similarity as a consequence of the weak reproductive barriers. This hypothesis is supported by an extensive hybridization of species that occurs in overlapping regions. However, from subsequent studies an alternative hypothesis was proposed: the high genetic similarity found in the genus is probably due to the relatively recent origin of these species (Bessega et al. 2000; Catalano et al. 2008).

Another study aimed to characterize the variability of the genus by combining morphological characters (through numerical taxonomy) and isozyme analysis was carried out by Joseau et al. (2007) in natural populations where four *Prosopis* species (*P. chilensis*, *P. flexuosa*, *P. alba*, *P. nigra*, and intermediate individuals, considered as possible hybrids between these four species) coexist, in the semi-arid Chaco. Six isozyme systems were analyzed. Numerical taxonomy constituted a fundamental tool for the morphological analysis and the correct identification of the materials used in that study. They found five morphological and genetically differentiated groups. The high genetic diversity that characterized all morphological groups was related to the fact that they are species with a wide geographical distribution, open crossing system, and with animal-mediated seed dispersal (Hamrick and Murawski 1991). Three morphogenetic groups were considered as representatives of pure species: group 1 = *P. alba*, group 2 = *P. chilensis*, and group 5 = *P nigra*. In addition, it was possible to identify two groups of intermediate morphology that at the same time had the greatest genetic diversity and the lowest differentiation with respect to the rest of the groups. The two groups with intermediate morphology were also the groups with the highest morphological and genetic variability, while group 1 (*P. alba*) was the least morphologically and genetically diverse.

The great diversity of this "species complex" would come from processes of introgression with more "specific" species adapted to narrower niches. In this way, the complex would maintain a high potential evolutionary capacity through the conservation of high variation in certain species and in hybrid swarms. At the same time, some species that conform the complex are highly adapted with capacity to occupy much more restricted niches. Thus, the complex as a whole should be considered as an evolutionary unit and the taxonomic species as part of it. The cohesion of this species complex is maintained through the genetic exchange among species, mediated by hybrid swarms.

9.3 Natural Interspecific Hybridization Processes Evaluated Through Morphological Traits and Molecular Markers

The capacity of algarrobos' species to exchange genetic information makes the study of these contact areas of great interest, both for taxonomic, evolutionary, and ecological studies. On one hand, they could lead to the formation of new species and on the other, these new genetic combinations could colonize new habitats allowing a rapid selective response to natural or anthropized unstable environments (open or hybridized environments) (Grant 1989; Arnold 1997). In this context, the hybridization processes that operate in secondary contact areas between species would be the key to understand the evolutionary mechanisms of the complex. Their study offers the basic knowledge necessary for the development of strategies in forest improvement and for the dynamic conservation (sensu Namkoong et al. 2000) of these genetic resources. Below, we present some studies carried out in contact areas of different algarrobos species characterizing morphological and genetic variation.

9.3.1   *Morphological and Molecular Characterization of a Hybrid Zone between* **Prosopis alba** *and* **P. nigra** *in the Chaco Region of Northwestern Argentina*

Due to its wide distribution, *P. alba* displays multiple zones of contact with different congeners throughout its geographic range. One of these congeneric species is *P. nigra*, which has good capacity to settle in highly degraded sites (Demaio et al. 2002), supporting drier soils than *P. alba*.

The study was carried out near Santa Victoria town, in northwestern Argentina (22° 16′ S; 62 42′ W). Three discontinuous sites were sampled: sites 1 and 2, where pure morphotypes of *P. alba* and *P. nigra*, were present, respectively, and site 3, a zone where both intermediate and pure morphotypes were found (i.e., a putative hybrid swarm). The three studied sites constitute patches of forests without continuous forested masses connecting them. This co-distribution area provides a suitable natural experimental design for understanding hybridization between *P. alba* and *P. nigra* and its consequences at the phenotypic level. In addition, the larger sample size included in our study (138 from the hybrid site) with respect to the previous one (it included just 7 hybrids, Vega and Hernandez 2005) allowed us to evaluate the spatial structure and the association between genetic and morphological variation, which has been very useful in the study of other hybrid swarms of *Prosopis* (Teich et al. 2015).

Through the morphological classification of leaves and fruits it was possible to differentiate the individuals belonging to the pure species of *P. alba* and *P. nigra* and to morphological groups formed by individuals with intermediate morphology to both species. In the individuals of the pure sites of *P. alba* and *P. nigra*, an evident morphological identity was found, where the leaf type characteristic of *P. alba* corresponds to the characteristic fruit of this species, occurring the same in *P. nigra*. However, in site 3 belonging to the hybrid swarm, it was observed that there is an uncoupling in the type of leaf and fruit, that is, individuals with leaves of *P. nigra* and fruit of *P. alba* appeared. According to the literature, this is the first time that this type of decoupling has been detected in fruit and leaf characters. In other hybridization studies carried out in *P. chilensis* and *P. flexuosa* (Verga 1995; Joseau 2006), the hybrids presented intermediate characteristics to both pure species, that is, the decoupling between fruit and leaf variables detected here was not observed.

In addition, a genetic characterization was performed through 6 microsatellite loci. Genetic differentiation among sites was significant (P < 0.001), with the greatest observed divergence between pure sites, site 1 and 2 ($F_{ST} = 0.305$), whereas the hybrid site showed an intermediate differentiation ($F_{ST} = 0.087$ and 0.092 between sites 1 and 3, and between sites 2 and 3, respectively).

It has been shown in different plant species that molecular and morphological assignments do not always correlate and that phenotypically intermediate individuals are the result of a wide intraspecific variation or convergent evolution (Rieseberg et al. 1993; Craft et al. 2002; Curtu et al. 2007; Moran et al. 2012) rather than of hybridization. In this study, the high correspondence between leaf morphology and genetic differentiation in the three sites strongly confirms genetic and morphological differences between the two parental species and the existence of intermediate morphotypes, which correspond to interspecific hybrids.

9.3.2 Genetic and Morphometric Markers are Able to Differentiate Three Morphotypes Belonging to Section Algarobia of Genus Prosopis

Through successive field trips and observations, it was detected in Sumampa, south of the Province of Santiago del Estero (29° 22′ S, 63° 23′ W to 29° 15′ S, 63° 10′ W), an area where *P. alba* and *P. ruscifolia* overlap and apparently hybridize producing fertile hybrids, as was suggested by the presence of morphotypes with intermediate morphology to both species. Thus, a study was conducted in that area along a transect which represent an environmental gradient going from the foothills of the Sumampa mountain ranges up to the salty lowlands of Laguna de Mar Chiquita basin. This gradient means a gradual phytosociological shift from the *Schinopsis lorentzii* ("quebracho") forests in the highlands ("quebrachales") with *P. alba* as a secondary species, to a halophyte community with *P. ruscifolia* as the dominant species ("vinalar").

The evidence of hybridization between these two species is derived from the combination of morphological and molecular analyses through RAPD molecular markers (Random Amplified Polymorphic DNA). Both analyses allow the recognition and significant differentiation of three groups, one corresponding to *P. alba*, another to *P. ruscifolia* and a third group to interspecific hybrids. These results indicate that the leaf morphology between these two species also constitutes a tool for the identification of interspecific hybrids. Surprisingly, the dispersion of the phenotypic values around the average was not different between the hybrid group and the putative parental species. This is not expected if we assume that the hybrid group is genetically more variable as a consequence of the addition of the set of genes from both parental species. One possible explanation would be that phenotypic variation is caused by environmental rather than genetic factors. If this were the case, the phenotypic variation would be the consequence of plasticity. However, a correlation analysis performed between phenotypic and genetic distances suggests that in almost all traits heritabilities were significant. The three groups studied showed high genetic variability. However, the hybrids showed significantly higher values than the determined morphotypes such as *P. alba* and *P. ruscifolia* (Table 9.1).

It is important to note that the hybrid morphotype was identified as *P. vinalillo* by Burkart (1976). In fact, the author suggested that this species could be a hybrid between *P. alba* and *P. ruscifolia*. Moreover, so far we have not detected a pure population of *P. vinalillo*; this species is always associated to the presence of both *P. alba* and *P. ruscifolia* (Ferreyra et al. 2013).

Table 9.1 Correlations (above diagonal) between morphological characters and RAPD distance matrices and their significance (below diagonal) according to Mantel tests based on (10,000 permutations)

	RAPD	LEW	LEA	LEX	LEX/LEA	LEF	LEL/LEW	LEL	PEL	PIL	NLP
RAPD		0.23	0.15	0.14	0.00	0.34	0.39	0.17	0.01	0.05	0.51
LEW	0.00		0.94	0.37	0.12	0.48	0.58	0.89	−0.01	0.17	0.40
LEA	0.00	0.00		0.27	0.07	0.28	0.42	0.87	−0.02	0.15	0.26
LEX	0.00	0.00	0.00		−0.02	0.15	0.40	0.29	−0.03	0.02	0.29
LEX/LEA	0.41	0.02	0.03	0.71		0.22	0.07	0.06	0.00	0.03	0.00
LEF	0.00	0.00	0.00	0.01	0.00		0.53	0.29	0.08	0.06	0.42
LEL/LEW	0.00	0.00	0.00	0.00	0.07	0.00		0.34	−0.04	0.26	0.65
LEL	0.00	0.00	0.00	0.00	0.07	0.00	0.00		−0.01	0.09	0.31
PEL	0.26	0.48	0.71	0.88	0.37	0.06	0.99	0.52		−0.03	−0.01
PIL	0.07	0.01	0.01	0.19	0.16	0.08	0.00	0.03	0.88		0.29
NLP	0.00	0.00	0.00	0.00	0.35	0.00	0.00	0.00	0.39	0.00	
NPI	0.00	0.00	0.00	0.00	0.54	0.00	0.00	0.00	0.51	0.01	0.00

9.3.3 Morphological and Genetic Differentiation of Disjunct Prosopis chilensis Populations across its Distribution Range

Prosopis chilensis (Molina) Stuntz emend. Burkart, locally known as "white algarrobo" or "Chilean algarrobo," belongs to section *Algarobia* and is a tree species with significant distant leaflets on the rachis as a distinctive feature. It has a disjunct geographical distribution: growing in southern Peru (Burghardt et al. 2010), in southeastern Bolivia, and in northwestern Argentina (Burkart 1976). The species also distributes in central Chile, although archaeobotanical and paleoecological data suggest it would not be native to Chile, appearing in the late Holocene and most likely associated to human activity (McRostie et al. 2017). According to its wide distribution, *P. chilensis* exhibits considerable variation in phenotypic characters. However, it is not clear to what extent the individuals from disjunct areas differ in morphological and genetic characters. Therefore, morphological and genetic studies were performed in *P. chilensis* in order to characterize Argentine and Bolivian populations (Aguilar et al. 2020).

First, from each tree (50 and 54 individuals from Argentina and Bolivia were collected, respectively), five totally expanded leaves were sampled from different places of the canopy and photographed; a total of eleven leaf characters were measured using the software HOJA 3.4. In a Principal Component Analysis, we could observe that morphological characters varied between the two provenances. The first two axis of the principal component analysis (PC1 and PC2) explained 65.5%

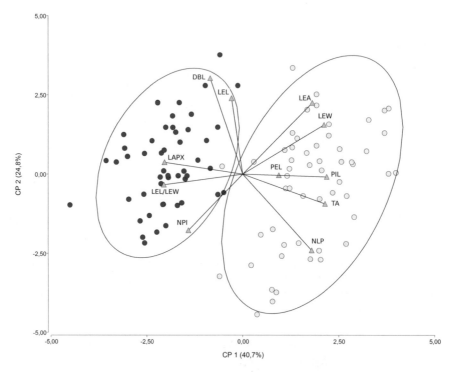

Fig. 9.4 Principal Component Analysis (PCA) based on morphological leave traits of *Prosopis chilensis* from Argentina and Bolivia. Blue dots represent each tree from Argentina while yellow dots represent individuals from Bolivia. Variables are represented with triangles

of the total variability among individuals (Fig. 9.4). PC1 (which accounted for 40.7% of the total variability) was positively correlated to the number of leaflets per pinna (NLP), pinna length (PIL), and leaflets width (LEW) and negatively correlated to leaflet shape (LEL/LEW) and to the leaflet apex (LAPX). PC2 was positively related to the distance between leaflets (DBL), leaflet length (LEL), width (LEW), and leaflet area (LEA) and negatively related to the number of pinnas (NPI) and the number of leaflets per pinna (NLP), accounting for 24.8% of the total variability (Fig. 9.4). *Prosopis chilensis* trees from Argentina have shorter pinnae, lower number of leaflets per pinnae and narrower and more elongated leaflets than the trees from Bolivia.

Additionally, using a Linear Discriminant Analysis (LDA) with leaf morphometric variables as predictors, we observed that all individuals from Argentina were classified correctly, while one Bolivian individual was assigned to the Argentinean group. Then, we explore which was the best variable to classify both groups, and the best splitter in a Classification and Regression Tree (CART; Breiman et al. 1984) was leaflet apex (LAPX). This predictor allowed to classify individuals belonging to the Argentine group (p = 0.86) when LAPX>0.155 cm (58 individuals) and to the Bolivian group when LAPX≤0.155 cm (54 individuals, p = 1).

Despite this clear morphological differentiation (expressed in their home habitats separately), we wanted to explore to what extent these disjunct areas of distribution differed genetically. Moreover, we wanted to get insights into the underlying historical process shaping this disjunct distribution. Noncoding fragments of the chloroplast genome are the most appropriate markers for phylogeographical studies due to their uniparental inheritance and their ability to detect neutral processes of evolution (Avise 2000). However, previous studies had shown that chloroplast DNA markers were not able to differentiate species of section *Algarobia* and used other genetics techniques (Ramírez et al. 1999; Vázquez-Garciadueñas et al. 2003; Ferreyra et al. 2004, 2007; Vega and Hernandez 2005; Catalano et al. 2008). Nevertheless, we performed a screening of multiple primers (*trn*Q-*rps*16, *trn*H-*psb*A, *rpl32*R-*ndh*F, *rpl32*F-*trn*L, *trn*D-*trn*T; Shaw et al. 2007) and found that the noncoding chloroplastic regions *ndh*F-*rpL32* and *rpl32*F-*trn*L retrieved enough variability at intraspecific level. The noncoding chloroplastic region *ndh*F-*rpL32* was sequenced in 52 individuals from 15 localities of Argentina and 24 localities of Bolivia, obtaining eight haplotypes. Haplotype genealogy was reconstructed using the *median-joining* algorithm implemented in Network v 5.0.0.1 (Bandelt et al. 1999). It showed a typical star-like topology with two very frequent and widespread haplotypes, one in Bolivia (H1, N = 17) and the other in Argentina (H6, N = 27), from which derived haplotypes of very low frequency (H2, H3, H4, H5, H7 and H8), separated from the central ones by few mutational steps (Fig. 9.5). Moreover, the presence of a past historical barrier between the two geographical groups was evidenced using the spatial analysis of the molecular variance implemented in SAMOVA v.2.0 (Dupanloup et al. 2002) retrieving an optimal partitioning genetic diversity with k = 2 (FCT = 0.84; p < 0.0001).

The conclusions from both morphological and molecular analyses are consistent and indicate that the tree species *P. chilensis* differs at least into two groups, accordingly with its disjunct geographic distribution. Argentinean *P. chilensis* trees have shorter and narrower leaves than the trees from Bolivia, which could be at least partially associated to climatic differences between both geographical regions. Also, genetic evidence reinforced that significant differences were detected between the disjunct populations.

9.3.4 Genetic-Adaptive Studies in Populations of Prosopis alba

Genetic diversity is the basis on which natural selection acts, giving organisms the ability to adapt to changes in their environment (Krutovsky and Neale 2005). Populations with little genetic variation are more vulnerable to the arrival of new pests or diseases, pollution, changes in climate, and habitat destruction due to human activities or other catastrophic events (Krutovsky and Neale 2005). Particularly in the face of the impact of climate change, natural populations can survive through migration to more favorable sites, or by phenotypic plasticity or local adaptations thus remaining in situ (Aitken et al. 2008). As plants, and more

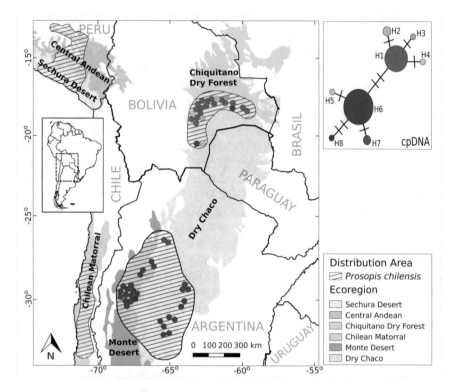

Fig. 9.5 (**a**) Spatial genetic diversity distribution of *Prosopis chilensis*. The circles represent individuals sampled from *P. chilensis*, colored according to their respective haplotype. Shaded areas indicate the different ecoregions (from Olson et al. 2001) and striped areas the geographical distribution of *P. chilensis*, respectively. (**b**) Network representing the genealogical relationships among the 8 cpDNA haplotypes (designated with numbers) found. The size of the circle is proportional to the frequency of haplotypes. The transverse lines indicate the amount of mutational changes between different haplotypes (modified from Aguilar et al. 2020)

specifically trees, are sessile organisms, and their rate of migration and generational time are slower than changes in climate (Aitken and Whitlock 2013), local adaptation and plasticity have a more decisive impact than migration. Natural selection is the driving force of local adaptation, and usually increases along environmental gradients (Grivet et al. 2011; Song et al. 2016).

The observed pattern of population differentiation is the product of mutation, migration, genetic drift, or selection (Wright 1951). While the first three are considered neutral processes, natural selection acts on certain traits and genes differently depending on the environmental conditions, which can vary both in space and time (Kawecki and Dieter 2004; Savolainen et al. 2013). Therefore, local adaptation occurs when the individuals of a given environment perform better than individuals from other places (Williams 1966).

The identification of genes under natural selection is key to understanding the genetic basis of adaptation to different environments and also useful for practical applications, such as the development of programs for the conservation of diversity and in genetic improvement assisted by molecular markers (Li et al. 2014; Song et al. 2016). The genomic regions involved in the selection can be identified in natural populations by detecting outlier loci that exhibit non-neutral variation patterns (Vitalis et al. 2001; Luikart et al. 2003). This is because local adaptation and directional selection reduce the genetic variation of these loci within populations and increase the differentiation between populations (Stinchcombe and Hoekstra 2008). By identifying these loci and excluding them from population analyses, better approximations of the historical population parameters and of the genetic structuring product of the neutral processes can be obtained (Vitalis et al. 2001).

In particular, *P. alba* is the species of algarrobo of greater geographic distribution of the genus in the Gran Chaco, and the one that has been most studied from the genetic point of view. So far there are only studies on genetic variability using neutral molecular markers (Bessega et al. 2009; Bessega et al. 2011), which reflect population dynamics and evolutionary forces such as genetic drift, mutation and migration, but can only be related to processes of selection indirectly. The recent sequencing of the transcriptome of *P. alba* and the analysis of candidate genes has enabled the use of polymorphic markers of genome-coding regions related to drought and salinity tolerance (Torales et al. 2013). These tools allow the direct study of the adaptive genetic diversity of the species.

In *Prosopis alba*, it was recently studied the genetic structure of three natural populations located in contrasting environmental conditions with respect to precipitation, using neutral molecular markers (Mottura et al. 2005; Bessega et al. 2013) as well as potentially adaptive markers (EST-SSRs) (Torales et al. 2013). From the analysis of F_{ST} for detection of outlier loci, a strong signal of divergent selection was detected in two markers, one EST-SSR (P73) and one genome microsatellite (gSSR) (GL8). Both markers had high differentiation rates (GL8, F_{ST} = 11.1 and P73, F_{ST} = 13.7%) (López Lauenstein 2019). The putatively neutral microsatellite marker (GL8) was developed by Bessega et al. (2013) from the massive DNA sequencing of three species of the Algarrobia section of the genus *Prosopis* (*P. pallida*, *P. velutina*, and *P. glandulosa* and putative hybrids). Since it comes from a massive sequencing, it is possible that the marker is not in a noncoding region or, alternatively, that is strongly linked to a coding region. The P73 marker, on the other hand, was developed from de novo sequencing of the *P. alba* transcriptome (Torales et al. 2013), so its location in a coding region is accurate, being its putative function associated to the pentatricopeptide repeat-containing protein (PPR) type. PPR proteins are a large family of the "RNA binding protein" type, involved in gene expression processes, mainly in organelles (Lurin et al. 2004), but also at the nuclear level. In plants, progress has been made particularly with the characterization in the control of processes where these proteins facilitate splicing, stability and transcription of RNAs. Key roles have been cited for PPR proteins in response to various stresses such as drought, salinity, and cold (Jiang et al. 2015) and also in general in response to biotic and abiotic stresses (Xing et al. 2018).

Population genetic structure analysis reveals a strong structure among the three natural populations studied (López Lauenstein 2019). However, when the two markers that are under selection were excluded from the analysis, no genetic structure was observed. This suggests that genetic differences and structuring observed among populations are mainly due to the processes of local adaptation and, to a lesser extent, to stochastic processes such as genetic drift. In this context, these processes of local adaptation should be taken into account when identifying the most appropriate seed sources best adapted to each particular condition where forest plantations, reforestation plants, or ecosystem restoration will be implemented (López Lauenstein 2019).

An alternative approach was also applied in order to detect evidences of local adaptation in *P. alba* (Bessega et al. 2015). As proposed by Merila and Crnokrak (2001), the univariate neutrality test analysis based on the comparison of the levels of phenotypic and molecular differentiation can contribute to discriminate selection from genetic drift effects over a group of quantitative traits. Following the classical interpretation, if $F_{ST} = Q_{ST}$, there is no need to invoke natural selection and differentiation is explained simply by gene flow-genetic drift interaction; if $Q_{ST} > F_{ST}$, positive directional natural selection favoring different phenotypes in different populations is suspected; and $Q_{ST} < F_{ST}$ suggests stabilizing selection favoring the same phenotype in different populations (Merila and Crnokrak 2001).

The study was conducted in an orchard established in 1991 in San Carlos (Province of Santiago del Estero) initiated from seeds collected from open pollinated trees (half-sib families) belonging to 8 sampling sites of northern Argentina: Añatuya, Castelli, Gato Colorado, Ibarreta, Pinto, Quimili, Rio Dulce Irrigation Zone, and Sumampa. A total of 142 individuals belonging to 32 different families, 7 replicates (blocks), and 4 trees per replicate (with a 4 × 4 m spacing) were sampled for the present study.

In order to apply the Q_{ST}-F_{ST} comparison tests, all the individuals were genotyped using 6 unlinked microsatellites markers (Mo05, Mo07; Mo08, Mo09, Mo13, and Mo16) and characterized morphologically. Three life history traits (height [H], basal diameter [BD], biomass [BMS]), 11 leaf characters (petiole length [PEL], number of pairs of leaflets per pinna [NLP], pinna length [PIL], spine length [SPL], number of pinnae [NP]), leaflet length [LEL], leaflet length/width [LEL/LEW], leaflet falcate [LEF], leaflet apex [LEX], and leaflet apex/total area [LEX/LEA]), and the spine length (SPL) were measured (see details in Bessega et al. (2009)). From the analysis of variance components, morphological differentiation indices (Q_{ST}) were estimated for each trait following Spitze (1993) approximation, assuming a half-sib model (Pressoir and Berthaud 2004), applying the following expression.

$$Q_{ST} = \frac{\sigma_O^2}{\sigma_O^2 + 2\left(4\sigma_f^2\right)},$$

where σ_O^2 is the variance among provenances, and σ_f^2 the variance among families. The coefficient F_{ST} was also estimated based on the molecular information.

The phenotypic differentiation (Q_{ST}) ranged from near cero for SPL and NLP to 0.36 for NPI, with an average over traits of 0.14. Molecular differentiation among origins was low but significant ($F_{ST} = 0.069$, p F_{ST} whereas for the remaining five traits (NLP, PIL, SPL, LEF, LEX/LEA) $Q_{ST} - F_{ST} < 0$. Cases where Q_{ST} was higher than the upper limit of F_{ST} confidence interval were considered as suggestive of directional selection while the cases where Q_{ST} was lower than the lower F_{ST} confidence interval limit are indicative of stabilizing selection. As discussed in Bessega et al. (2009), the univariate tests suggest heterogeneous selection over phenotypic traits with different optima in different populations. This result is in agreement with several authors that described the occurrence of ecotypes adapted to particular areas in several species of the genus *Prosopis* (Morello et al. 1971; Burghardt et al. 2004; Verga et al. 2009, 2014). The heterogeneous selection over phenotypic traits has important implications for gene resource management, suggesting that population source and traits with adaptive significance should be considered in breeding programs as a crucial point prior to any selecting decision.

9.4   Estimation of Mating System and Pollen Dispersal Parameters

Mating system and pollen dispersal parameters can be estimated indirectly in different populations based on molecular markers (Ritland 1990; Austerlitz and Smouse 2001; Smouse et al. 2001). An efficient method based on the mixed multilocus mating model was developed by Ritland (2002) and is implemented in the MLTR program (Ritland 2002). The method requires the grouping of the population samples into families (seeds collected from the same tree) and the multilocus genotyping of corresponding individuals. The comparison of the allelic frequencies in the pollen cloud that attends each mother plant allows, among other parameters, to establish the outcrossing rate (t), the biparental inbreeding ($tm - ts$), the correlated outcrossing rate (rt), the correlated paternity (rp), the pollen structure parameter (Φ_{ft}), and the effective number of pollen donors siring each mother plant (Nep).

On the basis of genetic data, it is also possible to indirectly assess the distance (in meters) that pollen is dispersed in populations. From the genotypes of the mother plants and their offspring (seeds collected individually from each tree), it is possible to estimate again the composition of the pollen cloud and indirectly estimate the pollen dispersion parameters using the computer program POLDISP 1.0 (Robledo-Arnuncio et al. 2007), applying two different algorithms: TwoGener (Smouse et al. 2001; Austerlitz and Smouse 2002) and Kindist (Robledo-Arnuncio et al. 2006). Both methods are based on the expected relationship between the physical distance between the sampled mother plants and the interclass genetic correlation (Φ_{FT}) between pairs of individuals (seeds) for the pollen set sampled in each mother. The coefficient Φ_{FT} can be considered an estimator of kinship, and it is expected that the kinship between the descendants of different mother plants is inversely proportional to the distance that separates them, and directly proportional to the dispersal capacity of the pollen. The difference between the estimates generated by the two algorithms is related to additional parameters required by both methods, such as the density of the study area, that in one case is predicted and the other has to be provided by the researcher and some assumptions such as floral asynchrony, uniformity in male fertility rates, etc. (see Austerlitz and Smouse 2002).

Mating system and pollen dispersal parameters were studied in a wild *P. alba* population (Bessega et al. 2011) and in a preserved *P. flexuosa* population (Bessega et al. 2017) using the indirect methods described above. These studies based on microsatellite markers showed that both species are mainly outcrossers (*tm* = 1 in *P. alba* and *tm* = 0.996 in *P. flexuosa*). However, in both populations inbreeding was detected (from the difference *tm-ts*) at low but significant frequency, which is the result of crosses between related individuals. The results showed that the number of pollen donors per mother plant (i.e., the number of trees that act as male parents of each family) was on average 6 individuals for *P. alba* and 3 for *P. flexuosa* (Table 9.2).

Additionally, studies that evaluated the internal constitution of each family, defined as the set of seeds collected from each mother plant, were carried out. This is equivalent to determining the proportion of half-sibs and full-sibs present in each fraternal group as a whole and considering the seeds that belongs to the same or different pods. In this case, the kinship was estimated according to the proportion of alleles shared among all the pairs of individuals of each fraternal group. In this way, it could be established that the proportion of full-sibs in *P. alba* is 64% when considering seeds from the same fruit, while this proportion drops to 10% when seeds of different fruits are compared. In *P. flexuosa,* the estimates were similar and vary between 71% and 24%, respectively. This information can be interpreted on the basis of the behavior of the pollinators and the anatomy of the inflorescence. The pollinating insects associated with these species usually focus their efforts on plants with higher floral density and limit their movement between plants to the nearest neighbors. As a consequence, each pollination event involves the pollen of a single or a few pollen donors fertilizing each inflorescence favoring the occurrence of full-sibs in each fruit.

The dispersal analysis allowed the description of the average dispersion distance of pollen in *P. alba* and *P. flexuosa* that ranges between 4 and 31 m, as estimated using the Kindist or TwoGener algorithms. Indirect estimation in these *Prosopis* species indicates that the dispersion of pollen and seeds would be limited. These results can be discussed in reference to the *algarrobo* management program as they provide valuable information to be considered in order to avoid the effects of inbreeding and genetic drift within populations as a consequence of the intensive use of these resources.

Table 9.2 Indirect estimations of mating system indices and pollen dispersal distances in *P. alba* and *P. flexuosa* populations

Model	Indices	*P. alba* estimate Bessega et al. 2011	*P. flexuosa* estimate Bessega et al. 2011
Mixed mating	Multilocus outcrossing rate (*tm*)	1.000*	0.996*
	Biparental inbreeding (*tm-ts*)	0.049*	0.193*
	Correlated outcrossing rate (*rt*)	0.107*	0.999[NS]
	Correlated paternity (*rp*)		
	Global	0.166*	0.312*
	Among fruits	0.098*	0.240*
	Within fruits	0.638[NS]	0.714*
Pollen pool structure	Φ_{ft} (pollen structure parameter)	0.10	0.17
	Effective number of pollen donors under pollen structure model (N_{ep})	5.11	2.93
	Kindist		
	Mean pollination distance (δ) Normal	5.36 m	4.56 m
	Mean pollination distance (δ) exponential	11.96 m	5.38 m
	TwoGener		
	Mean pollination distance (δ) Normal	13.74[a]/13.81[b] /23.74[c] m	20.12[a] /20.35[b] m
	Mean pollin. Distance (δ) exponential	15.5[a]/15.58[b] /30.92[c] m	17.83[a] /18.04[b] m

[a]Distance obtained from the global Φ_{ft} value, [b]pairwise Φ_{ft} values assuming the density equals the census density (*D*), [c]distance estimated obtained from the joint estimate of the effective density and pollen dispersal variance (σ^2). a, b, and c estimates were obtained using TwoGener. *and [NS] indicate significant (P < 0.05) and nonsignificant values, respectively

9.5 Phenological Analysis of a Stand and Its Application in a Hybrid Swarm of *Prosopis* spp.

Another useful method for the study of the mating system is to incorporate the floral phenology factor into the analysis. Expected genotype frequencies are obtained in the offspring that already involve the phenological behavior of the stand, empirically observed.

Accordingly, a study was performed (Córdoba and Verga 2006) in two periods (2002–2003 and 2003–2004) in a plot located in the Provincial Park Chancaní (Province of Córdoba), in the Arid Chaco (31° 19′ S- 65° 27′ W). The plot consisted of 112 algarrobos' trees and is located in an abandoned farm with 25 years of recolonization of the native forest. There come into contact both pure *P. chilensis* and *P. flexuosa* trees, and numerous individuals with intermediate morphological characteristics.

The work began with the classification of the individuals of the hybrid swarm from a morphological point of view. On the basis of this classification, the phenological behavior of each group of individuals was studied, and based on this behavior some descriptive parameters were estimated on the possible mating system that reigns in the hybrid swarm. The morphological classification of the trees was carried out taking into account leaf and fruit characters. Group 1, individuals with intermediate morphological characteristics; group 2, individuals of *P. flexuosa*; and group 3, individuals of *P. chilensis*. The individual flowering of each tree was monitored and the intensity of flowering (Flower Production Index) was estimated. The opening of the flowers and their receptive stage (at which time the flowers can be fertilized) were taken into account.

An important regularity was found regarding the number of days that elapsed between the beginning of the flowering of one group with respect to the others and the moment of the flowering peak. This regular behavior of the groups defined suggests that both the onset of flowering and the time at which the maximum production of flowers (flowering peak) occurs would be determined genetically and that these groups respond differentially in their interaction with the environment.

With all this information a first approximation was made to the description of the mating system of the hybrid swarm, trying to elucidate between which individuals and groups mating is possible and the frequency with which mating occurs. For this, the recorded data of the flowering were taken into account and some assumptions were made: (1) all the trees of the stand that have produced flowers can participate in the crosses; (2) two individuals can only cross if they bloom simultaneously; (3) it is considered that 10% of crosses come from self-fertilization; (4) the rest of the crosses can occur between individuals of the same or different groups; and (5) the amount or probability of crossing between trees of the same or different groups depends on (a) the number of flowers that the tree presents at that time, without considering those that will be fertilized by flowers of the same individual (equivalent to 10% of their flowers), and (b) the probability that at any given moment a tree A fecundates a tree B depends on the relative contribution of flowers of tree A with respect to the total number of flowers present at that moment in the stand.

Finally, the mating preference within and between the morphological groups was calculated. The trees of group 1 (individuals with intermediate morphological characteristics) show preference to mate with individuals of their own group, avoiding crossing with the trees of groups 2 (*P. flexuosa*) and 3 (*P. chilensis*). Mating between individuals of this group are 7 or 4.5 times more frequent than expected by random whereas the opposite occurs with groups 2 and 3, which mate less than expected by random. Group 2 also had a greater preference for mating with the trees of group 1, while mating between trees in the same group is similar to that expected at random. The number of crosses between the trees of the group will increase or decrease as their density increases or decreases, respectively. Individuals of group 3 have a greater preference for trees of group 1, than with those of their same group and repel or avoid mating with group 2.

The number of hybrids in the stand would tend to increase in each generation. It can be said that the trees have preference to mate with the individuals of group 1, since being group 1 only 7% of the stand trees, it participated in 47% of all the crosses. If we add to this, the low preference of groups 2 and 3 to mate with themselves, it would be expected that the offspring in the stand have a significant genetic component coming from group 1. According to these results, in the long term, a loss of the genetic identity of the *P. chilensis* population is expected as a genetically differentiable group.

With respect to pure species, hybrids seem to have adaptive advantages for their development in environments that have been modified by human action. Additionally, considering that *P. chilensis* populations have been affected by logging and that this species has limited its development in these increasingly extensive environments in the region, in the long term, the disappearance of the group as an entity could be expected, even without the negative effect on these populations as a result of anthropogenic action. To advance and deepen the knowledge of the behavior of the flowering and mating system of the hybrid swarm, controlled crosses are being carried out to effectively determine the possible crosses in the stand, complementing this study with the use of molecular markers (microsatellites).

References

Aguilar DL, Acosta MC, Baranzelli MC, Sérsic AN, Delatorre-Herrera J, Verga A, Cosacov A (2020) Ecophylogeography of the disjunct South American xerophytic tree species *Prosopis chilensis* (Fabaceae). Biol J Linn Soc 20:1–17

Aitken SN, Whitlock MC (2013) Assisted gene flow to facilitate local adaptation to climate change. Annu Rev Ecol Evol Syst 44:367–388

Aitken S, Yeaman S, Holliday J, Wang T, Sierra M (2008) Adaptation, migration or extirpation: climate change outcomes for tree populations. Evol Appl 1:95–111

Arnold M (1997) Natural hybridization and evolution. Oxford University press, Oxford

Austerlitz F, Smouse PE (2001) Two-generation analysis of pollen flow across a landscape. II. Relation between FFT, pollen dispersal and interfemale distance. Genetics 157:851–857

Austerlitz F, Smouse PE (2002) Two-generation analysis of pollen flow across a landscape. IV Estimating the dispersal parameter. Genetics 161:355–363

Avise JC (2000) Phylogeography: the history and formation of species. Harvard University Press, Cambridge, MA

Bandelt HJ, Forster P, Röhl A (1999) Median-joining networks for inferring intraspecific phylogenies. Mol Biol Evol 16:37–48

Barton NH (2001) The role of hybridization in evolution. Mol Ecol 10:551–568

Bessega C, Ferreyra L, Vilardi J, Saidman BO (2000) Unexpected low genetic differentiation among allopatric species of section Algarobia of *Prosopis* (Leguminosae). Genetica 109:255–266

Bessega C, Saidman BO, Darquier MR (2009) Consistency between marker- and genealogy-based heritability estimates in an experimental stand of *Prosopis alba* (Leguminosae). Am J Bot 96:458–465

Bessega C, Pometti CL, Ewens M, Saidman BO, Vilardi JC (2011) Strategies for conservation for disturbed *Prosopis alba* (Leguminosae, Mimosoidae) forests based on mating system and pollen dispersal parameters. Tree Genet Genome 8:277–288

Bessega C, Pometti CL, Miller JT, Watts R, Saidman BO, Vilardi JC (2013) New microsatellite loci for *Prosopis alba* and *P. chilensis* (Fabaceae). Appl. Plant Sci 1:1200324

Bessega C, Pometti C, Ewens M (2015) Evidences of local adaptation in quantitative traits in *Prosopis alba* (Leguminosae). Genetica 143:31–44

Bessega C, Pometti CL, Campos C, Saidman BO, Vilardi JC (2017) Implications of mating system and pollen dispersal indices for management and conservation of the semi-arid species *Prosopis flexuosa* (Leguminosae). For Ecol Manag 400:218–227

Breiman L, Friedman J, Stone CJ, Olshen RA (1984) Classification and regression trees. Routledge, New York, 368 pages

Burghardt AD, Palacios RA (1981) Caracterización electroforética de algunas especies de *Prosopis* (Leguminosae). 12th Congreso Argentino de Genética, p 11

Burghardt AD, Espert SM, Braun-Wilke SH (2004) Variabilidad genética en *Prosopis ferox* (Mimosaceae). Darwin 42:31–36

Burghardt AD, Brizuela MM, Mom MP, Albán L, Palacios RA (2010) Análisis numérico de las especies de *Prosopis* L. (Fabaceae) de las costas de Perú y Ecuador. Rev Peru Biol 17:317–324

Burkart A (1976) A monograph of genus *Prosopis* (Leguminosae subfam. Mimosidae). J Arnold Arbor 57:219–249

Carlson J, Adams C, Holsinger K (2016) Intraspecific variation in stomatal traits, leaf traits and physiology reflects adaptation along aridity gradients in a south African shrub. Ann Bot 117:197–207

Catalano SA, Vilardi J, Tosto D, Saidman BO (2008) Molecular phylogeny and diversification history of *Prosopis* (Fabaceae: Mimosoideae). Biol J Linn Soc 93:621–640

Córdoba A, Verga AR (2006) Método de análisis fenológico de un rodal: Su aplicación en un enjambre híbrido de *Prosopis* spp. Ciencia e Investigación Forestal (CIFOR) 14:92–109

Craft KJ, Ashley MV, Koenig WD (2002) Limited hybridization between *Quercus lobata* and *Quercus douglasii* (Fagaceae) in a mixed stand in central coastal California. Am J Bot 89:1792–1798

Curtu A, Gailing O, Finkeldey R (2007) Evidence for hybridization and introgression within a species-rich oak (*Quercus spp.*) community. BMC Evol Biol 7:218

Demaio P, Karlin UO, Medina M (2002) Árboles nativos del centro de Argentina. Buenos Aires. L.O.L.A (Literature of Latin America). Secretariat of the Organization of American States, Washington, DC

Dupanloup I, Schneider S, Excoffier L (2002) A simulated annealing approach to define the genetic structure of populations. Mol Ecol 11:2571–2581

Ferreyra LI, Bessega C, Vilardi J, Saidman BO (2004) First report on RAPD patterns able to differentiate some Argentinean species of section Algarobia (*Prosopis*, Leguminosae). Genetica 121:33–42

Ferreyra LI, Bessega C, Vilardi J, Saidman BO (2007) Consistency of population genetics parameters estimated from isozyme and RAPDs dataset in species of genus *Prosopis* (Leguminosae, Mimosoideae). Genetica 131:217–230

Ferreyra LI, Vilardi J, Tosto D, Julio N, Saidman B (2010) Adaptive genetic diversity and population structure of the "Algarrobo" [*Prosopis chilensis* (Molina) Stuntz] analysed by RAPD and isozyme markers. Eur J For Res 129:1011–1025

Ferreyra LI, Vilardi JC, Verga A, López V, Saidman BO (2013) Genetic and morphometric markers are able to differentiate three morphotypes belonging to section Algarobia of genus *Prosopis* (Leguminosae, Mimosoideae). Plant Syst Evol 299:1157–1173

Geber M, Griffen LR (2003) Inheritance and natural selection on functional traits. Int J Plant Sci 164:S21–S42

Grant V (1981) Plant speciation. Columbia University Press, New York

Grant V (1989) Especiación Vegetal. Editorial Limusa

Grivet D, Sebastiani F, Alía R, Bataillon T, Torre S, Zabal-Aguirre M et al (2011) Molecular footprints of local adaptation in two mediterranean conifers. Mol Biol Evol 28:101–116

Hamrick JL, Murawski DA (1991) Levels of allozyme diversity in populations of uncommon neotropical tree species. J Trop Ecol 7:395–399

Hunziker JH (1986) Hybridization and genetic variation of Argentina species of *Prosopis*. For Ecol Manag 16:301–305

Jiang S, Mei C, Liang Sh YY, Lu K, Wu Z, Wang X (2015) Crucial roles of the pentatricopeptide repeat protein SOAR1 in *Arabidopsis* response to drought, salt and cold stresses. Plant Mol Biol 88:369–385

Joseau MJ (2006) Caracterización morfológica y genética de poblaciones del género *Prosopis* del Chaco Semiárido del Norte de Córdoba y Sur de Santiago del Estero. Doctoral thesis, Fac. Cs. Agrs., Universidad Nacional de Córdoba, Argentina

Joseau MJ, Verga AR (2005) Caracterización morfológica y genética de poblaciones del género *Prosopis* del Chaco Semiárido del Norte de Córdoba y Sur de Santiago del Estero. XXXIV Congreso Argentino de Genética. Trelew, Chubut, 11–14 Sept. S-191

Joseau MJ, Verga AR, Diaz MP, Julio N (2007). Caracterización morfológica y genética de poblaciones del género *Prosopis* del Chaco Semiárido del Norte de Córdoba y Sur de Santiago del Estero. Ciencia e Investigación Forestal-Instituto Forestal /Chile 13(3):427–448

Joseau MJ, Verga A, Diaz MP, Julio N (2013) Morphological diversity of populations of the genus *Prosopis* in the semiarid Chaco of northern Córdoba and southern Santiago del Estero. Am J Plant Sci 4:2092–2111

Kawecki TJ, Dieter E (2004) Conceptual issues in local adaptation. Ecol Lett 7:1225–1241

Krutovsky K, Neale D (2005) Forest genomics and new molecular genetic approaches to measuring and conseving adaptive genetic diversity in forest trees. In: Geburek T, Turok J (eds) Conservation and managment of forest genetic resources in Europe. Arbora, Zvolen, Slovakia, pp 369–390

Li C, Sun Y, Huang HW, Cannon CH (2014) Footprints of divergent selection in natural populations of *Castanopsis fargesii* (Fagaceae). Heredity 113:533–541

López Lauenstein D (2019) Adaptación a la sequía en el contexto de cambio climático en poblaciones de *Prosopis alba* del Parque chaqueño: una aproximación desde la genética y la ecofisiología. Doctoral thesis: Fac. Cs. Exactas, Físicas y Naturales. Universidad Nacional de Córdoba (Argentina)

Luikart G, England PR, Tallmon D, Jordan S, Taberlet P (2003) The power and promise of population genomics: from genotyping to genome typing. Nat Rev Genet 4:981–994

Lurin C, Andrés C, Aubourg S, Bellaoui M, Bitton F, Bruyere C et al (2004) Genome-wide analysis of *Arabidopsis* pentatricopeptide repeat proteins reveals their essential role in organelle biogenesis. Plant Cell 16:2089–2103

McRostie VB, Gayo EM, Santoro CM, De Pol-Holz R, Latorre C (2017) The pre-Columbian introduction and dispersal of Algarrobo (*Prosopis*, section *Algarobia*) in the Atacama Desert of northern Chile. PLoS One 12:7

Merila J, Crnokrak P (2001) Comparison of genetic differentiation at marker loci and quantitative traits. J Evol Biol 14:892–903

Moran EV, Willis J, Clark JS (2012) Genetic evidence for hybridization in red oaks (*Quercus* sect. Lobatae, Fagaceae). Am J Bot 99:92–100

Morello JH, Crudelli NE, Saraceno M (1971) Los vinalares de Formosa, República Argentina. Serie Fitogeográfica 11, INTA

Mottura M, Finkeldey R, Verga AR, Gailing O (2005) Development and characterization of microsatellite markers for *Prosopis chilensis* and *Prosopis flexuosa* and cross-species amplification. Mol Ecol Notes 5:487–489

Namkoong G, Koshy MP, Aitken S (2000) Selection. In: Young A, Boshier D, Boyle T (eds) Forest conservation genetics. Principles and practice. CSIRO/CABI, Collingwood/Oxford, pp 101–111

Olson D, Dinerstein E, Wikramanayake E, Burgess N, Powell GC, Underwood E et al (2001) Terrestrial ecoregions of the world: a new map of life on earth. Bioscience 51:933–938

Palacios RA, Bravo LD (1981) Hibridización natural en *Prosopis* (Leguminosae) en la región chaqueña Argentina. Evidencias morfológicas y cromatográficas Darwiniana 23:3–35

Palacios R, Brizuela M (2005) Fabaceae, parte 13. Subfam. II. Mimosoideae, parte 4. Tribu VI. Mimoseae, parte B. Prosopis L. In: Anton AM, Zuloaga FO (eds) Flora Fanerogámica Argentina, vol 92, pp 3–25

Pasiecznik NM, Felker P, Harris PJC, Harsh LN, Cruz G, Tewari JC et al (2001) The *Prosopis juliflora-Prosopis pallida* Complex: a monograph. HDRA, Coventry, p 172

Petit C, Bretagnolle F, Felber F (1999) Evolutionary consequences of diploid-polyploid hybrid zones in wild species. Trends Ecol Evol 14:306–311

Pressoir G, Berthaud J (2004) Population structure and strong divergent selection shape phenotypic diversification in maize landraces. Heredity 92:95–101

Ramírez L, De la Vega A, Razkin N, Luna MV, Harris PJC (1999) Analysis of the relationships between species of the genus *Prosopis* revealed by the use of molecular markers. Agronomie 19:31–43

Reich PB, Wright IJ, Cavender-Bares J, Craine JM, Oleksyn J, Westoby M, Walters MB (2003) Evolution of functional traits in plants. Int J Plant Sci 164:143–164

Rieseberg LH (1997) Hybrid origin of a plant species. Ann Rev Ecol Syst 28:359–389

Rieseberg LH, Carney SE (1998) Plant hybridization. New Phytol 140:599–624

Rieseberg LH, Gerber D (1995) Hybridization in the Catalina island mountain mahogany (*Cercocarpus traskiae*): RAPD evidence. Cons Biol 9:199–203

Rieseberg LH, Ellstrand NC, Arnold M (1993) What can molecular and morphological markers tell us about plant hybridization. CRC Cr Rev Plant Sci 12:213–241

Ritland K (1990) A series of FORTRAN computer programs for estimating plant mating systems. J Hered 81:235–237

Ritland K (2002) Extensions of models for the estimation of mating systems using n independent loci. Heredity 88:221–228

Robledo-Arnuncio JJ, Austerlitz F, Smouse PE (2006) A new indirect method of estimating the pollen dispersal curve, independently of effective density. Genetics 173:1033–1045

Robledo-Arnuncio JJ, Austerlitz F, Smouse PE (2007) POLDISP: a software package for indirect estimation of contemporary pollen dispersal. Mol Ecol Notes 7:763–766

Saidman BO (1985) Estudio de la variación alozimica en el género *Prosopis*. PhD Thesis, Fac. Cs. Exactas y Naturales. Universidad de Buenos Aires

Saidman BO (1986) Isoenzymatic studies of alcohol dehydrogenase and glutamate oxalacetate transaminase in four Southamerican species of *Prosopis* and their natural hybrids. Silvae Genet 35:3–10

Saidman BO (1988) La electroforesis de isoenzimas para la medición de la variabilidad genética en especies de *Prosopis*. In: Prosopis en Argentina. Documento preliminar elaborado para el I Taller Internacional sobre Recurso Genético y Conservación de Gerrnoplasma en *Prosopis*. Fac. Cs. Agrarias (U. N. Córdoba) – FAO – PIRB, pp 107–118

Saidman BO (1990) Isoenzyme studies on hybrid swarms of *Prosopis caldenia* and sympatric species. Silvae Genet 39:5–8

Saidman BO (1993) Las isoenzimas en el estudio de la variación genética y las afinidades entre especies de *Prosopis*. Bol Genético Inst Fitotécnico Castelar 16:25–37

Saidman BO, Naranjo C (1982) Variación de esterasas en poblaciones de *Prosopis ruscifolia* (Leguminosae: Mimosoideae). Mendeliana 5:61–70

Saidman BO, Vilardi JC (1987) Analysis of the genetic similarities among seven species of *Prosopis* (Leguminosae: Mimosoidae). Theor Appl Genet 75:109–116

Saidman BO, Vilardi JC (1993) Genetic variability and germplasm conservation in the genus *Prosopis*. In: Puri S (ed) Nursery technology of forest tree species of arid and semiarid regions. Winrock-Oxford & IBH Publishing Co. PVT. Ltd., New Delhi/Bombay/Calcutta, pp 187–198

Saidman BO, Bessega C, Ferreyra LI, Vilardi JC (1998) Random Amplified Polymorphic DNA (RAPDS) variation in hybrid swarms and pure populations of genus *Prosopis* (Leguminosae). In: Bruns S, Mantell S, Tragardh C, Viana AM (eds) Recent advances in biotechnology for tree conservation and managment, pp 122–134

Saidman BO, Bessega C, Ferreyra LI, Julio N, Vilardi JC (2000) The use of genetic markers to assess population structure and relationships among species of the genus *Prosopis* (Leguminosae). Bol Soc Argent Bot 35:315–324

Savolainen O, Lascoux M, Merilä J (2013) Ecological genomics of local adaptation. Nat Rev Genet 14:807

Shaw J, Lickey EB, Schilling EE, Small RL (2007) Comparison of whole chloroplast genome sequences to choose noncoding regions for phylogenetic studies in angiosperms: the tortoise and the hare III. Am J Bot 94:275–288

Smouse P, Dyer RJ, Westfall RD, Sork VL (2001) Two-generation analysis of pollen flow across a landscape. I Male gamete heterogeneity among females. Evolution 55:260–271

Song Z, Li F, Zhang M, Weng Q (2016) Genome scans for divergent selection in natural populations of the widespread hardwood species *Eucalyptus grandis* (Myrtaceae) using microsatellites. Sci Rep 6:1–13

Spitze K (1993) Population structure in *Daphnia obtusa*: quantitative genetic and allozymic variation. Genetics 135:367–374

Stinchcombe JR, Hoekstra HE (2008) Combining population genomics and quantitative genetics: finding the genes underlying ecologically important traits. Heredity 100:158–170

Teich I, Mottura MC, Verga A (2015) Asociación entre variabilidad genética y fenotípica con ajuste por autocorrelación espacial en *Prosopis*. Basic Appl Genet 26:63–74

Teich I, Cosacov A, López Lauenstein D, Vega C, Sérsic A (2017) Variabilidad morfológica y diferenciación de entidades en un complejo de especies de *Prosopis* en el Gran Chaco Americano. Libro de II RABE:76

Torales SL, Rivarola M, Pomponio MF, Gonzalez S, Acuña CV, Fernández P et al (2013) De novo assembly and characterization of leaf transcriptome for the development of functional molecular markers of the extremophile multipurpose tree species *Prosopis alba*. BMC Genomics 14:705

Vázquez-Garcidueñas S, Palacios RA, Segovia-Quiroz J, Frías-Hernandez JT, Olalde-Portugal V, de la Vega O M et al (2003) Morphological and molecular data to determine the origin and taxonomic status of *Prosopis chilensis* var. riojana (Fabaceae, Mimosoideae). Can J Bot 81:905–917

Vega M, Hernandez P (2005) Molecular evidence for natural interspecific hybridization in *Prosopis*. Agrofor Syst 64:197–202

Vega C, Teich I, Acosta MC, Lopez Lauenstein D, Verga A, Cosacov A (2020) Morphological and molecular characterization of a hybrid zone between *Prosopis alba* and *P. nigra* in the Chaco region of northwestern Argentina. Silvae Genet 69:44–55

Verga AR (1995) Genetische Untersuchungen an *Prosopis chilensis* und *P. flexuosa* (Mimosaceae) im trockenen Chaco Argentiniens. Göttingen Research Notes in Forest Genetics. Doctoral thesis Abteilung für Forstgenetik und Forstpflanzenzüchtung der Universität Göttingen 19: 1–96

Verga AR (2015) Hoja 3.6. Programa de distribución gratuita. Instituto de Fisiología y Recursos Genéticos Vegetales (IFRGV). Instituto Nacional de Tecnología Agropecuaria (INTA). Camino 60 Cuadras, km 5.5 X5020ICA, Córdoba, Argentina. E-mail: verga.anibal@inta.gob.ar

Verga AR, Gregorius HR (2007) Comparing morphological with genetic distances between populations: a new method and its application to the *Prosopis chilensis – P. flexuosa* complex. Silvae Genet 56:45–51

Verga AR, López Lauenstein D, López C, Navall M, Joseau J, Gómez C et al (2009) Caracterización morfológica de los algarrobos (*Prosopis* sp.) en las regiones fitogeográficas Chaqueña y Espinal norte de Argentina. Quebracho (Santiago del Estero) 17:31–40

Verga AR, López C, Navall M, Joseau J, Gómez C, Royo O, et al. (2014) Caracterización morfológica, distribución geográfica y estimación de nichos ecológicos de algarrobos (*Prosopis* sp.) en las regiones fitogeográfica Chaqueña y Espinal norte de Argentina.http://redaf.org.ar/wp-content/uploads/2008/02/Distribucion-geografica-Prosopis.pdf

Vitalis R, Dawson K, Boursot P (2001) Interpretation of variation across marker loci as evidence of selection. Genetics 158:1811–1823

Williams GC (1966) Adaptation and natural selection. Princeton University Press, Princeton

Wright S (1951) The genetical structure of populations. Ann Eugenics 15:223–254

Xing H, Fu X, Yang C, Tang X, Guo L, Li C et al (2018) Genome-wide investigation of pentatricopeptide repeat gene family in poplar and their expression analysis in response to biotic and abiotic stresses. Sci Rep 8:2817

Chapter 10
Genetic Breeding of *Prosopis* Species from the "Great American Chaco"

Diego López Lauenstein, Ingrid Teich, Edgardo Carloni, Mariana Melchiorre, Mónica Sagadin, Javier Frassoni, and M. Jacqueline Joseau

10.1 Use and Domestication of "Algarrobos": A Set of Multipurpose Tree Species

In the Argentine Chaco, "algarrobos" wood (mainly *Prosopis alba* and *P. nigra*, but also *P. chilensis*, *P. flexuosa*, *P. hassleri*, and other *Prosopis* of minor or local relevance), gives rise to a sawing industry that supports a significant number of jobs (Kees and Michela 2016). In this region there are 678 sawmills, in which approximately 4500 people work and consume 419,000 m^3 of native wood per year, and 40% of this volume comes from different species of *Prosopis*. Other important native species that supply raw material to these sawmills are *Schinopsis balansae*, *Schinopsis lorentzii*, and *Aspidosperma quebracho-blanco*. More than 85% of the establishments are micro-companies, and the rest are medium-sized companies (Min. Agro. 2015). The raw material for these sawmills comes exclusively from the native forest, which has led to a progressive environmental degradation, and also to a decrease in log quality. Moreover, the average yield of the sawmills does not exceed 35%, partly due to its low level of technification, but also due to the low quality and unevenness of logs from unmanaged forests.

The *algarrobos* have been intensively used in Argentina for more than 50 years for making high-quality solid wood furniture. Also, their wood has been traditionally converted in charcoal by means of craft methods with hemispherical brick kilns. According to official statistics, 1663 ton of charcoal were made with *Prosopis*

D. López Lauenstein (✉) · I. Teich · E. Carloni · M. Melchiorre · M. Sagadin
Instituto de Fisiología y Recursos Genéticos Vegetales, Unidad de Estudios Agropecuarios (IFRGV, UDEA) INTA-CONICET, Córdoba, Argentina
e-mail: lopez.diego@inta.gob.ar

J. Frassoni · M. J. Joseau
Facultad de Ciencias Agropecuarias – Universidad Nacional de Córdoba, Córdoba, Argentina

© Springer Nature Switzerland AG 2021
M. J. Pastorino, P. Marchelli (eds.), *Low Intensity Breeding of Native Forest Trees in Argentina*, https://doi.org/10.1007/978-3-030-56462-9_10

wood in 2017 in Argentina (SAyDS 2019). However, the economic relevance of the *Prosopis* species from the Great American Chaco is not only related to its wood production. They are multipurpose trees, generally used for more than one product or service. They are suitable for silvopastoral systems, providing livestock with not only shade but also highly nutritious fruits for forage. In fact, their pods (regionally called "algarrobas") are also used for human consumption in many ways, from traditional food preparations (e.g., "patay") and beverages (e.g., "aloja" and "añapa") with economic relevance in the local markets, to food industry inputs traded regionally. A kind of flour is made by grinding the pods, which is used at home, in artisan bakeries and in cookie industries. A coffee substitute is also manufactured with ground pods. The official statistics have registered for 2017 the production in Argentina of 180 ton of algarrobas for human consumption, 67.63 ton of algarrobas for livestock forage, 30 ton of patay, 20 ton of algarroba flour and 2.5 ton of algarroba coffee (SAyDS 2019). Pods are collected from the natural forest, where each tree produces in average 60 to 120 kg per year, depending on the species, the site, and the year conditions (Galera 2000). Additionally, due to their tolerance to drought and salinity, and their ability to fix nitrogen in the soil, *Prosopis* species are regarded as a promising alternative for ecological restoration in degraded arid and semiarid environments.

The drastic reduction of *Prosopis* forests in recent decades has promoted interest in using the most productive species of the genus in afforestation, for a dual purpose of renewing the productive resource and recovering degraded ecosystems. Since the promotion of commercial plantations through the national subsidy granted by Law 25,080 for over 20 years, this interest has been translated into effective plantations (Fig. 10.1), that made *Prosopis alba* the second most planted native forest species in Argentina after *Araucaria angustifolia*. Currently, about 9000 ha are implanted with *P. alba* in Argentina, which are, in average, 12 years old with few having reached the rotation age. In most of this plantations silvicultural management was not done, and the origin of the genetic material (seeds) is unknown (Salto and Lupi 2019; Leandro Arce, personal communication).

These afforestation experiences have contributed to the domestication of the species by generating information on cultivation techniques and seedling production in the nursery. However, only in the last 10 years more specific silvicultural investigations begun in order to develop appropriate and modern methods for the cultivation of *P. alba*. Concerning seedling production on an industrial scale, the traditional system where seedling beds were at ground level, using polypots as "container" and substrates containing soil, is being replaced by modern technics with suspended seedling beds, using tray-pots and specific substrates appropriate for root development and optimal balance between root and aboveground biomass growth.

A mixture of composted pine bark, perlite, and vermiculite as substrate, in 125 cm^3 containers, resulted to be a good alternative for technical production in the nursery. Fertilization is necessary to produce seedlings under these conditions of substrate and containers. In this sense, a specific formulation has been developed for *P. alba*, which increases growth in diameter and height to reach the optimal size for transplant in 90 days (Salto et al. 2019a). This technological leap resulted in a

Fig. 10.1 Five-year-old afforestation of *P. alba* (Campo Durán provenance) located in Laguna Yema (Province of Formosa), with *Opuntia* sp. leaves drying between lines. (Photo: Diego López Lauenstein)

decrease of nursery time from 5 months with the traditional methods to 3 months with the modern technics, also reducing the volume of substrate, and the cost of both, production and movement of seedlings to the plantation. Information about edaphic characteristics for site quality determination for cultivating *P. alba* is also available. The best growth corresponds to soils with more than 50 cm of effective root exploration depth, good drainage, and light surface texture (loamy and/or silty sandy soils). On the contrary, in soils with somewhat poor drainage, shallow, heavy (clayey) soils, and that are in subnormal relief (depressed, waterlogged), the *algarrobos* are small in size (Kees et al. 2019). To initiate the plantation, site preparation depends on numerous factors, like previous use, type of soil, owner resources availability, and identification of possible soil limitations, among others. The usually used tools are a disc harrow, and if required, a subsoiler to improve the water infiltration. Planting should be done outside the frost period and coinciding with either the beginning or the end of the rainy season, which is concentrated in summer. The planting period can be longer, but always depending on water availability and favorable temperatures (MAGyP 2016).

In *P. alba*, the stem is formed by the succession of branching orders, of which three branches are generated, generally one aborts and one of the rest takes an orthotropic direction (upward) to continue with the trunk (Moglia and Giménez 2006). As a result of this branching model, young *algarrobo* plants without management show abundant branching, making it difficult to identify a main axis. Therefore, formative and lift pruning are essential to allow knot-free timber to form around the defect core. This management tends to be used for high-quality uses such as veneer, timber moldings, and furniture. On the other hand, planting density plays an essential role in branching degree, since denser plantations generate fewer branches and straighter stems. Zárate et al. (2019) tested different pruning treatments under a range of densities extending from 450 to 4500 plants/ha in a Nelder plot design, showing that growth up to 7 years in plantations at low densities (450, 560, and 750 pl/ha) have potential to achieve the highest bole volumes per plant and with few knots in the wood.

Another silvicultural treatment that should be used in any *Prosopis* plantation is thinning. The aim is to redistribute the growth of the stand in the best possible way, regulating the use of resources (water, light, and nutrients) in their growth space. The remaining individuals increase their growth in response to thinning, concentrating the stand growth on a smaller number of trees, but of higher quality, thus increasing the value of the produced wood. One of the limitations for decision-making is the opportune moment of thinning, which is carried out by evaluating the curves of current annual increment (CAI) and mean annual increment (MAI). For plantations in the Chaco region, the appropriate time for thinning ranges from 9 to 13 years. This variation is mainly due to the initial planting density (Gómez et al. 2019). Other factors that influence growth curves (CAI, MAI) are the seed provenance (genetics) and site quality (environment). Applying this silvicultural management, a rotation age of 25 years can be estimated, with diameters greater than 30 cm in the best quality sites (Kees et al. 2018).

10.2   Low-Intensity Breeding

Afforestation plans with *Prosopis* requires a large volume of seeds every year, and to supply this demand, fruits from the natural forest are collected. However, when collecting seeds, interspecific crosses that occur naturally within the genus must be avoided because can generate unevenness plantations, both in diameter and height growth, as well as great variation in morphological characters (e.g., multibranches or thorns). Like in all wild species, diversity is very large, which might constitute an advantage for breeding, but, at the same time, it forces to reduce variability in order to achieve discrete, more stable, and uniform genetic units. Therefore, the main objective of the low-intensity breeding of *Prosopis* is to assure the proper amount of seeds for seedling production, guaranteeing a high specific purity and uniformity of the seedlings. The genetic and ecophysiological knowledge available on *Prosopis chilensis*, *P. flexuosa*, and, more recently, *P. alba*, constitute the conceptual basis for advancing toward the implementation of seed stands for these purposes.

Given the alarming loss of forests that has occurred in Argentina during the last 35 years (estimated in 12 million ha in the Chaco region, e.g., Spensley et al. 2013, see Chap. 8), and the risk of losing biodiversity due to climate change, specific studies are required to identify and delimit seed production areas (SPA) for each species, and its subsequent transformation into seed stands. These actions aim to preserve existing variability and obtain base material with greater genetic uniformity and purity, avoiding hybridization between *Prosopis* species (Verga 2014 and see Chap. 9). At the same time, it seeks to respond to the suggestions of FAO (2014) to improve the sustainable use and management of forest genetic resources. Despite the intense degradation of *Prosopis* forests in Argentina, there are stands of good purity and silvicultural quality that, with a proper management, can be used in the short term as genetically stable seed stands. Accordingly, protocols were established together with the Argentine National Seed Institute (INASE; Res. 374/14) in order to systematize seed harvest for seedling production of the main *Prosopis* species. These protocols involve the morphological analysis of leaves of adult individuals, and the genetic analysis through biochemical markers (isoenzymes) of offspring (seeds) for the definition of purity thresholds (Verga 2014). So far 11 SPA (Fig. 10.2) of *Prosopis alba*, *P. flexuosa* and *P. chilensis* have been defined, which can be transformed into seed stands, some of which are already certified by INASE. This definition of SPA was a joint task of the Instituto Nacional de Tecnología Agropecuaria (INTA) and the National Germplasm Bank of *Prosopis* (BNGP) of the Facultad de Ciencias Agropecuarias de la Universidad Nacional de Córdoba (FCA-UNC). To identify these SPA, a joint analysis of information from databases of both institutions, bibliographic references, local informants, and satellite images was done. Then, field confirmation was performed accounting for several aspects: (1) specific

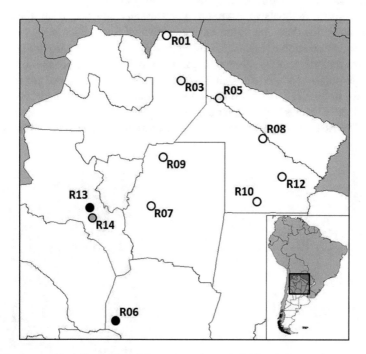

Fig. 10.2 Geographical distribution of *Prosopis* seed stands. White circles: *P. alba*; black circles: *P. flexuosa*; grey circle: *P. chilensis*. R01: Campo Durán, R03: La Unión, R05: Isla Cuba, R06: San Miguel, R07: Santiago del Estero, R08: Bermejito, R09: Chañar Bajada, R10: Villa Ángela. R12: Plaza, R13: Pipanaco, R14: Palampa

composition of the stand, i.e., predominance of the target species; (2) morphological characteristics and health status of the trees; (3) accessibility of the stand and (4) legal conditions and agreement of the landowner where the stand is located.

Considering the large interspecific hybridization rate of the genus and the need for the highest degree of purity for each seed source (i.e., regarding the target species), the first step to define a SPA is the accurate identification of the taxonomic status of the trees that compose the candidate stand. In this sense, there are two relatively simple and inexpensive methods to detect genetic variation of interspecific origin in the *Prosopis* complex: leaf morphology of adults and isoenzymatic analysis of seeds using the allozyme marker alcohol dehydrogenase (ADH). Due to the presence of species-specific alleles, the screening of the ADH isozyme in a sample of seeds allows to determine the proportion of interspecific crosses between the "white" algarrobos (*P. alba*, *P. hassleri*, *P. fiebrigii*, *P. chilensis*) and "black" algarrobos (*P. nigra*, *P. flexuosa*, *P. ruscifolia*) (Saidman 1986; Verga 1995). On the other hand, through morphological analysis of the leaves, it is possible to identify those algarrobos that are not the target species or that may presumably be of hybrid origin (Verga 2014). Both methods can be applied because algarrobos have a very strong correlation between their leaf morphological characteristics and their genetic basis, particularly that related to their specific origin (Saidman 1986; Verga 1995; Verga and Gregorius 2007; Ferreyra et al. 2013; Joseau et al. 2013).

The protocol establishes the sampling, at the selected stand, of all the *Prosopis* individuals with a diameter at the base greater than 5 cm. Each tree is identified, georeferenced and photographed to subsequently appreciate its morphological characteristics and estimate its allometric parameters. Leaves and fruits are harvested for morphological and genetic analysis (Figs. 10.3a and 10.3b), and for incorporation into the ACOR Herbarium (http://www.agro.unc.edu.ar/herbario) of the Facultad de Ciencias Agropecuarias, Universidad Nacional de Córdoba (FCA-UNC). The first step is performing the foliar morphology analysis on five scanned

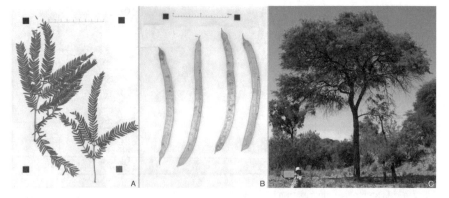

Fig. 10.3 Material used for the identification of trees in a natural stand for its transformation into a seed stand: (**a**) leaves, (**b**) fruits, and (**c**) a target tree of the seed stand of *P. alba* located in the Province of Salta. (Photos: Carmen Vega, Diego López Lauenstein)

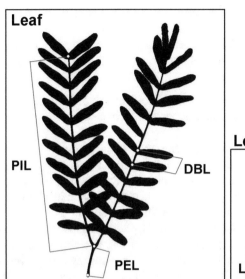
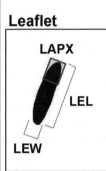

Fig. 10.4 Foliar morphological traits evaluated in *Prosopis alba* by means of the software Hoja 3.4. *PIL* pinna length, *PEL* petiole length, *DBL* distance between leaflets, *LAPX* leaflet apex, *LEL* leaflet length, *LEW* leaflet width

leaves for each tree (Fig. 10.3a) using the software Hoja 3.4 (Verga 2015). The most frequently used leaf traits are petiole length (PEL), number of pairs of pinnae (NPI), pinna length (PIL), number of pairs of leaflets per pinna (NLP), leaflet length (LEL), leaflet width (LEW), leaflet area (LEA), leaflet apex (LAPX), distance between leaflets (DBL), and the ratios LEL/LEW and LAPX/LEA (Fig. 10.4). Fruit morphometric characters usually include fruit length (FrL), fruit width (FrW), and fruit thickness (FrTh). *Prosopis* fruits are also characterized according to their shape (FSh), the fruit edge shape (FESh), and the color (FrC). To determine the taxonomical status a principal component analysis with the 13 leaf traits is carried out using each sampled tree as unit of analysis. Typical trees of each morphological group of *P. alba* (i.e., "chaqueño", "santiagueño" and "salteño," Verga et al. 2009) are included as reference. Once the species and the morphological group to which the stand belongs have been identified, a new principal component analysis is performed, and each tree is examined to detect possible outlier individuals. A threshold $(3, -3)$ is set on the principal axes N° 1 and N° 2. For doubtful individuals (i.e., possible hybrids) a visual appreciation of the fruit is used.

Subsequently, the specific purity of the offspring (seeds) is evaluated using the ADH allozyme marker. A pool of seeds is harvested from at least 20 trees and not less than 10% of the individuals, and 100 seeds are analyzed by starch gel electrophoresis. The following two maximum allele frequency thresholds are set:

1. The SPA will not be suitable for certification as a seed stand if the heterozygous individuals (seeds) exceeds 10%.
2. The SPA will not be suitable for certification as a seed stand if the frequency of the allele corresponding to the "algarrobo negro" (*P. nigra, P. flexuosa, P. ruscifolia*) in the seed pool exceeds 2%.

The two combined analyses (allozyme and leaf/fruit traits) allow identifying, within each SPA, individuals to thin with the purpose of increasing the specific purity of the seeds from the seed stand.

Seed production in *Prosopis*, as in many forest tree species, is cyclic, so seeds are not available every year. For this reason, it is important to stock up on seeds during high production years to ensure their availability for seedling production. Due to their thick seed coat, the *Prosopis* seeds stored at 4 °C maintain a high germination power (80%) for at least 30 years (Verzino and Joseau 2005). The National Germplasm Bank of *Prosopis* (BNGP) currently stores 1650 accessions at −18 °C, corresponding to seeds from 1106 trees or stands. Other ex situ conservation collections are botanical gardens and field trials such as progeny and provenance trials (e.g., López Lauenstein et al. 2016; Capello 2019; Verzino et al. 2020).

To assess the adaptability of some of the already mentioned seed stands, three provenance trials were installed in 2011 in three sites in the semiarid region of Chaco: Laguna Yema (24° 16′ 23.53″ S; 61° 14′ 5.28″ W; 160 m asl), Leales (27° 11′ 12.54″ S; 65° 14′ 37.86″ W; 320 m asl), and Fernández (27° 56′ 16.06″ S; 63° 52′ 26.10″ W; 160 m asl) (Fig. 10.5). The three sites differed in the mean annual precipitation and the mean annual temperature (Laguna Yema, 703 mm and 22.7 °C; Leales, 965 mm and 19.9 °C; and Fernández, 575 mm and 20.8 °C, respectively). The genetic material evaluated in these field trials was seven provenances of *Prosopis alba* from different seed stands, two *P. alba* selections based on the evaluation of three progeny trials (povenances 6 and 7), and one *P. chilensis* selection from a natural forest of the Arid Chaco (López Lauenstein et al. 2019).

The performance of this small network of trials was evaluated with periodic measurements of survival, diameter at breast height (DBH), and total height. At the time of this publication, the latest records are Leales, 2018, 7 years old; Fernández, 2019, 8 years old; and Laguna Yema, 2016, 5 years old. For comparing between sites, the mean annual increment (MAI) was calculated, dividing the last record, both DBH and height, by the number of years of each plantation. The main results show significant differences between provenances in both DBH and height growth. In this sense, provenance N° 1 *Prosopis alba* "Campo Durán" had the highest MAI both in height and diameter (Fig. 10.6), standing out mainly in Leales (12.1 cm DBH and 6.24 m in height at 7 years old) and Laguna Yema (12.62 cm DBH and 6.04 m high at 5 years old). The selection of *P. alba* from the progeny trials and the selection of individuals from the natural stand of *P. chilensis* served as reference for the natural provenances of *P. alba*. Provenance N° 7 stands out and corresponds to a selection of 10 open-pollinated families carried out from the evaluation of more than 200 *P. alba* families from the entire Chaco region, in 3 progeny trials (López Lauenstein

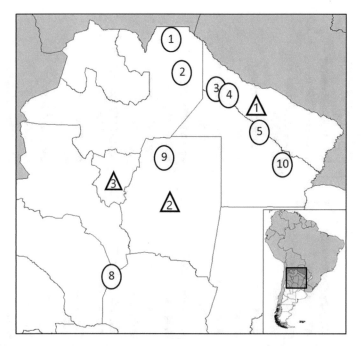

Fig. 10.5 Location of field trials (triangles) *1*, Laguna Yema; *2*, Fernández; *3*, Leales; and provenances (circles) *1*, Campo Durán; *2*, La Unión; *3*, Bolsa Palomo; *4*, Isla Cuba; *5*, Bermejito; *8*, *P. chilensis* selection; *9*, Chañar Bajada; and *10*, Plaza

et al. 2016). This evaluation was done 3 years after implanting the field trials and was based on growth and shape characters, highlighting the value of early selection and the high heritability of the chosen characters. It is worth noting that Campo Durán natural provenance, represented by open-pollinated seeds from unselected trees, performed better (or at least equivalently) than the progeny of the acute selection of 10 individuals out of 200 in a progeny trial. Likewise, provenance N° 8 (*P. chilensis* selection), as expected, showed the lowest growth in diameter and height in the three field trials. These results suggest a lack of adaptation of *P. chilensis* to the Semiarid Chaco region (it is a species of the Arid Chaco).

Although there is a clear (statistically significant) separation of treatments N° 1 and N° 8 (above and below, respectively) in terms of growth, the other provenances of *P. alba* did not show great differences between them in any of the three field trials. These provenances would have a high phenotypic plasticity to adapt to a wide variety of environmental conditions. In this sense, expanding the genetic base of recommended materials will increase the resilience of the plantations to climate change, e.g., different kinds of abiotic stress (saline, drought, thermal, etc.) and the occurrence of new pests or diseases.

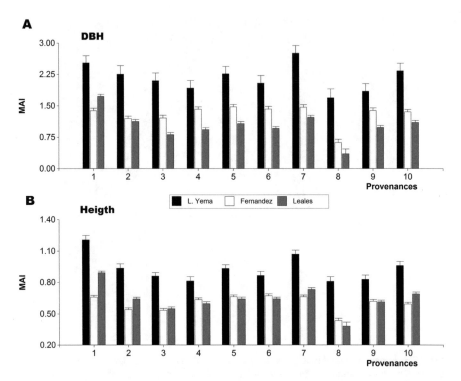

Fig. 10.6 Mean annual increment (MAI) in diameter at breast height (**a**) and height (**b**) for 10 *Prosopis* genetic materials (8 *P. alba* provenances, 2 *P. alba* selection, and 1 *P. chilensis* provenance; see Fig. 10.5) in three field trials: Fernández, Laguna Yema, and Leales

10.3 High-Intensity Breeding

The primary goal of *Prosopis* species breeding has traditionally been related to wood production. Growth traits (total height and diameter) and trunk shape and bole length have been the focus of improvement. However, as mentioned in the first section of this chapter, the objectives of different *Prosopis* breeding programs can be diverse. In addition to the wood volume yield, the objectives can be oriented to improve wood quality, fruits production and quality, resistance to pests and diseases, adaptability to marginal (e.g., dry) environments, or degraded (e.g., salinized) soils, among others. Improvement programs can contemplate one or more of the aforementioned goals, and in all cases, they must have the premise of maintaining a wide genetic diversity to guarantee that the genetic gains do not decrease, even in advanced generations.

Forest tree breeding has been successful at delivering genetically improved material for multiple traits based on recurrent cycles of selection, mating, and testing (Grattapaglia et al. 2018). Pedigree-based phenotypic selection, rather than genetic dissection approaches such as quantitative trait mapping and association genetics, is currently the main approach in *Prosopis* breeding programs of Argentina.

Initially, the genetic breeding strategy for *P. alba* was based on an adjustment of the "multiple populations" method proposed by Namkoong (1980), where genetic improvement and conservation of genetic resources are combined. This method consists of creating a base population conformed by the progeny of individuals from natural populations. Each base population is composed by subpopulations isolated from each other and located in different environments. These subpopulations are used as progeny trials and subsequently, after evaluation and selection, as seed orchards. In this way, each subpopulation is improved separately and, at the same time, passes through a process of differentiation from the others (due to the adaptation to the different environments). This allows for future recovering of genetic variability in stands arising by the mixture of seeds of the different subpopulations (Verga 2005).

Provenance and progeny trials of *Prosopis* in Argentina have been implemented since 1990 (Cony 1996; Felker et al. 2001; López Lauenstein et al. 2016), allowing the study of population performance in different environments and the estimation of genetic parameters related to traits such as height, diameter, stem shape, and growing rate. Progeny trials are a powerful tool in forest breeding programs (Zobel and Talbert 1984). Through the estimation of breeding values, family and/or individual rankings are made in order to carry out backward or forward selection. These progeny trials are then thinned according to the rankings (i.e., forward selection), to become seed orchards (Ruotsalainen and Lindgren 1998). Currently, *P. alba* breeding program in Argentina includes a network of three progeny trials established in 2008 (INTA net): Laguna Yema (24° 19′ 15.5″ S; 61° 17′ 31.7″ W; 161 m asl), Santiago del Estero (27° 56′ 45.1″ S; 64° 13′ 12.5″ W; 172 m asl), and Plaza (26° 56′ 3.6″ S; 59° 46′ 22.3″ W; 78 m asl). These trials include 217 open-pollinated families from ten different provenances, which cover a large part of the natural range of the species in the Argentine Chaco. This base population come from seeds collected during successive field campaigns between 2004 and 2007 (Verga et al. 2009) from phenotypically selected individuals (mother trees) in wild populations.

In this first phase of the program, the base population was constituted on the basis of the specific purity without considering differences by geographical origin. This decision was based on previous genetic studies in *P. chilensis* and *P. flexuosa* from the Arid Chaco region, where the main source of genetic variation was shown to come from hybridization processes (Verga 1995, 2005). In this species complex, there are no significant differences between populations that grow in dissimilar environments compared to the enormous variation within them as an effect of interspecific crosses and the presence of interspecific hybrids (Verga 1995). More recently, however, and based on leaf trait analysis and variation detected with molecular markers, three morphological groups within *P. alba* were determined, which could be considered subspecific taxonomic groups: *P. alba* "chaqueño," *P. alba* "santiagueño," and *P. alba* "salteño" (Verga et al. 2009, Verga 2014; Chap. 9). These groups show a separate geographic distribution and from their identification and evaluation through provenance trials, the breeding strategy was rethought. The new strategy considers dividing the current base populations into each subspecific taxonomic group and provenance. The objective is to advance in the installation of new progeny tests corresponding to each morphological group, developing each one as the base population of its own taxonomic group. In addition, and because there are traceability records of the seed used in the most recent plantations, plus trees are being evaluated and selected within the plantations to incorporate them into the base populations of each morphological group.

These two schemes are not excluding, and their simultaneous development can serve to face two challenges. On the one hand, with the original scheme (i.e., a mixture of provenances), a high genetic diversity is maintained improving adaptation to a large number of environments with high resilience to climate change. With the new scheme (i.e., separate morphological groups), on the other hand, the hope is to give a better response to the pursuit of specific objectives related to market demands and high growth rates. This new scheme began to be applied in particular to the outstanding population *Campo Durán*, which corresponds to the morphological group *P. alba* "salteño." Thus, in 2018, a progeny trial with 45 open-pollinated families from *Campo Durán* was installed in Sáenz Peña Experimental Station of INTA (26° 51′ 15.3″ S; 60° 25′ 16.8″ W). The experimental design includes the identification of the family and also of the seedlings from seeds of the same pod, since it is known that 64% of the seeds from the same fruit correspond to complete siblings (Bessega et al. 2012). This design allows to increase the accuracy in the estimation of breeding values from separately considering the treatments of complete siblings from those of half siblings.

Advanced forest genetic evaluation involves analyzing data from progeny tests using mixed linear models to estimate the best linear unbiased predictors (BLUPs) of tree breeding values (BVs). The high number of provenances and families in the INTA net not only allows these estimations but also contains ex situ and in vivo conservation material representative of the genetic variation of the species in the Argentine Chaco (Verga et al. 2009). In 2005, the first genetic markers for *Prosopis* species were published (Mottura et al. 2005); later, more molecular markers were developed (Suja et al. 2007; Bessega et al. 2013; Torales et al. 2013; Pomponio et al. 2015). These markers are mainly used for studies of genetic structure of the different *Prosopis* species and their interspecific hybrids (Chap. 9). Although the number of specific markers is still very limited, the challenge is to incorporate them into the breeding program through marker-assisted breeding and, at the same time, implement next-generation sequencing techniques for the development of large-scale markers (SNPs) and shorten the breeding cycle through genomic selection. With the current genomic resources, kinship matrices between individuals will be established, instead of using the theoretical values of the kinship relationships for the calculation of BVs, increasing the accuracy in their estimation (Marcucci Poltri and Gallo 2016; Grattapaglia et al. 2018).

As *P. alba* is a multipurpose resource (Galera 2000), its breeding process involves different strategies according to the product to be improved (wood quality, fruit production, etc.). One of the first evaluations of *P. alba* progeny trials considering its multiple uses included total biomass production, height, rate of pod production, and pod organoleptic characteristics (Felker et al. 2001). This study reported the evaluation of a 9-year-old progeny trial containing 57 open-pollinated families of 8 provenances from Northwestern Argentina established in 1990 in the Province of Santiago del Estero (27° 45′ S, 64° 15′ W, 200 m asl). The family-narrow-sense-heritability was 0.487 for height, 0.548 for aerial biomass production, and 0.244 for pod production. The mean values across families ranged from 10.7 kg/tree to 57.4 kg/tree for aerial biomass production, from 2.2 m to 3.6 m for height, and from 13 g/year to 874 g/year for pod production. Even though *P. alba* produces highly nutritious fruits for forage and human food with commercial value (Fagg and Stewart 1994), most studies have focused on traits related to wood production (Salto 2011; Cappa and Varona 2013; Carreras et al. 2016). Currently, ongoing studies seek to estimate breeding values at tree level for flowering and fruiting traits (flowering intensity, production of pods, nutritional quality of fruits) using the INTA net trials (Cisneros et al. in preparation).

The selection for wood yield requires improving traits like tree architecture, trunk shape, and bole length. After the progeny trials of INTA net were established in 2008, the families' early growth and shape were evaluated at 18 months of age. These studies reported significant interaction between provenance and environment in diameter (ranged from 1 to 8 cm) and total height (ranged from 1 to 3 m) and a negative correlation between tree form and the two growth traits (Salto 2011). This represents a constraint for the breeding program, as trees selected for fast growth would show low quality in terms of tree form. To overcome this limitation, Carreras et al. (2016) evaluated genetic parameters for several traits of economic importance in *P. alba*, including tree form (number of stems), height, diameter, and size increments using a multi-trait selection approach based on a selection index, with the goal of getting the maximum possible gain in all traits simultaneously (Bessega et al. 2015). They developed a breeding strategy for multiple trait selection taking into account the heritability of individual traits, the genetic correlation between them, and the increase of selected group kinship inherent to the selection process. This issue is important for breeding to preserve wide genetic diversity and prevent inbreeding depression in the following generations (Lindgren and Mullin 1997). They reported that although gain at individual trait level is reduced in comparison with the maximum potential, index selection allows significant gains for all three traits together and represents a suitable strategy to improve *P. alba* in order to establish clonal seed orchards. This study evaluated only one progeny trial of INTA net, and therefore no genotype x environment interaction was estimated. Currently, height, diameter, and tree form at different ages, measured in the three trials, are under analysis with individual tree mixed models (Borralho 1995) to predict breeding values and simultaneously estimate genetic and environmental effects (Cisneros et al. in preparation). Current efforts in *P. alba* breeding program aim at expanding the approaches to multiple uses (multipurpose selection) such as fruit production, forage, recovery of degraded areas, biomass production, and environmental services, among others.

10.4   Vegetative Propagation: A Useful Tool for Genetic Improvement

Agamic propagation constitutes a fundamental technique to be developed for the *P. alba* breeding genetic program, since it will allow the eventual installation of clonal seed orchards (López Lauenstein et al. 2016). In this way, ex situ conservation is also done by cloning representative samples of threatened natural populations. For these reasons, the INTA *P. alba* breeding program is developing asexual propagation methodologies through different in vitro tissue culture and grafting techniques. Also, different techniques of rejuvenation of vegetative material that allow the multiplication of *P. alba* through the use of mini-stakes are being tested.

The advantage of agamic propagation in forest genetic breeding is its ability to capture and transfer to the new individuals all the genetic potential of the mother plant, in a short period of time since it is not necessary to wait for the production of seeds to obtain their offspring. One of the vegetative propagation methods being tested is the in vitro culture of uninodal segments of 2-year-old seedlings from seeds of Campo Durán population (Fig. 10.7a, b). On the other hand, different types of grafts were tested from cuttings collected from the natural population and grafted on seedlings produced from seeds. The spike graft was the most appropriate in *P. alba*, in agreement with reports by other authors (Wojtusik and Felker 1993; Ewens and Felker 2003) (Fig. 10.7c–e). This technic allows us to propagate selected

Fig. 10.7 Vegetative propagation of *Prosopis alba*. (**a** and **b**), in vitro micropropagation from uninodal segments of seedlings obtained from seeds; (**c** and **d**), grafts of adult plants on juvenile seedlings; (**e** and **f**), mini-cuttings from Campo Durán population. (Photos: Edgardo Carloni)

individuals from the natural population and also adapt field materials to controlled conditions, looking to obtain uninodal segments to establish in vitro cultures.

Rooting of cuttings of rejuvenated material was also tested. Salto et al. (2012) selected 50 individuals of *P. alba* from the evaluation of a progeny trial installed in Laguna Yema. A multi-criteria selection index was calculated, considering the growth (diameter at the base and height) and the shape of the stem. The selection was made with the estimation of the individual improvement values at the third year of planting (Salto 2011). The 50 selected genotypes were coppiced in late winter (August) at a height of 20–25 cm, using a chainsaw. The stumps were sealed with a pruning bandage. The shoot harvesting campaigns took place at three moments in time (41, 61, and 110 days after coppicing). The length and diameter of shoots ranged from 15 to 70 cm and 2 to 6 mm, respectively. At the greenhouse, plant cuttings of 8–10 cm long were conditioned leaving two leaves reduced to a 50% in area. Then, the cutting bases were treated with indole-3-butyric acid at a 0.45 g/kg concentration as root inducer, using microbiological talc as vehicle. The rooting percentage of the plant cuttings brought from the field showed values ranging from 12.5% to 100% among genotypes (López Lauenstein et al. 2016).

Likewise, grafting adult plants on juvenile rootstocks provides rejuvenated material with the goal of establishing a propagation system for macro-cuttings. In *P. alba* and, as mentioned, in other species (Wendling et al. 2014), rejuvenated material is the most efficient way to achieve rooting of cuttings (de Souza and Felker 1986; Arce and Balboa 1991; Oberschelp and Marcó 2010). In recent years, de Souza et al. (2014) have implemented the mini-cuttings technique, which consists on the use of buds of mother plants under controlled conditions in order to generate a degree of rejuvenation and a lower lignification. These authors indicate that the mini-cutting have good rhizogenic capacity with rooting percentages ranged from 98% to 100%.

10.5   Beneficial Microorganisms

Soil microbiomes play important roles in terrestrial ecosystem regulation and functioning, impacting on productivity, diversity, and structure of plant communities. Different types of abiotic stress like drought, salinity, high temperatures, and low nutrient availability acting either alone or in combination have a strong influence on plant diversity, conditioning their survival. Usually plants coevolve with the biodiversity of soil microorganisms and mutual relationships between them exist. The use of rhizobia and mycorrhizal inoculants in nurseries is a strategy to improve the adaptation and survival of seedlings in transplanting to the field, both in wood production plantations and in restoration or ecosystem recovery plans. Therefore, isolation and characterization of specific strains of these microorganisms is highly valued in order to incorporate them into a technological package with the genetic breeding of plant germplasm.

10.5.1   Rhizobacteria for P. alba Cultivation

The symbiotic relationship between plants and rhizobacteria allows the incorporation of N_2 via biological fixation (NBF), inducing a set of systemic changes in the plant that contributes to efficient adaptive responses. Therefore, inoculation with rhizobacteria enhances plant survival during the development and establishment of forest plantations, especially in limiting environmental conditions (Zamioudis and Pieterse 2012).

In the Arid Chaco region of Argentina, the first described species with the ability to form root nodules in *P. alba* was *Mesorhizobium chacoense* (Velázquez et al. 2001). More recently, Chávez Díaz et al. (2013) reported that rhizobacteria belonging to *Mesorhizobium*, *Sinorhizobium* (Ensifer), and *Bradyrhizobium* genera, isolated from environments of five *P. alba* populations in the Chaco region, were also able to induce nitrogen-fixing nodules in this species. To obtain these isolates, soil was collected from *P. alba* seed stands in five localities in the Chaco: San Miguel-Córdoba (31° 45' 59" S; 65° 25' 39" W); Padre Lozano-Salta (23° 12' 51" S; 63° 50' 39" W), Isla Cuba-Formosa (24° 17' 31" S; 61° 51' 10" W), Bolsa Palomo-Formosa (24° 13' 15" S; 61° 57' 42" W), and Colonia Benítez-Chaco (27° 20' S; 58° 55' 60" W). Seeds were grown under these substrates to capture the diversity of rhizobacteria, and after 40 days, fixing nodules were harvested form roots (Chávez Díaz et al. 2013) (Fig. 10.8a). To isolate symbiotic microorganisms, nodules were superficially disinfected and macerated under sterile conditions. The macerate liquid was streaked on LMA culture medium (Vincent 1970), and the plates were incubated in an oven for 3 to 5 days at 28 °C (Fig. 10.8b).

From the developed colonies, DNA extractions and amplification of randomly repeated fragments were performed by rep-PCR, using the primer BOX-A1 (Versalovic et al. 1994). A total of 100 isolates were analyzed, and amplification

Fig. 10.8 Rhizobacteria nodules in *Prosopis alba* roots (**a**) and developed colonies in Petri dish (**b**). (Photos: Mariana Melchiorre)

patterns were grouped according to a cluster analysis (UPGMA) by locality. Based on this, representative isolates from each site whose similarity was less than 60% were selected for subsequent analyses using the Dice coefficient (Di Rienzo et al. 2012). In total 33 isolates were selected: 14 from Bolsa Palomo, 11 from Isla Cuba, 5 from Padre Lozano, 1 from San Miguel, and 2 from Colonia Benítez.

Tolerance to water deficit was evaluated by cultivating the selected isolates in hyperosmotic solutions of polyethylene glycol (PEG) testing water potentials of −0.6 MPa and − 2 MPa. It was observed that isolates that tolerated greater water deficit, correlated with greater indole compounds (IAA) production. From this set of isolates, the 16S rRNA gene was analyzed. Alignment, sequence analysis and construction of a phylogenetic tree were performed using the 16S rRNA gene fragment (1163 bp) with the ClustalW alignment tool of the MEGA software and the UPGMA method (Tamura et al. 2011). The best results were obtained with two isolates from Bolsa Palomo (*Mesorhizobium* and *Sinorhizobium* (Ensifer)), one from Colonia Benítez (*Mesorhizobium*), and two from Padre Lozano (*Bradyrhizobium* and *Sinorhizobium* (Ensifer)) (Chávez Díaz et al. 2013); all these isolates were found to promote growth in *P. alba*.

In another study, Pozzi Tay (2016) evaluated drought tolerance of *P. alba* in symbiosis with rhizobacteria under greenhouse conditions. Using three isolates [N° 2 and N° 63 from Bolsa Palomo, (KC-759691, *Mesorhizobium* spp. and KC-759695, *Sinorhizobium* spp.) and N° 53 from Padre Lozano (KC-759699, *Bradyrhizobium* spp.)], a biofertilizer was elaborated. Seeds from two contrasting *P. alba* provenances with respect to precipitations were used: Santiago (27° 52′ 44″ S, 64° 9′ 16″ O, 579 mm average annual precipitation) and Campo Durán (22° 12′ 01″ S, 63° 40′ 33″ W, 1054 mm average annual precipitation). A factorial design with three factors of two levels each was used: provenance (Santiago and Campo Durán), biofertilization (with and without) and water stress (control and drought). Growth variables and physiological responses to stress were evaluated (Pozzi Tay 2016). Under drought stress, plants of Santiago with biofertilization conserved significantly more leaves (node with leaf/total node) than the uninoculated plants of both provenances ($p \geq 0.05$). Furthermore, the highest values of chlorophyll and proline content were recorded in biofertilized plants of Santiago. Biofertilized plants of both provenances have similar values in number and weight of nodules, but under drought conditions, these parameters were reduced, without distinction of provenances.

Summing up, the analysis of growth parameters and physiological variables allowed to discriminate the behavior of two *P. alba* provenances under drought conditions in symbiosis with specific rhizobacteria from the Chaco region. Both biofertilized provenances maintained higher levels of proline and chlorophylls under drought compared to those not inoculated. This result suggests that the establishment of symbiosis also gives adaptive advantages to water restriction. Future studies will seek to identify the selected rhizobacterial species by sequencing other housekeeping genes as well as to analyze a possible synergistic effect of the mix of microorganisms in the biofertilizer or the prevalence of any of them for the colonization of nodules and biological nitrogen fixation.

10.5.2 *Mycorrhizae for* Prosopis alba

Mycorrhizae are classified according to the characteristics of the infection and the mutual organisms that establish it. The most important group is the endomycorrhizae, which has been subdivided into several groups, the most significant being the arbuscular mycorrhizal fungi (AMF). About 96% of plants form this type of mycorrhiza and are the most abundant group of mycorrhizal fungi. The absorption of phosphate through AMF results in an increase in the absorption of inorganic phosphorus (which is an almost immobile element in the soil) and therefore in the growth of plants. Furthermore, the AMF-plant symbiosis increases the stability of soil aggregates in natural systems, acting as an adherent, agglutinating soil particles into more stable aggregates, therefore increasing water retention. In arid and semiarid ecosystems, plant establishment is increased when mycorrhizal plants are used, which have greater protection and tolerance to adverse soil and climate conditions (Begum et al. 2019).

In order to isolate, characterize, and obtain mixed native AMF inoculants that confer tolerance to abiotic stress (i.e., drought, salinity) in *P. alba* seedlings, a microbiological study was carried out (Sagadin et al. 2018). Soil was collected from two *P. alba* pure stands with contrasting edaphoclimatic characteristics: Colonia Benítez (CB) 27° 20′ 00″ S, 58° 55′ 60″ W (1300 mm of mean annual precipitation) and Padre Lozano (PL) 23° 12′ 51″ S, 63° 50′ 39″ W (650 mm) (Cabrera 1976). Soil AMF species were identified using *Medicago sativa*, *Sorghum bicolor*, and *P. alba* as trap plants. The predominant presence of the Glomeraceae family was recorded in the isolated inoculum of trap plants. Species such as *Funneliformis mosseae* and *Rhizophagus intraradices* have been identified, as well as *Claroideoglomus etunicatum* (Claroideoglomeraceae), frequently reported in association with the vegetation of arid and semiarid ecosystems.

To evaluate the performance of inocula in nursery, a comparative test was carried out with the application of fertilizer (Salto et al. 2019b). The inoculation was applied at the moment of sowing by inoculating 20 g of each inoculum (PL and CB) and a mixture of them (MIX). The fertilization treatments were fertilization (100%), diluted fertilization (25%), and without fertilization (NF) according to Salto et al. (2016). After 120 days of sowing, the following variables were measured: diameter at the base (DAB), total height (H), and number of leaves (NL). Percentage of mycorrhizal colonization was determined in a sample of five seedlings per treatment. On the other hand, 15 seedlings per treatment (120-day old) were transferred to the greenhouse and acclimatized during 10 days. Then, irrigation was suspended to approximately 10% of the soil water content. The recovery was evaluated in 145-day seedlings by watering them to their maximum capacity for 10 days and measuring resprouting capacity as number of plants with new green leaves.

The addition of 100% and 25% of fertilization solution to AMF inoculated treatments did not promote significant differences ($p = 0.4561$) in the percentage of mycorrhizal colonization, suggesting that the addition of fertilizer did not alter the capacity of colonization nor inhibited the formation of the different characteristic structures of AMF, such as arbuscules, vesicles, and hyphae. Furthermore, in unfertilized plants, AMF inocula promoted growth. The increased leaf production in inoculated plants suggests that mycorrhizae stimulate ontogeny and delay leaf senescence. In addition, increasing the fertilizer to 100% improved growth in the treatment with the PL inoculum, and on the contrary, the MIX and CB inocula had less effect than fertilization of 100% in most of the variables of increase. On the other hand, the resprouting capacity of *P. alba* after drought stress conditions varied with different levels of fertilization and with different types of inoculum. In this context, the uninoculated and unfertilized treatment exhibited the lowest resprouting rate of all the treatments. On the contrary, the effect of the treatment (AMF × fertilization) was significant ($p = 0.0198$) increasing the resprouting capacity. Subsequent comparisons indicate that the proportion of resprouted plants in PL_0% and PL_ 25% treatments were significantly higher (Fig. 10.9). These results demonstrated that the AMF inoculum isolated from the semiarid regions of the study area, such as PL, may have the potential to mitigate drought stress in *P. alba* seedlings compared to the inoculum isolated from wet areas (Salto et al. 2019b).

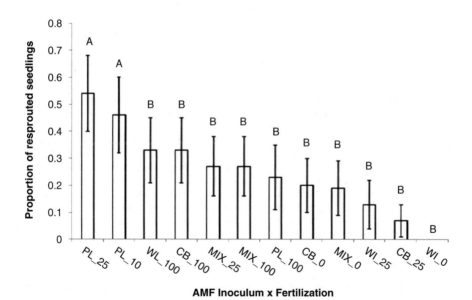

Fig. 10.9 Effect of different fertilization and AMF inoculation regimes on *P. alba* seedlings submitted to drought stress. Resprouting capacity is expressed as the proportion of resprouted plants after 15 days of drought stress. Mean values and their standard errors. Different letters indicate significant differences ($p \leq 0.05$) according to DGC's test (Salto et al. 2019b). *PL* Padre Lozano inoculum, *CB* Colonia Benitez inoculum, *MIX* mixture of both inocula, *WI* without inoculum, 100 fertilization without dilution, 25 fertilization diluted to 25%, 0 without fertilization

References

Arce P, Balboa O (1991) Seasonality in rooting of *Prosopis chilensis* cuttings and in-vitro micropropagation. For Ecol Manag 40:163–173

Begum N, Qin C, Ahanger M, Raza S, Khan M, Ashraf M et al (2019) Role of arbuscular mycorrhizal fungi in plant growth regulation: implications in abiotic stress tolerance. Front Plant Sci 10:1–15

Bessega C, Pometti C, Ewens M, Saidman BO, Vilardi JC (2012) Strategies for conservation for disturbed *Prosopis alba* (Leguminosae, Mimosoidae) forests based on mating system and pollen dispersal parameters. Tree Genet Genomes 8:277–288

Bessega C, Pometti C, Miller J, Watts R, Saidman BO, Vilardi JC (2013) New microsatellite loci for *Prosopis alba* and *P. chilensis* (fabaceae). Appl Plant Sci 1:1–4

Bessega C, Pometti C, Ewens M, Saidman B, Vilardi J (2015) Evidences of local adaptation in quantitative traits in *Prosopis alba* (Leguminosae). Genetica 143:31–44

Borralho N (1995) The impact of individual tree mixed models (BLUP) in tree breeding strategies hardwood forestry. In: Proceedings of the CRC-IUFRO conference: Eucalypts plantations: Improving fiber yield and quality, 12–24 February, 1995, Hobart, pp 141–145

Cabrera AL (1976) Regiones Fitogeográficas Argentinas. Enciclopedia argentina de agricultura y jardinería, Tomo 2, fasc. 1. ACME, Buenos Aires

Capello R (2019) Banco de Germoplasma Forestal. Manejo de los Recursos Genéticos Forestales Nativos. Dirección de Recursos Naturales. Gobierno de la Provincia de Formosa

Cappa E, Varona L (2013) An assessor-specific Bayesian multi threshold mixed model for analyzing ordered categorical traits in tree breeding. Tree Genet Genomes 9:1423–1434

Carreras R, Bessega C, López C, Saidman B, Vilardi J (2016) Developing a breeding strategy for multiple trait selection in *Prosopis alba* Griseb., a native forest species of the Chaco Region in Argentina. Forestry 90:199–210

Chávez Díaz L, González P, Rubio E, Melchiorre M (2013) Diversity and stress tolerance in *rhizobia* from Parque Chaqueño region of Argentina nodulating *Prosopis alba*. Biol Fertil Soils 49:1153–1165

Cony M (1996) Genetic potential of *Prosopis* in Argentina for its use in other countries. Semiarid fuelwood and forage tree building consensus for the disenfranchised. Center for Semi-Arid Forest Resources, Texas, pp 6–3. 6–24

De Souza S, Felker P (1986) The influence of stock plant fertilization on tissue concentrations of N, P and carbohydrates and the rooting of *Prosopis alba* cuttings. For Ecol Manag 16:181–190

De Souza J, Bender A, Tivano J, Barroso D, Moroginski L, Vegetti A, Felker P (2014) Rooting of *Prosopis alba* mini-cuttings. New For 45:745–752

Di Rienzo J, Casanoves F, Balzarini M, Gonzalez L, Tablada M, Robledo C (2012) InfoStat versión 2012. Grupo InfoStat, FCA, Universidad Nacional de Córdoba, Argentina

Ewens M, Felker P (2003) The potential of mini-grafting for large-scale production of *Prosopis alba* clones. J Arid Environ 55:379–387

Fagg C, Stewart J (1994) The value of *Acacia* and *Prosopis* in arid and semi-arid environments. J Arid Environ 27:3–25

FAO (2014) Plan de acción mundial para la conservación y la utilización sostenible y el desarrollo de los recursos genéticos forestales p comisión de recursos genéticos para la alimentación y la agricultura, 35 pp

Felker P, López C, Soulier C, Ochoa J, Abdala R, Ewens M (2001) Genetic evaluation of *Prosopis alba* (algarrobo) in Argentina for cloning elite trees. Agrofor Syst 53:65–76

Ferreyra L, Vilardi J, Verga A, López V, Saidman B (2013) Genetic and morphometric markers are able to differentiate three morphotypes belonging to section Algarobia of genus *Prosopis* (Leguminosae, Mimosoideae). Plant Syst Evol 299(6):1157–1173

Galera F (2000) Los algarrobos: Las especies del género *Prosopis* (algarrobos) de América Latina con especial énfasis en aquellas de interés económico. Graziani Gráfica, Córdoba, 269 pp

Gómez A, Rossi F, Bravo S (2019) Determinación del momento oportuno de raleo de algarrobo blanco en plantaciones: técnica alternativa para la medición de anillos de crecimiento. En: Avances en el conocimiento y tecnologías productivas de especies arbóreas nativas de Argentina / [Compiladoras]: Carla S. Salto; Ana María Lupi. Ediciones INTA, Buenos Aires, 78 pp

Grattapaglia D, Silva-Junior O, Resende R, Cappa E, Müller B, Tan B, El-Kassaby Y (2018) Quantitative genetics and genomics converge to accelerate forest tree breeding. Front Plant Sci 9:1693. https://doi.org/10.3389/fpls.2018.01693

Joseau J, Verga A, Días MP, Julio N (2013) Morphological diversity of populations of the genus Prosopis in the semiarid Chaco of Northern Cordoba and Southern Santiago del Estero. Am J Plant Sci 4:2092–2111

Kees S, Michela J (2016) Aspectos de la producción primaria y el mercadeo del algarrobo en Chaco, Argentina. Presidencia Roque Sáenz Peña, Argentina

Kees S, Ferrere P, Lupi A, Michela J, Skoko J (2018) Producción y crecimiento de las plantaciones de *Prosopis alba* Griseb. en la provincia Chaco. Revista de Investigaciones Agropecuarias 44:113–120

Kees S, López A, Zurita J, Brest E, Roldan M, Rojas J (2019) Identificación de las características edáficas determinantes de la calidad de sitio para el cultivo de *Prosopis alba*. En: Avances en el conocimiento y tecnologías productivas de especies arbóreas nativas de Argentina / [Compiladoras]: Carla S. Salto; Ana María Lupi. Ediciones INTA, Buenos Aires, 78 pp

Lindgren D, Mullin T (1997) Balancing gain and relatedness in selection. Silvae Genet 46:124–129

López Lauenstein D, Vega C, Luna C, Sagadin M, Melchiorre M, Pozzi E et al (2016) Subprograma *Prosopis*. In: Marcó MA, Llavallol CE (eds) Domesticación y mejoramiento de especies forestales. Unidad para el Cambio Rural. Ministerio de Agroindustrias. Presidencia de la Nación Argentina, Buenos Aires, pp 113–136

López Lauenstein D, Vega C, Verga A, Fornes L, Saravia P, Feyling M et al (2019) Evaluación de diez orígenes de Algarrobo para establecer sistemas silvopastoriles en el Chaco semiárido argentino. In: Proceedings of X Congreso Internacional sobre Sistemas Silvopastoriles. Asunción, 24–26 September, Editorial CIPAV, Cali

MAGyP (2016) Avances en la silvicultura del algarrobo blanco. https://www.agroindustria. gob.ar/sitio/areas/ss_desarrollo_foresto_industrial/biblioteca_forestal/publicaciones/_archivos/000000_Avances%20en%20la%20silvicultura%20del%20algarrobo%20blanco.pdf

Marcucci Poltri S, Gallo L (2016) Herramientas moleculares. In: Marcó MA, Llavallol CE (eds) Domesticación y mejoramiento de especies forestales. Unidad para el Cambio Rural. Ministerio de Agroindustrias. Presidencia de la Nación Argentina, Buenos Aires, p 201

Min. Agro (2015) Informe Nacional del Relevamiento Censal de Aserraderos. Ministerio de Agroindustria. República, Argentina. https://www.agroindustria.gob.ar/sitio/areas/ ss_desarrollo_foresto_industrial/censos_inventario/

Moglia G, Giménez A (2006) Resultados preliminares de la arquitectura vegetal de *Prosopis alba* y *Prosopis nigra*. En: II Jornadas Forestales de Santiago del Estero: Forestación y Aprovechamiento Integral del Algarrobo. Santiago del Estero, 5 p

Mottura M, Finkeldey R, Verga A, Gailing O (2005) Development and characterization of microsatellite markers for *Prosopis chilensis* and *Prosopis flexuosa* and cross-species amplification. Mol Ecol Notes 5:487–489

Namkoong G, Barnes R, Burley J (1980) Philosophy of breeding strategy for tropical forest trees, Tropical Forest Paper N°16. Commonwealth Forestry Institute, Oxford, 67 pp

Oberschelp G, Marcó M (2010) Efecto del ácido 3-indolbutírico sobre el enraizamiento adventicio y la altura de plantines clonales de *Prosopis alba* Grisebach. Quebracho Revista de Ciencias Forestales 18(1–2)

Pomponio MF, Acuña C, Pentreath V, López Lauenstein D, Marcucci Poltri S, Torales S (2015) Characterization of functional ssr markers in *Prosopis alba* and their transferability across *Prosopis* species. For Syst 24:2013–2016

Pozzi Tay E (2016) Simbiosis en *Prosopis alba* como estrategia para la mejora de su tolerancia a estrés hídrico. Graduate thesis, Fac. Cs. Exactas, Físicas y Naturales. Universidad Nacional de Córdoba

Ruotsalainen S, Lindgren D (1998) Predicting genetic gain of backward and forward selection in forest tree breeding. Silvae Genet 47:42–50

Sagadin M, Monteoliva M, Luna C, Cabello M (2018) Diversidad e infectividad de hongos micorrícicos arbusculares nativos provenientes de algarrobales del Parque Chaqueño argentino con caracteríticas edafoclimáticas contrastantes. Agriscientia 35:19–33

Saidman B (1986) Isoenzymatic studies of alcohol dehydrogenase and glutamate oxalacetate transaminase in four South American species of *Prosopis* and their natural hybrids. Silvae Genet 35:3–10

Salto C (2011) Variación genética en progenies de polinización abierta de Prosopis alba griseb. de la región chaqueña. Tesis de Maestría. Facultad de Ciencias Agrarias. Universidad Nacional de Rosario, Argentina, 75 pp

Salto C, Lupi A (2019) Avances en el conocimiento y tecnologías productivas de especies arbóreas nativas de Argentina / [Compiladoras]: Carla S. Salto; Ana María Lupi. Ediciones INTA, Buenos Aires, 78 pp

Salto C, Oberschelp J, Harrand L (2012) Recolección, acondicionamiento y transporte de material vegetal de Prosopis alba Griseb. para propagación vía estacas. En: Reunión Nacional del Algarrobo, 13–14 noviembre 2012, Córdoba, Argentina

Salto C, Harrand L, Oberschelp J, Ewens M (2016) Crecimiento de plantines de *Prosopis alba* en diferentes sustratos, contenedores y condiciones de vivero. Bosque (Valdivia) 37:527–537

Salto C, Harrand L, Oberschelp J, Ewens M (2019a) Efecto del tamaño de envase y calidad del sustrato utilizado sobre la calidad del plantín de *Prosopis alba*. En: Avances en el conocimiento y tecnologías productivas de especies arbóreas nativas de Argentina / [Compiladoras]: Carla S. Salto, Ana María Lupi. Ediciones INTA, Buenos Aires, 78 pp

Salto CS, Sagadin M, Luna C, Oberschelp G, Harrand L, Cabello M (2019b) Interactions between mineral fertilization and arbuscular mycorrhizal fungi improve nursery growth and drought tolerance of *Prosopis alba* seedlings. Agrofor Syst 94:103–111

SAyDS (2019) Anuario de Estadística Forestal de Especies Nativas 2017–2018. https://www.argentina.gob.ar/ambiente/bosques/estadistica-forestal

Spensley J, Sabelli A, Buenfil J (2013) Estudio de Vulnerabilidad e Impacto Del Cambio Climático En El Gran Chaco Americano. Ed. Programa de las Nacionales Unidas para el Medio Ambiente – PNUMA

Suja A, Venkataraman G, Parida A (2007) Identification of stress-induced genes from the drought-tolerant plant *Prosopis juliflora* (Swartz) D. C. Through analysis of expressed sequences tags. Genome 50:470–478

Tamura K, Peterson D, Peterson N, Stecher G, Nei M, Kumar S (2011) MEGA5: molecular evolutionary genetics analysis using maximum likelihood, evolutionary distance and maximum parsimony methods. Mol Biol Evol 28:2731–2739

Torales S, Rivarola M, Pomponio MF, Gonzalez S, Acuña C, Fernández P et al (2013) De novo assembly and characterization of leaf transcriptome for the development of functional molecular markers of the extremophile multipurpose tree species *Prosopis alba*. BMC Genomics 14:705

Velázquez E, Igual J, Willems A, Fernández MP, Muñoz E, Mateos PF et al (2001) Mesorhizobium chacoense sp., a novel species that nodulates *Prosopis alb*a in the Chaco Arido region (Argentina). Int J Syst Evol Microbiol 51:1011–1021

Verga A (1995) Genetic study of *Prosopis chilensis* y *Prosopis flexuosa* (Mimosaceae) in the dry Chaco of Argentina. Doctoral thesis, Abteilung für Forstgenetik und Forstpflanzensüchtung der Universität Göttingen, Alemania, 96 pp

Verga A (2005) Recursos Genéticos, Mejoramiento y Conservación de Especies del Género Prosopis. En: Mejores Árboles Para Más Forestadores: El Programa de Producción de Material de Propagación Mejorado y el Mejoramiento Genético en el Proyecto Forestal de Desarrollo Edición: Carlos A. Norberto. SAGPyA-BIRF

Verga A (2014) Rodales semilleros de *Prosopis* a partir del bosque nativo. Quebracho 19:125–138

Verga A (2015) "Hoja 3.6." Programa de distribución gratuita. Instituto de Fisiología y Recursos Genéticos Vegetales (IFRGV). Instituto Nacional de Tecnología Agropecuaria (INTA). Camino 60 Cuadras, km 5.5 X5020ICA, Córdoba, Argentina. E -mail: verga.anibal@inta.gob.ar

Verga A, Gregorius H (2007) Comparing morphological with genetic distances between populations: a new method and its application to the *Prosopis chilensis – P. flexuosa* Complex. Silvae Genet 56:45–51

Verga A, López Lauenstein D, López C, Navall M, Joseau J, Gómez C et al (2009) Caracterización morfológica de los algarrobos (*Prosopis* sp.) en las regiones fitogeográficas Chaqueña y Espinal norte de Argentina. Quebracho 17:31–40

Versalovic J, Schneider M, de Bruijn F, Lupski J (1994) Genomic fingerprinting of bacteria using repetitive sequence-based polymerase chain reaction. Methods Mol Cell Biol 5:25–40

Verzino G, Joseau MJ (eds) (2005) El Banco Nacional de Germoplasma de *Prosopis*. Conservación de recursos forestales nativos en Argentina, 172 pp

Verzino G, Frassoni J, Joseau M, Clausen G, Navarro C (2020) Conservación ex situ, circa situ e in situ realizada por el Banco Nacional de Germoplasma de *Prosopis*, Córdoba, Argentina. Revista Nexo (NA-V7N1-9) (in press)

Vincent J (1970) A manual for the practical study of the root-nodule bacteria. Blackwell, Oxford

Wendling I, Trueman S, Xavier A (2014) Maturation and related aspects in clonal forestry. Part II: Reinvigoration, rejuvenation and juvenility maintenance. New For 45:473–486

Wojtusik T, Felker P (1993) Interspecific graft incompatibility in *Prosopis*. For Ecol Manag 59:329–340

Zamioudis C, Pieterse C (2012) Modulation of host immunity by beneficial microbes. Mol Plant Microbe Intract 25:139–150

Zárate M, Gyenge J, Gómez A (2019) ¿Cuál es el efecto de la poda y la densidad de plantación en el crecimiento de los algarrobos? In: Avances en el conocimiento y tecnologías productivas de especies arbóreas nativas de Argentina / [Compiladoras]: Carla S. Salto; Ana María Lupi. Ediciones INTA, Buenos Aires, 78 pp

Zobel B, Talbert J (1984) Applied forest tree improvement. Wiley, New York, 505 pp

Chapter 11
Species Without Current Breeding Relevance But High Economic Value: *Acacia caven*, *Acacia aroma*, *Acacia visco*, *Prosopis affinis*, *Prosopis caldenia* and *Gonopterodendron sarmientoi*

Carolina Pometti, Gonzalo A. Camps, María Cristina Soldati,
Teresa Velasco Sastre, Gregorio Gavier, Noga Zelener, Aníbal Verga,
Mauricio Ewens, Beatriz O. Saidman, Alicia N. Sérsic, and Andrea Cosacov

11.1 Botanical, Ecological and Usage Features

All the species included in this chapter have been utilized directly from the natural forest since the time of European pre-colonization. The cultural multipurpose use that the original inhabitants of the region had done turned to an industrial exploitation when the Spaniards settled and imposed. Since then, the hard wood of all these species has been utilized for rural carpentry, poles, charcoal and firewood. Although their wood plays a relevant role in the local economy, the official statistics does not reflect it properly, since most of their commerce is informal.

The genus *Acacia* (Fabaceae) includes over 1450 species distributed in tropical and subtropical regions of the Americas, Australia, Africa and southern Asia (Guinet

C. Pometti (✉) · B. O. Saidman
Laboratorio de Genética en Especies Leñosas, Departamento de Ecología, Genética y Evolución (DEGE), Facultad de Ciencias Exactas y Naturales, Universidad de Buenos Aires., Buenos Aires, Argentina
e-mail: cpometti@ege.fcen.uba.ar

G. A. Camps
Instituto de Fisiología y Recursos Genéticos Vegetales, Unidad de Estudios Agropecuarios (IFRGV, UDEA) INTA-CONICET, Córdoba, Argentina

Laboratorio de Ecología Evolutiva-Biología Floral, IMBIV (UNC-CONICET), Córdoba, Argentina

M. C. Soldati · G. Gavier
Instituto de Recursos Biológicos (IRB), Centro de Investigaciones de Recursos Naturales (CIRN), INTA. Hurlingham, Buenos Aires, Argentina

© Springer Nature Switzerland AG 2021
M. J. Pastorino, P. Marchelli (eds.), *Low Intensity Breeding of Native Forest Trees in Argentina*, https://doi.org/10.1007/978-3-030-56462-9_11

and Vassal 1978; Ross 1981; Luckow 2005). In northern and Central Argentina, *Acacia* is represented by 21 woody species. Most of them are trees or shrubs 2–6 m high, although some species reach up to 20 m.

According to recent taxonomy, the circumscription of the genus *Acacia* is currently controversial, since it may be treated as a single genus or as comprising multiple genera. Considerations about this subject can be found in Orchard and Maslin (2005), Smith et al. (2006), Rijckevorsel (2006) and Moore et al. (2011). Following Vassal's treatment (1972; Polhill et al. 1981), *Acacia* is considered as a single genus with three subgenera (*Acacia*, *Aculeiferum* and *Phyllodineae*). The Native American species of *Acacia* belong to two subgenera: *Acacia* and *Aculeiferum*. In the present chapter, for the benefit of the users, the authors recognize this last classification although they will use the name *Acacia s. l.*

Acacia caven (Mol.) Mol. (syn. *Vachellia caven*, Seigler and Ebinger, 2006) belongs to the subgenus *Acacia*, and it is native to Argentina, Chile, Paraguay, Uruguay, Brazil and Bolivia (Aronson 1992), where this 6 m tall tree (or shrub) is known by the common name "espinillo" or "churqui". In Argentina, it is widely distributed from the central to the north-eastern regions of the country (Fig. 11.1), reflecting its remarkable climate tolerance and ecological adaptability to a wide range of environmental conditions. As such, this species is suitable for colonizing sites degraded by human activities (i.e. intensive agriculture, cattle raising and fires, among others). Due to its great plasticity, it is recommended for reforestation of degraded ecosystems.

It has very small and perfumed yellow flowers arranged in compact spherical inflorescences with a short peduncle between 4 and 18 mm. The fruit is a cylindrical, woody pod, dark brown, between 4 and 7 cm long (Fig. 11.2). The seeds are greenish, hard, approximately 6 mm in diameter, and about 1000 seeds can be counted per kilogram of fruit, for which a germination power of 85.7% has been reported (Karlin et al. 1997).

T. Velasco Sastre
Instituto de Fisiología y Recursos Genéticos Vegetales, Unidad de Estudios Agropecuarios (IFRGV, UDEA) INTA-CONICET, Córdoba, Argentina

N. Zelener
Centro de Investigación de Recursos Naturales (CIRN), INTA. Hurlingham, Buenos Aires, Argentina

A. Verga
Instituto de Fisiología y Recursos Genéticos Vegetales, Unidad de Estudios Agropecuarios (IFRGV, UDEA) INTA-CONICET, Córdoba, Argentina

INTA AER La Rioja, La Rioja, Argentina

M. Ewens
Estación Experimental Fernández-UCSE (Prov. Santiago del Estero- Universidad Católica Santiago del Estero), Santiago del Estero, Argentina

A. N. Sérsic · A. Cosacov
Laboratorio de Ecología Evolutiva y Biología Floral - Instituto Multidisciplinario de Biología Vegetal (IMBIV) UNC - CONICET. Facultad de Ciencias Exactas, Físicas y Naturales, Universidad Nacional de Córdoba., Córdoba, Argentina

Fig. 11.1 Geographic distribution in Argentina of *Acacia aroma, A. caven, A. visco, Prosopis affinis* and *P. caldenia*. (Taken from Funes et al. 2007; Dimitri and Biloni 1973; Leonardis and Milanesi 1975, and Mariani et al. 2012, respectively)

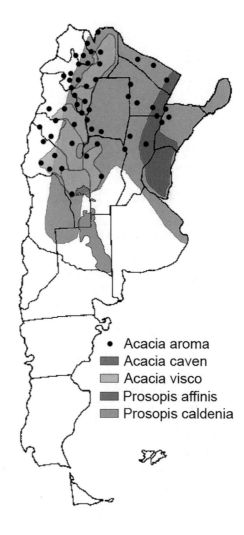

- • Acacia aroma
- ■ Acacia caven
- ■ Acacia visco
- ■ Prosopis affinis
- ■ Prosopis caldenia

As in many *Acacia* species, their fruits have fodder aptitude. A common practice in the north of Chile is to cut the branches to feed the cattle along the dry season, meanwhile the remaining parts of the tree are used as fuel or to construct fences (Karlin et al. 1997). Its wood is hard and heavy, being its density from 0.8 to 0.98 g/cm^3, providing material for charcoal, firewood and poles (Tortorelli 1956).

Acacia aroma Gillies ex Hook. & Arn. (syn. *Vachellia aroma*, Seigler and Ebinger, 2006) also belongs to the subgenus *Acacia* and is distributed in Central and Northern Argentina, Bolivia, Western Paraguay, Peru and Southern Ecuador (Ebinger et al. 2000). This species grows as a tree or shrub 4–6 m in height, which is not only ecologically but also economically important due to multiple uses and functions. From an ecological point of view, it is a foundation species, and the roots are good nitrogen fixers, what highlights the interest of the species to be used in restoration programmes. Economically, the fruits and leaves provide forage for cat-

Fig. 11.2 Fruits and flowers of *Acacia caven* var. caven. (Photo **a** from (c) Dick Culbert Creative Commons Attribution 4.0 International Public License; Photo **b**: Carolina Pometti)

tle and goats, fruits and bark are rich in tannins, the flowers are useful in the perfume industry and also have importance for honey production, the seeds have medicinal uses, and its wood has high density and durability (density 0.78 g/cm³), which makes it appropriate to be used for floors, canes and charcoal production (Pensiero and Gutiérrez 2005; Pometti et al. 2009).

Acacia visco Lor. ap. Griseb (syn. *Senegalia visco*, Seigler et al. 2006; *Parasenegalia visco* (Lorentz ex Griseb.) Seigler & Ebinger) belongs to the subg. *Aculeiferum*. It is native to Chile, Bolivia and Argentina, and it has been introduced to Africa and naturalized in Europe. In this last continent, it was introduced as ornamental in roads, parks and gardens (Romero-Zarco and Tormo Molina 2016). In Argentina it grows in the north-western provinces of Salta, Jujuy, Tucumán, Catamarca, La Rioja, San Juan and San Luis (Fig. 11.1), where it is commonly known as "viscote" or "arca". It is used for ornamental purposes due to its abundant and scented yellow flowers, and its wood is utilized in carpentry, bodywork and parquet due to its hardness and durability (density 0.8 to 0.9 g/cm³), characteristics for which it is also used for fence posts (Tortorelli 1956). The official statistics register for 2017 the commerce of 57 tonne of roundwood of this species and 15 tonne of poles for rural fences (SGAyDS 2019). Methanol extracts of leaves and bark of *A. visco* has been shown to have short- and long-term anti-inflammatory effects in mice (Pedernera et al. 2007). Among the compounds identified from the leaves of this 6 to 12 m tall tree, the triterpenoid lupeol, α-amyrin and β-amyrin may be mainly responsible for the pharmacological activities (Pedernera et al. 2010).

Prosopis affinis Spreng., locally known as "ñandubay", is a tree of 2.5 to 10 m in height, 50 to 60 cm in diameter, and slow growth, native to Argentina, Brazil, Uruguay and Paraguay (GBIF 2019). In Argentina, it is distributed in the northeastern part of the country (Fortunato et al. 2008) (Fig. 11.1) and defines a phytogeographic district of the Espinal ecoregion (Cabrera and Willink 1973). It differs from other species of the genus by vegetative (flat topped tree, armed stems with geminate spines, leaflets up to 2 mm long, opposite to imbricate, nerves of the hypophyllum prominent), floral (racemes mainly spiciform) and carpological traits (legume moniliform with margin undulate) (Palacios and Brizuela 2005). The ñandubay is a dominant species of the forest, since it determines its structure and functioning (SAyDS 2007). At the same time, it is an important resource for local communities for its medicinal and chemical properties (i.e. dyes, tanning) and its importance as a honey-bearing species. It has fruits (pods) of high nutritive value and palatability that are used as fodder and food for humans, whereas its wood is utilized in numerous applications, determining the possible use of the species in silvopastoral and agroforestry systems (Roig 1993). *P. affinis* has significantly decreased the size of its populations in Argentina due to deforestation and timber overexploitation, especially in the easternmost part of its distribution, where it used to be a widespread species and nowadays only remains as relict forests (SAyDS 2007; Matteucci 2012).

Prosopis caldenia L. is an endemic species from Argentina that characterizes the phytogeographic district of the Espinal ecoregion known as the Caldenal (Fig. 11.1) (Cabrera and Willink 1973), according to its common name: "caldén". The present state of this ecosystem is far from being a continuous forest. The current patches of caldén woodlands are the result of fragmentation and degradation (Fig. 11.3), where the remaining forest represents 18% of the original area (Collado et al. 2002; Gabutti et al. 2009; SAyDS 2006). They have different conservation status and varying sizes and represent variations of the same type of vegetation (Boyero 1985; Zinda et al. 2005). The natural range of *P. caldenia* extends over the Chaco-Pampa Plain, covering 48,872 km^2 (Morello et al. 2012; Oyarzabal et al. 2018) of the southern part of the Espinal ecoregion. The temperate – semiarid climate of the Caldenal has gradually increased its mean precipitation from 560 to 900 mm.year^{-1} in the last 50 years

Fig. 11.3 Lone *Prosopis caldenia* tree in a deforested area, with cattle under its shadow. (Photo: Teresa Velasco)

(Contreras et al. 2011; Risio et al. 2014), thus raising the deforestation area used for crops and shifting westward the agricultural frontier (Barbosa 2005).

Caldén is a heliophilous and deciduous tree, 12–15 m high, with a superficial radical system. The flowers that are arranged in spikes are hermaphroditic, actinomorphic and yellow (Burkart 1976). Flowering period occurs in spring but can be aborted by late frosts or rains. A second smaller flowering can happen in January. The fruit is an indehiscent helical legume containing up to 35 oval seeds, and the fructification occurs at the end of December and January (Steibel 2003; Lell 2004). It is a diploid species with a somatic number of $2n = 28$ and small chromosomes (Burkart 1976). The dispersion of the Caldén is mostly endozoic, although it also sprouts vegetatively from the stump (Tortorelli 1956; Boyero 1985; Lell 2004). The bark is persistent and resistant to fire. Their specimens can be very long-lived, reaching up to 318 years (Scarone et al. 2000). Caldén has a high-quality wood, used for furniture, parquet flooring and fuel with a high calorific value. It has a density of 0.750 kg/dm^3 in the dry state (Bogino 2006). The official statistics registered in 2017 the production in Argentina of 229 tonne of roundwood, 722 tonne of poles, 22 tonne of charcoal and 58,504 tonne of firewood (SGAyDS 2019). In addition, the fruit is a fodder highly desired by native fauna and livestock, because of its carbohydrates wealth. Nevertheless, their inter-annual production and their spatial distribution are very variable (Menvielle and Hernandez 1985; Risio et al. 2016). As a heliophyte species, open areas favour its growth (Velasco et al. 2018), and it responds vigorously to early thinning treatments and pruning (Amieva 2013). Mean radial growth values estimations of 2.47 to 3.47 mm.year^{-1} are cited in the Province of San Luis (Bogino and Villalba 2008) and 1.88 to 4.04 mm.year^{-1} in the Province of La Pampa (Peinetti et al. 1994; Velasco et al. 2018). Silvicultural rotation is between 60 and 100 years in natural forests without management (Bogino and Villalba 2008), and the natural regeneration period has been calculated as 4 years in unperturbed sites under optimal light conditions.

Gonopterodendron sarmientoi (Lorentz ex Griseb.) A.C. Godoy-Bürki (syn. *Bulnesia sarmientoi* Lorentz ex Griseb.) (Zygophyllaceae) is a ~ 20 m high tree distributed in Central-Northern Argentina, South-Eastern Bolivia, Western Paraguay and marginally in south-western Brazil (Zuloaga et al. 2008; Waller et al. 2012) (see Fig. 11.7). This species, known as "palo santo", occurs mainly in semiarid areas (Dry Chaco), under strong seasonality conditions; few natural populations of the species were found marginally in the ecotone with humid climate, such as in Chiquitano Dry Forest, the Pantanal or the Humid Chaco. Across its distribution range, the average annual precipitation varies from 500 to 1200 mm and the average annual temperature from 22 to 26 °C. The species is associated with temporarily floodable soils with high clay content (Adamoli et al. 1972), particularly in the forests where palo santo is dominant (called "palosantales"), such as in those around the Pilcomayo and Bermejo rivers.

Regarding the structure and population dynamics of palosantales, *G. sarmientoi* trees (Fig. 11.4) show a diametric distribution with a preponderance of individuals in the lower classes (diameter <20 cm) being the species density around 286 adult trees/ha. In forests where the species does not dominate, it co-exists with typical

Fig. 11.4 *Gonopterodendron sarmientoi* tree at the border of the natural forest. (Photo: Gonzalo Andrés Camps)

trees of the region such as *Tabebuia nodosa, Aspidosperma quebracho-blanco*, and *Schinopsis lorentzii*; in this type of forests, the diameter distribution of the species is more regular, and the species density is between 25 and 38 trees/ha (Loto et al. 2018). Giménez et al. (2007) studied the wood anatomy and annual growth rings of the species, reporting that 100 years of growth correspond to a basal diameter of 45 cm, approximately. Loto et al. (2018) registered an annual periodic increase (period 2007–2012) of 1.14 mm/year as a mean for a total of 126 specimens.

Exploitation of *G. sarmientoi* has been reported in communities of Argentina (Filipov 1994; Scarpa 2000; Martínez 2011) as well as in Bolivia and Paraguay (Arispe and Rumiz 2002; Benítez et al. 2008; Quiroga et al. 2009); the timber is used for home and urban construction and for handicrafts and tools; the resin is used in medicine or as repellent (Mereles and Pérez de Molas 2008). Palo santo is exploited at industrial level to produce charcoal, floor boards or to extract an essential oil ("guaiacol"), and it is also exported as roundwood (Jacobs 1990; SAyDS 2007; Mereles and Pérez de Molas 2008; Waller et al. 2012). The Argentinian official statistics register the production of 3282 tonne of roundwood, 821 tonne of poles and 277 tonne of other products derived from the species wood (SGAyDS 2019).

11.2   Genetic Diversity by Means of Molecular Markers

Genetic diversity is the result of long-term evolution and represents the evolutionary potential of a species. Surviving in a varying environment requires the accumulation of genetic variation in order to increase adaptability (Li et al. 1999). Accurate estimates of genetic diversity are useful for optimizing sampling strategies and for conserving and managing the genetic diversity of trees (Hamrick and Godt 1996).

Six botanical varieties were described for *A. caven* (var. *caven*, var. *dehiscens*, var. *sphaerocarpa*, var. *stenocarpa*, var. *microcarpa* and var. *macrocarpa*) based on both morphological traits (Aronson 1992; Pometti et al. 2007) and RAPD markers (Pometti et al. 2010). Argentina is the only country where all six varieties are present (Aronson 1992).

The genetic diversity of the species was studied for 15 Argentinian populations belonging to the six mentioned varieties by means of RAPD and AFLP techniques. With RAPD, genetic diversity (expected heterozygosity: H_E) ranged from 0.16 to 0.33 (mean $H_E = 0.254$) and percentage of polymorphic loci *PPL* varied from 27.7% to 97.9% (mean *PPL* = 78.73%), evaluating 47 discernible bands (Pometti et al. 2010). In the case of AFLP technique, 217 neutral loci were evaluated, and *PPL* ranged from 65% to 93.5% (mean *PPL* = 78.2%), meanwhile H_E ranged from 0.206 to 0.353 (mean $H_E = 0.278$) (Pometti et al. 2012).

In the case of *A. aroma*, six Argentine populations were studied where H_E varied from 0.18 to 0.24 (mean $H_E = 0.21$), and *PPL* varied from 56.4% to 67.1% (mean *PPL* = 62.1%). These values were obtained with 401 AFLP loci (Pometti et al. 2018).

In *A. visco*, *PPL* ranged from 43.4% to 74.2% with a mean of 60.89% over the seven Argentine populations analysed. Estimations of H_E varied from 0.16 to 0.26 with a mean of 0.20. These indices were obtained by means of 431 AFLP loci (Pometti et al. 2016). The estimates of genetic diversity for the three species of *Acacia* mentioned here were similar to other widely distributed African and American acacias like *A. albida*, *A. senegal*, *A. raddiana*, *A. farnesiana and A. curvifructa* (Shrestha et al. 2002; Chiveu et al. 2008; Pometti et al. 2015).

A total of 132 individuals of *P. affinis* distributed in 16 fragments of native forest within an agricultural intensification gradient in the West-Centre of the Province of Entre Ríos (Central-Eastern Argentina) were examined. They were assessed by means of the patterns obtained for ten polymorphic SSR markers transferred from phylogenetically close species (Soldati et al. 2018). Descriptive parameters of genetic diversity were estimated. Unbiased expected heterozygosity levels were moderate (from 0.621 to 0.816), with an average value of U $H_E = 0.734$. Numerous exclusive alleles (*Ea*) were found, distributed within the fragments of native forest, in a range of 0 to 4. Nei's genetic distance among fragments was low to moderate, with an average value of 0.238. Higher levels of genetic variability were found in the forest fragments sampled in areas of the gradient with greater agricultural expansion.

The knowledge about *P. caldenia* genetic diversity and mating system is scarce, and specific studies using DNA markers such as RAPD, ISSR or SSR are restricted to the Province of San Luis (Central Argentina). It has been suggested that the Caldén have not suffered marked genetic erosion yet, and a great phenotypic variability has been reported. As for the rest of the species of *Prosopis* genus, swarms of hybrids have been observed in zones with sympatry between *P. caldenia* and *P. flexuosa*. Absence of barriers of reproductive isolation and sympatry have facilitated hybridization and introgression in the genus *Prosopis* (e.g. Palacios and Bravo 1981; Naranjo et al. 1984; Hunziker et al. 1986) and have contributed to the morphological and genetic variability observed in pure populations of Caldén.

High percentage of polymorphism in *P. caldenia* was reported by means of microsatellites transferred from other *Prosopis* species (Sherry et al. 2011; Bessega et al. 2013). The pioneer study by Pérez Díaz (2014) allowed the genetic characterization of two Caldén populations through different molecular markers. As a result, SRR and ISSR were more valuable to detect polymorphic loci among individuals of *P. caldenia* than those obtained by RAPD. According to the general expectation for the Algarobia section (Bessega et al. 2005), diversity in *P. caldenia* was high (He = 0.54) (Velasco 2018). Nei's genetic distances of *P. caldenia* conspecific populations (0.080) are similar to those obtained among *Prosopis nigra* populations (0.096) and *Prosopis ruscifolia* (0.081) (Ferreyra 2000).

Currently, there is only one publication referred to the genetic diversity of *G. sarmientoi* (Camps et al. 2018). In this study the authors characterized the genetic diversity across almost the entire geographical range of the species (sampling natural populations from Argentina, Bolivia and Paraguay) using two non-coding fragments of chloroplast DNA (cpDNA). The retrieved haplotype network revealed three distinct phylogroups which overlapped geographically. At the species level, haplotype diversity was $h = 0.685$ (SD: 0.039), and nucleotide diversity was $\pi = 0.0012$ (SD: 0.0007). The highest value of haplotype diversity per population was $h = 0.867$ (SD: 0.1291), while the lowest value was $h = 0.333$ (SD: 0.2152). For the nucleotide diversity, values varied between $\pi = 0.000220$ (SD: 0.000290) and $\pi = 0.002332$ (SD: 0.001591). The highest levels of genetic diversity were found in the northwest, centre and southeast areas of the distribution (see Fig. 11.7). Exclusive haplotypes were found throughout the spatial range, although they were more frequent in sites located at the edge of the geographical distribution.

11.3 Mating System

The information on population structure and mating system is paramount for developing strategies for rational use and conservation programmes of native species as they contribute to define the breeding units in the wild.

For *Acacia* species the experimental evidence indicates that most are predominantly outcrossers, sometimes presenting self-incompatibility systems (Kenrick and Knox 1985; Sedgley et al. 1992). In the study of mating system of four populations of *A. caven* using isozymes (Pometti et al. 2011), the estimate for the multilocus outcrossing rate (*tm*) was high (≥0.957) in all populations, indicating that this species is predominantly an outcrosser. Moreover, this study showed a high probability that the individuals within each progeny array are half sibs. On the other hand, a similar study on three populations of *A. aroma* also using isozymes, similarly revealed a high multilocus outcrossing rate *tm* ≥ 0.914, although with a high probability that individuals within progeny arrays are full rather than half sibs (Casiva et al. 2004). These last results should be taken carefully due to a reduced sample size. In the case of *A. visco*, a study of the mating system was assessed by means of the AFLP technique in three different natural scenarios. The estimate of multilocus outcrossing rate (*tm*) was high (≥0.971) in all populations. Moreover, in the three populations studied, the progenies of open pollination were constituted mainly by half-sibs (94.3%) (Pometti et al. 2013). Nowadays, seeds of all South American *Acacia* species are mainly dispersed by cattle, so they likely have a long-distance dispersal rate (Aronson 1992).

Prosopis affinis is pollinated by insects (primarily bees) and has endozoochory dispersion through wild and domestic mammals (Burkart 1976; Abraham de Noir et al. 2002; Martín 2014). There are no specific data about its mating system; however, the species of Algarobia section are considered mostly cross-pollinated (Simpson 1977; Simpson et al. 1977), having been reported low levels of autogamy in some species of the genus (Bessega et al. 2000; Saidman et al. 2000). There are few studies about distances of seed and pollen dispersal for the genus. It has been estimated that the seeds can be dispersed about 50 m by small rodents or may be dispersed by cattle at greater distances (between 4 and 6 km) (Reynolds 1954; Keys 1993; Keys and Smith 1994). On the other hand, the entomophilic condition of *Prosopis* species determines that pollen is usually unable to migrate large distances (Bessega et al. 2000), but it is estimated that can be dispersed among 5.36 and 30.92 m (Bessega et al. 2012).

Very little is known about the reproduction of *Gonopterodendron sarmientoi*. There is only a mention of a *Meliponini* bee species, *Geotrigona argentina* (Vossler 2014) pollinating the species. The fruit, dispersed by autochory, consists of a three-winged capsule where each mericarp contains one seed (Abraham de Noir et al. 2002). Regarding the reproduction by seeds, information related to the germination requirements is scarce (but see de Viana et al. 2014). Field observations (Camps et al. unpublished information) revealed asexual reproduction by sprouting roots, being this kind of reproduction very frequent in the monospecific forests ("palosantales").

11.4   Population Structure and Landscape Genetic Structure

Plant populations are not randomly arranged assemblages of genotypes but are usually structured in space and time. Diverse approaches implemented in computer programmes like STRUCTURE, GENELAND and BAPS allow studying the genetic structure of natural populations to a fine scale. This genetic structure may be manifested among geographically distinct populations, within a local group of plants, or even in the progeny of individuals. Ecological factors affecting reproduction and dispersal are likely to be particularly important in determining genetic structure. Also, spatial and genetic patterns are often assumed to result from environmental heterogeneity and differential selection pressures (Loveless and Hamrick 1984).

The spatial structure of the genomic variation among natural populations constitutes a central topic in evolutionary biology. The structure is primarily influenced by the population density, breeding system and environmental heterogeneity, among other factors. For plants, the ability to extend the geographical distribution and maintain genetic variability within populations depends on the gene flow mediated by seed movement and pollen dispersal (Peakall et al. 2003; Moran and Clark 2011). These mechanisms influence the structuration of genetic diversity within and between populations, which is usually referred to as spatial genetic structure (SGS) (Vekemans and Hardy 2004).

A study of landscape genetic structure in the six varieties of *A. caven* by means of AFLP, showed that the 15 populations analysed were significantly structured using both the Wright's approach ($F_{ST} = 0.315$) and AMOVA ($\Phi_{ST} = 0.315$), despite a significant proportion of genetic variation (about 68.5%) existed within populations (Pometti et al. 2012). When software STRUCTURE was applied, the optimal number of *K* genetic clusters was 11, what is almost the same number of analysed populations, which is indicative of a high differentiation (Fig. 11.5a). The populations analysed in this study were sited in five ecoregions of Argentina: Pampa, Puna, Espinal, Wet Chaco and Dry Chaco. When regions were studied separately, Wet Chaco showed the lowest values, while Puna showed the highest values of genetic structure, both in terms of F_{ST} and Φ_{ST} (Pometti et al. 2012).

A similar study was done in *A. visco* with AFLP markers in seven populations within two subregions of Argentina: Puna and Chaco. A significant amount of genetic differentiation among populations was observed using both the Wright's approach ($F_{ST} = 0.126$) and AMOVA ($\Phi_{ST} = 0.23$), and a large proportion of genetic variation (about 77.4%) existed within populations. The analysis of molecular variance showed that the variance between subregions was relatively low (2.1%) but highly significant. The analysis with STRUCTURE showed that the optimal number of *K* clusters was 6 (Fig. 11.5b). These results indicated that populations belonging to Puna subregion were more differentiated from the rest in comparison to those belonging to Chaco (Pometti et al. 2016).

In a study of six natural populations of *A. aroma* in the Argentinean Chaco, the analysis of population structure by means of Wright's F_{ST} statistic was high ($F_{ST} = 0.42$) and significant. The analysis of molecular variance indicated that the largest component of genetic diversity (60.7%) was found within populations as usual, but a great part of it (39.3%) was found between populations. The analysis with STRUCTURE showed that the optimal number of clusters (*K*) was 3, what was interpreted to be caused by the geographical proximity of some populations (Pometti et al. 2018) (Fig. 11.5c). Significant SGS was detected in short to medium distances

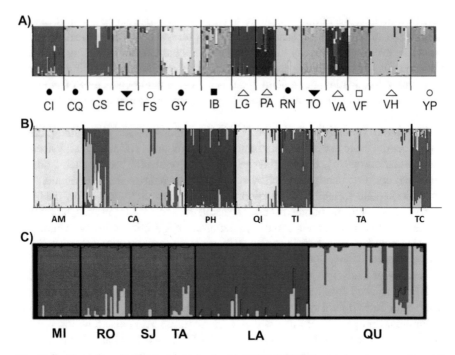

Fig. 11.5 Clustering of individuals obtained with STRUCTURE for (**a**) *Acacia caven* ($K = 11$) (Taken from Pometti et al. 2012); (**b**) *Acacia visco* ($K = 6$) (Taken from Pometti et al. 2016); and (**c**) *Acacia aroma* ($K = 3$) (Taken from Pometti et al. 2018). Each individual is represented by a vertical coloured line. Same colour in different individuals indicates that they belong to the same cluster

in three of the six populations (up to 530 m). The neighbourhood size ranged from 15.2 to 64.3 individuals. The estimation of gene dispersal (σ_g) was conducted considering four different effective densities in populations where SGS was significant and ranged from 67 m to 864 m.

In *Prosopis affinis* genetic differentiation among the analysed fragments of populations by AMOVA was moderate (F_{ST}: $0.080 - p \leq 0.001$). Through Bayesian methods (STRUCTURE), it was possible to identify two genetic groups heterogeneously distributed.

So far, there are very few populations of *P. caldenia* studied through genetic markers; therefore the results of genetic diversity of the species are preliminary. Three populations of *P. caldenia* were studied in the Province of San Luis, 20 km apart from each other, resulting very low differentiated ($F_{ST} = 0.014$) (Velasco 2018). Pérez Díaz (2014) also showed low interpopulation variation of this species by sampling two populations located in the north and in the south of Villa Mercedes City. Despite a general degradation scenario, which includes overgrazing, fragmentation due to the advance of agriculture over native forests, deforestation and fires, this study showed a high intrapopulation genetic variability, which is of great value for the conservation of the species. Velasco (2018) found hybrid swarms in sympatric populations of *P. caldenia* and *P. flexuosa*. Interspecific Nei's genetic distance between populations of *P. caldenia*, *P. flexuosa* and interspecific hybrids varied between 0.099 and 1.534 and reflected that interspecific hybrids were more similar genetically to *P. caldenia* (Velasco 2018). Structural analysis of *P. caldenia*, *P. flexuosa* and interspecific hybrids populations showed that genetic differentiation between populations of these three species is moderate ($0.15 < F_{ST} < 0.25$), thus indicating that populations of *P. caldenia* have genetic coherence and preserve certain genetic isolation from *P. flexuosa*, although they hybridize in sympatry. These results could reveal a kind of fragmentation within *P. caldenia* populations marked by the presence of *P. flexuosa* in the stand (Velasco 2018).

In *Gonopterodendron sarmientoi*, Camps et al. (2018) did not find a significant population structure when analysed the spatial distribution of the retrieved cpDNA haplotypes. This pattern was also detected in other plants from the studied region, e.g. the herb *Turnera sidoides* (Speranza et al. 2007), and the trees *Astronium urundeuva* (Caetano et al. 2008; Caetano and Naciri 2011) and *Geoffroea spinosa* (Caetano and Naciri 2011). Although across the distribution area of *G. sarmientoi*, the Pilcomayo and Bermejo rivers are potential barriers to gene flow, it seems that they were not determinant in the dispersion of its seeds throughout the geographical range of the species (Fig. 11.6). Genetic variation was also not associated with the climatic gradient (Pearson correlation values <0.15; $P > 0.05$), as was evidenced in correlation analyses between genetic diversity indices (h and π) and climate

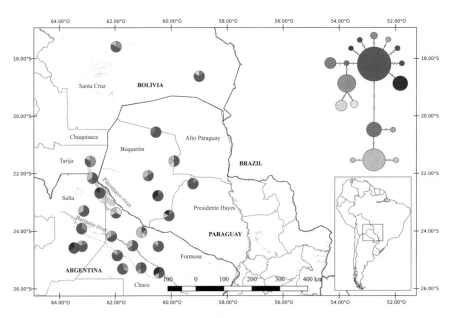

Fig. 11.6 Genealogical relationships among haplotypes recovered from 24 populations of *Gonopterodendron sarmientoi*. On the map, pie charts show the haplotype frequency in each population. Haplotype colours correspond to those shown in the network on the right. In the network, circle sizes are proportional to haplotype frequency. (Figure modified from Camps et al. 2018)

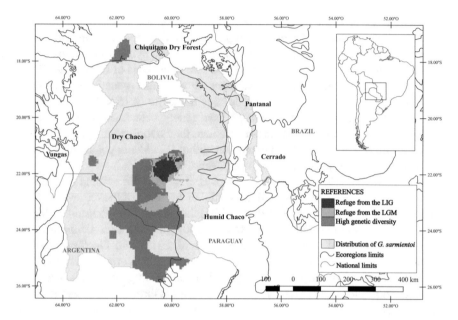

Fig. 11.7 Spatial distribution of high genetic diversity of *Gonopterodendron sarmientoi* (haplo-type diversity >0.73) retrieved interpolating haplotype diversity values of 24 localities. Refuges (*LIG* last interglacial, *LGM* last glacial maximum) proposed for *G. sarmientoi* based on genetic information (cpDNA) and ecological niche modelling are shown. Ecoregions (Chiquitano Dry Forest, Pantanal, Cerrado, Yungas, Dry Chaco and Humid Chaco) correspond to those of Olson et al. (2001). (Figure modified from Camps et al. 2018)

(considering those climatic variables of biological importance, such as mean temperature of most humid quarter, specific humidity mean of coldest quarter, annual mean specific humidity and annual mean temperature). Since no current environmental factors seem to influence the distribution of the genetic variation in *G. sarmientoi*, the observed genetic patterns could be associated to historical demographic fluctuations (may be related to past climatic changes). In fact, phylogeographic analyses suggest that glaciation periods promoted pulses of demographic expansion, while interglacial periods promoted the diversification of the lineages (Fig. 11.7) (Camps et al. 2018).

11.5   Recommendations for Management and Conservation Programmes

The overall genetic diversity of a taxon has great implications for its long-term survival and evolution (Frankel et al. 1995; Avise and Hamrick 1996). Therefore, knowledge of the levels and patterns of genetic diversity is important for designing conservation and management strategies for endangered species (Hamrick 1983; Francisco-Ortega et al. 2000).

The Chaco and Espinal ecoregions of Argentina are under a strong use pressure intending to convert these woodlands in agricultural crops (mainly soybean) or extensive pasturelands. The subtropical dry forest of the humid Chaco walks its way to agriculture, moving the traditional cattle production carried out there to the dry Chaco and Espinal (and even to the Monte) ecoregions, which used to hold this practice but at a lower intensity level. This sequence in rangeland use gives place to a severe degradation risk. Anywhere grass forage tends to decrease or disappear; cattle increases browsing pressure on shrubs and tree seedlings. In areas strongly degraded by overstocking, instead of removing cattle, farmers usually prefer to replace them by goats, which are capable of browsing on almost any type of available plant growth. This kind of decisions can determine disappearance of both the wood characterizing the climax situation and the herbaceous cover in many locations (Fernández and Busso 1997). The situation is extremely fragile, and conservation and restoration practices need to be considered. Planting the key species of these ecosystems must be planned taking into account the genetic factor.

The studies of *A. caven* populations showed that since differentiation in ecoregions was highly significant (Pometti et al. 2012), an adequate management strategy for the use of this species in active restoration plans would be using local sources for seed provision, namely, not moving seeds long distances, in order to prevent maladaptation and genetic contamination processes. Likewise, populations from the Puna showed the highest level of structuration while those from the humid Chaco the lowest. In this regard, the use of local seeds should be considered more rigorously in the former.

In the case of *A. visco,* differentiation is lower than in *A. caven*, but still significant (Pometti et al. 2016), and the use of local seed sources is also recommendable for restoration purposes. On the other side, the differentiation estimated among *A. aroma* populations was the highest among these three acacias (42%), making it essential to use local seed sources in case a restoration programme is conducted with this species.

The findings of evidence of SGS and the estimation of gene dispersal (σ_g) for *A. aroma* (Pometti et al. 2018) suggest that seed sampling for restoration programmes should consider a minimal distance between mother trees of 50 to 150 m to minimize genetic relatedness among them. In cases of disturbed populations with low density, the minimum and maximum distances should be much larger (270/870 m). These results are important for managing and conserving the extant trees and populations of the species in the fragmented landscapes (Pometti et al. 2018).

Controlled crosses are currently being performed between trees with and without spines in order to conduct genetic studies about the determinism of this trait in *A. aroma* (Mauricio Ewens, personal communication). This research is conducted in the frame of a collaboration between the Santiago del Estero Province – Universidad Católica de Santiago del Estero agreement and the Universidad de Buenos Aires.

There is scarce information about cultivation in the three species of *Acacia* studied here; however, there is some information about germination and dormancy of the seeds in *A. caven* and *A. aroma*, and the results showed that these species have physical dormancy imposed by an impermeable seed coat (Funes and Venier 2006). Moreover, scarified seeds have proven to germinate equally in light and darkness at all temperature regimes studied (Funes and Venier 2006). In a study of possible nitrogen-fixing species that could be used for ecological rehabilitation of degraded ecosystems in Chile, seeds of Argentinean populations of *A. caven*, *A. aroma* and *A. visco* were assessed for survival after 6 years, plant growth and nutritive value. *A. caven* populations presented >80% survival at the end of the 6 year of growth. In contrast, *A. visco* and *A. aroma* showed survival rates <25% (Arredondo et al. 1998). These three species failed to attain even 1 m in height after 6 years. About nutritive values, the pods of *A. aroma* contain relatively high protein levels (10.4%). Moreover, the leaves of *A. visco* and *A. caven* appear rich in protein with 32 and 18.5%, respectively (Arredondo et al. 1998). The results of these trials support that new trees, shrubs (*A. caven* and *A. visco* in particular) and microorganisms can be identified for artificial introduction to the "Espinal" region's agroecosystems (Arredondo et al. 1998).

The preliminary genetic results in *P. affinis* could indicate that, in the area with higher agricultural intensification levels, remnant native forest fragments have a high conservation value because of their high levels of genetic variability. However, further sampling and analyses are required to better understand the observed patterns and to find the conservation threshold of the landscape that will allow preserving the genetic variability of the species in the area. Seedling production for active restoration with this species appears as a possible task. In a germination study under controlled conditions, 100% of germinated seeds were obtained in only 4 days with a pre-germinative treatment of sulfuric acid 95% during 20 min (Castillo et al. 2015). Likewise, an experimental restoration with interesting results can be mentioned (Casermeiro et al. 2015). Sowing of pretreated seeds and planting of 80-cm high seedlings were assayed in a grassland and in a secondary *Prosopis* sp. forest, yielding a survival rate lower than 20% for the seeds at 3 months, but up to 68% for the seedlings after 2 years.

For *P. caldenia* management and conservation, studies on genetic diversity, breeding system and pollination distance should be expanded to the entire distribution range of the species, since the existing analyses are limited to restricted areas. Hybrid swarms should be taken into account in areas of sympatry between *P. caldenia* and *P. flexuosa*. Although it has been achieved that the hybrid does not have more vigour than its parentals (Velasco 2018), it will be necessary to deepen the knowledge of these systems in order to be able to make decisions about the management and conservation of *P. caldenia* populations.

An interesting study of assisted regeneration was conducted in a degraded forest of the species (Bistolfi 2016). Different treatments were compared in a field trial to test the artificial recruitment of *P. caldenia* and *P. flexuosa* in weeded plots within a productive ranch with livestock. Germinated seeds, 1-month-old seedlings and 1-year-old seedlings, were planted in early autumn and late spring. The first two treatments resulted clearly inappropriate, since the seedlings died soon due to drought stress or cattle trampling. The cattle was also a problem for the large seedlings, as they browsed them. With a recruitment of 32 saplings per ha and year, the control treatment (natural regeneration) was the best, which turned out to be an endozoochoric dispersion facilitated by the cattle. Therefore, paradoxically, livestock was essential for the best of the options essayed for the regeneration of this degraded forest.

The exploitation of *Gonopterodendron sarmientoi* as a logging resource is carried out using exclusively native forest trees, thus, being environmentally and economically unsustainable. The International Union for the Conservation of Nature (IUCN) considers *G. sarmientoi* in the category "endangered" (Barstow 2018). The species conservation status in Paraguay is classified as endangered too, because of habitat loss and selective logging. In Argentina, the species is not included in a list of endangered taxa, but there are laws and normative that regulate the exploitation of native flora in general and in particular the palo santo logging. Currently, our country is writing a Comprehensive Management Plan for the species, which includes the revision of the existing regulations, the management and control of exploitation, the generation of information on distribution, genetic variability and growth dynamics of the species. In Bolivia, the species is considered vulnerable due to a restricted geographical range and a decrease in the size of the subpopulation (De la Barra and Guillén 2018). For Brazil, no specific regulations were found.

As a first step in proposing tools that contribute to the conservation and management of palo santo, Camps et al. (2018) performed a synthetic map (Fig. 11.7) based on genetic information (cpDNA) and species distribution modelling across time to reveal zones where high genetic diversity values overlaps with climatically stable areas (refugees). This approach highlights three main relevant areas that must be considered to conserve the genetic resources of the species: one is an extended area in the centre of the range (Paraguayan Chaco) with stable climate since the last interglacial maximum (120 ka), and the other two zones (Paraguayan Chaco and the triple boundary between Argentina, Bolivia and Paraguay) are stable since the Last Glacial Maximum (20 ka). Because climatic refugees are areas that can preserve the evolutionary history of lineages (Keppel et al. 2012; Tzedakis et al. 2013), these sites are of conservation interest. In addition, Chiquitano Dry Forest and Southeast zone are of conservation value because of its high levels of genetic diversity, the presence of exclusive haplotypes, and because they are localities where the species is adapted to extreme environmental conditions.

Cultivation of *G. sarmientoi* has not been developed so far. Preliminary results obtained from germination experiments at three different thermoperiods (35/15 °C, 25/15 °C and 20/10 °C) showed that the highest percentages of germination (90%) were obtained at the intermediate temperature, 25/15 °C, both in light (12/12 h) and permanent darkness (Venier, Ferreras and Funes, unpublished data). Some planting experiences at small scale have been carried out. For example, in the Centro de Validación de Tecnologías Agropecuarias (CEDEVA), in Laguna Yema (Province of Formosa), native seedlings at 7 years of age show an approximate height of 1.5 to 2 m (Hector Taboada, personal communication), a preliminary observation that proves the slow growths of the species, but also its aptitude for cultivation.

References

Abraham de Noir F, Bravo S, Abdala R (2002) Dispersal mechanisms in some woody native species of Chaco Occidental and Serrano. Quebracho 9:140–150

Adamoli J, Neumann R, De Colina ADR, Morello J (1972) El Chaco aluvional salteño. Revista de Investigaciones Agropecuarias 9:165–237

Amieva R (2013) Respuesta temprana del crecimiento de *Prosopis caldenia* y *Prosopis flexuosa* sometidos a raleos y podas en la provincia de la pampa. Graduation thesis Universidad Nacional de San Luis, Argentina

Arispe R, Rumiz D (2002) Una estimación del uso de los recursos silvestres en la zona del bosque chiquitano, cerrado y pantanal de Santa Cruz. Rev Bol Ecol 11:17–36

Aronson J (1992) Evolutionary biology of *Acacia caven* (Leguminosae, Mimosoideae): Infraspecific variation in fruit and seed characters. Ann Mo Bot Gard 79:958–968

Arredondo S, Aronson J, Ovalle C, del Pozo A, Avendaño J (1998) Screening multipurpose legume trees in Central Chile. For Ecol Manag 109:221–229

Avise JC, Hamrick JL (1996) Conservation genetics, case histories from nature. Chapman and Hall, New York, pp 1–9

Barbosa O (2005) Descripción del ecosistema. In: Barbosa O.A. y Privitello, M.J.L. (Eds.), Caracterización Ecológica y utilización del caldenal de San Luis (Argentina). FICES, Universidad Nacional de San Luis (Argentina)

Barstow M (2018) *Bulnesia sarmientoi*. The IUCN red list of threatened species 2018: e.T32028A68085692. https://doi.org/10.2305/IUCN.UK.20182.RLTS.T32028A68085692.en. Downloaded on 28 Nov 2019

Benítez B, Bertoni S, González F, Céspedes G (2008) Uso artesanal de especies vegetales nativas en Tobatí, Paraguay. Aspectos botánicos y socioeconómicos. Rojasiana 8:10–25

Bessega C, Ferreyra L, Julio N, Montoya S, Saidman B, Vilardi JC (2000) Mating system parameters in species of genus *Prosopis* (Leguminosae). Hereditas 132:19–27

Bessega C, Saidman BO, Vilardi JC (2005) Genetic relationships among American species of *Prosopis* (Leguminosae) based on enzyme markers. Genet Mol Biol 28:277–286

Bessega C, Pometti CL, Ewens M, Saidman BO, Vilardi JC (2012) Strategies for conservation for disturbed Prosopis alba (Leguminosae, Mimosoidae) forests based on mating system and pollen dispersal parameters. Tree Genet Genomes 8:277–288

Bessega CF, Pometti CL, Miller JT, Watts R, Saidman BO, Vilardi JC (2013) New microsatellite loci for *Prosopis alba* and *P. chilensis* (Fabaceae). Appl Plant Sci 1:1200324

Bistolfi NM (2016) Análisis comparativo de métodos de reforestación en bosque de caldén, *Prospis caldenia* Burkart, en un contexto de rehabilitación ecológica. Graduation thesis, Facultad Cs. Exactas y Naturales, Universidad Nacional de La Pampa. http://www.biblioteca.unlpam.edu. ar/rdata/tesis/x_bisana024.pdf

Bogino SM (2006) Revalorizar al calden. Una especie nativa única de la Argentina y el mundo. SAGPyA Forestal

Bogino SM, Villalba R (2008) Radial growth and biological rotation age of *Prosopis caldenia* Burkart in Central Argentina. J Arid Environ 72:16–23

Boyero MA (1985) *Prosopis caldenia* Burk, en Argentina. Segundo Encuentro Regional CIID América Latina y el Caribe, Santiago de Chile. Forestación en zonas áridas y Semiáridas:270–323

Burkart A (1976) A monograph of the genus *Prosopis* (Leguminosae subfam. Mimosoideae). J Arnold Arbor 57:219–525

Cabrera ÁL, Willink A (1973) Biogeografía de América Latina. Monografía 13. Serie de Biología. Secretaría General de la Organización de los Estados Americanos. EE.UU, Washington, DC, 120 pp

Caetano S, Naciri Y (2011) The biogeography of seasonally dry tropical forests in South America. In: Dirzo R, Young HS, Mooney HA, Ceballos G (eds) Seasonally dry tropical forests. Island Press, Washington, DC, pp 23–44

Caetano S, Prado D, Pennington RT, Beck S, Oliveira- Filho A, Spichiger R, Naciri Y (2008) The history of Seasonally Dry Tropical Forests in eastern South America: inferences from the genetic structure of the tree *Astronium urundeuva* (Anacardiaceae). Mol Ecol 17:3147–3159

Camps GA, Martínez-Meyer E, Verga AR, Sérsic AN, Cosacov A (2018) Genetic and climatic approaches reveal effects of Pleistocene refugia and climatic stability in an old giant of the Neotropical Dry Forest. Biol J Linn Soc 125:401–420

Casermeiro J, Spahn E, De Petre A, Prand M, Ronconi AP, Rosenberger J, Martínez MH, Apaulaza J, Casermeiro L, Meza D, Müller A (2015) Enriquecimiento de sistemas forestales degradados del distrito Ñandubay con especies nativas leñosas. Ciencia, Docencia y Tecnologia (Suplemento) 5:N°5

Casiva PV, Vilardi JC, Cialdella AM, Saidman BO (2004) Mating system and population structure of *Acacia aroma* and *A. macracantha* (Fabaceae). Am J Bot 91:58–64

Castillo D, Bennadji Z, Alfonso M (2015) Evaluación de métodos de quiebra de dormancia en *Prosopis affinis* Spreng. In: Reunión Anual, International Seed Testing Association (ISTA), Montevideo, 15 al 18 de junio, 2015. http://www.ainfo.inia.uy/digital/bitstream/item/5255/1/POSTER-ISTA-2015-Bennadji.pdf

Chiveu CJ, Dangasuk OG, Omunyin ME, Wachira FN (2008) Genetic diversity in Kenyan populations of *Acacia senegal* (L.) willd revealed by combined RAPD and ISSR markers. Afr J Biotech 7:2333–2340

Collado AD, Chuvieco E, Camarasa A (2002) Satellite remote sensing analysis to monitor desertification processes in the crop-rangeland boundary of Argentina. J Arid Environ 52:121–133

Contreras S, Jobbágy EG, Villagra PE, Nosetto MD, Puig de fábregas J (2011) Remote sensing estimates of supplementary water consumption by arid ecosystems of Central Argentina. J Hydrol 397:10–22

De la Barra N, Guillén R (2018) Libro rojo de la flora amenazada de Bolivia. Volumen II. Centro de Biodiversidad y Genética. Tierras Bajas, Cochabamba

de Viana ML, Morandini MN, Urtasun MN, Giamminola EM (2014) Caracterización de frutos y semillas de cuatro especies arbóreas nativas del Noroeste Argentino para su conservación ex situ. Lhawet 3:41–48

Dimitri MJ, Biloni JS (1973) Esencias forestales indígenas de la Argentina de aplicación ornamental. Celulosa Argentina, 104 pp

Ebinger JE, Seigler DS, Clarke HD (2000) Taxonomic revision of South American species of the genus *Acacia* subgenus Acacia (Fabaceae: Mimosoideae). Syst Bot 25:588–617

Fernández OA, Busso CA (1997) Arid and semi-arid rangelands: two thirds of Argentina. Rala report N0 200, p 20

Ferreyra LI (2000) Estudio de la variabilidad y la diferenciación genética por medio de técnicas de lsoenzjmas y RAPD en poblaciones naturales de especies e híbridos de Género *Prosopis* (Leguminosas). Doctoral thesis, Facultad Cs. Exactas y Naturales Universidad de Buenos Aires

Filipov A (1994) Medicinal plants of the Pilagá of Central Chaco. J Ethnopharmacol 44:181–193

Fortunato RH et al (2008) Fabaceae (Leguminosae) Catálogo de las Plantas Vasculares del Cono Sur (Argentina, Sur de Brasil, Chile, Paraguay y Uruguay). Monographs in Systematic Botanic from the Missouri Botanical Garden 107(2–3):2078–2319. Saint Louis

Francisco-Ortega J, Santos-Guerra A, Kim SC, Crawford DJ (2000) Plant genetic diversity in the Canary Islands: a conservation perspective. Am J Bot 87:909–919

Frankel OH, Brown AHD, Burdon JJ (1995) The conservation of plant biodiversity. Cambridge University Press, Cambridge

Funes G, Venier P (2006) Dormancy and germination in three *Acacia* (Fabaceae) species from Central Argentina. Seed Sci Res 16:77–82

Funes G, Venier P, Galetto L, Urcelay C (2007) *Acacia aroma* Gillies ex Hook. & Arn. Kurtziana 33:55–65

Gabutti EG, Maceira NO, Gómez Hermida V, Leporati JL (2009) Superficie remanente y patrones de fragmentación del bosque de calden en la provincia de San Luis, Argentina. http://www.fices.unsl.edu.ar/cga/Caldenal1.pdf

GBIF Secretariat (2019) Prosopis affinis Spreng. GBIF Backbone Taxonomy. Checklist dataset. https://doi.org/10.15468/39omei. Accessed via GBIF.org on 2020-04-08

Giménez AM, Hernández P, Geréz R, Spagarino C (2007) Anatomía de leño y anillos de crecimiento de Palo Santo (*Bulnesia sarmientoi* Lorenz ex. Griseb Zygophyllaceae). Quebracho 14:23–35

Guinet P, Vassal J (1978) Hypotheses on the differentiation of the major groups in the genus *Acacia* (Leguminosae). Kew Bull 32:509–527

Hamrick JL (1983) The distribution of genetic variation within and among natural plant populations. In: Schonewald-Cox CM, Chambers SM, McBryde B, Thomas WL (eds) Genetics and conservation. Benjamin/Cummings, Menlo Park, pp 335–348

Hamrick JL, Godt MJW (1996) Conservation genetics of endemic plant species. In: Avise JC, Hamrick JL (eds) Conservation genetics. Chapman & Hall, New York, pp 281–304

Hunziker JH, Saidman BO, Naranjo CA, Palacios RA, Poggio L, Burghardt AD (1986) Hybridization and genetic variation of Argentine species of *Prosopis* (Leguminosae, Mimosoidae). For Ecol Manag 16:301–315

Jacobs H (1990) Vegetations analytische und strukturelle Untersuchungen einer regengrünen Trockenwald vegetation im östlichen Bereich des zentralen Chacos unter Berücksichtigung des Einflusses der Viehweide. Doctoral thesis, Forstlichen Fakultät, Georg-August-Universität Göttingen, 113 pp

Karlin OU, Coirini RO, Catalan L, Zapata R (1997) Acacia caven. In: Oficina Regional de la FAO para America Latina y el Caribe (ed) Especies arbóreas y arbustivas para las zonas áridas y semiáridas de América Latina, 157–167. Zonas áridas y semiáridas, 12. FAO/PNUMA, Santiago

Kenrick J, Knox RB (1985) Self-incompatibility in the nitrogen-fixing tree, Acacia retinodes: quantitative cytology of pollen tube growth. Theor Appl Genet 69:481–488

Keppel G, Van Niel KP, Wardell-Johnson GW, Yates CJ, Byrne M, Mucina L, Schut AG, Hopper SD, Franklin SE (2012) Refugia: identifying and understanding safe havens for biodiversity under climate change. Glob Ecol Biogeogr 21:393–404

Keys RN (1993) Mating system and pollination biology of velvet mesquite (*Prosopis velutina* Wooton). PhD dissertation, The University of Arizona

Keys RN, Smith S (1994) Mating system parameters and population genetic structure in pioneer populations of *Prosopis velutina* (Leguminosae). Am J Bot 81:1013–1020

Lell J (2004) El caldenal: una visión panorámica del mismo enfatizando en su uso. In: Arturi MF, Frangi JL, Goya JF (eds) Ecología y manejo de los bosques de Argentina. UNLP

Leonardis RFJ, Milanesi CA (1975) Esencias forestales indígenas de la Argentina de aplicación industrial. Celulosa Argentina, 144 pp

Li F, Xiong ZT, Li FM, Zhu YG (1999) Genetic diversity and divergence between populations of *Hemerocallis lilioasphodelus* L. from Henan and Hunan Province. Wuhan Univ J Nat Sci 45:849–851

Loto DE, Gasparri I, Azcona M, García S, Spagarino C (2018) Estructura y dinámica de bosques de palo santo en el Chaco Seco. Ecol Austral 28:064–073

Loveless MD, Hamrick JL (1984) Ecological determinants of genetic structure in plant populations. Annu Rev Ecol Syst 15:65–95

Luckow M (2005) Tribe Mimoseae. In: Lewis G et al (eds) Legumes of the world, pp 163–183

Mariani D, Urioste M, Betelu M, Fantini M, Saravia V, Titarelli F (2012) El caldén, símbolo de nuestra identidad cultural. Ecología para Todos N°4. Subsecretaría de Ecología, Provincia de La Pampa, 40 pp. https://ambiente.lapampa.gob.ar/images/stories/Imagenes/Archivos/Ecologia_para_todos_el_calden.pdf

Martín GO (2014) Técnicas de Refinamiento y Recuperación de Pastizales, Serie Didáctica N°85. San Miguel de Tucumán: Universidad Nacional de Tucumán. https://docplayer.es/10033296-Tecnicas-de-refinamiento-y-recuperacion-de-pastizales.html

Martínez GJ (2011) Use of medicinal plants in the treatment of waterborne diseases in a toba (qom) community of the "Impenetrable" (Chaco, Argentina): an ethnoecological and sanitary perspective. Bonplandia 20:329–352

Matteucci SD (2012) Ecorregión Espinal. En: Morello J, Matteucci SD, Rodríguez AF, Silva ME (eds) Ecorregiones y Complejos Ecosistémicos Argentinos, pp 349–390

Menvielle EE, Hernandez OA (1985) El Valor Nutritivo de las Vainas de Caldén (*P. caldenia* Burk). Revista Argentina de Producción Animal 5:7–8

Mereles F, Pérez de Molas L (2008) *Bulnesia sarmientoi* Lorentz ex Griseb. (Zygophyllaceae): estudio de base para su inclusión en el Apéndice II de la Convención CITES. WWF Paraguay, Lambaré, p 15

Moore G, Smith GF, Figueiredo E, Demissew S, Lewis G, Schrire B, Rico L, van Wyk AE (2011) Acacia, the 2011 nomenclature section in Melbourne, and beyond. Taxon 59:1188–1195

Moran EV, Clark JS (2011) Estimating seed and pollen movement in a monoecious plant: a hierarchical Bayesian approach integrating genetic and ecological data. Mol Ecol 20:1248–1262

Morello J, Matteucci SD, Rodriguez AF, Silva ME, Mesopotámica P, Llana P, Medanosa P (2012) Ecorregiones y complejos Ecosistémicos de Argentina. Orientación Gráfica Editora, Buenos Aires

Naranjo CA, Poggio L, Zeiger SE (1984) Phenol chromatography, morphology and cytogenetics in three species and natural hybrids of *Prosopis* (Leguminosae-Mimosoideae). Plant Syst Evol 144:257–276

Olson DM, Dinerstein E, Wikramanayake ED, Burgess ND, Powell GVN, Underwood EC, D'amico JA, Itoua I, Strand HE, Morrison JC, Loucks CJ, Allnutt TF, Ricketts TH, Kura Y, Lamoreux JF, Wettengel WW, Hedao P, Kassem KR (2001) Terrestrial ecoregions of the world: a new map of life on Earth. Bioscience 51:933–938

Orchard AE, Maslin BR (2005) The case for conserving Acacia with a new type. Taxon 54:509–512

Oyarzabal M, Clavijo J, Oakley L, Biganzoli F, Tognetti P, Barberis I, Matur HM, Aragón R, Campanello PI, Prado D, Oesterheld M, León RJC (2018) Unidades de vegetación de la Argentina. Ecología Austral 28:40–63

Palacios RA, Bravo LO (1981) Hibridación natural en *Prosopis* (Leguminosas) en la región chaqueña argentina. Evidencias morfológicas y cromatográficas. Darwiniana 23:3–35

Palacios R, Brizuela MM (2005) Prosopis. In: Anton AM, Zuloaga FO (eds) Flora Fanerogámica Argentina 92. Museo Botánico – IMBIV, Córdoba, pp 1–25

Peakall R, Ruibal M, Lindenmayer DB (2003) Spatial autocorrelation analysis offers new insights into gene flow in the Australian bush rat, *Rattus fuscipes*. Evolution 57:1182–1119

Pedernera AM, Garcia Aseff S, Guardia T, Guardia Calderón CE, Pelzer LE (2007) Study of acute toxicity of *Acacia visco* methanolic extract in mice. Sociedad Argentina de Farmacología experimental, XXXVIII annual scientific meeting, November 1–3, 2006. Biocell 31:75–112

Pedernera AM, Guardia T, Guardia Calderón CE, Rotelli AE, de la Rocha NE, Saad JE, Lopez Verrilli MA, Garcia Aseff S, Pelzer LE (2010) Anti-inflammatory effect of *Acacia visco* extracts in animal models. Inflammopharmacology 18:253–260

Peinetti R, Dussart E, Boninsegna J (1994) Análisis dendroecológico preliminar de la tendencia de edad en caldén (*Prosopis caldenia* Burk.). In: Proceedings of international meeting of the IAWA. Mar del Plata

Pensiero JF, Gutiérrez HF (2005) Flora vascular de la provincia de Sante Fe: claves para el reconocimiento de las familias y géneros: catálogo sistemático de las especies. Ciencia y técnica. Ed. Universidad Nacional del Litoral, pp 22; 251

Pérez Díaz JP (2014) Análisis de la variabilidad genética en *Prosopis caldenia* (Burkart) mediante el uso de marcadores moleculares. Graduation thesis, Universidad Nacional de San Luis (Argentina)

Polhill RM, Raven PH, Stirton CH (1981) Evolution and systematics of the Leguminosae: 1–26. In: Polhill RM, Raven PH (eds) Advances in legumes systematics, vol 1. Royal Botanic Gardens, Kew

Pometti CL, Cialdella AM, Vilardi JC, Saidman BO (2007) Morphometric analysis of varieties of *Acacia caven*: (Leguminosae, Mimosoideae): taxonomic inferences in the context of other Argentinean species. Plant Syst Evol 264:239–249

Pometti CL, Pizzo B, Brunetti M, Macchioni N, Ewens M, Saidman BO (2009) Argentinean native woods species: physical and mechanical characterization of some *Prosopis* species and *Acacia aroma* (Leguminosae; Mimosoideae). Bioresour Technol 100:1999–2004

Pometti CL, Vilardi JC, Cialdella AM, Saidman BO (2010) Genetic diversity among the six varieties of *Acacia caven* (Leguminosae, Mimosoideae) evaluated at molecular and phenotypic levels. Plant Syst Evol 284:187–199

Pometti CL, Vilardi JC, Saidman BO (2011) Mating system parameters and genetic structure in Argentinean populations of *Acacia caven* (Leguminosae, Mimosoideae). Plant Syst Evol 292:25–32

Pometti CL, Bessega CF, Vilardi JC, Saidman BO (2012) Landscape genetic structure of natural populations of *Acacia caven* in Argentina. Tree Genet Genomes 8:911–924

Pometti CL, Bessega CF, Vilardi JC, Saidman BO (2013) Comparison of mating system parameters and genetic structure in three natural scenarios of *Acacia visco* (Leguminosae, Mimosoideae). Plant Syst Evol 299:761–771

Pometti CL, Bessega CF, Vilardi JC, Cialdella AM, Saidman BO (2015) Genetic diversity within and among two Argentinean and one Mexican species of *Acacia* (Fabaceae). Bot J Linn Soc 177:593–606

Pometti CL, Bessega CF, Vilardi JC, Ewens M, Saidman BO (2016) Genetic variation in natural populations of *Acacia visco* (Fabaceae) belonging to two sub-regions of Argentina using AFLP. Plant Syst Evol 302:901–910

Pometti C, Bessega C, Cialdella A, Ewens M, Saidman B, Vilardi J (2018) Spatial genetic structure within populations and management implications of the South American species *Acacia aroma* (Fabaceae). PLoS One 13(2):e0192107. https://doi.org/10.1371/journal.pone.0192107

Quiroga R, Arrázola S, Tórrez E (2009) Medicinal flora diversity and useful local in the village of Weenhayek of Gran Chaco Province, Tarija, Bolivia. Rev Bol Ecol y Cons Amb 25:25–39

Reynolds HG (1954) Some interrelations of the Merriam kangaroo rat to velvet mesquite. J Range Manag 7:176–180

Risio L, Herrero C, Bogino SM, Bravo F (2014) Aboveground and belowground biomass allocation in native *Prosopis caldenia* Burkart secondaries woodlands in the semi-arid Argentinean pampas. Biomass Bioenergy 66:249–260

Risio L, Calama R, Bogino SM, Bravo F (2016) Inter-annual variability in *Prosopis caldenia* pod production in the Argentinean semiarid pampas: a modelling approach. J Arid Environ 131:59–66

Roig FA (1993) Aportes a la etnobotánica del género *Prosopis*. In: Unidades de Botánica y Fisiología Vegetal, IADIZA (eds) Contribuciones Mendocinas a la Quinta Reunión Regional para América Latina y el Caribe de la Red de Forestación del CIID, pp 99–121

Romero-Zarco C, Tormo Molina R (2016) Acacia visco Lor. ex Griseb. (Mimosaceae), árbol mal identificado en la flora ornamental española. Bouteloua 23:111–117

Ross JH (1981) An analysis of the African *Acacia* species: their distribution, possible origins and relationships. Bothalia 13:389–413

Saidman BO, Bessega CF, Ferreyra L, Julio N, Vilardi JC (2000) Estudios evolutivos y poblacionales en el género *Prosopis* utilizando marcadores bioquímicos y moleculares. Multequina 9:81–93

SAyDS (2006) Primer Inventario Nacional De Bosques Nativos. Secretaria de Ambiente y Desarrollo Sustentable de la Nación. https://www.argentina.gob.ar/sites/default/files/primer_inventario_nacional_-_informe_nacional_1.pdf

SAyDS (2007) Primer Inventario Nacional de Bosques Nativos. Informe Regional Parque Chaqueño. Buenos Aires: Secretaría de Ambiente y Desarrollo Sustentable de la Nación.

https://www.argentina.gob.ar/sites/default/files/primer_inventario_nacionalinforme_regional_parque_chaqueno_0.pdf

Scarone M, Lell J, Giunchi A, Viroletti M (2000) Respuesta de *Prosopis caldenia* a la destrucción de su parte aérea por fuego. Multequina 9:161–164

Scarpa GF (2000) Plants employed in traditional veterinary medicine by the criollos of the Northwestern Argentine Chaco. Darwin 38:253–265

Sedgley M, Harbard J, Smith R-MM, Wickneswari R, Griffin AR (1992) Reproductive biology and interspecific hybridization of *Acacia mangium* and *A. auriculiformis* A. Cunn. ex Benth. (Leguminosae: Mimosoideae). Aust J Bot 40:37–48

Seigler DS, Ebinger JE (2006) Mimosaceae *Vachellia aroma*. Phytologia 87(3):143

Seigler DS, Ebinger JE, Miller JT (2006) The genus *Senegalia* (Fabaceae: Mimosoideae) from the New World. Phytologia 88:38–93

SGAyDS (2019) Anuario de Estadística Forestal 2017–2018. Buenos Aires, 175 pp. https://www.argentina.gob.ar/ambiente/tierra/bosques-suelos/manejo-sustentable-bosques/programa-nacional-estadistica-forestal

Sherry M, Smith S, Patel A, Harris P, Hand P, Trenchard L, Henderson J (2011) RAPD and microsatellite transferability studies in selected species of *Prosopis* (section Algarobia) with emphasis on *Prosopis juliflora* and *P. pallida*. J Genet 90:251–264

Shrestha MK, Golan-Goldhirsh A, Ward D (2002) Population genetic structure and the conservation of isolated populations of *Acacia raddiana* in the Negev Desert. Biol Conserv 108:119–127

Simpson BB (1977) Breeding system of dominant perennial plants of two disjuncts warm desert ecosystems. Oecologia 27:203–226

Simpson BB, Neff JL, Moldenke AR (1977) *Prosopis* flowers as a resource. In Simpson BB (ed) Mesquite: Its biology in two desert ecosystems, vol 5. US/IBP. Syntesis Series Dowden Hutchinson & Ross, pp 84–107

Smith GF, van Wyk AE, Luckow M, Schrire B (2006) Conserving *Acacia* Mill. with a conserved type. What happened in Vienna? Taxon 55:223–225

Soldati MC, Gavier G, Morales M, Solari LM, Suares RP, Zelener N (2018) Diversity and genetic structure of *Prosopis affinis* in a fragmented landscape, an SSR analysis. https://www.sbg.org.br/pt-br/anais-eletronicos

Speranza PR, Seijo JG, Grela IA, Solís Neffa VG (2007) Chloroplast DNA variation in the *Turnera sidoides* L. complex (Turneraceae): biogeographical implications. J Biogeogr 34:427–436

Steibel P (2003) Flora y vegetación de la Provincia de La Pampa. Class notes. Cátedra de Botánica, Facultad de Agronomía, Univ. Nacional de La Pampa

Tortorelli LA (1956) Maderas y Bosques Argentinos. Ed, Acme. Bs. As, 190

Tzedakis PC, Emerson BC, Hewitt GM (2013) Cryptic or mystic? Glacial tree refugia in northern Europe. Trends Ecol Evol 28:696–704

Van Rijckevorsel P (2006) Acacia: what did happen at Vienna? Anales Jard Bot Madrid 63:107–110

Vassal J (1972) Apport des recherches ontogéniques et séminologiques a l'étude morphologique, taxonomique et phylogénique du genre Acacia. Bull Soc Hist Nat Toulouse 108:1–115

Vekemans X, Hardy OJ (2004) New insights from fine-scale spatial genetic structure analyses in plant populations. Mol Ecol 13:921–935

Velasco Sastre T (2018) Dinámica evolutiva de enjambres híbridos entre Prosopis caldenia y Prosopis flexuosa en el Espinal, distrito del Caldenal. Doctoral thesis, Universidad Nacional de Córdoba

Velasco Sastre T, Vergarechea M, Tapia A, Dussart E, Leporati J, Bogino S (2018) Growth dynamics and disturbances along the last four centuries in the *Prosopis caldenia* woodlands of the Argentinean pampas. Dendrochronologia 47:56–66

Vossler FG (2014) Small pollen grain volumes and sizes dominate the diet composition of three South American subtropical stingless bees. Grana 54:68–81

Waller T, Barros M, Draque J, Micucci P (2012) Conservation of the Palo Santo tree, *Bulnesia sarmientoi* Lorentz ex Griseb, in the South America Chaco Region. Med Plant Conserv 15:4–9

Zinda R, Adema E, Rucci T (2005) Relevamiento fisonómico de la vegetación en el área del Caldenal. INTA Publicación Técnica 60, p 24

Zuloaga OF, Morrone O, Belgrano MJ (eds) (2008) Catálogo de Las Plantas Vasculares Del Cono Sur: Argentina, Sur de Brasil, Chile, Paraguay y Uruguay. Missouri Botanical Garden Press, St. Louis

Part III
Subtropical Rainforests

Chapter 12
Subtropical Rainforests: The Yungas and the Alto Paraná Rainforest

Luis F. Fornes

Argentinean subtropical rainforests include two formations, which are highly significant in terms of magnitude, biological diversity, and advanced degree of environmental degradation. These are the Tucumano-Oranense or Yungas Rainforest, located at the Northwest part of Argentina (NWA), and the Alto Paraná Rainforest, in the Northeast (NEA).

The historical background of the forestry activity that has developed in the Argentine subtropical rainforests had a purely extractive basis. Abinzano (1985) speaks about an "extractive front," and defines it as the predominant form of occupation and organization of the economic circuits of the nineteenth and early twentieth centuries. Then, forestry activity has been associated since the time of colonization, and even before, with logging practices. The "extractive front" involved a model of land use and occupation, and simultaneously a productive system of low investment and destruction of nonrenewable resources in the short term. Moreover, this productive system was inserted in a market system regulated from outside the region by the tradable sectors located in the port of Buenos Aires, which soon developed the monopoly of the definitive industrialization of the products and its mass marketing.

Logging history is an important factor that affects the genetic diversity and structure of many forest tree species. Several findings have demonstrated that overexploitation of valuable timbers puts genetic diversity at risk (Hall et al. 1994, 1996; Lee et al. 2002). Genetic effects result in reduction of gene flow, loss of genetic diversity within populations and species, changes in the genetic structure, genetic drift, and inbreeding (Young and Boyle 2000). In addition, habitat disturbance affects diversity, abundance, and behavior of pollinators with changes on mating patterns of species (Aizen and Feinsinger 2002; Ward et al. 2005; Inza et al. 2012). Genetic diversity is needed in order to ensure acclimation and evolutionary processes

L. F. Fornes (✉)
INTA EEA Famaillá, Tucumán, Argentina
e-mail: fornes.luis@inta.gob.ar

© Springer Nature Switzerland AG 2021
M. J. Pastorino, P. Marchelli (eds.), *Low Intensity Breeding of Native Forest Trees in Argentina*, https://doi.org/10.1007/978-3-030-56462-9_12

of adaptation of forest genetic resources to changing environmental conditions and social requirements (Young et al. 1996, Bawa and Dayanandan 1998).

12.1 The Yungas Rainforest: Geographical Location, Logging Background, Conservation Status, and Trade

The Yungas, a subtropical montane rainforest from Northwest of Argentina (NWA), is a hotspot of biodiversity in this country, especially on the Upper Bermejo River Basin (UBRB). It is known that the Yungas plays an important role in providing vital environmental services such as watershed protection. This ecosystem is between 400 and 2300 m asl, and it is located on the southernmost limit of the neotropical cloud forests, a major system in Latin America (Brown et al. 2001). It occurs along a narrow and discontinuous belt throughout the Argentinean provinces of Salta, Jujuy, and Tucumán, from 22° to 28° 15′ S. The Yungas has been ecologically subdivided into Northern, Central, and Southern sectors according to the north–south orographic pattern of mountain ranges of Sierras Subandinas and Sierras Pampeanas that precede, from east to west, the Andean Cordillera in Argentina (Brown et al. 2001). The Northern sector covers the mountainous formations of Sierras de Santa Victoria and Zenta; the center sector covers the Sierras de Lumbrera, Santa Barbara, Centinela, and Maíz Gordo, and the Southern sector covers the Sierras de Metán, Aconquija, and Medina (De la Sota 1972; Brown et al. 2001). These latitudinal sectors are separated by Chaco Serrano and show a clear decrease in taxonomic diversity with an increasing latitude that has been principally associated with less benign weather conditions toward Southern Yungas (De la Sota 1972; Cabrera 1994; Morales et al. 1995; Brown et al. 2001; Juárez et al. 2007).

In general, the climate is defined as humid subtropical, with a marked dry season (May–October) (Hurtado et al. 2013). The altitudinal extension hosts a great heterogeneity of environmental characteristics, which is reflected in the composition and specific richness of the vegetation decreasing from the basal point toward the summits. As for trees, their distribution in the altitudinal gradient responds first to changes in precipitation and temperature, and second to local factors related to topography and disturbances (Brown et al. 2001; Blundo et al. 2012). This vegetation type expands across a large altitudinal gradient, where tree species turnover promotes the occurrence of three altitudinal belts with differentiable physionomic and floristic characteristics. The basal zone is occupied by the "Selva Pedemontana," approximately between 400 and 900 m asl in the piedmont and low altitude mountains, occupying 1217 km^2, with annual average rainfall of 820 mm (550–1400 mm) and annual average temperature of 21.5 °C (maximum mean = 27.6 °C, minimum mean = 15.4 °C). The tree species of highest commercial value such as *Cedrela balansae*, *Pterogyne nitens*, and *Amburana cearensis* have practically disappeared; on the contrary a notorious increase in the abundance of species like *Anadenanthera colubrina* var. *cebil*, which is not palatable for livestock, is detected. Of the total number of adult trees per hectare (187), only 23.5% (44 individuals) corresponds to species of commercial value; the remainder (149 individuals) has a DBH of less than 30 cm, and the vast majority corresponds to species disregarded by the market. Natural regeneration has a significant number of individuals (1320/ha), but only 6.9% corresponds to species of actual commercial value. The current state of the environment of the Selva Pedemontana reflects the process of irrational logging throughout the region, by which tree specimens of noble species disappeared (Del Castillo et al. 2005).

At mid-altitudes it is found the "Selva Montana," on the slopes of the mountains between 900 and 1600 m asl, with annual average rainfall of 1800 mm (1100–2300 mm). In the highest altitude is the "Bosque Montano," between 1600 and 2300 m asl as continuous forest, with annual average rainfall of 1100 mm (800–1400 mm). The last two altitudinal floors occupy 19,721 km^2 and represent, broadly speaking, what is referred to as cloudy forests (Fig. 12.1a), in which fog can contribute with water up to 100% additional to vertical precipitation, and the average annual temperature is 11.7 °C (maximum mean = 21.8 °C, minimum average = 8.8 °C) (Malizia et al. 2012). Deciduous tree species are dominant at the extremes of the gradient, and semi-deciduous and evergreen species are dominant at mid-elevations. On the other hand, wind and gravity-dispersed tree species characterize the lower part of the gradient, while animal-dispersed trees characterize the upper part of the gradient.

In the Selva Montana (Fig. 12.1b), the process of logging began later than in the piedmont area and was very selective, being extracted only *Cedrela saltensis*, an emblematic species of this ecosystem. There is a higher density of adult trees (225), of which 62.8% corresponds to individuals with DBH lower than 30 cm and only 22.2% have commercial value. Forest regeneration reaches only 500 individuals/ha,

Fig. 12.1 (a) View of the cloudy forest from the alpine meadows, descending from Santa Ana in the Puna to Valle Colorado in the Yungas. (b) Giant of the forest emerging from the general canopy, in the Selva Montana in Calilegua National Park. (Photos: Mario Pastorino)

including all species; of which only 31% are commercially valuable species (*Handroanthus lapacho*, *Juglans australis*, *Cedrela saltensis*, *Patagonula americana*, and *Parapiptadenia excelsa*). The Bosque Montano is mainly represented by *Podocarpus parlatorei* and *Alnus jorullensis* and to a lesser extent *Juglans australis*. It has a large area without forest exploitation due to the difficult accessibility and low value of the woods. However, these forests are subjected to grazing pressure throughout the year (Del Castillo et al. 2005).

Physical and biological changes along the altitudinal gradient determine uses and conservation strategies (Malizia et al. 2012). In addition, it has been proposed that the climatic history of the region with glacial refuges in the Northern Yungas during the Pleistocene (Pennington et al. 2000) would have contributed to configurate the current pattern of biological diversity revealed from the geographic distribution of endemic species (Brown et al. 2001, 2006).

The main environmental service of the Yungas is the water provision for irrigation and human consumption (Balvanera 2012), having an influence over approximately 400,000 ha of crops and more than 2 million people. Regarding good's provision, native forests continue to be the main source of quality wood in the NWA. The Yungas is the traditional supplier of the domestic hardwood market and even has a surplus that is traded abroad, but unplanned logging is a threat to a sustained supply. The main problem is the informal trade of native woods: according to estimations, 40% of the declared volumes comprises illegal timber. The problem is due to the reduction of natural stocks because of selective extraction and the significant increase in illegal logging (Fornes et al. 2016). The overexploitation of the Yungas along the second half of the last century has been particularly alarming and has caused habitat forest disturbance (Grau and Brown 2000; Brown and Pacheco 2006). Since 1960, mechanization led to overexploitation of the Yungas. Thus, continuous selective logging for 50 years in a regional context of unsustainable management and lack of control and transparency of the local markets led to a deep ecological impact on accessible forests (Grau and Brown 2000).

The Argentinean Yungas' transformation into crops and urbanization achieved to 30% of its original surface in the year 2010, while in 1970 the grade of transformation was 18% (Malizia et al. 2012). The highest levels of landscape transformation in South America between 2001 and 2010 were recorded at the transitional area between the Chaco forest and the basal altitude of the Yungas (Volante et al. 2012; Aide and Grau 2004). It is estimated that the remaining area of the Yungas is 3,726,835 ha (Malizia et al. 2012). The Selva Pedemontana is the altitudinal floor of the Yungas most affected by transformations, mainly in flat lands below 5% slope, where it is estimated that more than 80% of the original area has been already transformed (Gutiérrez Angonese and Grau 2014).

The forestry exploitation of the Yungas Rainforest, although with very low impact intensity, began at the time of the colony, mainly in the Province of Tucumán, where "the abundance of excellent wood made it easier for its inhabitants to build good carts" (Levene 1940). The carriage industry became a relevant economic activity at that time in NWA. The arrival of the railroad to Tucumán in 1876 promoted the development of the timber exploitation at an important commercial pace. The introduction of steam-driven sawing machines and the cheap access to faraway markets (especially Buenos Aires) were determinant for this development, which began to soar with immigrant income. The further advance of the railway network from Tucumán toward the North, allowed the expansion of the timber exploitation at its side. In 1891 the railway network reached the cities of Salta and Jujuy; and at the end of the decade of the 1930s, the lines were completed. At that time, each railway station from the south of Tucumán to the north had one or two sawmills (some up to 10 or 15, like Orán and Tartagal). In all the railway stations, the forest was the initial generator of wealth and source of work, giving origin to the formation of villages (Del Castillo et al. 2005).

Nowadays, the highest percentage of the remaining piedmont rainforest is located in the Northern sector of the Yungas, while the Southern sector has been almost completely transformed into agricultural areas by the end of the nineteenth and early twentieth century (Brown and Malizia 2004). One of the last relics of Selva Pedemontana in the Southern sector is kept in the Provincial Park and Flora and Fauna Reserve "La Florida" (Tucumán), which was created in 1936, being the first protected area in provincial jurisdiction of the entire Argentina (Inza et al. 2012; Lomáscolo et al. 2014).

Currently, approximately 11% of the Argentine Yungas are legally designated as protected areas (Malizia et al. 2012). However, many of these protected areas are not representative of the main forest ecosystems or are poorly implemented. These facts have led to judge that the existing protection is insufficient to guarantee the conservation of the Yungas biodiversity, especially in the Selva Pedemontana (Brown and Malizia 2004; Pidgeon et al. 2015). Protected areas are generally located in sectors where losses due to opportunity costs are low (Pancel 2015). The incorporation of important regions for biodiversity conservation into the system of protected areas is not always feasible. However, it is possible that these regions can be managed recognizing biodiversity conservation as a goal (Blendinger et al. 2009).

Different kinds of natural reserves exist in the Yungas (Fig. 12.2). The National Park Administration is responsible for the national parks Baritú, El Rey, Calilegua, and Los Alisos. In addition, three provinces are currently managing reserves and parks distributed into Yungas environments: Jujuy (2), Salta (3), and Tucumán (7). On the other hand, there are academic institutions and companies that also have some flora and fauna reserves of their own. However, the degree of protection and management is lower than in national parks.

Focusing on the exploitation of the natural forest, a falling trend is evident in the last years due to the resource depletion, in terms of intervened surface and also of harvested volumes. In fact, a change in the target species have been occurring, particularly a decrease in traditional species such as cedars (*Cedrela balansae*, *C. saltensis*, and *C. angustifolia*), *Handroanthus impetiginosus* and *Myroxylon peruiferum*, and an increase in the case of *Anathenantera colubrina var. cebil* (Eliano et al. 2009). Many of the species that used to be less important for the domestic market are currently in the top spots in terms of extraction due to their abundance, in detriment of the quality required for manufacturing products with high added value

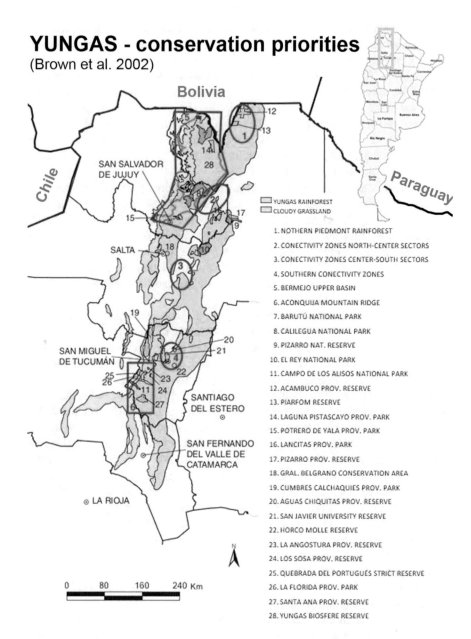

YUNGAS - conservation priorities
(Brown et al. 2002)

YUNGAS RAINFOREST
CLOUDY GRASSLAND

1. NOTHERN PIEDMONT RAINFOREST
2. CONECTIVITY ZONES NORTH-CENTER SECTORS
3. CONECTIVITY ZONES CENTER-SOUTH SECTORS
4. SOUTHERN CONECTIVITY ZONES
5. BERMEJO UPPER BASIN
6. ACONQUIJA MOUNTAIN RIDGE
7. BARUTÚ NATIONAL PARK
8. CALILEGUA NATIONAL PARK
9. PIZARRO NAT. RESERVE
10. EL REY NATIONAL PARK
11. CAMPO DE LOS ALISOS NATIONAL PARK
12. ACAMBUCO PROV. RESERVE
13. PIARFOM RESERVE
14. LAGUNA PISTASCAYO PROV. PARK
15. POTRERO DE YALA PROV. PARK
16. LANCITAS PROV. PARK
17. PIZARRO PROV. RESERVE
18. GRAL. BELGRANO CONSERVATION AREA
19. CUMBRES CALCHAQUIES PROV. PARK
20. AGUAS CHIQUITAS PROV. RESERVE
21. SAN JAVIER UNIVERSITY RESERVE
22. HORCO MOLLE RESERVE
23. LA ANGOSTURA PROV. RESERVE
24. LOS SOSA PROV. RESERVE
25. QUEBRADA DEL PORTUGUÉS STRICT RESERVE
26. LA FLORIDA PROV. PARK
27. SANTA ANA PROV. RESERVE
28. YUNGAS BIOSFERE RESERVE

Fig. 12.2 Argentinean protected areas system in the NWA region. (Taken from: http://faunayflor-adelargentinanativa.blogspot.com/2017/10/selva-yungas.html)

and specialized labor. Cedars continue to be among the highest value species in the forest market products.

According to the official forest statistics corresponding to 2016, the timber formal trade from Yungas Rainforest amounts to around 52,000 m³ per year, with Salta Province contributing 90% (MAyDS 2018). However, a parallel trade from illegal logging, of equal or greater magnitude than formal trade, increases considerably the extracted volume. In fact, the illegal use of forests is a historical practice and, in a sense, commonly accepted by the local society. This has led to the development of a much-uncontrolled informal economy, which is claimed as the main cause of degradation of the Yungas (https://www.argentina.gob.ar/ambiente/sustentabilidad/planes-sectoriales/bosques).

The official statistics point out that the main species exploited in 2016 were *Anadenanthera colubrina* var. *cebil* (10,230 m³/year), *Phyllostylon rhamnoides* (5200 m³/year), *Handroanthus impetiginosus* (4850 m³/year), cedars (2800 m³/year), *Calycophyllum multiflorum* (2250 m³/year), *Myroxylon peruiferum* (800 m³/year), and *Chlorophora tinctoria* (730 m³/year). On a smaller scale (<200 m³/year per species), *Cordia trichotoma*, *Juglans australis*, *Tipuana tipu*, and *Enterolobium contortisiliquum* can be mentioned (MAyDS 2018).

Domestic market of wood from the native forest depends mainly on supply and demand. The price of the highest valuable wood ranges from US$ 200 to 480/m³, such as *Handroanthus impetiginosus* (US$ 330–480/m³), *Cedrela angustifolia* (US$ 300–400/m³), *Cedrela balansae*, and *Cordia trichotoma* (US$ 200–300/m³). Timber of lower value but greater abundance such as *Anadenanthera colubrina* var. *cebil* and *Tipuana tipu* have a commercial value between US$ 130 and 200/m³. In the international market the main timber sold are Spanish cedar (*Cedrela* spp.) and Ipé (*Handrohanthus impetiginosus*), from Peru and Brazil mainly, with values that triple the price of the domestic market.[1]

According to the latest Argentinean forest plantations inventory (2019), the provinces of Tucumán, Salta, and Jujuy (NWA) represent only 2% (25,000 ha) of the national planted area, which mainly includes species grown for industrial purposes such as pines, eucalypts, poplars, and willows. However, the "others" category includes native cultivated species, in which *Cedrela* and *Prosopis* genus are the most planted in the NWA, representing 15% of the native forest tree species cultivated in Argentina, thus marking a regional trend.[2]

Currently, the NWA has about 900 ha of *Cedrela* spp. cultivated in commercial plantations. The plantations' features change based on the species, the site, the cultivation system, and the age; different modalities were found with *Cedrela* species: clump plantation and trails enrichment with one, two, or three strips (Fornes et al. 2016).

---

[1] Newsletter #21: Forest Industry of Northwest Argentina, Nov. 2018.

[2] https://datos.agroindustria.gob.ar/dataset/inventario-nacional-plantaciones-forestales-por-superficie/archivo/147acbc6-2048-4d2b-9cd7-df13efe328fa

12.2 Alto Paraná Rainforest: Geographical Location, Logging Background, and Conservation Status

The Alto Paraná Rainforest, locally called Selva Misionera, concentrates the largest plant biodiversity of Argentina in less than 0.5% of the national territory, with more than 2800 vascular plant species (Zuloaga et al. 1999). However, it is probably the least studied unit of forest vegetation in the country and is currently affected by the fastest rate of urbanization. It extends to the two neighboring countries (Southwest Brazil and East Paraguay) and consists mainly of multistratified subtropical forests, marginal forests or riverbanks in large rivers, and even savannas in the south of the Province of Misiones. It is located at the Northeast end of Argentina and covers almost 30,000 km². It has a humid subtropical climate, with annual rainfall of 1700–2400 mm, distributed evenly during the year. The average temperature of January is 25 °C, with an absolute maximum of 39 °C, while the average of the coldest month is 15 °C (Burgos 1970; Margalot 1985).

Physiographic variations make it possible to divide this rainforest into five geomorphological regions: Serrana, Paraná Coastal, Uruguay Coastal, Southern Plains, and Plains Region. It is possible to establish a physiognomic correspondence between these geomorphological regions and the phytogeographic units proposed by Martínez Crovetto (1963), who divides the Selva Misionera into five Districts: Laureles' District, Palo Rosa's District, Tree Fern's District, Urunday's District, and Fluvial District (Paranaense and Paraguayense sub-districts). However, Cabrera (1971) divides the Alto Paraná Rainforest into two zones called (1) "Selvas Mixtas" District, which occupies the North of the Province of Misiones, including the coastal communities along the Paraná and Uruguay rivers, and (2) "Campos" District in the South of the Province of Misiones and the North of the Province of Corrientes, which has an edaphic character related to shallowness (Martínez Crovetto 1963) and is constituted by herbaceous communities with predominant presence of the tree *Myracrodruon balansae*.

The "Selvas Mixtas" District includes three tree communities (sub-districts): "Laurel, Guatambú, and Palo Rosa Rainforest," "Laurel, Guatambú, and Paraná Pine Rainforest," and "Urunday Forest." The first sub-district is characterized by the ostensible presence of two species with greater thermal and water requirements, which are very affected by the overexploitation: *Aspidosperma polyneuron* (Palo Rosa) and *Euterpe edulis* (Palmito), a graceful stipe palm with edible inner core and growing bud (heart of palm), whose extraction causes its death. They are accompanied by *Holocalyx balansae* and numerous exclusive species of this particular and limited ecosystem.

The tree ferns community occurs in the valleys of the tributaries of the Uruguay River and in the lower parts of the central mountains. It is characterized by the abundance of several species of tree ferns, which dominate the undergrowth in the few relics of this formation that have still managed to persist in pristine state. The "Laurel, Guatambú, and Paraná Pine" sub-district occupies the Northeast end of Misiones, in the highest altitude regions and with a slightly colder climate. Martínez

Fig. 12.3 (**a**) View of "Laurel, Guatambú, and Paraná pine" District; (**b**) *Cedrela fissilis* tree selected in the Manuel Belgrano Forest Station of INTA. (Photos: Luis Fornes)

Crovetto (1963) called this area "Sector Planaltense" (Fig. 12.3a). These Paraná pine trees (*Araucaria angustifolia*) are frequently associated with "yerba mate" trees (*Ilex paraguariensis*) (Martínez Crovetto 1981). Besides these two species, the main Planalto's tree species are *Balfourodendron riedelianum*, *Nectandra megapotamica*, *Patagonula americana*, *Diatenopteryx sorbifolia*, *Holocalyx balansae*, *Cedrela fissilis*, *Cabralea canjerana*, *Myrocarpus frondosus*, *Parapiptadenia rigida*, *Ocotea puberula*, *Enterolobium contortisiliquum*, *Cordia trichotoma*, *Peltophorum dubium*, *Apuleia leiocarpa*, *Handroanthus heptaphyllus*, and *Bastardiopsis densiflora*, among others.

Another forest formation in the "Selvas Mixtas" District is the "Urunday forest." The Urunday (*Myracrodruon balansae*) communities form an irregular strip in the South of Misiones, in the ecotone between the "Laurel and Guatambú" rainforest and the "Campos" District. In the rocky hills of the south of Misiones, the forest is smaller and poorer in species, sometimes forming pure forests of Urunday on the slopes of the nearby hills and plains (Martínez Crovetto 1963). Also in this region occurs *Anadenanthera colubrina* var. *cebil* and others, such as *Acacia caven*, *Lithraea molleoides*, *Cereus uruguayanus*, and *Celtis pubescens*.

Throughout the Argentine Mesopotamian region (provinces of Misiones, Corrientes, and Entre Ríos) appears a forest formation with particular elements: the riparian forest. These forests include the vegetation of the Paraná and Uruguay rivers shores and their tributaries (Fluvial District according to Martínez Crovetto 1963). In general, they constitute a very narrow strip forming a forest in gallery along the rivers, which includes many trees such as *Handroanthus heptaphylla*, *Enterolobium contortisiliquum*, *Parapiptadenia rigida*, and *Peltophorum dubium*, mainly.

The climax of "Selvas Mixtas" has different strata: a herbaceous and muscinal undergrowth, a layer of bamboos and shrubs and a tree stratum, where a layer of lianas and epiphytes can be distinguished. In the tree stratum, there is an average density of 482 trees per ha in a primary forest, and 66% are deciduous and semi-deciduous species.

The Alto Paraná Rainforest has been affected by intense logging. In the search for the most valuable forest species, the entire rainforest of the Province of Misiones has been explored and exploited. In most cases, this process has developed without respecting the minimum concepts of sustainable use that would have preserved the productive capacity of these forests (Milkovic 2012). This purely extractive action has led to a permanent decrease in the volume and quality of the harvested timber. Among the most required and exploited species due to their quality wood, we can mention *Cedrela fissilis*, *Araucaria angustifolia*, *Cordia trichotoma*, *Peltophorum dubium*, and *Handroanthus heptaphylla* (Tortorelli 2009).

The forest history in the Alto Paraná Rainforest also has an extractivist genesis, which acquired greater emphasis in the 1920s, with the arrival of the main immigration streams. The immigrant settler became the social actor responsible for transforming natural borders into productive territories. During most of the twentieth century, the rainforest was seen as a brake on progress. In the 1930s, since a local industry was emerging, changes that influenced the productive organization of forestry activity at territorial level were introduced. At the end of that decade, Misiones aimed to produce sawn timber and other derived products, still rustic, with native wood. Around the production of these products, factories and worker's villages were installed (Scalerandi 2012). Subsequently, the native wood sawing industry and forest camps declined with the degradation of the forest resources (Kraustofl 1991; Scalerandi 2012; Ramírez 2017).

The transition from a purely extractivist model to one with local basic industrialization fell within the broader framework of national change from the agro-export model to the import substitution industrialization model, which, in the case of timber, was aimed at the provision of finished products and intermediate inputs to the domestic market based on the development of forest resources. The forestry-industrial model extended from the middle of the twentieth century to the end of the 1980s and was part of the government developmental perspective. In this context, in the Alto Paraná area, there was a continuous and persistent increase in forest plantations with industrial purpose, which substantially modified the landscape of the Northwest of Misiones.

The impetus for the installation of the pulp industrial plants in Misiones was not only due to the ecological conditions that led to the rapid growth of the implanted species (the raw material for the pulp mills) but also to the developmental paradigms of that time. Forestry was perceived as more industrial than the agricultural model, since industry was considered to generate more added value. Thus, more jobs were invested in industrial projects as an engine of development.

The Province of Misiones leads the production of long fiber pulp of Argentina, and its production systems provide 70 to 85% to the Argentine timber market. In 2018, Misiones production was at 10 million m³ of raw material, mainly of yellow pines (*Pinus elliottii* var. *elliottiii* and *Pinus taeda*). The total consumption of the 438 registered sawmills is around 3.3 million m³ of wood, while pulp, paper, and board factories consumed a similar value of wood (Laharrague 2018). According to the provincial inventory (2015), the area cultivated with trees in the Province of Misiones is 405,824 ha. It is noted that 92% of the local consumption of wood comes from local plantations, led by yellow pines (87%) and followed by *Eucalyptus* spp. (9%), Paraná pine (2%), and others (*Toona, Paulownia,* and *Melia* among the main genus). Similarly, 86% of the raw material used by sawmills comes from the province itself (MinAgro 2018).

Among the native tree species of the Alto Paraná Rainforest, *Araucaria angustifolia* is the most important one destined to commercial forest plantations. In recent years, it has seen a sustained growth in demand, which has led to an improvement in prices compared to previous cycles. Compared to exotic resinous pines, today it is possible to pay double or more for the ton of wood of planted Paraná pine in the local market. There are 16,000 ha planted with Paraná pine at present (Pinazo 2013). The Instituto Nacional de Tecnología Agropecuaria (INTA) is currently carrying out a genetic improvement program of this species and is in charge of providing seeds to the productive sector (see Chap. 15).

With respect to native species, the activity remains clearly extractive and the average annual volume is 289,274 m³, being the main species: *Parapiptadenia rigida, Patagonula americana, Ocotea puberula, Balfourodendron riedelianum,* and *Cordia trichotoma,* a group that provides 15% of the volume. A second group provides the 3.5% of the volume marketed: *Luehea divaricata, Cedrela fissilis, Cabralea canjerana, Apuleia leiocarpa,* and *Myrocarpus frondosus.* The rest is made up of a large series of minor species.

The consolidation of the forestry-industry model is characterized by the establishment of paper industries in the Province of Misiones, the growth of plantations for industrial purposes and the formation of villages and colonies around sawmills and manufacturing enterprises. The technology used in the new factories exceeded the rudimentary ones of the extractivist model; thus, the Alto Paraná began to be considered as a forestry-industrial pole (Ramírez 2017). However, Argentina produces less than 1 million ton/year of pulp and 1.7 million ton/year of paper. The factories are mostly obsolete and those that are being modernized are slowly doing so. The scale of production of all factories is far below the international average (Laharrague 2018).

The average annual deforestation of the Province of Misiones was estimated a few years ago as 6700 ha per year (Milkovic 2012), which has decreased significantly from previous data (18,780 ha/year for the period 1989–2004, Guerrero Borges et al. 2007). Natural forests of the emblematic *A. araucana* were reduced considerably through exploitation since the early twentieth century (Fahler 1981) and are now "critically endangered" according to the red list of species (IUCN 2015).

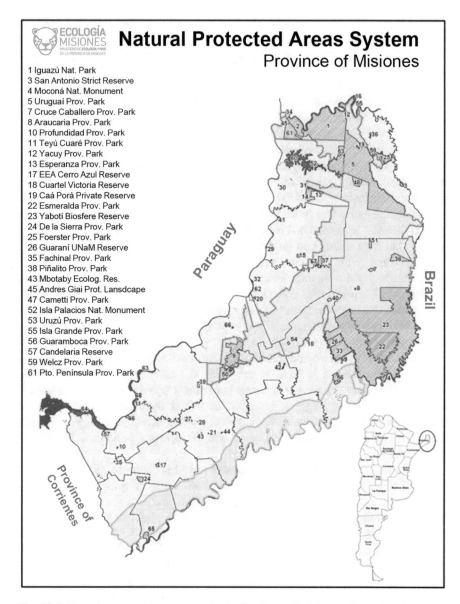

Fig. 12.4 Natural protected areas system in the Province of Misiones. (From: http://ecologia. misiones.gov.ar/ecoweb/images/eventgallery/mapas/areas-naturales-protegidas.jpg)

Today, approximately 1,490,000 ha of the total provincial surface are included within some natural protected area (Fig. 12.4), representing 50% of the original area covered with forest (http://www.ecologia.misiones.gov.ar/ecoweb/index.php/anp-descgen/sistema-de-areas-naturales-protegidas). However, the conservation status of this surface is highly variable. The System of Protected Natural Areas (SANP) created by Provincial Law XVI-N° 29 (formerly 2932) is formed by: 22 Provincial Parks – (157,567 ha), 2 Natural Monuments (416 ha), 3 Ictic Reserves (963 ha), 2 Natural Cultural Reserves (15,695 ha), 3 Multiple Use Reserves (500 ha), 2 Protected Landscapes (8012 ha), 22 Private Reserves (10,450 ha), 5 Municipal Natural Parks (2029 ha), and the Yaboti Biosphere Reserve of provincial jurisdiction (YBR, 236,313 ha). The YBR includes the following reserves: Provincial Parks Esmeralda, Moconá and Caá Yarí, Experimental Reserve Guaraní (Universidad Nacional de Misiones), and the Cultural Reserve Papel Misionero. In addition, there are two areas of provincial jurisdiction that intends to avoid uses of deep ecological impacts: the "Corredor Verde" (1,108,000 ha) and Riverside Parkway (260,082 ha). They constitute a biological corridor that crosses the Alto Paraná Rainforest, covering a mosaic of landscapes that includes protected areas, private properties of diverse use, agricultural colonies with varied socioeconomic situations, indigenous communities, and areas of land use and tenure conflicts. It involves 22 municipalities and about 37% of the provincial surface.

National jurisdiction is restricted to the Iguazú National Park (67,620 ha) and the Strict Natural Reserve San Antonio (400 ha). The Iguazú National Park was designated a World Heritage Site by UNESCO in 1984 and the Iguazú Falls one of the new wonders of the world (https://www.parquesnacionales.gob.ar/areas-protected/region-noreste/pn-iguazu).

12.3 General Considerations from a Breeding Point of View for a Sustainable Management of the Argentinean Rainforests

Both subtropical rainforests are the southern edge of larger formations from neighboring countries and are separated by a huge phytogeographic region called "Great American Chaco" (see Chap. 8). However, the Yungas and the Alto Paraná Rainforest share some tree genus of trade importance, such as *Cedrela, Cordia, Jacaranda, Enterolobium, Handroanthus, Anadenanthera, Parapiptadenia*, and *Chlorophora* (Tortorelli 2009), with species having disjunct ranges, such as *Cordia trichotoma, Anadenanthera colubrina* var. *cebil, Patagonula americana*, and *Enterolobium contortisiliquum*, among others.

There is a considerable confusion regarding the taxonomy of *Cordia* genus. Variations in climate, soil, and altitude within their extensive natural distributions contribute to great differences in the flowering phenology and fruit production and also on morphological features such as the size of flowers and leaves. Therefore, there are doubts about whether *Cordia trichotoma*, which grows in Brazil and Argentina, is really a different species or just a variety of *Cordia alliodora* (Liegel and Stead 1990). In the same sense, Gottschling (2005) studied the congruence of a phylogeny of Cordiaceae (Boraginales) inferred from ITS1 sequence data with morphology, ecology, and biogeography issues, finding a great genetic similarity between *C. alliodora* and *C. trichotoma*, as shown by the neighbor-joining tree of *Cordia* into the *Sebestena* clade.

In the *Cedrela* case, both rainforests do not share the same species. For South American Cedrela genus, the Amazonian Rainforest of Peru constitutes a center of species diversity, although they are certainly not a center of old *Cedrela* diversity. The ancestral habitat preference of *Cedrela* could be inferred by morphological adaptations. Deciduous habit, shoot apices protected by a cluster of bud scales, and capsular fruits with dry, winged, and wind-dispersed seeds of extended viability are features that point to a long evolutionary history of Cedrela in dry forest habitats (Muellner et al. 2010).

Natural hybridization between closely related forest tree species is a common event in nature and has been documented in several economically important genera (Zelener et al. 2016), such as *Quercus* (Whittemore and Schaal 1991), *Eucalyptus* (Potts and Wiltshire 1997), *Nothofagus* (Stecconi et al. 2004), *Prosopis* (Vega and Hernández 2005), and *Populus* (Di Fazio et al. 2011). Hybrid zones under forest harvesting could impact in the wood trade somehow. In 2007, the company GMF Latino Americana was harvesting mainly *Cedrela balansae* in a management unit of the natural forest in Northwest of the Province of Salta. Surprisingly, *C. balansae*'s wood from this area was rejected on the domestic market. The genetic relationships between individuals was determined through molecular markers (AFLP and ITS), recognizing the presence of individuals with ancestry of both *C. balansae* and *C. saltensis*. This result uncovered strong support for the occurrence of natural hybridization between both *Cedrela* species. Additionally, hybrid zones were identified in areas of sympatry (at both the Calilegua National Park and the San Andrés farm) and in transition zones from 820 to 1100 m asl (localities of Pintascayo and Acambuco). That is an interesting case, because the researchers do not usually put the results of their work into practice, even when the primary purpose of their research is the preservation of biodiversity (Gallo et al. 2009).

Likewise, evidence could not be found of hybridization with *C. angustifolia*, neither of *C. balansae* nor of *C. saltensis* (Zelener et al. 2016). Although Cedrela wood from hybrid areas is not desirable in the domestic market, from an evolutionary point of view, these hybrid zones may be important due to their potential adaptive value to the new climatic scenarios (see Chap. 13). The correct characterization of Cedrela's germplasm has direct implications for enrichment and restoration plans (specific purity) and the implementation of appropriate conservation strategies and genetic improvement.

Muellner (2010) places *Cedrela balansae* (one of the cedars from the Yungas) and *C. fissilis* (the cedar from the Alto Paraná Rainforest) phylogenetically very close, although geographically separated, which would imply a greater affinity for hybridization and a risk of genetic contamination in the event of seed or seedlings transfers. However, this factor is not sufficient since a phenological synchrony is required in the flowering and action of pollinators. *Cedrela fissilis* has been used for enrichment of degraded forests in the Southern sector of the Yungas, whose latitude coincides with the Alto Paraná Rainforest, but so far no evidence of entities of the genus *Cedrela* was found as invasive forms (Trapani, personal communication).

However, several exotic invasive species began to appear in expansion the last 30 years (Malizia et al. 2017; Montti et al. 2017; Powell and Aráoz 2018), mainly of the genus *Ligustrum*, *Morus*, and *Leucaena*.

Many of the remaining Yungas rainforests diminished their productive potential due to unplanned forest use, but, in addition, extensive livestock farming has negatively affected the regeneration of some forest species, encouraging the advance of invasive species (Lorenzatti 2014). To reverse the economic and ecological degradation of many of the Yungas' stands, changes have been proposed in forestry techniques to ensure sustainable forest management based on the natural dynamics of species (Balducci et al. 2012). These proposed changes in Yungas management are based on a Reduced Impact Logging (RIL) scheme – an adaptive management framework. Implementing low impact forest management, including seed tree selection and future tree protection, is important because conventional forest harvesting produces, among other negative impacts, damage to approximately 30 trees for each extracted tree (Uhl 1990). This last strategy is suggested especially for the Northern sector of the basal Yungas rainforest.

The differentiation of geographic sectors in the latitudinal gradient of Yungas (North–Central–South) related mainly to species diversity and climatic and topographic conditions (Brown and Kappelle 2001; Blundo et al. 2012) is consistent with the genetic zones identified by Inza et al. (2012) for *Cedrela angustifolia*. This differentiation suggests diverse approaches for the Northern and Central–Southern sectors related to the forest management and restoration methods (Chap. 13). The same trend is observed in *Handroanthus impetiginosus* (see Chap. 16). The most noticeable change, considering specific biodiversity, forest structure, and genetic diversity, occurs at a latitude of around 24° S. Between 24° and 28°15′ S, the Argentinean Yungas shows in general an impoverishment in terms of the number of commercially valuable species, but also in the magnitude of the population genetic diversity of emblematic species.

Given the above conditions, enrichment with native species of high commercial value is a valid and sustainable alternative in places where land use change is not legally feasible (Yellow Zones, according to the National Law 26,331) either through the opening of narrow trails leaving strips of secondary forest or in agroforestry systems, especially in the Central and South sectors.

Regarding the legal framework, National Law 26,331 of Minimum Requirements for the Environmental Protection of Native Forests regulates the extraction of native timber and the expansion of the agricultural frontier. This law represents a major opportunity for the conservation of natural forests, as it restricts spaces in which logging and land use changes are permitted. Additionally, it encourages the recovery and improvement of degraded forests, which helps to maintain the supply of these types of hardwood to the market, based on sustainable production (Yellow Zones), and, consequently, the ecosystem services offered by these forests. According to forest management maps prepared by the provinces, at present, the NWA region (Salta, Jujuy, and Tucumán) and Misiones represent approximately 1.5 million hectares suitable for enrichment through hardwood species in Yellow Zones (Fornes 2012).

In 2006, INTA formally initiated the "Program for Domestication, Conservation and Use of Native Forest Tree Genetic Resources," in articulation with the National Parks Administration (APN), several national universities, and private companies with forest management certification. In order to prioritize species to domesticate among 40 high value wooden species of the Argentinean rainforests (Eliano et al. 2009), a survey was conducted to know the opinion of the potential beneficiaries/ users, considering the value of their wood, their growth rate, their potential surface cultivation, the level of existing knowledge, and their ecological vulnerability. Selected species in order of importance for the NWA were *Cedrela balansae*, *Cedrela angustifolia*, *Cordia trichotoma*, *Handroanthus impetiginosus*, *Enterolobium contortisiliquum*, *Tipuana tipu*, *Juglans australis*, *Myroxylon puruiferum*, *Anadenanthera colubrina* var. *cebil*, *Amburana cearensis*, *Pterogyne nitens*, and *Jacaranda mimosifolia.* The domestication program began with the first four mentioned species. The same procedure was done in the NEA, being the selected species *Cedrela fissilis*, *Cordia trichotoma*, *Enterolobium contortisiliquum*, *Handroanthus heptaphylla*, *Peltophorum dubium*, *Parapiptadenia rígida*, *Bastardiopsis densiflora*, *Myrocarpus frondosus*, *Cabralea oblongifoliola*, *Astronium urundeuva*, and *Luehea divaricata,* of which the domestication process of the first two was initiated (Fornes et al. 2016).

Experiences of restoration of degraded native forests in Argentinean Sub-tropical Rainforest are not abundant. An example is the experience conducted by Grance and Maiocco (1995), who performed enrichment strips in the Alto Paraná Rainforest with *Bastardiopsis densiflora* and recorded a 60% survival rate in the third year with plants coming from natural regeneration. They also noted that the open paths favored the installation of natural regeneration. Eibl et al. (1996) assessed regeneration in different forest harvesting situations, concluding that in the disturbed areas the quantity of seedlings was greater than under forest coverage, since most tree species would be favored by the luminosity increment. Montagnini et al. (1997) conducted enrichment experiences with commercially valuable native species in an overexploited forest where they evaluated ten species including the palm *Euterpe edulis*. After 4 to 7 years of planting, the species that showed the highest mean height and diameter at breast height (DBH) were *Bastardiopsis densiflora*, *Enterolobium contortisiliquum*, *Nectandra lanceolata*, *Ocotea puberula*, and *Peltophorum dubium*. In addition, *Cordia trichotoma* and *Balfourodendron riedelianum,* appreciated as timber, were recommended for enrichment despite their relatively slow growth. Likewise, *Euterpe edulis*, despite its low survival rate, presented individuals with good development. Another important aspect is the differences between species in the requirements for their regeneration, which plays a crucial role in the structuring of communities in general (Pickett and White 1985) and in tropical forests in particular (Denslow 1987). Thus, the requirements for regeneration in the different phases of the gaps contribute to the maintenance of the diversity of the tropical forests (Denslow 1980).

Crechi et al. (2010) reported another experience including a silvicultural field trial with three native species assayed under pine canopy and at open sky. They assessed the total height and DBH up to 12 years of age, showing a mean survival above 70%, but differences arose between species and coverage conditions. *Balfourodendron riedelianum* stood out in open sky with 96% survival and showed an 81% survival under pine canopy. Likewise, *Enterolobium contortisiliquum* showed the best average growth in DBH (14 cm) and height (7 m) in both situations, although the crowns were opened resulting in a poor forestry shape. Comparatively, *Cordia trichotoma* and *B. riedelianum* presented monopodial growth and straight stem, resulting in promising species for the logging industry. Another study (Barth et al. 2008) evaluated some trials and plantations trying to identify native species that could adapt to plantation systems under three different conditions: open sky, under a protective cover into "capueras" (degraded forest) and in agroforestry systems (with cattle rearing and without it). The purpose was to search alternatives to recover degraded forestry ecosystems. The evaluated species were *E. contortisiliquum*, *Cedrela fissilis*, *Gleditsia amorphoides*, *Peltophorum dubium*, *Myrocarpus frondosus*, *Handroanthus heptaphylla*, *Cordia trichotoma*, *Lonchocarpus muehlbergianum*, *B. riedelianum*, and *Myracrodruon balansae*, planted in trails within "capueras." Survival was good in the test sites, except for *L. muehlbergianum* and *P. dubium* in mixed plantation that did not surpass a survival of 56%. *E. contortisiliquum* stood out for having the best growth in the most degraded site, suggesting its use for restoration of poor soils. However, the susceptibility of this species to the attack of the beetle *Epicauta adspersa* raised doubts about the possibility of its establishment under open sky planting conditions, although the intense attacks only occurred at the site of lower degradation and affected the trees in the initial stages. In other experiments in the region, *E. contortisiliquum* showed good growth in enrichment trails. *H. heptaphylla* showed satisfactory behavior in the least degraded site. The high mortality of *M. balansae* and *P. dubium* may have been explained by the soil compaction, which did not allow the roots to explore in depth for water supply under drought conditions. Both species exhibited basal-stem regrowth of approximately 20% of the dead standing specimens, with growth of 1.5 m on average per year. This suggests the possibility of silvicultural conduction of both species. *B. riedelianum* did not adapt to any of the sites, despite being replaced for a total of 3 years. The mixed plantations were beneficial for all the species considered in this work. *E. contortisiliquum* and *M. balansae* combined showed the greatest survival, while *P. dubium* in mixed plantation showed greater diameter and height. In terms of predictive models of performance analysis, logarithmic ones proved to be suitable but constrained to an age limit of 5 years, since at lower ages negative values would be obtained. The implementation of a productive diversification makes up a promising tool for the recovery of areas that suffer from a species impoverishment and difficulties in the process of natural forest succession, adding to the degradation of physical aspects like soil and water.

Moretti et al. (2019) recently reported another experience of rainforest restoration, in San Antonio (Misiones). The work focused on a species that has high timber and ecological value: *Cabralea canjerana*, and the aim was (again) to evaluate its establishment under different coverage conditions. Two experiments were carried out: a pot experiment in which survival and growth under two extreme natural coverages were evaluated, and a field experiment in which this species was planted in a gradient of coverage produced by natural gaps. In both experiments, more than 90% of the plants were initially established in a wide gradient of coverage. In the pot experiment, plants at full sun exposure were damaged by frost in winter. Four years after plantation, 70% of the plants survived in the field experiment, and mortality was observed mostly in microenvironments with lower light incidence. They conclude that *C. canjerana* can establish and survive in many microenvironmental conditions within the rainforest, it acclimates in few months to sudden changes in coverage and, 1 year after planting, its growth rises with higher availability of light. Thus, planting seedlings under cover conditions, and the subsequent removal of the understory (gap opening), is a good strategy for the installation of the species.

Similarly, Balducci et al. (2009) experimented in the Yungas Rainforest with 20 native species in pure and mixed plantations. Among the main species evaluated were *Tipuana tipu, Cedrela balansae, Cordia trichotoma, Handroanthus impetiginosus, Jacaranda mimosifolia, Pterogyne nitens, Enterolobium contortisiliquum*, and *Astronium urundeuva*, with genetic material from the Northern Sector of Yungas Rainforest. The trials were installed between the years 2000 and 2002 in Valle Morado (Province of Salta) located in the UBRB environment. The mortality rates varied between less than 5% and values close to 30%. The species with the highest diameters were *E. contortisiliquum, T. tipu* and *C. balansae* with average DBH above 10 cm at the fifth year of planting, showing *T. tipu* the best forest shape. *C. trichotoma, J. mimosifolia, P. nitens, A. urundeuva*, and *H. impetiginosus* followed in growth, with an average DBH between 6 and 8 cm. The combination of *C. trichotoma* and *H. impetiginosus* is interesting given the good stem shapes that were observed. As for DBH, the species with best total height was *T. tipu*, with an average of more than 9 m at the age of 5 years. It was followed by *C. balansae, J. mimosifolia, E. contortisiliquum*, and *P. nitens*, with a total height between 5 and 7 m. Most of the species evaluated generally showed a high tendency to the formation of lateral branches when planted at open sky, thus losing a good forestry shape, although this factor does not have a significant impact on the plant survival.

Del Castillo et al. (2005) performed some planting experiences with native and exotic Meliaceae in Yuto (Jujuy), at 350 m asl. They compared three valuable wood species: *Cedrela balansae, Toona ciliata* var. *australis*, and *Melia azedarach*, in open sky plantation at age 14, obtaining a wood yield of 6.7, 9.7, and 5.6 m³/ha/year, respectively.

The natural dynamics of the rainforest generate changes in the canopy coverage and, consequently, in the quality and quantity of light that reaches the understory. Similar effects can be achieved by the anthropogenic selective extraction of trees. The microsites produced by the disturbances represent a continuum of environmental conditions that ensures the coexistence of many species. The knowledge of the requirements of the species and the evaluation of the optimal microenvironments to plant growth are useful in decision-making to restore degraded areas.

The presented trials and experiences of plantations point out the feasibility of cultivation of the main forest tree species of the Argentinean rainforests, both with commercial and conservation purposes. Given the current attention on mitigation of the effects of the climate change, probably the main concern is related to active restoration of degraded forests. This finality encourages the conduction of low-intensity breeding programs for these species. The state of the art in this regard is presented in detail in the following four chapters.

References

Abinzano R (1985) Procesos de Integración en una sociedad multiétnica: la provincia Argentina de Misiones. Doctoral thesis, Departamento de Antropología y Etnología de América, Universidad de Sevilla

Aide TM, Grau HR (2004) Globalization, migration, and Latin American ecosystems. Science 305:1915–1916

Aizen MA, Feinsinger P (2002) Bees not to be? Responses of insect pollinator faunas and lower pollination to habitat fragmentation. Ecol Stud 162:111–129

Balducci E, Arturi M, Goya J, Brown A (2009) Potencial de plantaciones forestales en el pedemonte de las Yungas. Ediciones del Subtrópico, p 33

Balducci ED, Eliano P, Iza HR, Sosa I (2012) Bases para el manejo sostenible de los bosques nativos de Jujuy. Incotedes, Jujuy

Balvanera P (2012) Los servicios ecosistémicos que ofrecen los bosques tropicales. Ecosistemas 21:136–147

Barth S, Eibl B, Montagnini F (2008) Adaptabilidad y crecimiento de especies nativas en áreas en recuperación del noroeste de la provincia de Misiones. XIII Jornadas Técnicas Forestales y Ambientales UNaM. Eldorado, June 5–7, 2008

Bawa KS, Dayanandan S (1998) Global climate change and tropical forest genetic resources. Clim Chang 39:473–485

Blendinger PG, Rivera LO, Álvarez ME, Nicolossi G, Politi N (2009) Selección de áreas prioritarias para la conservación de las aves en la Selva Pedemontana de Argentina y Bolivia. In: Brown AD, Blendinger PG, Lomáscolo T, García Bes P (eds) Selva Pedemontana de las Yungas: historia natural, ecología y manejo de un ecosistema en peligro. Ediciones del Subtrópico, Tucumán

Blundo C, Malizia LR, Blake JG, Brown AD (2012) Tree species distribution in Andean forests: influence of regional and local factors. J Trop Ecol 28:83–95

Brown AD and Kappelle M (2001) Introducción a los bosques nublados del neotrópico: una síntesis regional. In: Kappelle M and Brown A (eds) Bosques nublados del neotrópico 2001. Instituto Nacional de Biodiversidad, Costa Rica, pp 25-40.

Brown AD, Malizia LR (2004) Las Selvas Pedemontanas de las Yungas: en el umbral de la extinción. Ciencia Hoy 14:52–63

Brown AD, Pacheco S (2006) Importancia del género *Cedrela* en la conservación y desarrollo sustentable de las Yungas australes. In: Pacheco S, Brown A (eds) Ecología y Producción de cedro (género *Cedrela*) en las Yungas australes. LIEY-ProYungas, Tucumán, pp 9–18

Brown AD, Pacheco S, Lomáscolo T, Malizia L (2006) La situación ambiental en los bosques andinos yungueños. In: Brown A, Martínez Ortiz U, Acerbi M, Corchera J (eds) La situación ambiental argentina 2005. Fundación Vida Silvestre Argentina, Buenos Aires, pp 52–71

Brown AD, Grau HR, Malizia LR, Grau A (2001) Bosques nublados del neotrópico: Argentina. In: Kappelle M, Brown AD (eds) Bosques Nublados del Neotrópico. Instituto Nacional de Biodiversidad. Santo Domingo, Heredia, pp 623–659

Burgos JJ (1970) El clima de la región noreste de la Rep. Argentina en relación con la vegetación natural y el suelo. Bol Soc Arg Bot 11(Supl):37–101

Cabrera AL (1971) Fitogeografía de la República Argentina. Boletín de la Sociedad Argentina de Botánica 14:1–2

Cabrera AL (1994) Enciclopedia Argentina de Agricultura y Jardinería. Regiones Fitogeográficas Argentinas. ACME, Tomo II, Fascículo 1, Buenos Aires

Crechi E, Hennig A, Keller A, Hampel H, Domecq C y Eibl B (2010) Crecimiento de 3 especies latifoliadas nativas a cielo abierto y bajo dosel de pino hasta los 12 años de edad, en Misiones Argentina (*Cordia trichotoma* Vell. Arrab. ex Steudel, *Balfourodendron riedelianum* Engl., *Enterolobium contortisiliquum* Vell. Morong.). XIV Jornadas Técnicas Forestales y Ambientales. Facultad de Ciencias Forestales, UNaM. Eldorado, 10–12 June, 2010

De la Sota ER (1972) Sinopsis de las Pteridófitas del Noroeste de Argentina. Darwiniana 17:11–103

Del Castillo E, Zapater MA, Gil MN, Tarnowski C (2005) Selva de Yungas del Noroeste Argentino (Jujuy, Salta, Tucumán). Recuperación Ambiental y Productiva. Lineamientos Silvícolas y Económicos para un Desarrollo Forestal Sustentable. Estación Experimental de Cultivos Tropicales INTA Yuto. Documento Técnico N°1, p 47

Denslow JS (1980) Patterns of plant species diversity during succession under different disturbance regimes. Oecologia 46:18–21

Denslow JS (1987) Tropical rainforest gaps and tree speciesd diversity. Ann Rev Ecol Syst 18:431–451

Di Fazio SP, Slavov GT, Joshi CS (2011) *Populus*: a premier pioneer system for plant genomics. In: Joshi CP, Di Fazio SP (eds) Genetics, genomics and breeding of poplar. Science Publishers, Enfield, pp 1–28

Eibl B, Montagnini F, Woodward C, Szczipanski L, Rios R (1996) Evolución de la regeneración natural en dos sistemas de aprovechamiento y bosque nativo no perturbado en la provincia de Misiones. Yvyraretá 7:63–78

Eliano PM, Badinier C, Malizia LR (2009) Manejo forestal sustentable en Yungas: protocolo para el desarrollo de un plan de manejo forestal e implementación en una finca piloto. Ediciones del Subtrópico, Tucumán

Fahler J (1981) Variación geográfica entre y dentro de orígenes de *Araucaria angustifolia* (Bert. O. Ktze.) a los ocho años de edad en la provincia de Misiones, Argentina. MSc dissertation. Universidade Federal do Paraná, Curitiba

Fornes L (2012) Domesticación de Especies de Alto Valor de las Selvas Subtropicales. Producción Forestal 2:28–31

Fornes L, Zelener N, Gauchat M, Inza MV, Soldati MC, Ruíz V, Meloni D, Grignona J, Barth S, Ledesma T, Tapia S, Tarnowski C, Eskiviski E, Figueredo I, González P, Leiva N, Rodríguez G, Alarcon P, Cuello R, Gatto M, Rotundo C, Giannoni F, Alonso FM, Saravia P, Trápani A (2016) Subprograma Cedrela. In: Domesticación y Mejoramiento de Especies Forestales. PROMEF. UCAR, pp 136–159

Gallo L, Marchelli P, Chauchard L, Gonzalez Peñalba M (2009) Knowing and doing: research leading to action in the conservation of forest genetic diversity of Patagonian Temperate Forests. Conserv Biol 23:895–898

Gottschling JSM, Weigend M, Hilger HH (2005) Congruence of a phylogeny of Cordiaceae (Boraginales) inferred from ITS1 sequence data with morphology, ecology, and biogeography. Ann Missouri Bot Gard 92:425–437

Grance LA, Maiocco DC (1995) Enriquecimiento del bosque nativo con *Bastardiopsis densiflora* (Hook et Arn.) Hassl., cortas de mejoras y estímulo a la regeneración natural en Guaraní, Misiones. Yvyraretá 6:29–44

Grau A, Brown AD (2000) Development threats to biodiversity and opportunities for conservation in the mountain ranges of the upper Bermejo River Basin, NW Argentina and SW Bolivia. Ambio 29:445–450

Guerrero Borges V, Cotti Alegre J, Sarandón R (2007) Cambios en la cobertura del Bosque Atlántico Argentino durante el periodo 1989–2004. Congreso de la Asociación Española de Teledetección, septiembre de 2007

Gutiérrez Angonese J, Grau HR (2014) Assessment of swaps and persistence in land cover changes in subtropical a periurban region, NW Argentina. Landsc Urban Plan 127:83–93

Hall P, Chase MR, Bawa KS (1994) Low genetic variation but high population differentiation in a common forest species. Conserv Biol 8:471–482

Hall P, Walker S, Bawa K (1996) Effects of forest fragmentation on genetic diversity and mating system in a tropical tree *Pithecellobium elegans*. Conserv Biol 10:757–768

Hurtado R, Fernandez Long M, Serio L, Portal M, Vakdiviezo Corte M (2013) Estudio de las precipitaciones en la Región Noroeste de la Argentina. Agraria VII(14):69–73

Inza MV, Zelener N, Fornes L, Gallo LA (2012) Effect of latitudinal gradient and impact of logging on genetic diversity of *Cedrela lilloi* along the Argentine Yungas Rainforest. Ecol Evol 2:2722–2736

IUCN (2015) Red list of threatened species. www.iucnredlist.org. Accessed on July 15

Juárez A, Ortega-Baes P, Sühring S, Martin W, Galíndez G (2007) Spatial patterns of dicot diversity in Argentina. Biodivers Conserv 16:1669–1677

Kraustofl E (1991) Condiciones de trabajo y calidad de vida de los peones forestales de bosque nativo de Misiones. Depto. de Antropología Social, Universidad Nacional de Misiones

Laharrague N (2018). http://www.argentinaforestal.com/2018/07/13/resultados-del-censo-nacional-de-aserraderos-en-misiones-sobre-438-industrias-refuerzan-la-potencialidad-de-proyectos-de-inversion-con-biomasa-forestal-para-energia-limpia/

Lee CT, Wickneswari R, Mahani MC, Zakri AH (2002) Effect of selective logging on the genetic diversity of Scaphium macropodum. Biol Conserv 104:107–118

Levene R (1940) Historia Argentina (Programa) UNLP. FaHCE. Facultad de Humanidades y Ciencias de la Educación

Liegel LH, Stead JW (1990) *Cordia alliodora* (Ruiz & Pav.) Oken. Laurel, capá prieto. In: Burns RM, Honkala BH (eds) Silvics of North America: 2. Hardwoods. Agric. Handb. 654. U.S. Department of Agriculture, FS, Washington, DC, pp 270–277

Lomáscolo T, Grau A, Brown AD (2014) Guía de Áreas Protegidas de la Provincia de *Tucumán*. Ediciones del Subtrópico, Argentina

Lorenzatti S (2014) Efecto de la ganadería sobre la estructura del bosque y regeneración de especies forestales en las Yungas Argentinas. MSc thesis, Facultad de Agronomia, Univesidad de Buenos Aires

Malizia L, Pacheco S, Blundo C, Brown AD (2012) Caracterización altitudinal, uso y conservación de las Yungas subtropicales de Argentina. Ecosistemas 21:53–73

Malizia A, Osinaga-Acosta O, Powell PA, Aragón R (2017) Invasion of *Ligustrum lucidum* (Oleaceae) in subtropical secondary forests of NW Argentina: declining growth rates of abundant native tree species. J Veget Sci 28:1240–1249

Margalot JA (1985) Geografía de Misiones. Buenos Aires, Argentina, 236 pp

Martínez Crovetto R (1963) Esquema Fitogeográfico de la Provincia de Misiones. Bonplandia 1:171–223

Martínez Crovetto R (1981) Composición florística de yerbales naturales de los alrededores de San Pedro (Misiones). Publ. Técn. N° 3, Fac. Cs. Agrarias, UNNE, Misiones, Argentina, 13 pp

MAyDS (2018) Anuario de Estadística Forestal – Especies Nativas 2016. https://www.argentina.gob.ar/ambiente/bosques/estadistica-forestal

Milkovic M (2012) Mapa de cobertura forestal de la Provincia de Misiones 2010. Informe de consultoría a Fundación Vida Silvestre Argentina

MinAgro (2018) Censo Nacional de Aserraderos. Informe del relevamiento censal en la provincia de Misiones. https://www.agroindustria.gob.ar/sitio/areas/ss_desarrollo_foresto_industrial/censos_inventario/_archivos/censo//000000_Provincia%20de%20Misiones%20(Marzo%20 2018).pdf

Montagnini F, Eibl B, Grance L, Maiocco D, Nozzi D (1997) Enrichment planting in overexploited subtropical forest of the Paranaense region of Misiones, Argentina. For Ecol Manag 99:237–246

Montti L, Carrillo VP, Gutiérrez-Angonese J, Gasparri NI, Aragón R, Grau HR (2017) The role of bioclimatic features, landscape configuration and historical land use in the invasion of an Asian tree in subtropical Argentina. Landsc Ecol 32:2167–2185

Morales JM, Sirombra M, Brown AD (1995) Riqueza de árboles en las Yungas argentinas. In: Brown AD, Grau HR (eds) Investigación, Conservación y Desarrollo en Selvas Subtropicales de Montaña. LIEY, Fac. de Cs. Nat. e Inst. Miguel Lillo. Univ. Nacional de Tucumán, Tucumán, pp 163–174

Moretti A, Olguin F, Pinazo M, Gortari F, Vera Bahima J, Graciano C (2019) Supervivencia y crecimiento de un árbol nativo maderable bajo diferentes coberturas de dosel en el Bosque Atlántico, Misiones, Argentina. Ecol Austral 29:99–111

Muellner AN, Pennington TD, Koecke AV, Renner SS (2010) Biogeography of Cedrela (Meliaceae, Sapindales) in Central and South America. Am J Bot 97:511–518

Pancel L (2015) Nature conservation in the tropics. In: Pancel L, Köhl M (eds) Tropical forestry handbook. Springer, Berlin/Heidelberg

Pennington RT, Prado DE, Pendry CA (2000) Neotropical seasonally dry forest and Quaternary vegetation changes. J Biogeogr 27:261–273

Pickett STA, White PS (1985) The ecology of natural disturbance and patch dynamics. Academic, Orlando

Pidgeon AM, Rivera L, Martinuzzi S, Politi N, Bateman B (2015) Will representation targets based on area protect critical resources for the conservation of the Tucuman Parrot? Condor 117:503–517

Pinazo M (2013). https://inta.gob.ar/noticias/araucaria-el-arbol-insignia-de-misiones-es-tambien-una-importante-alternativa-productiva

Potts BM, Wiltshire RJE (1997) Eucalypt genetics and genecology. In: Williams J, Woinarski J (eds) Eucalypt ecology: individuals to ecosystems. Cambridge University Press, Cambridge, pp 56–91

Powell PA, Aráoz E (2018) Biological and environmental effects on fine-scale seed dispersal of an invasive tree in a secondary subtropical forest. Biol Invasions 20:461–473

Ramírez D (2017) Un abordaje histórico de la actividad forestal en Misiones: del frente extractivo al agronegocio forestal. Folia Historica del Nordeste N° 30, Sept.-Dic. 2017. IIGHI – IH-CONICET/UNNE, pp 29–49

Scalerandi V (2012) La Fábrica en Cabure-í Trabajadores, campesinos y foresto industria en el Noreste de Misiones 1930–1970. MSc thesis, Fac. Humanidades y Cs. Sociales, Universidad Nacional de Misiones

Stecconi M, Marchelli P, Puntieri J, Picca P, Gallo L (2004) Natural hybridization between a deciduous (Nothofagus antarctica, Nothofagaceae) and an evergreen (N. dombeyi) forest tree species: evidence from morphological and isoenzymatic traits. Ann Bot 94:775–786

Tortorelli L (2009) Maderas y Bosques Argentinos. Tomo I. 515 pp

Uhl C (1990) Deforestation, fire susceptibility, and potential tree responses to fire in the Eastern Amazon. Eology 71:437–449

Vega MV, Hernández P (2005) Molecular evidence for natural interspecific hybridization in Prosopis. Agrofor Syst 64:197–202

Volante JN, Alcaraz-Segura D, Mosciaro MJ, Viglizzo EF, Paruelo JM (2012) Ecosystem functional changes associated with land clearing in NW Argentina. Agric Ecosyst Environ 154:12–22

Ward M, Dick CW, Gribel R, Lòwe AJ (2005) To self, or not self. A review of outcrossing and pollen-mediated gene flow in neotropical trees. Heredity 95:246–254

Whittemore AT, Schaal BA (1991) Interspecific gene flow in sympatric Oaks. Proc Natl Acad Sci U S A 88:2540–2544

Young AG, Boyle TJ (2000) Forest fragmentation. In: Young AG, Boshier D, Boyle TJ (eds) Forest conservation genetics: principles and practice. CSIRO, CABI, Australia, pp 123–134

Young A, Boyle T, Brown T (1996) The population genetic consequences of habitat fragmentation for plants. Trees 11:413–418

Zelener N, Tosto D, de Oliveira LO, Soldati MC, Inza MV, Fornes LF (2016) Molecular evidence of hybrid zones of Cedrela (Meliaceae) in the Yungas of Northwestern Argentina. Mol Phylogenet Evol 102:45–55

Zuloaga F, Morrone O, Rodríguez D (1999) Análisis de la biodiversidad en plantas vasculares de la Argentina. Kurtziana 27:17–167

Chapter 13
Patterns of Neutral Genetic Variation for High-Value Cedar Species from the Subtropical Rainforests of Argentina

Noga Zelener, María Cristina Soldati, María Virginia Inza, Leonardo A. Gallo, and Luis F. Fornes

13.1 The Genus *Cedrela* in Argentina: Distribution Area, Ecological Niches, and the Current State of Its Genetic Resources

Cedrela (Meliaceae) is distributed in Central and South America, from Mexico (24° N) to Argentina (28° S). This genus includes tropical and subtropical highly valuable species for the domestic and international forestry sector (Kageyama et al. 2004; Pennington and Muellner 2010). In Argentina, *Cedrela* occurs in the Tucumano-Oranense Rainforest or Yungas Rainforest (Yungas hereafter), in the Northwestern Argentina (NWA), and in the Paraná Atlantic Forest also known as Paranaense Rainforest or Alto Paraná Rainforest (name used hereafter), located in Northeastern Argentina (NEA) (Fig. 13.1; also see Chap. 12).

In the Yungas of NWA, four altitudinal strata are distinguished, three of which host forest species: (i) Temperate Montane Forest (TMF), (ii) Montane Rainforest

N. Zelener (✉)
Centro de Investigación de Recursos Naturales (CIRN), INTA. Hurlingham, Buenos Aires, Argentina
e-mail: zelener.noga@inta.gob.ar

M. C. Soldati · M. V. Inza
Instituto de Recursos Biológicos (IRB), Centro de Investigaciones de Recursos Naturales (CIRN), INTA. Hurlingham, Buenos Aires, Argentina

L. A. Gallo
Instituto de Investigaciones Forestales y Agropecuarias Bariloche (IFAB) INTA-CONICET, San Carlos de Bariloche, Río Negro, Argentina

L. F. Fornes
INTA EEA Famaillá, Famaillá, Tucumán, Argentina

© Springer Nature Switzerland AG 2021   343
M. J. Pastorino, P. Marchelli (eds.), *Low Intensity Breeding of Native Forest Trees in Argentina*, https://doi.org/10.1007/978-3-030-56462-9_13

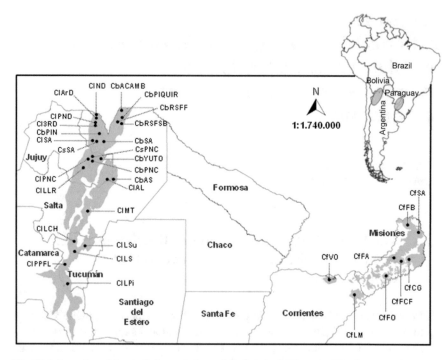

Fig. 13.1 Location of the studied populations of *Cedrela* in the Yungas of Northwestern Argentina (left side of the map) and in the Alto Paraná Rainforest of Northeastern Argentina (right side of the map). Labels indicate populations (codes are as in Table 13.1)

(MR), and (iii) Piedmont Rainforest (PR) (Brown et al. 2001; Grau et al. 2006; SAyDS 2007). Both TMF and PR strata concentrate species of interest for forestry (Brown and Pacheco 2006). Furthermore, the Yungas is ecologically subdivided into northern, central, and southern sectors according to the north-south orographic pattern (Brown et al. 2001). These latitudinal sectors exhibit a clear decrease of taxonomic diversity with increasing latitude (Brown et al. 2001; Juárez et al. 2007). In addition, glacial refuges in northern Yungas during the Pleistocene (Pennington et al. 2000) would have contributed to shape the current pattern of biological diversity, revealed from the geographic distribution of endemic species in this region (Brown et al. 2001, 2006).

The Yungas plays an important role in providing many environmental services; even so, in the second half of the last century, the Yungas has faced depletion processes. Most Piedmont Rainforest surface has been modified by anthropogenic intervention (e.g., Brown et al. 2006; SAyDS 2007, 2008; Seghezzo et al. 2011). The historical overexploitation of Yungas, selective logging, illegal cutting, and lack of control and transparency of local market led to habitat disturbance of accessible forests. Additionally, urban expansion and conversion to agricultural use led to increasing depletion and habitat loss processes (Aide and Grau 2004; Muellner

Table 13.1 Study populations of *Cedrela*, with locality, population code, sample sizes (N, number of samples in AFLP analyses; N′, number of samples in SSR analyses; N″, number of samples in ITS sequence analyses), geographic coordinates, and mean altitude of the population (m asl). The superscript (a) indicates the populations of the Yungas included in a study that combined dataset of AFLP markers and ITS sequence data

Argentine Yungas Rainforest				
Cedrela angustifolia				
Locality	Code	N/N′/N″	Latitude (S)/longitude (W)	Altitude (m asl)
El Nogalar-Natural Reserve	ClND	15/0/0	22°16′32.8″/64°43′06.6″	1825
Argencampo	ClArD	9/0/0	22°20′26.7″/64°43′13.1″	1728
Baritú-National Park	ClPND[a]	15/0/2	22°29′ 6.9″/64°44′53.8″	1704
Empresa 3R	Cl3RD	11/0/0	22°32′15.8″/64°45′12.4″	1713
San Andrés	ClSA[a]	15/0/0	23°04′54.5″/64°50′50.6″	1744
Calilegua-National Park	ClPNC[a]	9/0/2	23°40′59.8″/64°53′46.1″	1642
La Ramada	ClLLR[a]	12/0/0	23°57′54.1″/65°09′32.9″	1656
El Arenal	ClAL[a]	9/0/0	24°20′26.2″/64°20′31.1″	1111
Metán	ClMT[a]	10/0/0	25°23′04.4″/65°00′37.6″	1102
Choromoro	ClLCH	12/0/0	26°22′09.2″/65°28′06.5″	1457
Sunchal	ClLSu[a]	9/0/3	26°30′55.5″/65°06′07.2″	1468
El Siambón	ClLS[a]	15/0/0	26°41′51.3″/65°27′19.0″	1389
La Florida-Provincial Park	ClPPFL[a]	10/0/3	27°07′19.6″/65°47′09.9″	1328
Los Pizarros	ClLPi[a]	10/0/3	27°45′20.6″/65°41′42.3″	1113
Cedrela saltensis				
Calilegua-National Park	CsPNC[a]	9/0/6	23°06′31.3″/64°42′17.8″	1069
San Andrés	CsSA[a]	13/0/10	23°41′14.6″/64°51′07.1″	837
Cedrela balansae				
Acambuco	CbACAMB[a]	16/16/11	22°05′2.05″/63°56′08.6″	911
Piquirenda	CbPIQUIR	15/15/0	22°19′13.4″/63°50′52.6″	852
Río Seco FF	CbRSFF[a]	10/10/4	22°27′57.5″/63°59′49.5″	665
Río Seco FSB	CbRSFSB[a]	17/17/6	22°31′20.0″/63°55′57.5″	699
Pintascayo	CbPIN[a]	11/ 0/8	22°51′15.2″/64°36′36.7″	926
San Andrés	CbSA[a]	12/12/9	23°06′47.4″/64°27′35.7″	471
Yuto	CbYUTO[a]	16/16/6	23°40′15.4″/64°33′29.4″	416
Calilegua-National Park	CbPNC[a]	12/12/6	23°44′59.7″/64°51′08.7″	707
Apolinario Saravia	CbAS[a]	9/9/2	24°20′15.8″/64°12′04.9″	589
Argentine Alto Paraná Rainforest				
Cedrela fissilis				
Locality/province	Code	N/N′/N″	Latitude (S)/longitude (W)	Altitude (m asl)
Puerto Bossetti	CfFB	0/16/0	25°48′28.5″/54°15′54.5″	315
INTA San Antonio	CfSA	0/15/0	26°02′51.2″/53°46′10.4″	535
El Alcázar	CfFA	0/14/0	26°52′43.5″/54°43′25.7″	250

(continued)

Table 13.1 (continued)

Argentine Yungas Rainforest				
Campo Guaraní	CfCG	0/19/0	26°55′31.6″/54°13′0.40″	482
Eldorado	CfFCF	0/9/0	26°59′12.2″/54°28′49.6″	217
Oberá	CfFO	0/13/0	27°28′11.4″/54°58′57.5″	350
Villa Olivari	CfVO	0/11/0	27°36′20.4″/56°53′37.3″	73
Las Marías	CfLM	0/10/0	28°05′30.2″/56°03′00.6″	123

et al. 2009; Seghezzo et al. 2011; FAO 2016). Today, native forests remain the main source of high-quality wood for local forestry development (Fornes et al. 2016).

In the NEA, the Alto Paraná Rainforest is invaluable as biogeographic relict and reserve of biodiversity, due to its genetic resources and endemic species. Despite this, the fragmentation and degradation processes, mainly caused by the patterns of land subdivision, land use with exotic species plantations, spread of agriculture, and unsustainable management of forest species, led to a severe reduction in the surface occupied by forests; at current times only 6–7% of the original surface remains. Along the twentieth century, timber species present in the remnant forest have been overexploited, in a regional context of lack of planning and legal control (Placci 2000; Rau 2005; SAyDS 2007; Inza et al. 2018).

In both Yungas of NWA and Alto Paraná Rainforest of NEA (Fig. 13.1), *Cedrela* is among the most important forest resources because of the mechanical properties of its wood, color, veining, and easiness of work in carpentry. Consequently, it has undergone intense processes of logging, which affected the genetic diversity and structure of the populations (Inza et al. 2012; Soldati et al. 2013; Zelener et al. 2016).

Three endangered species of *Cedrela* (*C. angustifolia*, *C. saltensis*, and *C. balansae*) follow latitudinal and altitudinal patterns of distribution in the Yungas, with transition zones along elevation where the species co-occur.

Cedrela angustifolia (syn = *C. lilloi*; Pennington and Muellner 2010) locally called "cedro coya" is the most valuable timber species of cedar in the Yungas. It occurs in Ecuador, Peru, Brazil, Paraguay, Bolivia, and Argentina, where it can be found in the upper level of the MR and the TMF strata, in an altitude range of 900–2500 m above sea level (m asl), from the border with Bolivia at 22° S to 28° 15′ S, building the southernmost edge of *Cedrela*'s range. It is a canopy, long-lived, and deciduous hardwood species, whose height ranges from 30 to 40 m. The populations of *C. angustifolia* show individuals at low density and aggregated distribution with a gap-dependent recruitment (Grau 2000; Grau et al. 2003; Zamora Petri 2006), which is consistent with the spatial distribution and regeneration patterns described for the genus and family in the Neotropics (Mostacedo and Fredericksen 1999; Van Rheenen et al. 2004). Little is known about the species' mating system; however, most other *Cedrela* species, as well as the Meliaceae family species, are monoecious and predominantly outcrossing, with high self-incompatibility and small unisexual flowers pollinated by small insects (Kageyama et al. 2003; Ward et al. 2005; Aschero 2006). Because of selective logging, the presence of individuals with high commercial value is low, suggesting genetic erosion of the species.

Currently, *C. angustifolia* is included in the IUCN (International Union for Conservation of Nature and Natural Resources) Red List "endangered" category (2020-1, https://www.iucnredlist.org/species/32989/9741887). In addition, it is also included in Appendix III of CITES (Convention of International Trade in Endangered Species of Wild Fauna and Flora 2017; http://cites.org/sites/default/files/esp/app/2017/S-Appendices-2017-10-04.pdf).

Cedrela saltensis, in Argentina, is known as "cedro salteño" and occurs in the MR (from 700 to 1100 m asl) on a mountain strip on the northwest of the country (see top-right panel in Fig. 13.1). It extends between the border with Bolivia at 22° S and its southern limit at 24° 40' S, within restricted zones of sympatry between *C. angustifolia* and *C. balansae* (Brown et al. 2001; Zapater et al. 2004; Malizia et al. 2006). Regarding the taxonomic identity of *C. saltensis*, it was rejected the hypothesis of its hybrid origin from *C. balansae* and *C. angustifolia*, and therefore it needs to be treated as a separate taxonomic entity (Premoli et al. 2011).

Cedrela balansae, also known as "cedro orán" (Zapater et al. 2004), typically grows in the PR (from 300 to 700 m asl) on flat land and low hills due to its susceptibility to low temperatures and humidity requirements. The PR altitudinal stratum has been the most intensely exploited forests because of its proximity to cities and routes and consequently exhibits the highest levels of degradation. The distribution of *C. balansae*, in the NWA, is restricted to a small latitudinal range (22° to 24° 30' S). It is a canopy, long-lived, and deciduous tree species, up to 20 m high, with a dark brown or gray deeply fissured bark. Like other species of the Meliaceae family, *C. balansae* occurs naturally at low densities per hectare and presents an aggregated spatial pattern (Grau et al. 2003; Zamora Petri 2006). Its wood has excellent properties such as fast drying, stability, high durability, and high insect resistance. It also features good aesthetic characteristics like aroma, color, and grain (Rivera 2006) turning *C. balansae* as a highly rated species on the timber market. In addition, the easy accessibility to its populations and the development of agriculture under benign climatic conditions resulted in the conversion of 70–80% of the piedmont surface to agricultural activities (Brown and Malizia 2004; SAyDS 2007). This, in turn, has led to the fragmentation of the remaining *C. balansae* populations, thus increasing their susceptibility to further degradation.

In the Alto Paraná Rainforest of NEA, even though two *Cedrela* species were reported (*C. odorata* and *C. fissilis*; Zapater et al. 2004), only populations of *C. fissilis*, which is commonly known as "cedro misionero," have been found. This cedar species is widespread throughout South America (Pennington and Muellner 2010), although it is experiencing a significant decline because of large-scale forest clearance across its range. Unfortunately, this anthropogenic process is expected to continue, leading to a 30% population decline over the next century (Barstow 2018). In Argentina, the species occurs from 25° 40' S (on the border with Brazil) to 28°S, where reaches its southern limit of distribution. In the northern area of its distribution, almost 60% of the Alto Paraná Rainforest has been transformed (Placci and Di Bitetti 2006). *Cedrela fissilis* has timber attributes that placed it among the species of greatest economic value in the local and international market. Accordingly, the species turned as one of the most affected forest species of the Alto Paraná Rainforest

(Brown et al. 2006); currently being assessed as "vulnerable" in IUCN Red List (https://www.iucnredlist.org/species/33928/68080477; Barstow 2018); also it is found on CITES Appendix III (2017).

13.2 Conservation and Breeding Program of *Cedrela* in Argentina: Molecular Genetic Bases

Since 2006, the National Institute of Agricultural Technology (INTA) of Argentina develops a national program to address the domestication and conservation of high-value native species from subtropical forests. Increasing the production of hard-wood and recovering degraded areas, as well as preserving forest productive function and the environmental services that forests provide, are the main goals of this program. In this context, the genus *Cedrela* was selected as promising genetic resource for breeding purposes due to its high productive potential, its value in the forest products market, and the relatively fast growth rates of their species (Fornes et al. 2016) not requiring large planting areas to ensure an attractive economic return (Brown et al. 2006).

The initial purpose was the characterization of the conservation status of *Cedrela* in Argentina, with the aim of outlining breeding and conservation strategies. Molecular genetics is a useful tool to know the extent and distribution of the cur-rently available genetic variability, which is required for genetic resources manage-ment (Fornes et al. 2016). Thus, we sampled 398 individuals from 33 populations that spanned most of *Cedrela*'s geographical range in the Yungas of NWA and in the Alto Paraná Rainforest of NEA (Table 13.1; Fig. 13.1). Nuclear molecular systems were then used in order to (i) identify suitable sets of markers for unambiguous identification of genetic patterns in *C. balansae, C. saltensis, and C. fissilis* (Soldati et al. 2014a); (ii) describe geographic distribution patterns of *C. angustifolia* genetic variability and quantify the incidence of anthropic disturbance on genetic diversity (Inza et al. 2012); (iii) investigate hybrid zones of *Cedrela* (Zelener et al. 2016); (iv) assess the effect of fragmentation on gene flow in *C. balansae* and *C. fissilis* (Soldati et al. 2013, 2014b); (v) identify priority *Cedrela* populations for conservation pur-poses; and (vi) describe the genetic materials to be used in the breeding programs (Fornes et al. 2016).

13.2.1 Molecular Genetic Variability of Cedrela angustifolia in the Yungas

Genetic variability pattern of *C. angustifolia* (14 populations, 160 individuals; Table 13.1; Fig. 13.1) was recently addressed to explain the effect of the latitudinal gradient and the logging history (disturbance) through the assessment of 293 polymorphic AFLP (amplified fragment length polymorphism) markers. High-priority areas for conservation and domestication purposes, both from protected areas and private fields, were identified (Inza et al. 2012).

13.2.1.1 Genetic Diversity and Differentiation

The average value of genetic diversity of *C. angustifolia* in Argentina is low (Ht = 0.135, He = 0.087; GenAlEx6, Peakall and Smouse 2006), which is consistent with the fact of being the southernmost edge of its continental range (Table 13.1, Fig. 13.1). Additionally, *C. angustifolia* populations are isolated, with low density of individuals per population and different levels of disturbance, all features of species with low levels of genetic diversity (Hamrick et al. 1992; Young and Boyle 2000).

Several indices, such as mean expected heterozygosity (He), percentage of polymorphic *loci* (PPL), and the number of exclusive bands (EB), were used to assess genetic population diversity. The results showed that genetic population variability decreased with increasing latitude and disturbance level (Table 13.2); genetic diversity is much higher in the northern populations than in central and southern populations, though northern populations display similar or greater levels of disturbance than those located in the south. However, La Florida Provincial Park, which is located in the south (Table 13.1; Fig. 13.1), was a singular population displaying genetic diversity levels comparable with the level found in populations located in the north (Table 13.2).

Table 13.2 Genetic diversity of 14 *C. angustifolia* populations in the Yungas with locality, population code, latitudinal sectors (LS: *SS* southern sector, *CS* central sector, *NS* northern sector) of Yungas (Brown et al. 2001), disturbance levels (DL: *UP* undisturbed populations, *LDP* low disturbed populations, *DP* disturbed populations). *He* means expected heterozigosity, *PPLp* percentage of polymorphic loci per population, *EB* number of exclusive bands

Locality	Code	LS	DL	He	PPL$_P$ (%)	EB
El Nogalar-Natural Reserve	ClND	NS	DP	0.119	41.0	9
Argencampo	ClArD	NS	DP	0.113	32.4	7
Barítú-National Park	ClPND	NS	UP	0.168	65.5	36
Empresa 3R	Cl3RD	NS	DP	0.107	33.5	5
San Andrés	ClSA	NS	LDP	0.088	30.7	5
Calilegua-National Park	CLPNC	CS	DP	0.062	17.8	1
La Ramada	ClLLR	CS	LDP	0.071	21.2	2
El Arenal	ClAL	CS	LDP	0.059	17.1	0
Metán	ClMT	SS	DP	0.043	12.3	0
Choromoro	ClLCH	SS	LDP	0.056	17.1	2
Sunchal	ClSu	SS	LDP	0.063	19.5	1
El Siambón	ClLS	SS	LDP	0.057	18.1	0
La Florida-Provincial Park	ClPPFL	SS	UP	0.157	41.3	16
Los Pizarros	ClLPi	SS	LDP	0.049	14.3	0

Most of the genetic variation (87%) was observed within populations (AMOVA; GenAlEx 6.2, Peakall and Smouse 2006), as expected for long-lived, woody, and outcrossing species (Hamrick et al. 1992; Nybom 2004). In addition, a moderate genetic differentiation among populations ($\Phi_{PT} = 0.130$, $F_{ST} = 0.115$) was found. Taking into account the extent of population distribution and the features of latitudinal sectors described for the Yungas (Brown et al. 2001), gene flow restriction associated with spatial isolation of sampled populations (~600 km) was suggested. The discontinuous orography of the Yungas and habitat disturbance caused by logging modified pollination patterns, which would also increase the genetic structure (Aizen and Feinsinger 2002).

13.2.1.2   Genetic Variation and Latitude

A combined data analysis showed a close relationship between the genetic variability patterns of C. angustifolia in the Yungas of NWA and the latitude (Table 13.2, Fig. 13.2). Intrapopulation genetic diversity (He, PPL) tested against latitude was significant and showed a clear decrease with increasing latitude (InfoStat 2008, Fig. 13.2a). In addition, genetic structure was in strong agreement with latitudinal sectors of the Yungas (Brown et al. 2001). Three genetic clusters (best K value according to Evanno et al. 2005 was $K = 3$) were identified by Bayesian modeling (Structure 2.2.3, Pritchard et al. 2000, Fig. 13.2b), and three UPGMA dendrogram' groups (NTSYS 2.0, Rohlf 1998; data shown in Inza et al. 2012) were in correspondence with northern, central, and southern Yungas. Additionally, diversity (He, PPL, and EB) was estimated for each latitudinal sector, and, as proposed by Brown et al. (2001, 2006), populations were grouped into northern, central, and southern subregions of Yungas (Table 13.2). To minimize the effect of disturbance, only populations with some disturbance level (LDP and DP) were compared.

In agreement with the latitudinal pattern of biological diversity of the Yungas (Brown et al. 2001; Juárez et al. 2007), the highest genetic diversity was detected in the north (He, 0.181; PPL, 89.3%; and EB, 68), decreasing towards the south (He, 0.093; PPL, 38.6%; and EB, 11; Inza et al. 2012). When only undisturbed populations were compared, genetic diversity of Baritú National Park in the north was higher than La Florida Provincial Park in the south (Table 13.2). A significant genetic differentiation among latitudinal sectors ($\Phi_{RT} = 0.07$, $p = 0.001$) and undisturbed populations of Baritú National Park and La Florida Provincial Park ($\Phi_{PT} = 0.10$, $p = 0.001$) support these results (AMOVA, GenAlEx 6.2, Peakall and Smouse 2006). Furthermore, the association of genetic diversity with latitude, which was calculated for each disturbance group (InfoStat 2008), was high, negative, and significant (Fig. 13.2a).

Latitudinal patterns of biological diversity in the Yungas have been explained by less benign climate towards the southern region (Brown et al. 2001, 2006; Juárez et al. 2007). We hypothesize that the warmth-prone habit of C. angustifolia (Villalba et al. 1992) may have led to larger populations and greater persistence of genetic variability in northern Yungas. In long-term periods, according to the climatic

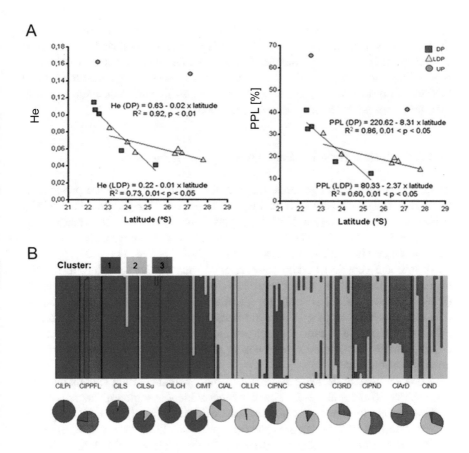

Fig. 13.2 Genetic variability patterns of *C. angustifolia* populations in Yungas. (**a**) Regression of genetic diversity (He, left; PPL, right) on latitude (°S) for each disturbance level. *He* expected heterozygosity, *PPL* percentage of polymorphic *loci, DP* disturbed populations, *LDP* low disturbed populations, *UP* undisturbed populations. (**b**) Genetic structure of *C. angustifolia* populations inferred by Bayesian clustering (Structure). Populations are presented from left to right in decreasing order of latitude (codes are as in Table 13.2). Each vertical bar represents an individual. Pie charts represent the percentages of assignment of populations to clusters 1 (blue), 2 (green), and 3 (red)

history of the region (Pennington et al. 2000), northern Yungas may have been a potential thermal refuge during glacial times for *C. angustifolia* in the Pleistocene, thus expecting lower diversity in more recently colonized southern Yungas because of genetic drift processes occurring during the species migration (Martin and McKay 2004). Nevertheless, high genetic diversity of La Florida Provincial Park in the south cannot be explained either by the environmental gradients or by the historical temperature refuges of Yungas. Taking into account humid instead of thermal conditions, historical humid refuges in extremely dry periods of the Pleistocene in the region can be postulated as a sound explanation for the entire genetic pattern found. The Upper Bermejo River Basin (UBRB) in the northern edge of Yungas and the wet slopes of Sierra de Aconquija towards the south may have been historical refuges of the species. They hold the highest rainfall of the region due to their orography. This is in agreement with the two areas that have been suggested as potential biodiversity refuges in Yungas due to its high levels of endemisms associated with long-term ecosystem stability (Brown et al. 2006).

13.2.1.3 Genetic Variation and Anthropogenic Disturbance

In addition to latitudinal patterns, logging history of *C. angustifolia* populations could also explain the genetic variability of the species in Yungas (Inza et al. 2012). A loss of genetic diversity in forest tree species was observed with increasing logging intensity, and bottlenecks in disturbed populations are suggested as a possible explanation (White et al. 1999; Kageyama et al. 2004; André et al. 2008).

To evaluate genetic variation patterns of *C. angustifolia* according to logging history, its populations were classified into three different disturbance levels (Table 13.2; see Inza et al. 2012 for classification criteria), and genetic diversity (He, PPL, and EB) was then calculated for each disturbance group. To minimize the effect of latitude, the disturbance groups were assessed within each latitudinal sector, and, finally, only pairs of neighboring populations with different disturbance levels were compared.

A general trend of decrease in genetic diversity with increasing logging intensity was observed (Inza et al. 2012). This was more evident in southern Yungas with the *undisturbed* group showing the highest genetic diversity (He = 0.314, PPL = 82.3%, EB = 22), followed by the *low disturbed* group (He = 0.138, PPL = 55.1%, EB = 11) and the *disturbed* group showing the lowest values (He = 0.086, PPL = 24.5%, EB = 2). Only in this subregion, genetic differentiation among *disturbed* group (Φ_{RT}, AMOVA, GenAlEx 6.2 Peakall and Smouse 2006) was significant (Table 13.3); the fact that overexploitation of *C. angustifolia* began on southern Yungas (Minetti 2006) could be the explanation. Additionally, genetic erosion processes at this latitude could have a greater impact due to the lower initial levels of genetic diversity (Inza et al. 2012). This is consistent with the assessment of neighboring populations that showed lower genetic diversity for populations with higher disturbance for all compared pairs (data shown in Inza et al. 2012) and a significant genetic differentiation (Φ_{PT}) among them (high in the south, low-moderate in the north and the center, Table 13.3). For neighboring populations no selection pressure is expected, and therefore, it was suggested that genetic drift processes are associated with reduced effective population sizes (Young and Boyle 2000; Degen et al. 2006), which could explain the loss of genetic diversity in logged populations (Inza et al. 2012). No genetic differentiation between Baritú National Park and Empresa 3R pair of populations was observed, but the proximity between them and the unlogging period of Empresa 3R population for more than 20 years may explain it.

Table 13.3 Analysis of molecular variance (AMOVA) according to disturbance. Genetic variation among different disturbance levels (Φ_{RT} or Φ_{PT}) is indicated. A, within each latitudinal sector; B, pairs of neighboring populatios with different disturbance levels. *UP* undisturbed populations, *LDP* low disturbed populations, *DP* disturbed populations

Comparative assessment	Compared disturbance groups or compared populations	Φ Statistics	South	Center	North
A: Within each latitudinal sector of Yungas	UP, LDP, DP	Φ_{RT}	0.15[b]	–	0.00 ns
	DP, LDP		0.05[a]	0.00 ns	0.02 ns
	Unlogging (UP), logged (DP + LDP)		0.22[b]	–	0.00 ns
B: Neighboring populations with different disturbance levels	La Florida Provincial Park, El Siambón	Φ_{PT}	0.21[b]	–	–
	La Ramada, Calilegua National Park		–	0.08[b]	–
	Baritú National Park, 3R		–	–	0.01 ns
	Baritú National Park, Argencampo		–	–	0.10[b]
	Baritú National Park, El Nogalar		–	–	0.03[a]

[a] $p \leq 0.01$; [b] $p \leq 0.001$; ns, not significant

13.2.2   Molecular Genetic Variability of Cedrela balansae in the Yungas

Little information is available about the genetic diversity of *C. balansae*, despite its ecological and economic importance. In order to provide information about the genetic variation of the species as a prerequisite to start breeding and conservation programs, genomic analyses combining both SSRs (simple sequence repeats) and AFLP markers were carried out (Soldati et al. 2013).

Young leaves from 107 adult trees randomly chosen were collected throughout the north and center of the Yungas in NWA, representing eight remnant natural populations covering the distribution area of the species in Argentina. For genomic DNA extraction and molecular marker procedures, see Soldati et al. (2013). Since no SSR markers have been reported in *C. balansae* so far, we tested the transferability of 45 SSRs developed in phylogenetically close species belonging to the Meliaceae family: (i) 21 primer pairs of SSRs developed for *C. odorata*; (ii) 6 primer pairs developed for *C. fissilis*; (iii) 10 primer pairs developed for *Swietenia macrophylla*; and (iv) 8 primer pairs developed for *Swietenia humilis*. Thirty primer pairs successfully amplified putatively homologous SSR *loci* in *C. balansae*, 12 of them were pre-selected based on clearness of resolution patterns and polymorphism levels. Null alleles were found in 10 of the 20 markers, but only reached values greater than 0.05 in 5 *loci*. The remaining 7 *loci* (i.e., Ced2, Ced41, Ced44, Ced61a, Ced95, CF66, and CF78) showed no significant proportion of null alleles (P < 0.05), and none of them revealed significant deviations from HWE (*Hardy-Weinberg*

Table 13.4 Genetic diversity parameters estimated by SSR and AFLP markers for *C. balansae* populations distributed in the Yungas. Locality, population code, *Ea* exclusive alleles, *Ho* observed heterozygosity, *He* expected heterozygosity, *NPLp* number of polymorphic *loci* per population, *PPLp* percentage of polymorphic *loci* per population, *NEM* number of exclusive markers, *SD* standard deviation

Locality	Code	SSRs			AFLPs			
		Ea	Ho	He	NPLp	PPLp	NEM	He
Acambuco	CbACAMB	3	0.723	0.664	327	80.63%	13	0.259
Piquirenda	CbPIQUIR	1	0.703	0.620	245	51.05%	2	0.181
Río Seco FF	CbRSFF	3	0.643	0.622	241	48.43%	1	0.203
Río Seco FSB	CbRSFSB	0	0.647	0.645	263	55.24%	1	0.211
San Andrés	CbSA	1	0.618	0.663	299	71.20%	8	0.254
Yuto	CbYUTO	2	0.602	0.645	298	70.94%	5	0.229
Calilegua-National Park	CbPNC	2	0.726	0.683	288	66.23%	2	0.258
Apolinario Saravia	CbAS	1	0.619	0.575	220	40.05%	1	0.180
Mean	–	1.625	0.660	0.643	272.6	60.47%	4.125	0.222
SD	–	1.408	0.029	0.026	36.2	4.90%	4.356	0.033

equilibrium) genotypic proportions (average Fis = 0.021) nor a significant genotyping linkage disequilibrium (P > 0.05). Sixty-two alleles were detected in the 107 individuals analyzed through SSR markers. Allelic multiplicity showed an average of 8.85 alleles per *locus* (SD = 4.09) and ranged from 5 to 16 alleles. Also, a total of 13 exclusive alleles were identified in 7 out of 8 analyzed populations (Table 13.4). Observed heterozygosity (Ho) values across populations ranged from 0.602 to 0.726 (mean = 0.660, SD = 0.029). Genetic diversity (Nei) ranged from 0.575 to 0.683 for Apolinario Saravia and Calilegua National Park populations, respectively, with an average He of 0.643 (SD = 0.026) (Soldati et al. 2013).

Two out of six AFLP primer combinations were selected expressing a clear and highly informative genetic pattern, as well as the representativeness of the genome of the species, detecting 382 polymorphic *loci*. Genetic diversity showed an average value of 0.222 (SD = 0.033), ranging from 0.180 to 0.259. The number of polymorphic *loci* per population ranged from 220 to 327, whereas the percentage of polymorphic *loci* per population ranged from 40.05% to 80.63% (Table 13.4). Exclusive markers were observed in all populations, ranging between 1 and 13 with an average value of 4.125 (Table 13.4).

A good congruence was found between both SSR and AFLP results, since a high and significant correlation (r = 0.902; P < 0.01) among He population values obtained from both marker types was estimated. To visualize geographic patterns of genetic diversity obtained through SSRs and AFLPs, grid-based spatial analyses were carried out using 2.5 min grid cells (~5 km at the equator), following van Zonneveld et al. (2012). Construction of the circular neighborhoods and posterior bootstrap sample bias correction were performed in R statistical package version 2.14 (R Development Core Team 2011). The grid-based spatial analyses show, once again, that Calilegua National Park, San Andrés, and Acambuco populations are the areas with the highest levels of genetic diversity and Apolinario Saravia population exhibits the lowest values (Fig. 13.3a).

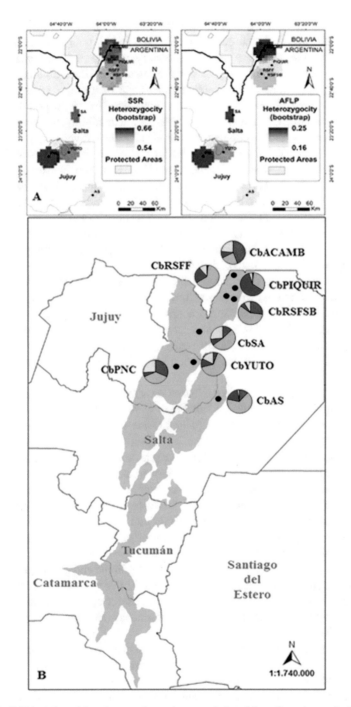

Fig. 13.3 Grid-based spatial analyses and genetic groups inferred from Bayesian methods in eight natural populations of *C. balansae*. (**a**) Average He values for SSRs and AFLPs in each 2.5 min grid cell with nine or more trees applying a 20° circular neighborhood. (**b**) Pie charts represent distribution of genetics groups in each population. Labels near pie charts indicate populations (codes are as in Table 13.4). Black dots indicate the exact location of populations

In summary, genetic diversity of *C. balansae* populations in the Yungas of NWA was moderate, with a similar general trend displayed by both markers, making the results comparable. Levels of genetic diversity observed in *C. balansae* might be associated with the species' marginal distribution area and the latitudinal position of the studied populations on the southernmost edge of a major ecosystem as the Neotropical Cloud Forests. The intense exploitation, to which populations have been subjected due to its hardwood qualities, easy accessibility, and species' spatial distribution pattern, also accounts for its levels of genetic diversity.

Population genetic differentiation, by SSR and AFLP data, was estimated using θp and Fst statistics. Historical gene flow (Nm) was then indirectly calculated, according to Crow and Aoki (1984). Bayesian clustering approach was implemented for both markers, to resolve the optimal number of genetic clusters (K) and population assignment across the identified clusters. The no-admixture model was used in order to detect subtle structure (Structure 2.3.3 software, Pritchard et al. 2000; Falush et al. 2007). Low but significant genetic differentiation among populations (SSRs: $\theta p = 0.049$, CI95: [0.025; 0.077]; AFLPs: Fst $= 0.041$, P ≤ 0.001) was observed. Additionally, considerable historical gene flow was detected among *C. balansae* populations, with Nm values of 3.71 and 4.47, for SSRs and AFLPs, respectively. Regarding population structure, four genetic clusters ($K = 4$, according to Evanno et al. 2005 method) homogeneously distributed were distinguished using Bayesian methods. No pattern of clustering between the populations was observed, and there were no populations assigned exclusively to a given genetic cluster (Fig. 13.3b).

A cluster analysis of individuals based on similarity matrices was also implemented using UPGMA method. Mantel tests were applied and cophenetic correlation coefficients (r) were computed; significance testing was achieved with 1000 random permutations using NTSYS pc.2.0 software. The 107 individuals analyzed could be discriminated using the AFLPs markers, whereas the SSR set did not allow to distinctively fingerprint 7 pairs of individuals (data not shown). No grouping by population nor by geographic location was observed in either case, being consistent with genetic cluster distribution. Moderate to high values were obtained for the cophenetic correlation coefficients (r = 0.676 and r = 0.884 for SSRs and AFLPs, respectively). Very low genetic distances between populations were observed, being the largest 0.047 and 0.070 for SSRs and AFLPs (r = 0.780 and r = 0.740), respectively. Finally, a very low correlation between geographic and genetic distances was observed for both molecular markers (SSRs, r = 0.197, P = 0.01; AFLPs, r = 0.056, P = 0.01), and no spatial pattern of distribution was detected. The weak grouping of individuals was further confirmed by cluster analysis and the extremely low pairwise genetic distances recorded.

Taking into account the short geographic distance among *C. balansae* populations in the Yungas (~250 km between the northernmost and the southernmost ones), it is likely that they are in a potential reproductive contact which is being reflected in a weak population structure and clustering of individuals based on similarity indices. In addition, the high levels of historical gene flow detected among populations would suggest that the 8 populations of *C. balansae* studied behave practically as a homogeneous genetic unit in the piedmont of the Yungas.

13.2.3   Molecular Evidence of Hybrid Zones of Cedrela in the Yungas

The inclusion on molecular genetics studies of the third species of *Cedrela* (*C. saltensis*) with natural distribution in the Yungas allowed us to investigate whether natural hybridization took place among *Cedrela* species in NWA, thus supporting the existence of hybrid zones between them (Zelener et al. 2016).

Species delimitation in *Cedrela* is difficult because of the lack of unique qualitative morphological diagnostic characters. They can be defined only by reference to six widely variable and overlapping characters (Pennington and Muellner 2010). Therefore, a combined dataset of AFLP markers and ITS (internal transcribed spacers) sequences data were used to test hybridization and/or introgressive hybridization in *Cedrela* species from the Yungas of NWA. For this study, 210 samples were available, including the three *Cedrela* species that occur in Yungas: (i) 104 samples from 10 populations of *C. angustifolia*; (ii) 84 samples from 8 populations of *C. balansae*; and (iii) 22 samples from 2 populations of *C. saltensis* located in zones of sympatry between *C. saltensis* and *C. angustifolia* and between *C. saltensis* and *C. balansae* in Calilegua National Park and San Andrés farm, respectively (Table 13.1; Fig. 13.1). For detailed information about AFLP and ITS procedures and data analyses, see Zelener et al. (2016).

13.2.3.1   Population Structure of *Cedrela* in the Yungas

AFLP fingerprinting of the 210 samples of *Cedrela* from the Yungas resulted in a matrix of 577 well-scorable polymorphic markers, which was analyzed by 2 complementary methods: (i) a multivariate approach as Principal Coordinates Analysis (PCoA) using GenAlEx 6.3 (Peakall and Smouse 2006) and (ii) a statistical inference based on the Bayesian clustering method of Structure 2.3.4 (Pritchard et al. 2000; Falush et al. 2007).

The PCoA showed that most of the samples grouped according to the species: there was a main cluster for each of the three species (*C. angustifolia*, *C. balansae*, and *C. saltensis*). The first principal coordinate (73.1% of the total variation) clearly distinguished samples of *C. angustifolia* from samples of the other two species. The second principal coordinate (14.8% of the total variation) grouped most of the samples of either *C. saltensis* or *C. balansae* in separate groups. However, some samples of these two species showed intermediate positions between *C. saltensis* and *C. balansae* clusters. The scattering was most noticeable for samples from Pintascayo, San Andrés, Acambuco, and Calilegua National Park populations of *C. balansae*. Remarkably, most samples of Pintascayo rested near the *C. saltensis* cluster (complete data shown in Zelener et al. 2016).

In Structure, a preliminary analysis using data from the three species (20 populations) revealed that the 210 samples of *Cedrela* split into two clusters (best $K = 2$). One of the clusters was congruent with a taxonomical-delimited species, since grouped all samples of *C. angustifolia*, while the other cluster brought together samples of *C. balansae* and *C. saltensis* (*Cbal-Csal* hereafter). PCoA had indicated analogous results. Next, we carried out analyses on Structure for *C. angustifolia* and *Cbal-Csal*, independently. Both analyses showed two clusters (best *K* = 2).

To display the latitudinal distribution of the genetic clusters estimated by both independent Bayesian analyses, we constructed a first map representing each of the 20 populations (Table 13.1) with the proportion of assignment to each genetic cluster (Fig. 13.4a). In this map, pie charts at sampled locations show the proportion of membership of populations to *C. angustifolia*, *C. balansae*, and *C. saltensis*. The two clusters of *C. angustifolia* (cluster I depicted in violet and cluster II depicted in lilac; Fig. 13.4a) were broadly represented among the populations, which suggests that populations behaved as metapopulations associated with latitudinal and altitudinal distributions across the Yungas (Table 13.1). There was a tendency to find populations located at higher latitudes and lower altitudes (Los Pizarros, La Florida Provincial Park, El Siambón, Sunchal, Metán, and El Arenal) with a higher percentage of membership in cluster I (74–100%). Populations located towards lower latitudes and higher altitudes (La Ramada, Calilegua National Park, San Andrés, and Baritú National Park) showed substantial increase in the percentage of membership in cluster II (43–100%) while exhibiting varying levels of admixture between clusters I and II (Fig. 13.4a); these results are in agreement with those presented above concerning patterns of genetic diversity found in *C. angustifolia* and reported by Inza et al. (2012) when this species was addressed independently.

To a certain extent, the two clusters of *Cbal-Csal* were congruent with taxonomical-delimited species: cluster I (depicted in red; Fig. 13.4a) brought together samples of *C. balansae*, while cluster II (depicted in green; Fig. 13.4a) brought together samples of *C. saltensis*. Accordingly, Río Seco FF, Río Seco FSB, Yuto, and Apolinario Saravia populations constitute genetically pure populations of *C. balansae*. In contrast, four populations that were defined *a priori* as *C. balansae* (Calilegua National Park, Acambuco, San Andrés, and Pintascayo) contained varying levels of admixture between clusters I and II (Fig. 13.4a). These results suggest a hybrid or introgressed origin for these four populations. The two populations of *C. saltensis* showed percentage of assignment in cluster II >97% for most of the samples, with the exception of four samples belonging to Calilegua National Park and San Andrés populations, which exhibited levels of admixture of clusters I and II >45%. In addition, Pintascayo – a population sampled as *C. balansae* according to morphological characteristics – showed percentage of membership >83% in cluster II, the *C. saltensis* cluster. Actually, 4 out of 11 samples of Pintascayo showed percentage of membership ≥99% in the *C. saltensis* cluster, which suggests that these four samples detained a large contribution of the *C. saltensis* genome.

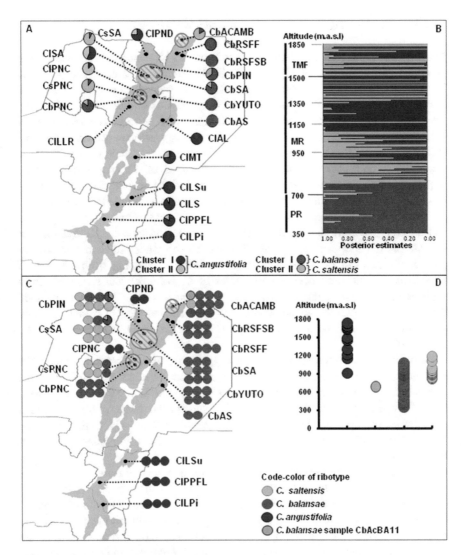

Fig. 13.4 Hybrid zones in the Yungas of Northwestern Argentina. Latitudinal distribution (**a**) and altitudinal distribution (**b**) of genetic clusters inferred from Structure for *Cedrela*; the strata are indicated as TMF (Temperate Montane Forest), MR (Montane Rainforest), and PR (Piedmont Rainforest). Spatial distribution (**c**) and altitudinal distribution (**d**) of the four ribotypes of ITS of *Cedrela*. Populations of hybrid origin display circles of two or more colors; multicolored circles represent samples harboring intragenomic polymorphism for ITS. Labels indicate populations (codes are as in Table 13.1). Hatched areas in the maps depict locations where *C. balansae* and *C. saltensis* form hybrid zones

13.2.3.2   Phylogenetic Relationships of *Cedrela* Species from the Yungas

The aligned ITS dataset consisted of a matrix of 131 sequences. Direct sequencing produced ITS sequences of *C. angustifolia* (13 samples), *C. balansae* (43), and *C. saltensis* (12), while sequencing of bacterially cloned amplicons yielded 60 clones from 9 samples of *C. balansae* and 4 samples of *C. saltensis* (Table 13.1). Three additional sequences from GenBank were used as reference sequences of *C. saltensis*, *C. balansae*, and *C. angustifolia*. Phylogenetic analysis based on this matrix returned a tree that grouped the sequences into two main clades. The first clade (90% bootstrap support) grouped sequences of *C. angustifolia*, while sequences of *C. balansae* and *C. saltensis* occupied the second clade (95% boot-strap support). Within this second clade, there were two sub-clades. The first one grouped most of the sequences of *C. balansae* obtained through direct sequencing and through bacterial cloning from samples that belong to either *C. balansae* or *C. saltensis* (91% bootstrap support). The second sub-clade (100% bootstrap sup-port) included most of the sequences from direct sequencing from samples of *C. saltensis*, some sequences from samples of *C. balansae* and sequences from samples classified as either *C. balansae* or *C. saltensis* that had been obtained through bacterial cloning. In agreement with the results of the AFLP analyses, both sub-clades contained sequences of both *C. balansae* and *C. saltensis* (Zelener et al. 2016).

Network analysis further assessed the genealogical relationships among the sequences, which were mostly congruent with the phylogenetic analysis (data shown in Zelener et al. 2016). Four ribotypes were found among the 128 sequences. One ribotype was present exclusively in *C. angustifolia*; a second ribotype was a singleton, which appeared in a sample of *C. balansae* (CbAcBA11); a third ribotype was present in sequences from samples of *C. balansae*; and a fourth one was present in *C. saltensis*.

In order to compare results achieved by a Bayesian method from AFLP markers and those from ITS analyses, we constructed a second map to display the latitudinal distribution of the ITS ribotypes that were identified in each of the 15 populations (Table 13.1; Fig. 13.4c). Each circle in this map represents the ribotype of a sample. Interestingly, ribotypes 3 and 4 appeared in bacterially cloned sequences obtained from both a sample of Pintascayo population of *C. balansae* and a sample of San Andrés population of *C. saltensis* (data shown in Zelener et al. 2016). In addition, some samples of *C. balansae* – from Pintascayo population – and *C. saltensis*, from San Andrés population, retained ITS paralogs that had originated in either *C. balan-sae* or *C. saltensis*, as expected in recent hybrids (Nieto Feliner and Rosselló 2007; Coleman 2009).

13.2.3.3 Natural Hybridization Between *C. balansae* and *C. saltensis* in the Yungas

The existence of samples possessing ITS paralogs with dual origins provides a strong evidence for the occurrence of natural hybridization between *C. balansae* and *C. saltensis* in the Yungas. For hybridization to occur among closely related species, several events should take place simultaneously. The first of such requirements is the overlapping distribution range of the target species (Rajora and Mosseler 2001; Lepais et al. 2009); shared pollinators and synchronicity of flowering are also essential factors (Carney et al. 2000).

In the Yungas of NWA, the distribution range of *C. angustifolia* partly overlaps with the range of *C. saltensis* at lower elevation of the MR, while there is no overlap of ranges of *C. angustifolia* and *C. balansae*. The two genetic clusters of *C. angustifolia* derived from Bayesian analysis did not occur in any sample of *C. saltensis*. Therefore, range overlapping does not seem to be enough to trigger hybridization between *C. angustifolia* and *C. saltensis* or between *C. angustifolia* and putative *C. saltensis*-*C. balansae* hybrids. According to phylogenetic studies, the clade that contains *C. angustifolia* is sister to the clade that contains *C. saltensis* and *C. balansae* (Muellner et al. 2009; Koecke et al. 2013). This phylogenetic distance may account for the absence of hybridization involving either *C. angustifolia* and *C. saltensis* or *C. angustifolia* and *C. balansae*. Studies on cross-species transferability of SSR markers provide further insights about the phylogenetic relationships among species of *Cedrela* in the Yungas (Soldati et al. 2014a).

The overlapping distribution ranges of *C. saltensis* and *C. balansae* follow a downstream transect in the same altitudinal stratum. Studies of reproductive biology revealed that species of *Cedrela* may share pollinators (Bawa et al. 1985; Kageyama et al. 2004; Ward et al. 2005; Aschero 2006; Pennington and Muellner 2010). In the Yungas, the three species of *Cedrela* exhibit synchronicity of the flowering period (Zapater et al. 2004). If *C. saltensis* and *C. balansae* share pollinators, as postulated previously, and there is some degree of interfertility between these two species, hybridization is likely to take place in sites where they co-occur.

The detection of interspecific hybrids by means of morphological characters generally assumes that trees of hybrid origin will be phenotypically intermediate between parental species. However, this assumption is often not valid (Allendorf et al. 2001). For *Cedrela*, morphometric characters did not discriminate between introgressant hybrids and trees harboring the expected morphology of *C. balansae*. When *C. balansae* was studied regardless of the other two species, as was described in this chapter and reported by Soldati et al. (2013), all samples were considered as belonging to *C. balansae*. Only a joint assessment by molecular genetics of samples, *a priori* considered either *C. balansae* or *C. saltensis*, allowed us to detect introgressant hybrids between them from those samples belonging to pure species. These findings suggest that an extensive range of phenotypic plasticity exists and that it may obscure morphological differences between *C. balansae* and *C. saltensis*. The highest levels of genetic diversity found in Calilegua National Park, Acambuco, and San Andrés populations *a priori* considered as *C. balansae*

(Table 13.4; and Soldati et al. 2013) reinforce the presence of introgressant hybrids into these populations since hybridization can lead to an increase in genetic diversity at population level (Donoso et al. 2004; Hoffmann and Agro 2011).

13.2.3.4 Delimitation of Hybrid Zones of *C. balansae* and *C. saltensis* in the Yungas

The abundance of samples with mixed ancestry in both *C. balansae* and *C. saltensis* corroborated the existence of hybrid contact zones in the Yungas. These hybrid zones exist in areas of sympatry as well as at intermediate elevations where *C. balansae* and *C. saltensis* co-occur. To display altitude ranges of hybrid contact zones, we addressed the altitudinal distribution of genetic clusters from AFLP at sample level. In Fig. 13.4b, each sample is represented by a horizontal colored bar; the extension of the color in each bar indicates the probability of belonging to the inferred genetic cluster in that sample. In addition, it was addressed the altitudinal distribution of ribotypes showing the correspondence between ribotype assignment and the altitude where the sample was obtained (Fig. 13.4d). The geographic locations of hybrid zones are depicted as hatched areas on the maps (Fig. 13.4a, c). Hatched areas on the first map included all populations with ancestry in more than one species, from Bayesian clustering analyses of AFLP markers (Fig. 13.4a). Hatched areas on the second map included the populations that showed ITS ribotypes from distinct species (Fig. 13.4c); each circle in this map represents the ribotype of a sample; populations of hybrid origin display ribotypes encoded with different color; multicolored circles represent samples harboring intragenomic polymorphism.

The changes in the frequencies of AFLP markers and ITS ribotypes were strongly congruent along the Yungas and strata elevation. Intermediate genotypes displayed ancestry in both species at sympatric areas that occurred in the Calilegua National Park and the San Andrés farm (Fig. 13.4a, c). Moreover, increasing proportions of ancestry admixture occur towards altitude ranges approximately from 820 to 1100 m asl (Pintascayo and Acambuco localities) (Fig. 13.4b, d). Additionally, pure populations of *C. saltensis* were genetically identified towards increasing altitudes of the MR. Meanwhile, pure populations of *C. balansae* were widely represented in the PR, in accordance with vegetation strata described for the Yungas (Fig. 13.4b, d) (Brown et al. 2001; Zapater et al. 2004; Malizia et al. 2006). In addition, genetically pure samples of either *C. balansae* or *C. saltensis* were found together with hybrids and introgressant forms between these species in Calilegua National Park, San Andrés, and Pintascayo localities.

We did not find any genetically pure sample of *C. balansae* in the Acambuco population; it is worth mentioning that an accession of *C. saltensis* was reported in Tarija (South of Bolivia) at 22° 16' S, 64° 30' W, and 1100 m asl (James C. Solomon collector, No.10066, herbaria LPB and MO; http://www.tropicos.org), not far from locality of Acambuco. The existence of *C. saltensis* nearby may explain the presence of introgressant forms of *C. balansae* in Acambuco population.

13.2.4   Clonal Seed Orchards of Cedrela balansae and Cedrela angustifolia: Molecular Genetic Bases

Seed orchards are the most common and cost-effective means of making available a stable supply of genetically improved seeds (Ipinza and Vergara 1998; Varghese et al. 2000). The optimal function of clonal seed orchards (CSO) depends on many factors; the genetic purity of the clones to be included in the orchard is a crucial starting point.

As was described in the present chapter, our work allowed us to detect genetically pure *C. balansae* populations from those containing a significant proportion of trees that harbor genetic contributions of both *C. balansae* and *C. saltensis* parental species. Consequently, Acambuco and Pintascayo populations were discarded to comprise the *C. balansae* CSO. In addition, all clones to be included should have a purity $\geq 99\%$, according to genetic assignment analysis by AFLP markers.

In long-term breeding programs, the intensity of selection applied over breeding populations to build up seed orchards with superior genotypes restricts the number of genotypes involved in the final orchard, thereby decreasing genetic diversity and increasing the risk of inbreeding depression over successive generations. To balance genetic gains and diversity, genomic diversity parameters through the employment of molecular markers represent a valuable tool as selection criteria (Marcucci Poltri et al. 2003; Zelener et al. 2005). Accordingly, for the optimal design of *C. balansae* CSO, a genome analysis of 51 superior individuals selected from *C. balansae* breeding population – showing a purity $\geq 99\%$ – was carried out (Soldati et al. 2015). To characterize the levels of genetic diversity, to estimate the rates of inbreeding (Fis), and to determine the genetic similarities and relationships between selected individuals, seven polymorphic SSRs (Soldati et al. 2014a) and two AFLPs markers combinations (Soldati et al. 2013) were used.

Genetic diversity was moderate to high (He = 0.716 from SSR and He = 0.269 from AFLP markers) and similar to average genetic diversity found in natural populations of the species (He = 0.618 from SSR and He = 0.222 from AFLP markers; Soldati et al. 2013) suggesting that selected individuals for the CSO are representative samples of the breeding population. In addition, no significant grouping of individuals either by families or by geographical origins was observed, which is consistent with results obtained for the natural populations studied; as was previously mentioned, *C. balansae* populations could behave as a homogeneous genetic unit in the Yungas. However, low average similarity index (0.229 and 0.521 for SSR and AFLP markers, respectively) between individuals and low rates of inbreeding (Fis = 0.12, SSRs) were detected. Therefore, inbreeding levels for the proposed CSO could be considered as acceptable.

The absence of interspecific hybridization between *C. angustifolia* and *C. saltensis* or between *C. angustifolia* and *C. balansae*, described in this chapter and reported by Zelener et al. (2016), suggests that all individuals of *C. angustifolia* breeding population are genetically pure; therefore, no CSO purification is required. Accordingly, a genome analysis of 44 superior individuals from breeding population was performed using 140 polymorphic AFLP markers. The genetic diversity of *C. angustifolia* CSO was moderate (He = 0.204) and greater (more than double) than the average genetic diversity of natural populations of the species in the Yungas of NWA (He = 0,087, Inza et al. 2012). Eleven trees with exclusive *loci* at individual level were identified. Genetic structure inferred by Bayesian clustering (Structure 2.2.3, Pritchard et al. 2000) was mainly associated with latitudinal origin of the trees, following the patterns described for the species in previous sections. The UPGMA dendrogram (NTSYS 2.0, Rohlf 1998) showed a homogeneous general distribution of genetic relationships among trees (r = 0.84) with some trees of northern Yungas more distant. The average similarity index (0.62) was considered adequate to avoid inbreeding depression processes. Finally, AFLP markers allowed the unmistakable identification of the total number of evaluated individuals.

13.2.5 Genetic Variability of Cedrela fissilis Populations in the Alto Paraná Rainforest

Cedrela fissilis in Argentina, was subjected to intense and selective overexploitation (best phenotypes were harvested), resulting its current natural populations in fragmented stands of small size lacking reproductive adult individuals with desirable economic characteristics. Despite the high value of *C. fissilis*, information on the genetic diversity of this species is scarce. In order to assess the current levels of genetic diversity and its distribution throughout its natural range in Argentina, we performed a genomic analysis using SSR markers (Soldati et al. 2014b). This knowledge will be used to sustain conservation strategies for the remnant populations and to identify future sources of genetic material for breeding programs.

We collected 107 individuals belonging to 8 natural populations of the species in the Alto Paraná Rainforest of NEA, covering the natural distribution area of *C. fissilis* in the country. Total genomic DNA was extracted from dried leaf material following the procedure described by Hoisington et al. (1994) with minor modifications. Concentration and integrity of DNA was determined by comparison with reference standards in agarose gels. For this species, the analysis was performed with ten validated SSR markers: (i) seven SSRs transferred from *C. odorata*; (ii) one SSR transferred from *Swietenia macrophylla*; and (iii) two SSRs developed for *C. fissilis* (Lemes et al. 2002; Hernandez et al. 2008; Gandara 2009; Soldati et al. 2014a). These were amplified by PCR according to optimized protocols for other species of the genus (Soldati et al. 2014b).

One hundred ninety-six alleles were detected, with an average number of 24.5 alleles for the 8 populations assessed. Forty-nine exclusive alleles (Ea) were found (Table 13.5). The effective number of alleles (ENa) was variable in a range of 3.713 to 6.476, with an average value of 5.774 (SD = 0.726). Observed heterozygosity (Ho) values ranged from 0.744 to 0.867 (mean = 0.820, SD = 0.016). Expected heterozygosity (He – genetic diversity of Nei) levels were high, with an average

Table 13.5 Genetic diversity parameters obtained by SSR markers for *C. fissilis* populations distributed in the Alto Paraná Rainforest in Northeastern Argentina. Locality, population code, *Ea* exclusive alleles, *Ho* observed heterozygosity, *He* expected heterozygosity, *SD* standard deviation

		SSRs		
Locality	Code	Ea	Ho	He
Puerto Bossetti	CfFB	9	0.828	0.841
INTA San Antonio	CfSA	4	0.833	0.818
El Alcázar	CfFA	11	0.834	0.830
Campo Guaraní	CfCG	4	0.842	0.824
Eldorado	CfFCF	7	0.867	0.871
Oberá	CfFO	6	0.831	0.834
Villa Olivari	CfVO	4	0.781	0.745
Las Marías	CfLM	4	0.744	0.818
Mean		6.125	0.820	0.822
SD[d]		2.695	0.016	0.012

value of 0.822 (Table 13.5). A difference among He values could be observed when comparing populations of the two different provinces (Misiones and Corrientes; see Fig. 13.1). Population genetic differentiation was estimated using θp statistics. Historical gene flow (Nm) was calculated indirectly through this statistics, according to Crow and Aoki (1984). Bayesian clustering approach was implemented to resolve the optimal number of genetic clusters (*K*) and the proportion of population assignment by the identified clusters. A no-admixture model was used in order to detect subtle structure (Structure 2.3.3 software, Pritchard et al. 2000; Falush et al. 2007).

An AMOVA (GenAlEx 6.2, Peakall and Smouse 2006) analysis showed a moderate and highly significant population genetic differentiation, with a θp value of 0.06 (P ≤ 0.001). This pattern of diversity distribution is expected for long-lived species, predominantly cross-pollinated, perennial, and woody plants. The Bayesian analysis allowed us to identify four genetic clusters (Fig. 13.5), heterogeneously distributed among populations; one of these clusters was predominant (~93%) in Villa Olivari population, located in the southwest of the distribution of the species (Province of Corrientes). This result is consistent with the lower genetic diversity levels observed in this population in comparison with the remaining ones. Considerable historical gene flow was detected among *C. fissilis* populations, with an Nm value of 3.004, which is consistent with a number of traits that favor cross-pollination and long-distance gene flow (Carvalho 1994).

To visualize any grouping patterns among populations, population genetic distances (Nei 1972) and its clustering (UPGMA) were determined using GDA software v1.1 (Lewis and Zaykin 2001). No grouping by geographic location was observed. However, those populations located in the southern portion of the range exhibited higher genetic distances when compared with those located in the northeast. These results are consistent, particularly for Villa Olivari population (CfVO in Fig. 13.5), because of the contrasting distribution of genetic clusters found in this

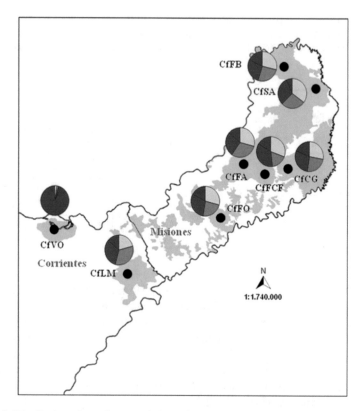

Fig. 13.5 Distribution of genetic groups inferred from Bayesian methods, in eight natural populations of *C. fissilis* in the Alto Paraná Rainforest of Northeastern Argentina. Pie charts represent
distribution of genetics groups in each population. Labels near pie charts indicate populations
(codes are as in Table 13.5). Black dots indicate the exact location of populations

population compared to the other populations, showing a clear separation of this
population (data shown in Soldati et al. 2014b).

The results (Soldati et al. 2014b) indicate that the variation in genetic variability
that was observed among populations from northwest and southeast (Provinces of
Corrientes and Misiones) could be associated with the phytogeographic districts in
which populations are located: Mixed Forests District and Open Woodlands District
(Cabrera 1976). Further research is necessary to better understand this pattern; fortunately, new analyses using AFLP markers are currently being carried out.

13.3 Contributions from Research to Conservation and Use of *Cedrela* Genetic Resources in Argentina

The studies we report here help to understand the current state of the genetic
resources of *Cedrela* genus in Argentina and thus to address suitable strategies for
conservation, to guide germplasm collection, and to select proper germplasm for

breeding programs. Definitions of management will mainly depend on the different gene pool of each species, the population accessibility in private lands, and their protection status under protected area systems or the absence of such condition.

Cedrela angustifolia still maintains important populations in many streams, in places with low accessibility. In addition, some populations are preserved within national or provincial natural protected areas, such as Calilegua National Park, Baritú National Park, El Nogalar National Reserve, and La Florida Provincial Park, respectively. The researches described in this chapter allowed us to propose priority areas to be preserved for safeguard the remnant gene pool of *C. angustifolia* (Inza et al. 2012) and also to provide a guide for appropriate sources of propagation material compatible with domestication and breeding purposes. Populations of *C. angustifolia* were assigned to three different priority areas for conservation, according to its contribution to the total genetic diversity and uniqueness as suggested by Petit et al. (1998). Baritú and La Florida populations were considered as *maximum-priority* populations because they have the highest diversity and exclusive genetic variants not present in the remaining Argentine populations. Fortunately, Baritú is a national park nowadays, but La Florida as a provincial park has a lower conservation status. As first conclusion, the preservation status of La Florida should be increased, and the use of both Baritú and La Florida protected areas should be restricted to research purposes. Second, we assigned northern populations of the Yungas, located in UBRB, to a *high-priority* conservation area due to their high levels of genetic diversity and differentiation. Thus, it is advisable to promote the protection status of El Nogalar National Reserve (including educational and tourism purposes), as well as to turn to sustainable forest management those forests located on private lands like San Andrés locality.

Maximum and high-priority areas are in agreement with the two areas proposed by Brown et al. (2006) for biodiversity conservation of the Yungas, and the efforts of species conservation should focus here. Eight remaining populations of central and southern Yungas were considered as belonging to a *medium-priority* area of conservation because of their moderate to low genetic diversity. However, they should have proper management considering their genetic differentiation (between them and with northern populations) and their critical locations. Los Pizarros is located on the southern edge of the natural range of *C. angustifolia*, and Metán population can act as stepping stone for gene flow between central and southern Yungas.

the increase of the protection status of La Florida Provincial Park and El Nogalar National Reserve, as well as, the use restriction of Baritú National Park only for research should be considered. Nevertheless, conservation only with protected areas is not enough. Thus, a sustainable forest management on private lands is required to preserve remaining *C. angustifolia* gene pool, which ensures evolutionary processes for species adaptation to changing environmental conditions and social requirements for conservation and domestication purposes.

Cedrela balansae exhibits a more critical situation than *C. angustifolia*. The species occurs in the lower altitudinal stratum of the Yungas, which facilitates the accessibility to its populations. Moreover, most of the remaining populations are located on private lands at marginal locations of protected areas; only one of the eight populations here described and reported by Soldati et al. (2013) is protected within the Calilegua National Park. Accordingly, INTA recently initiated activities for the *ex situ* conservation of these populations, in clonal banks and clonal seed orchards. This strategy includes populations with low genetic diversity, like Apolinario Saravia, whose value lies in its geographical location at the extreme southeastern range of the species.

Gene flow that likely occurred within *C. balansae* and *C. saltensis* contact zones suggests that barriers to the gene exchange remained weak for an extended period in the Yungas of NWA. In addition, climate changes affect the colonization and evolutionary dynamics of the species (e.g., Aitken et al. 2008; Hoffmann and Agro 2011; Koecke et al. 2013; Köcke et al. 2015). Genetic field trials conducted in the Yungas provided evidence of better adaptive behavior (survival and height in the third year) in hybrids between *C. balansae* and *C. saltensis* (e.g., population of Pintascayo) than in provenances of genetically pure populations of both species in transition zones between them (Grignola, personal communication). This evidence suggests that combination of introgression and selection may further contribute to novel allelic associations enhancing adaptation of introgressed forms. Thus, habitat forest degradation assembled with effects of climate changes may be potent evolutionary forces that contribute to structuring hybrid zones in the Yungas.

The genetic identity of *Cedrela* germplasm has implications for restoration planning and provides valuable baseline information for commercial forestry-related activities; genetic improvement programs are currently under development in order to supply the demand for high-quality genetic seed. In addition, most of the geographical area we sampled belongs to private land, where forestry productive activities should be developed according to the Law No. 26,331. The domestic market of forest products applies differential pricing to different *Cedrela* wood, punishing the round logs that do not adjust to the expected timber features. The molecular characterization of the populations can ensure the genetic provenance of the wood offered to the forestry sector.

References

Aide TM, Grau HR (2004) Globalization, migration, and Latin American ecosystems. Science 305:1915–1916

Aitken SN, Yeaman S, Holliday JA, Wang T, Curtis-McLane S (2008) Adaptation, migration or extirpation: climate change outcomes for tree populations. Evol Appl 1:95–111

Aizen MA, Feinsinger P (2002) Bees not to be? Responses of insect pollinator faunas and lower pollination to habitat fragmentation. Ecol Stud 162:111–129

Allendorf FW, Leary RF, Spruell P, Wenburg JK (2001) The problems with hybrids: setting conservation guidelines. Trends Ecol Evol 16:613–622

André T, Lemes MR, Grogan J, Gribel R (2008) Postlogging loss of genetic diversity in a mahogany (*Swietenia macrophylla* King, Meliaceae) population in Brazilian Amazonia. For Ecol Manage 255:340–345

Aschero V (2006) Biología reproductiva e importancia de la polinización en *Cedrela lilloi*. In: Pacheco S, Brown A (eds) Ecología y Producción de cedro (género *Cedrela*) en las Yungas australes. LIEY-ProYungas, Tucumán, pp 41–50

Barstow M (2018) *Cedrela fissilis*. The IUCN red list of threatened species 2018: e.T33928A68080477. https://doi.org/10.2305/IUCN.UK.2018-1.RLTS.T33928A68080477.en

Bawa KS, Bullock SH, Perry DR, Coville RE, Grayum MH (1985) Reproductive biology of tropical lowland rain forest trees. II. Pollination systems. Am J Bot 72:346–356

Brown AD, Malizia LR (2004) Las Selvas Pedemontanas de las Yungas: en el umbral de la extinción. Ciencia Hoy 14:52–63

Brown AD, Pacheco S (2006) Importancia del género *Cedrela* en la conservación y desarrollo sustentable de las Yungas australes. In: Pacheco S, Brown A (eds) Ecología y Producción de cedro (género *Cedrela*) en las Yungas australes. LIEY-ProYungas, Tucumán, pp 9–18

Brown AD, Grau HR, Malizia LR, Grau A (2001) Argentina. In: Kappelle M, Brown AD (eds) Bosques Nublados del Neotrópico. Instituto Nacional de Biodiversidad, Santo Domingo de Heredia, pp 623–659

Brown AD, Pacheco S, Lomáscolo T, Malizia L (2006) La situación ambiental en los bosques andinos yungueños. In: Brown AD, Martínez Ortiz U, Acerbi M, Corchera J (eds) La situación ambiental argentina 2005. Fundación Vida Silvestre Argentina, Buenos Aires, pp 52–71

Cabrera AL (1976) Regiones fitogeográficas argentinas. In: Kugler WF (ed.). Enciclopedia Argentina de Agricultura y Jardinería. Tomo 2. 2da edición. Acme, Buenos Aires, Argentina. Fascículo 1, pp 1–85

Carney SE, Wolf DE, Rieseberg LH (2000) Hybridization and forest conservation. In: Boyle TJB, Young A, Boshier D (eds) Forest conservation genetics: principles and practice. CSIRO, CABI, Collingwood, pp 167–182

Carvalho PER (1994) Espécies Florestais Brasileiras. Recomendações Silviculturais, Potencialidades e Uso da Madeira. EMBRAPA-CNPF, Brasília

Coleman AW (2009) Is there a molecular key to the level of "biological species" in eukaryotes? A DNA guide. Mol Phylogenet Evol 50:197–203

Crow JF, Aoki K (1984) Group selection for a polygenic behavioral trait: estimating the degree of population subdivision. Proc Natl Acad Sci U S A 81:6073–6077

Degen B, Blanc L, Caron H, Maggia L, Kremer A, Gourlet-Fleury S (2006) Impact of selective logging on genetic composition and demographic structure of four tropical tree species. Biol Conserv 131:386–401

Donoso C, Premoli A, Gallo L, Ipinza R (2004) Variación intraespecífica en especies arbóreas de los bosques templados de Chile y Argentina. Editorial Universitaria S.A, Santiago de Chile

Evanno G, Regnaut S, Goudet J (2005) Detecting the number of clusters of individuals using the software STRUCTURE: a simulation study. Mol Ecol 14:2611–2620

Falush D, Stephens M, Pritchard JK (2007) Inference of population structure using multilocus genotype data: dominant markers and null alleles. Mol Ecol Notes 7:574–578

FAO (2016) El estado de los Bosques del Mundo. http://www.fao.org/3/a-i5850s.pdf

Fornes L, Zelener N, Gauchat M, Inza MV, Soldati MC, Ruíz V, et al (2016) Subprograma *Cedrela*. In: Domesticación y Mejoramiento de Especies Forestales. PROMEF. UCAR. pp 136–159 http://forestoindustria.magyp.gob.ar/archivos/biblioteca-forestal/domesticacion-y-mejoramiento-de-especies-forestales.pdf

Gandara FB (2009) Diversidade genética de populações de Cedro (*Cedrela fissilis* Vell. Meliaceae) no Centro-Sul do Brasil. PhD. thesis. Escuela Superior de Agricultura, Universidad de San Pablo

Grau HR (2000) Regeneration patterns of *Cedrela lilloi* (Meliaceae) in northwestern Argentina subtropical montane forests. J Trop Ecol 16:227–242

Grau HR, Easdale TA, Paolini L (2003) Subtropical dendroecology – dates disturbances and forest dynamics in northwestern Argentina montane ecosystems. For Ecol Manage 177:131–143

Grau A, Zapater MA, Neumann RA (2006) Botánica y distribución del género *Cedrela* en el noroeste de Argentina. In: Pacheco S, Brown A (eds) Ecología y Producción de cedro (género *Cedrela*) en las Yungas australes. LIEY-ProYungas, Tucumán, pp 19–30

Hamrick JL, Godt MJW, Sherman-Broyles SL (1992) Factors influencing levels of genetic diversity in woody plant species. New For 6:95–124

Hernández LG, Buonamici A, Walker K, Vendramin GG, Navarro C, Cavers S (2008) Isolation and characterization of microsatellite markers for *Cedrela odorata* L. (Meliaceae), a high value neotropical tree. Conserv Genet 9:457–459

Hoffmann AA, Agro CM (2011) Climate change and evolutionary adaptation. Nature 470:479–485

Hoisington DA, Khairallah MM, Gonzales de Leon D (1994) Laboratory protocols. CIMMYT applied molecular genetics laboratory. CIMMYT, Hisfoa

InfoStat (2008) Grupo InfoStat, FCA, Universidad Nacional de Córdoba, Argentina

Inza MV, Zelener N, Fornes L, Gallo LA (2012) Effect of latitudinal gradient and impact of logging on genetic diversity of *Cedrela lilloi* along the Argentine Yungas rainforest. Ecol Evol 2:2722–2736

Inza MV, Aguirre NC, Torales SL, Pahr NM, Fassola HE, Fornes LF, Zelener N (2018) Genetic variability of *Araucaria angustifolia* in the Argentinean Parana Forest and implications for management and conservation. Trees 32:1135–1146

Ipinza Carmona R, Vergara Lagos R (1998) Diseños de Huertos semilleros. In: Ipinza Carmona R, Gutiérrez B, Emhart V (eds) Mejora Genética Forestal Operativa, pp.129–151

Juárez A, Ortega-Baes P, Sühring S, Martin W, Galíndez G (2007) Spatial patterns of dicot diversity in Argentina. Biodivers Conserv 16:1669–1677

Kageyama PY, Sebbenn AM, Ribas LA, Gandara FB, Castellen M, Perecim MB, Vencovsky R (2003) Diversidade genética em espécies arbóreas tropicais de diferentes estágios sucessionais por marcadores genéticos. Scientia Forestalis 64:93–107

Kageyama P, Caron D, Gandara F, do Santos JD (2004) Conservation of Mata Atlántica forest fragments in the state of São Paulo, Brazil. In: Vicenti B, Amaral W, Meilleur B (eds) Challenges in managing forest genetic resource for livelihoods: examples from Argentina and Brazil. IPGRI, Rome

Köcke AV, Muellner-Riehl AN, Cáceres O, Pennington TD (2015) *Cedrela ngobe* (Meliaceae), a new species from Panama and Costa Rica. Edinburgh J Bot 72:225–233

Koecke AV, Muellner-Riehl AN, Pennington TD, Schorr G, Schnitzler J (2013) Niche evolution through time and across continents: the story of Neotropical *Cedrela* (Meliaceae). Am J Bot 100:1800–1810

Lemes MR, Brondani RP, Grattapaglia D (2002) Multiplexed systems of microsatellite markers for genetic analysis of mahogany, *Swietenia macrophylla* king (Meliaceae), a threatened neotropical timber species. J Hered 93:287–291

Lepais O, Petit RJ, Guichoux E, Lavabre JE, Alberto F, Kremer A, Gerber S (2009) Species relative abundance and direction of introgression in oaks. Mol Ecol 18:2228–2242

Lewis PO, Zaykin D (2001) GDA 1.1, genetic data analysis: computer program for the analysis of allelic data

Malizia LR, Blundo C, Pacheco S (2006) Diversidad, estructura y distribución de bosques con cedro en el noroeste de Argentina y sur de Bolivia. In: Pacheco S, Brown A (eds) Ecología y Producción de cedro (género *Cedrela*) en las Yungas australes. LIEY-ProYungas, Tucumán, pp 83–104

Marcucci Poltri SN, Zelener N, Rodriguez Traverso J, Gelid P, Hopp HE (2003) Selection of a seed orchard of *Eucalyptus dunnii* based on genetic diversity criteria calculated using molecular markers. Tree Physiol 23:625–632

Martin PR, McKay JK (2004) Latitudinal variation in genetic divergence of populations and the potential for future speciation. Evolution 58:938–945

Minetti JM (2006) Aprovechamiento forestal de cedro en las Yungas de Argentina. In: Pacheco S, Brown A (eds) Ecología y Producción de cedro (género *Cedrela*) en las Yungas australes. LIEY-ProYungas, Tucumán, pp 143–154

Mostacedo BC, Fredericksen TS (1999) Regeneration status of important tropical forest tree species in Bolivia: assessment and recommendations. For Ecol Manage 124:263–273

Muellner AN, Pennington TD, Chase MW (2009) Molecular phylogenetics of Neotropical Cedreleae (mahogany family, Meliaceae) based on nuclear and plastid DNA sequences reveal multiple origins of "*Cedrela odorata*". Mol Phylogenet Evol 52:461–469

Nei M (1972) Genetic distance between populations. Am Nat 106:283–292

Nieto Feliner G, Rosselló JA (2007) Better the devil you know? Guidelines for insightful utilization of nrDNAITS in species-level evolutionary studies in plants. Mol Phylogenet Evol 44:911–919

Nybom H (2004) Comparison of different nuclear DNA markers for estimating intraspecific genetic diversity in plants. Mol Ecol 13:1143–1155

Peakall RP, Smouse E (2006) GenAlEx6: genetic analysis in Excel. Population genetics software for teaching and research. Mol Ecol Notes 6:288–295

Pennington TD, Muellner AN (2010) A monograph of *Cedrela* (Meliaceae). Milborne Port: DH Books, Sherborne

Pennington RT, Prado DE, Pendry CA (2000) Neotropical seasonally dry forest and Quaternary vegetation changes. J Biogeogr 27:261–273

Petit RJ, El Mousadik A, Pons O (1998) Identifying populations for conservation on the basis of genetic markers. Conserv Biol 12:844–855

Placci G (2000) El desmonte en Misiones: impactos y medidas de mitigación. In: Bertonatti C, Corcuera J (eds) Situación Ambiental Argentina 2000. Fundación Vida Silvestre Argentina, Buenos Aires, pp 349–354

Placci G, Di Bitetti M (2006) Environmental situation in the ecoregion of Atlantic Forest of Alto Paraná (Atlantic Forest). [In Spanish.]. In: Brown A, Ortiz UM, Acerbi M, Corcuer J (eds) *The environmental situation Argentina*. Fundación Vida Silvestre Argentina, Buenos Aires, pp 195–209

Premoli AC, Souto CP, Trujillo SA, del Castillo RF, Quiroga P, Kitzberger T et al (2011) Impact of forest fragmentation and degradation on patterns of genetic variation and its implication for forest restoration. In: Newton AC, Tejedor N (eds) Principles and practice of forest landscape restoration: case studies from the drylands of Latin America. IUCN, Gland, pp 205–228

Pritchard JK, Stephens M, Donnelly P (2000) Inference of population structure using multilocus genotype data. Genetics 155:945–959

Rajora OP, Mosseler A (2001) Challenges and opportunities for conservation of forest genetic resources. Euphytica 118:197–212

Rau MF (2005) Land use change and natural Araucaria forest degradation Northeastern Misiones, Argentina. Doctoral thesis, Faculty of Forestry, Albert-Ludwigs-University Freiburg in Breisgau, Germany

Rivera SM (2006) Características y usos de las maderas de *Cedrela* de las Yungas de Argentina. In: Pacheco S, Brown A (eds) Ecología y Producción de cedro (género *Cedrela*) en las Yungas australes. LIEY-ProYungas, Tucumán, pp 51–58

Rohlf FJ (1998) NTSYS-PC numerical taxonomy and multivariate analysis system version 2.0. Exeter Software, Setauket

SAyDS (2007) Primer Inventario Nacional de Bosques Nativos: Informe Regional Selva Tucumano-Boliviana, first ed. Secretaría de Ambiente y Desarrollo Sustentable de la Nación, Bs. As., Argentina

SAyDS (2008) El avance de la frontera agropecuaria y sus consecuencias. Secretaría de Ambiente y Desarrollo Sustentable. Subsecretaría de Planificación y Política Ambiental, Dirección Nacional de Ordenamiento Ambiental y Conservación de la Biodiversidad, Bs. As., Argentina

Seghezzo L, Volante JN, Paruelo JM, Somma DJ, Buliubasich EC, Rodríguez HE et al (2011) Native forests and agriculture in Salta (Argentina): conflicting visions of development. J Environ Dev 20:251–277

Soldati MC, Fornes L, van Zonneveld M, Thomas E, Zelener N (2013) An assessment of the genetic diversity of *Cedrela balansae* (Meliaceae) in Northwestern Argentina by means of combined use of SSR and AFLP molecular markers. Biochem Syst Ecol 47:45–55

Soldati MC, Inza MV, Fornes L, Zelener N (2014a) Cross transferability of SSR markers to endangered *Cedrela* species that grow in Argentinean subtropical forests, as a valuable tool for population genetic studies. Biochem Syst Ecol 53:8–16

Soldati MC, Fornes L, Barth S, Eskiviski E, Zelener N (2014b) Diversidad genética en poblaciones remanentes de *Cedrela fissilis* Vell. en la Selva Paranaense. In: Libro de Trabajos Técnicos del Grupo GEMFO, IV Reunión, pp 107–110. ISBN: 978987-521-4842

Soldati MC, Alonso FM, Fornes L, Zelener N (2015) Estrategias de selección para el diseño de Huertos Semilleros Clonales (HSC): un caso de estudio en *Cedrela balansae* CD.C. In: Proceedings of Congreso Nacional de Viveros Cítricos, Forestales y Ornamentales. Misiones. Argentina. 4–6 August. https://redforestal.conicet.gov.ar/download/congresos/DISERTACIONES.pdf

van Rheenen HMPJB, Boot RGA, Werger MJA, Ulloa M (2004) Regeneration of timber trees in a logged tropical forest in North Bolivia. For Ecol Manage 200:39–48

van Zonneveld M, Scheldeman X, Escribano P, Viruel MA, Van Damme P, Garcia W et al (2012) Mapping genetic diversity of cherimoya (*Annona cherimola* Mill.): application of spatial analysis for conservation and use of plant genetic resources. PLoS One 7:e29845

Varghese M, Nicodemus A, Nagarajan B, Siddappa KRS, Bennet SSR, Subramanian K (2000) Seedling seed orchards for breeding tropical trees. Scroll press, Coimbatore

Villalba R, Colmes RL, Boninsegna JA (1992) Spatial patterns of climate and tree growth variations in subtropical northwestern Argentina. J Biogeogr 19:631–649

Ward M, Dick CW, Gribel R, Lowe AJ (2005) To self, or not self. A review of outcrossing and pollen-mediated gene flow in neotropical trees. Heredity 95:246–254

White GM, Boshier DH, Powell W (1999) Genetic variation within a fragmented population of *Swietenia humilis* Zucc. Mol Ecol 8:1899–1909

Young AG, Boyle TJ (2000) Forest fragmentation. In: Young AG, Boshier D, Boyle TJ (eds) Forest conservation genetics: principles and practice. CSIRO, CABI, Collingwood, pp 123–134

Zamora Petri M (2006) Influencia de la ganadería trashumante y la apertura de claros en la supervivencia y el crecimiento de *Cedrela lilloi* en Tariquía, Bolivia. In: Pacheco S, Brown AD (eds) Ecología y Producción de cedro (género *Cedrela*) en las Yungas australes. LIEY-ProYungas, Tucumán, pp 131–142

Zapater MA, del Castillo EM, Pennington TD (2004) El Género *Cedrela* (Meliaceae) en la Argentina. Darwin 42:347–356

Zelener N, Marcucci Poltri SN, Bartoloni N, López CR, Hopp HE (2005) Selection strategy for a seedling seed orchard design based on both, trait selection index and genomic analysis by molecular markers: a case study for *Eucalyptus dunnii*. Tree Physiol 25:1457–1467

Zelener N, Tosto D, de Oliveira LO, Soldati MC, Inza MV, Fornes LF (2016) Molecular evidence of hybrid zones of *Cedrela* (Meliaceae) in the Yungas of Northwestern Argentina. Mol Phylogenet Evol 102:45–55

Chapter 14
Breeding Strategy for the *Cedrela* Genus in Argentina

Josefina Grignola, Ezequiel Balducci, Adrián Trápani, Pablo Saravia, Mario Hernán Feyling Montero, Miguel Gatto, Roberto Cuello, Gonzalo Antonio Perez, Liliana Ríos de Gonzalez, Verónica Eugenia Ruiz, Diego Meloni, Julio Victor Saez, and Mario J. Pastorino

14.1 Domestication of *Cedrela* sp. in Argentina

Trees of the genus *Cedrela* are among the native forest species of Argentina with the highest wood price in the domestic market. At the end of 2018, average prices of round wood on trucks in the forest were 240 U\$S/m^3 for *Cedrela balansae* and

J. Grignola (✉) · P. Saravia · M. Gatto · G. A. Perez · L. Ríos de Gonzalez · J. V. Saez
INTA EEA Famaillá, Famaillá, Argentina
e-mail: grignola.josefina@inta.gob.ar

E. Balducci
INTA EECT Yuto, Yuto, Argentina

A. Trápani
Fac. Agronomía y Zootecnia, Univ. Nac. Tucumán, Tucumán, Argentina

M. H. F. Montero
Dir. Nac. Desarrollo Foresto Industrial – MAGyP, Buenos Aires, Argentina

R. Cuello
Dir. Flora, Fauna Silvestre y Suelos de Tucumán – SAAyA, Tucumán, Argentina

V. E. Ruiz
ICiAgro Litoral (CONICET – Univ. Nac. Litoral), Santa Fe, Argentina

D. Meloni
Univ. Nac. Santiago del Estero, Santiago del Estero, Argentina

M. J. Pastorino
Instituto de Investigaciones Forestales y Agropecuarias Bariloche (IFAB) INTA-CONICET, Bariloche, Argentina

© Springer Nature Switzerland AG 2021
M. J. Pastorino, P. Marchelli (eds.), *Low Intensity Breeding of Native Forest Trees in Argentina*, https://doi.org/10.1007/978-3-030-56462-9_14

270 U$S/m^3 for *C. saltensis* in the Yungas.[1] Likewise, for *C. fissilis*, in the Alto Paraná Rainforest, average prices of 61 U$S/m^3 have been reported.[2] The wood of this genus is known as Spanish Cedar in the world market, originally referring to *Cedrela odorata*, a species exploited since colonial periods in Central America. According to the International Tropical Timber Organization (ITTO), the price of Spanish Cedar wood dried in a kiln and placed in the Port of Callao, Peru, is over 950 U$S/m^3.

Noteworthy, so far, all *Cedrela* wood sold in Argentina comes from natural forests. Its domestic market is concentrated in the large urban centers of Buenos Aires, Córdoba, and Santa Fe, where it is mainly processed into carpentries for the production of furniture and openings. It is also used in uprights, crossbars, wooden or parquet floors, and the manufacture of various domestic utensils. It is highly appreciated for its excellent aesthetics and workability. It is straight-grained with a medium texture and a subdued luster. When first cut, the heartwood is pale creamy, turning to pinkish brown upon exposure to light and air. The density varies according to the species from lightweight (*C. balansae* 450 kg/m^3) to moderately heavy (*C. fissilis* 550 kg/m^3) (Tortorelli 2009). It also presents a pleasant perfume and tolerance to insect attack once cut (Grau et al. 2006), as well as a good durability. This is endorsed by many still standing colonial churches in Central America and the Andes, whose doors, ceilings, and indoor decoration are made of *Cedrela* sp. wood (Pennington and Muellner 2010).

Due to its history, wood quality and price, together with the reduction of natural forests resources, the interest in the commercial cultivation of native *Cedrela* species has increased in Argentina during the last three decades. The experience of cultivation of *C. odorata* in block plantations in various tropical regions around the world, such as Central and South America, South Africa, and Southeast Asia, represents an outstanding example to imitate. Cozzo (1995) has reported that *C. odorata* plantations in Ecuador, with a density of 1125 trees/ha, reached 18 m in height and 24 cm in DBH on average at 6 years of age, and 25 m in height and 50 cm in DBH at 18 years, with a yield of 21 m^3/ha/year.

However, block plantations in the subtropical region exhibit failures due to the occurrence of frost. In the Yungas, the most sheltered sites of the piedmont region are used for short-cycle and high-yield agricultural crops and are scarcely available for forest cultivation. Thus, it would be useful to cultivate these species in natural forest enrichment systems, especially in forest degraded by overexploitation and extensive cattle grazing. In such sites, the remaining vegetation provides the seedlings of *Cedrela* sp. protection against eventual frosts during the first years, allowing their establishment.

Following this reasoning, the Instituto Nacional de Tecnología Agropecuaria (INTA) proposed, for both the Yungas and the Alto Paraná Rainforest, the cultivation of *Cedrela* native species in enrichment systems in areas of intermediate

---

[1] Personal Communication: National Direction of Industrial Forestry Development

[2] Personal Communication: Misiones College of Forest Engineers

conservation value ("yellow areas") according to Law No. 26,331. About 1.5 million hectares of "yellow areas" have been identified in the provinces of Jujuy, Tucumán, Salta, and Misiones with suitable conditions for *Cedrela* sp. cultivation under enrichment systems (Fornes 2012).

Two types of enrichment are proposed: (1) in clumps, without tree cover, and (2) in strips, with tree side cover (Fig. 14.1a and b). The first case involves forests implanted in clearings of 500 to 1000 m², previously made for crops production (later abandoned) or by fire. If the surface of the clearings exceeds 1000 m², it is difficult to achieve frost protection. In the second case, the strips are opened with machetes, axes, and chainsaws in the natural vegetation by removing the undergrowth and trees of species of low commercial value. They are about 5 m wide, with a length determined by the characteristics of the area, such as the slope, exposure, soil, and vegetation. One or two lines are planted in each strip according to their width, with a distance of 4–5 m between seedlings within the line. The natural seedlings and saplings of tree species of high commercial value are preserved. The distance between strips is 10–20 m in order to achieve a general density of about 200–250 trees/ha. At the time of plantation, the optimum size of the seedlings is from 0.80 to 1.50 m high, which is reached in the nursery 12–15 months after sowing. Planting should be done during the rainy season (November to March). The use of individual tree shelters is recommended to avoid damage from herbivores such as rabbits and hare. Fertilization is usually unnecessary in these kinds of environments. Weeding must be carried out two or three times a year to prevent the invasion of weeds in the strips (Del Castillo et al. 2004).

Fig. 14.1 Cedar enrichment in: (**a**) block plantation: 4-year-old trees of *C. fissilis* in Alpachiri, Tucumán; (**b**) strips: 11-year-old trees of *C. fissilis* in Potrero de las Tablas, Tucumán. (Photos: Adrián Trápani and Miguel Gatto, respectively)

According to the Chamber of Forest Industry of the Province of Tucumán,[3] the estimated cost for the production system in strips is 1142 U$S/ha, with the 52% of expenses concentrated in the first year and 25% corresponding to the task of strip opening. Of the total amount, 80% is commonly covered by the National State in accordance with Law No. 25,080 on investment in cultivated forests, which reimburses the money 1 or 2 years after planting, upon verification of the survival of 200 seedlings per hectare. There are also provincial laws that provides some indirect benefits, such as the Tucumán Law No. 7021, which exempts the payment of the tax on the gross income of these productive activities, the real estate tax, and the service of irrigation of forested lands, among others.

The technology for seedling production of Argentinian *Cedrela*'s species has been adjusted efficiently. The fruits are capsules and mature in the forest between June and August depending on the species. When they acquire a light-brown color, it is the adequate moment for collection, before their dehiscence. For *C. angustifolia*, it should be taken into consideration that it presents long periods without fruit production (Aschero 2006); therefore, seed supply can be a "bottleneck." To extract the seeds, the capsules are placed under the sun in a dry environment until they open (Lorenzi 1992). The seeds are winged and of an easy handling size, 1000 seeds weight 18–33 g for *C. balansae*, 19.5 g for *C. saltensis*, and 42.5 g for *C. fissilis*, and they have a germination rate of over 88% for the first two species and 59% for the third one (Del Castillo et al. 2002). The viability of the seeds decreases up to 50% from 1 year to the next, so it is convenient to use new seeds every year (Fornes et al. 2016). Using as substrate an inert mixture of perlite, coconut fiber, and blonde peat, with slow release fertilizer, seedlings of adequate size for planting are obtained in 3 months. However, it is convenient to raise them for 12–15 months in order to reach the field with larger and more vigorous plants. To do this, when they reach 5 cm in height and have adult leaves, the seedlings are placed in pots 20–25 cm high and 6–7 cm in diameter, filled with soil and forest litter in a ratio of 2:1 (Monteverde 2006). The most vigorous seedlings should be selected in the nursery to increase the survival in the field (Grignola 2014).

The first forest enrichment plantations in Argentina date from the early 1980s. Actually, in the Northwest of Argentina, there are about 1000 ha intervened with this productive system, mostly with species of the genus *Cedrela*, but also with other native and exotic species, such as *Jacaranda mimosifolia*, *Tipuana tipu*, *Enterolobium contortisiliquum*, and *Toona ciliata* var. *australis*. There have been reported 530 ha in Jujuy, 345 ha in Tucumán, and 74 ha in Salta enriched with *Cedrela* sp. specifically (Fornes et al. 2016). The good results obtained are evidence of the success of this production system. For example, in an experimental plantation of *C. balansae* located in Yuto (Province of Jujuy), the plants reached a DBH of 31 cm and a height of 14 m at 15 years of age. Likewise, in an experimental plantation of *C. angustifolia* in El Naranjo (Province of Tucumán), DBH of 36 cm and

---

[3] Personal Communication: Chamber of Forest Industry of the Province of Tucumán

heights of 19 m were obtained at the age of 31 years.[4] It is important to know at which age each *Cedrela* species form the heartwood, since the value of its wood resides in that portion of the trunk. It has been reported the early formation of heartwood in individuals of *C. angustifolia*, *C. balansae*, and *C. fissilis* in plantations, with values between 80 and 97% of the trunk at the ages between 12 and 30 years.

An alternative for small- and medium-sized producers is the development of agroforestry systems, where high-value timber species are combined with consociated crops that allow a profitability anticipated to the cutting cycle of the trees. Through the combination of agricultural and forestry production, the best functions of the production of forests and food crops can be achieved. There are environmental as well as socioeconomic advantages of integrated systems over agriculture and/or tree monocultures (Wiersum 1981). In Alto Paraná Rainforest, native forest species are combined with *Ilex paraguariensis*, *Carica papaya*, or *Manihot esculenta*. In Yungas, a combination of *Cedrela* sp. can be carried out with native fruit species such as *Solanum betaceum*, *Physalis* spp., and *Eugenia uniflora*, among others. These species also require half shade, especially in the juvenile stages. As they grow, openings in the canopy should be made, allowing more light to enter.

14.2   A Tiny Enemy Threatens the Cultivation of *Cedrela* sp.

The moth known as "Meliaceae borer," *Hypsipyla grandella* Zeller, (order Lepidoptera, family *Pyralidae*) is the main pest affecting cedar species, limiting their potential in commercial forestations. It is specific of *Cedrela* and *Swietenia* genus in the American continent. In Argentina, it affects all native cedar species. The incidence of this pest is verified in the native forest, but in plantations (either in blocks or enrichment strips), it becomes more notorious. It can also affect plants in nurseries.

The main damage is caused by the larva, which destroys the main and lateral terminal buds, drilling the apices and making tunnels in the young stems. In the affected plants, the loss of apical dominance and its subsequent sprouts result in numerous lateral branches and consequently malformed trees, which lose their value for the production of wood (Fig. 14.2).

The biology of *H. grandella* has been well studied in Argentina (Tapia 2012). It lays the eggs externally both on the leaves and on the stem of the host plant. Between 48 and 72 h later, the larvae are born and penetrate the axillary bud first and move by boring the terminal non-lignified stem, the axillary bud, or the tip of the branches (Briceño Vergara 1997). The feeding continues generally in the marrow, consuming the bark, phloem, and leaf. Inside the plant, the insect meets between 5 and 7 larval stages and pupa until the adult emerges, which has nocturnal habits and a wingspan

---

[4] Personal Communication: Adrián Trápani and Miguel Gatto, Programa de Domesticación y Mejoramiento de Especies Forestales Nativas e Introducidas para Usos de Alto Valor (PROMEF)

Fig. 14.2 (**a**) *Hypsipyla grandella* adult. (**b**) *H. grandella* larva within a *Cedrela* sp. stem. (Photos: Verónica Baca)

of 22–23 mm. The average life cycle is 5 weeks, depending on weather conditions and food availability.

Knowing the duration of the first generation of the year of *H. grandella* is of great importance to understand the population dynamics of this insect and implement strategies aimed at not allowing the exponential growth of the population over time. Baca et al. (2013) determined in the north of the Province of Salta the temporal distribution of *H. grandella* attacks in *C. balansae* saplings. They observed the first adult in September and the second in mid-November, although the largest adult capture was recorded during December and January. The first attacks were observed 8 weeks after the beginning of cedar foliation. The largest number of attacks was observed in mid-December, coinciding with the adult population peak. The first 10% of the observed attacks were concentrated in the term of 22 days (November–December), while the remaining 90% was concentrated within 71 days (December–February). These observations are important because there is a period of 20–30 days since the presence of the first adults of *H. grandella* and the first damages.

On the other hand, Baca et al. (2013) determined that a generation of *H. grandella* in field conditions lasted 44 ± 8 days. These results agree with the information obtained in laboratory conditions, where the species completed the total life cycle in 4–7 weeks (average 5 weeks), depending on the conditions and availability of food. The knowledge of life cycle duration allows determining the appropriate control time (Tapia 2012).

There are numerous studies regarding chemical control, mainly in Central America. In Argentina, different evaluated insecticides have proven to be effective. Generally, the effective products are hazardous and require numerous applications. Tapia (2012) conducted a series of control trials with chemicals. The experiments were carried out in commercial plantations of *C. balansae*, *C. angustifolia*, *C. salt-ensis*, and *C. odorata* in Salta, Jujuy, and Tucumán, with biweekly and monthly application frequencies. The products evaluated were Imidacloprid (SC 35% at

10‰), Alphacypermethrin (SC 6% at 20‰), Deltamethrin mixture (EC 10% at 10‰) plus Methyl Ethyl Azinphos (SC 36% at 6‰), and Azadirachtin (EC1% at 5‰), an organic farming insecticide extracted from neem seeds (*Azadirachta indica*). They obtained near-total control with Alphacypermethrin and the mixture of Deltamethrin and Methyl Ethyl Azinphos (Tapia 2012). These results are consistent with those obtained under nursery conditions by Eskiviski et al. (2010). However, the World Health Organization restricted the use of organophosphorus compounds Methyl Ethyl Azinphos (highly dangerous, Class Ib according to WHO). In contrast, Alphacypermethrin (as well as Deltamethrin) is a low toxicity pyrethroid (WHO Class II), making it the recommended choice (Tapia 2012).

It is important to understand the insect population dynamics in order to decide the appropriate moments of action. Likewise, knowing its natural enemies could be relevant, since this will allow the exploration of research lines based on biological controllers, as part of a pest integrated management. In this regard, Baca et al. (2013) were able to isolate from soil samples a strain of *Beauveria* sp., an entomopathogenic fungus capable of parasitizing the insect. It presents a good rate of sporulation, being possible its multiplication at large scale on solid substrate. This fungus could be used as a biological control strategy against *H. grandella*. Similarly, a microhymenoptera of the *Chalcididae* family was found parasitizing the pupae of *H. grandella*. The emergence of this parasitoid was observed in pupae collected in the field in the north of the Province of Salta, which is the first report of its presence on *H. grandella* in Argentina. It was identified as *Brachymeria subconica* (Aquino et al. 2015). This opens the door to further research on its potential and effectiveness as a biological controller.

An alternative to reduce the effect of *H. grandella* attacks would be to explore plants genetic resistance. The effect of provenance was proved to be highly significant in *C. odorata* and *Swietenia macrophylla* for the susceptibility to *H. grandella*. The mean number of attacks per tree ranged from 0.8 to 2.4 and from 0.6 to 1.3 depending on the provenances, in both species, respectively (Newton et al. 1999), that is, some provenances had triple the attacks than others. Likewise, a different susceptibility was observed among cedar species in a field trial carried out in Yuto Experimental Station of INTA, where 100% of *C. fissilis* saplings were affected while only 38% and 28% of *C. saltensis* and *C. odorata* saplings, respectively, were attacked by *H. grandella* (Del Castillo and Tapia 2005).

So far, chemical treatments with Alphacypermethrin and Deltamethrin are the most effective and with less toxicity. They must be conducted during the first 3 years of planting, and involve 15% of the total costs of cultivating *Cedrela* spp. (Tapia 2012). Applications should be made from the first year of planting to obtain a trunk length free of deformation of 3.5–4 m. The appropriate moment to start this treatment is defined by the insect population dynamics, and according to the abovementioned studies, it is commonly recommended from the second half of December until February.

14.3 Variation in Quantitative Traits

First approaches to unravel the variation of the *Cedrela* genus in quantitative traits concentrated on the species level, particularly considering physiological traits. An assay of thermic stress tolerance was carried out with seedlings of *C. fissilis* and *C. saltensis* (Meloni and Martínez 2011). One-month seedlings were transferred to pots containing Hoagland nutrient solution under greenhouse with photoperiod adjusted (12 h) and controlled temperature conditions (10 °C or 25 °C). Modulated fluorescence emission measurements were made daily between 8:30 am and 10:30 am, and concentrations of quercitol (a cyclitol) and catechin (a phenol), two natural compounds with likely cryoprotective action on cell membranes, were determined (Orthen and Popp 2000).

The modulated fluorescence variables evaluated in this study had different behavior in both species. Temperatures of 10 °C produced a significant increase of non-photochemical quenching (NPQ) and decrease F_v/F_m ratio, but the effect resulted more marked in *C. fissilis* than in *C. saltensis* (Fig. 14.3). These results indicate that temperatures of 10 °C generated a decrease of the proportion of photosystem II active reaction centers and increased the loss of energy absorbed through dissipation in the form of heat.

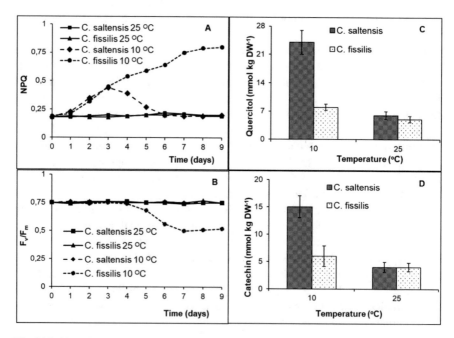

Fig. 14.3 Non-photochemical quenching (NPQ), Fv/ Fm ratio, and quercitol and catechin concentrations in *C. saltensis* and *C. fissilis* seedlings, grown at 10 °C and 25 °C. (From Orthen and Popp 2000)

The F_v/F_m ratio represents the maximum efficiency of the photosystem II; a low value of F_v/F_m indicates an inefficient use of the absorbed energy and photoinhibition (Murchie and Lawson 2013). The F_v/F_m ratio decreased in *C. fissilis* plants grown at 10 °C and remained with no changes in *C. saltensis* (Fig. 14.3). These results indicate that although *C. fissilis* increased energy dissipation in the form of heat, in the long term, this mechanism was not enough to compensate the deleterious effects of stress. Consequently, irreversible damage occurred in the photochemical stage of photosynthesis and photoinhibition.

From the results obtained on the modulated fluorescence variables, it can be inferred that *C. saltensis* was able to acclimate at low temperatures, whereas *C. fissilis* was not. This behavior agrees with the climatic features of the home sites of both species. It suggests that photosynthesis of *C. saltensis* from Yungas (Calilegua provenance), with minimum temperatures of 12 °C, is more tolerant to low temperatures than photosynthesis of *C. fissilis* from Alto Paraná Rainforest (San Antonio provenance), with minimum temperatures of 16 °C.

The leaves of *C. saltensis* grown at 10 °C showed higher concentrations of quercitol and catechin than those grown at 25 °C. Contrary to *C. saltensis*, in response to low temperatures, *C. fissilis* kept the concentrations of quercitol and catechin with no changes (Fig. 14.3).

It can be concluded that *C. saltensis*, with a higher altitudinal niche than *C. fissilis*, is more tolerant to low temperatures, since it has a stable photosynthesis and the ability to synthesize cryoprotectants in these environmental conditions.

Interspecific variation of water stress tolerance in the genus was also studied (Ruiz et al. 2013). It is well known that the seedling stage is the most critical for trees development (Garkoti et al. 2003), because its limited root system is more vulnerable to water shortage. The understanding of seedling adaptive physiological responses to drought is relevant to predict potential areas of cultivation. The aim of this work was to study the physiological response in greenhouse conditions of seedlings of *C. balansae*, *C. balansae* × *C. saltensis* hybrid (average annual rainfall below 1000 mm at their home sites), and *C. fissilis*, (average annual rainfall above 2000 mm) under different simulated water regimes. Two provenances of *C. balansae* (Río Seco and Yuto), one provenance of the hybrid (Pintascayo), and two provenances of *C. fissilis* (San Antonio and Guaraní) were submitted to four simulated rainfall treatments: 600 mm/year, 800 mm/year, 1000 mm/year, and 1200 mm/year. This factorial trial was installed in Famaillá Experimental Station of INTA (27° 3' S, 65° 25' W, 450 m asl) with a completely randomize design of 15 replications (N = 300). One seed per pot was sown in January; pots were 13 cm in diameter and 45 cm in deep and were filled with a local loamy soil. After sowing, the pots were transferred to a greenhouse in order to exclude the natural rainfall. The pots were maintained close to field capacity until the beginning of simulated rainfall treatments and were rotated regularly in their positions to avoid confounding effects of light and temperature gradients. With seedlings emerged, rainfall treatments were applied from March to December, and, from August onward, leaf relative water content (RWC) and water potential (Ψ_w) were measured. The RWC was calculated according to the following equation:

$$RWC = \left[\left(FW - DW \right) / \left(TW - DW \right) \right] \times 100.$$

where FW is fresh weight, DW is dry weight, and TW is turgid weight.

The Ψ_w was determined with a Schölander chamber at midday, and at the end of the trial, total biomass production and shoot height were measured.

Even though leaf RWC (Fig. 14.4A) showed no clear differentiation among provenances, it was possible to determine different water adjustment responses between species under the tested treatments. Significant differences in the Ψ_w were detected between the species from Yungas and that from Alto Paraná Rainforest (Fig. 14.4B). The provenances of *C. balansae* and *C. balansae* × *C. saltensis* hybrid were less susceptible to severe water deficit than *C. fissilis* provenances. Although interspecific differences were found in the physiological responses, it was not possible to separate the behaviors according to their provenances, indicating an intraspecific stability. It is important to highlight that *C. balansae* Río Seco provenance showed the best behavior under severe stress situation (600 mm).

Nevertheless, the biometric parameters total biomass production (dry weight) (Fig. 14.5a) and shoot height (Fig. 14.5b) showed significant differences among provenances or simulated rainfall treatments. The dry weight (Fig. 14.5a) of *C. fissilis* Guaraní was significantly affected under higher hydric deficit treatments (800

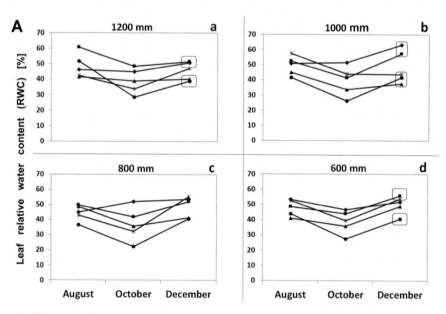

◆ *C. balansae* (Río Seco) - ■ *C. balansae* (Yuto) - ▲ *C. balansae* × *C. saltensis* (pintascayo)
● *C. fissilis* (San Antonio) - ✳ *C. fissilis* (Guaraní)

Fig. 14.4A Leaf relative water content of *Cedrela* seedlings growing under four simulated annual rainfall regimes: (**a**) 1200 mm/year, (**b**) 1000 mm/year, (**c**) 800 mm/year, and (**d**) 600 mm/year. Values are means of ten different measurements. (From Ruiz et al. 2013)

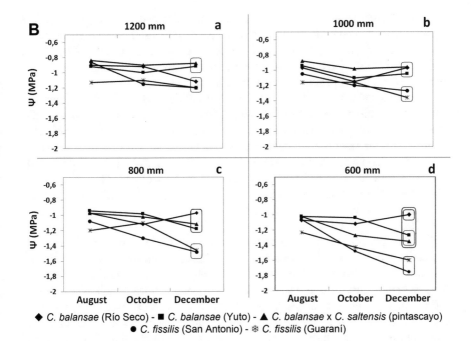

Fig. 14.4B Midday leaf water potential (Ψw) throughout the final phase of the experiment in *Cedrela* seedlings growing under four simulated annual rainfall regimes: (**a**) 1200 mm/year, (**b**) 1000 mm/year, (**c**) 800 mm/year, and (**d**) 600 mm/year. Values are means of ten different measurements. (From Ruiz et al. 2013)

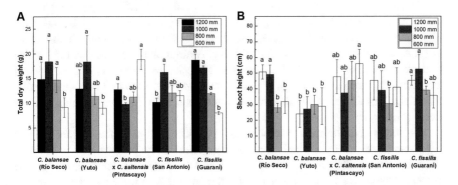

Fig. 14.5 (**a**) Total dry weight and (**b**) shoot height at the end of the experiment in *Cedrela* seedlings growing under four simulated annual rainfall regimes. Values are means of ten different measurements. Vertical bars represent ± standard error (≤ 0.05). (From Ruiz et al. 2013)

and 600 mm), while the provenance of *C. balansae* Río Seco was only significantly affected under 600 mm. An opposite behavior was observed in the provenance of *C. balansae* × *C. saltensis* hybrid (Pintascayo), with a significant increase of the dry weight when exposed to 600 mm, with respect to those under 1200 mm treatment.

The shoot height (Fig. 14.5b) of *C. balansae* Río Seco and *C. fissilis* Guaraní was significantly affected under 600 mm with respect to the 1200 mm treatment. A different behavior was observed in *C. balansae* × *C. saltensis* hybrid (Pintascayo), which was not affected by the water deficit, since it presented no significant differences between 600 mm and 1200 mm treatments.

The water balance was more efficient in seedlings from natural environments with lower rainfall regimes. Besides, the efficiency was also manifested in the water use, since the provenances from natural drier environments were able to grow more when the water regime treatment was lower. Differences on drought responses among tree ecotypes growing in tropical and subtropical rainforests may be attributed to genetic differences in physiological and morphological adaptive responses (Engelbrecht and Kursar 2003). Plant responses to water deficit, when they are outside their natural environments, depend on the rainfall regimes of their native habitats (Otieno et al. 2005).

In a subsequent trial, all Argentinian species of the genus were compared with respect to growth traits (Grignola 2014). It included five provenances of *C. balansae*, two of *C. fissilis*, two of the hybrids between *C. saltensis* and *C. balansae*, two of *C. saltensis*, and, since it was not possible to obtain enough seeds of *C. angustifolia* from the same provenance, a pool of several sources was used to represent this species. The trial was established on 100 cm^3 individual pots filled with inert substrate and slow-release fertilizer in the greenhouse of Famaillá Experimental Station (27° 3' S, 65° 25' W, 450 m asl), with seeds from trees selected in the natural forest according to growth, health, and shape criteria. The hybrid character of adult plants of *C. saltensis* × *C. balansae* was confirmed through genetic markers.

Seventy days after germination, the total height and collar diameter of 20 randomly chosen plants per treatment were measured (N = 240). Significant differences among species and among provenances were found for both variables, *C. fissilis* and *C. angustifolia* being the species with the tallest but thinnest seedlings. Among the provenances, *C. fissilis* San Antonio and Guaraní had the highest average heights of the trial (33.0 cm and 27.7 cm, respectively) and the lowest diameters (5.5 mm on average), while *C. balansae* Río Seco and Ledesma (undifferentiated) exhibited the highest mean diameters, with 7.8 and 8.0 mm, respectively (Fig. 14.6).

The status of carbohydrate reserves is a physiological parameter useful to infer vigor and health of a plant and might have an effect on its performance in the field (Birchler et al. 1998). In turn, the adaptation of a species to low temperatures can be evaluated by testing the resistance of its cell membranes to freezing. Such a test can be performed indirectly by subjecting its fresh leaves to freezing, then immersing them in deionized water, and finally measuring the electrical conductivity of the solution formed by the water and solutes released by the leaves. The measured conductivity is a function of the breakage of the cell membranes by freezing (Dexter et al. 1932; Rodríguez-Rey et al. 2000).

From the previous trial, seven seedlings per treatment were transplanted into 3000 cm^3 pots 70 days after germination and were grown for 1 year within a greenhouse. In autumn (temperatures between 10 °C and 15 °C), winter (& lt; 10 °C), and

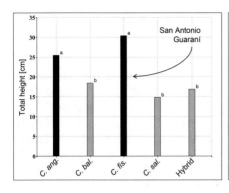 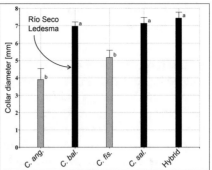

Fig. 14.6 Average values and standard errors of total height and collar diameter at 70 days of age of different *Cedrela* species. Different letters indicate significant differences (p > 0.05)

summer (~25 °C), leaves of these plants were sampled, and subsequently their content of simple soluble sugar (glucose + fructose) was measured. In addition, another sampling was done in autumn on four dates characterized by a temperature gradient of 5°, from 18.3 °C to 13.6 °C. In each sampling, six leaflets per plant were collected: three were subjected to −10 °C in a test tube for 1 h and then immersed in deionized water for 2 h at 25 °C. The other three leaflets were used for the control treatment. The electrical conductivity was measured in both solutions, thus obtaining the value of index of tissue damage (ITD) as the ratio between the conductivity of the control treatment and the conductivity of the −10 °C treatment (Grignola 2014).

Sugar concentration for the species and provenances studied varied according to the period of the year. For autumn-winter season, significant differences in sugar concentration were observed, but not in the summer season, differentiating *C. fissilis* from *C. balansae*, *C. angustifolia*, *C. saltensis*, and the hybrid. The highest value was recorded in Guaraní provenance of *C. fissilis* (22.5 ± 0.7 mg of glucose + fructose per gram of dry leaves) (Grignola 2014). Regarding the damage of the cell membranes due to freezing, significant differences among species were observed at the four sampling dates, with *C. fissilis* differing from *C. balansae*, *C. saltensis*, and the hybrid in three of the four sampling dates. The species from the Alto Paraná Rainforest had the most negative value of ln-ITD, which means that it was the one with the lowest solute release, namely, the one with the least damage to cell membranes. In addition, differences between the origins were demonstrated for the four sampling dates. For the two lowest temperature dates, the two *C. fissilis* provenances were different from the other ten tested. These results, besides differentiating *C. fissilis* from the Yungas species, would be indicative of a lower sensitivity to cold conditions for the species of the Alto Paraná Rainforest, at least at the seedling stage (Grignola 2014).

14.4   First Steps Toward the Breeding of the Genus in Argentina

The first progeny trial of the genus in Argentina corresponds to *C. balansae* and was installed in April 2001 in Valle Morado (23° 28′ 11.8″ S, 64° 25′ 52″ W, 385 m asl) by Fundación Proyungas (Balducci et al. 2009). It includes 42 open-pollinated families obtained from mass selected trees in the natural forest according to forestry shape and sanitary criteria. Seeds were collected from five provenances: Ingenio Ledesma, Yuto, Campo Chico, Los Naranjos, and Orán. The trial has an area of 2 ha and an experimental design of incomplete blocks, with 20 repetitions of single-tree plots (40 treatments per block; N = 800). It was installed without any coverage.

First evaluations were made at 4 years of age (Horlent and Monteverde 2006). Significant differences were demonstrated for survival: families without mortality were observed throughout the trial, as well as families with up to 47% mortality. No significant differences were found between families in terms of shoot borer resistance. The percentage of plants attacked per family varied between 35% and 79%. Accordingly, differences could not be shown for the stem shape either, a variable closely related to the *H. grandella* attack. Similarly, families did not show significant differences for the "burning of the bark." This damage is caused by direct insolation in the north face of the stem. It begins as spots in the bark and progresses to wounds that reach the cambium, which can reduce the value of the wood and allow the entry of pathogens.

At the age of 5 years, height and diameter increments were evaluated (Balducci et al. 2009), and differences between families were demonstrated. The best ten families had a mean DBH of 14.4 cm and a mean height of 7.1 m. Maximum mean annual increment was 3.89 cm for DBH and 1.67 m for height. In addition to the information that it provides, this trial is important as a source of genetically proven material for further advances in the improvement of the species.

In 2006, INTA formally began a breeding program for this genus. The planned strategy was to create clonal seed orchards from mass selected trees from the natural forest, while different genetic trials were carried out. In December 2008, a small network of three trials was established, in order to test species, provenances, and progenies. They were composed by 110 open-pollinated families coming from seeds collected from trees selected in the natural forest according to growth, health, and shape criteria (Grignola 2014). All Argentinian species of the genus were assayed: five provenances of *C. balansae,* two of *C. fissilis,* two of *C. saltensis,* three of the hybrid (*C. saltensis* × *C. balansae*), and a pool of several provenances for *C. angustifolia* as in the previous trial. Seed pools from *C. odorata* and *Toona ciliata* (a Meliaceae from Australia) were used as controls. The test sites were La Moraleja (24° 17′ 59″ S, 64° 1′ 19″ W, 375 m asl), La Fronterita (26° 58′ 18″ S, 65° 30′ 7″ W, 655 m asl), and El Siambón (26° 42′ 49″ S, 65° 26′ 35″ W, 1170 m asl), with strong altitudinal and rainfall contrasts (mean annual rainfall: 850 mm, 1400 mm and 1200 mm, respectively). The experimental design was in randomized complete blocks, with 16 repetitions of single-tree plots in the first two sites and with 8 repetitions in El Siambón. All trials were without coverage, with a distance of 3 m × 3 m between plants. The number of provenances and families tested at each site varied according to seedlings availability (Grignola 2014).

A strong mortality was observed at the third year (Grignola 2014), with greater incidence in El Siambón, the site of highest altitude and lowest minimum temperatures. *C. balansae*, *C. fissilis*, and the hybrid did not differentiate, with survival values between 20% and 32% depending on the trial. The highest survival rate in El Siambón was for *C. fissilis* San Antonio provenance, with 43%. In La Moraleja (the other altitudinal extreme), the highest survival rate was for *C. balansae* Ledesma provenance, with 36%. The planting system without coverage was a determining factor to explain this high mortality.

Species by site interaction was observed for height at the third year, and significant differences between species were found (Fig. 14.7). The tallest plants belonged to the undifferentiated group of *C. balansae* (147.5 cm ± 6) and the hybrid (136.7 cm ± 10) in La Moraleja, the hybrid (124.4 cm ± 10.5) in La Fronterita, and *C. fissilis* (102 cm ± 11) in El Siambón. The highest average height was observed for the plants from *C. balansae* Ledesma, with 153 cm ± 8 in La Moraleja.

Another evaluation was performed at the eighth year (Grignola, unpublished data) in which interaction between trial sites and provenances was confirmed (La Moraleja trial unfortunately could not be continued). In El Siambón, the two tested provenances of *C. fissilis* were those with the highest annual height growth rate

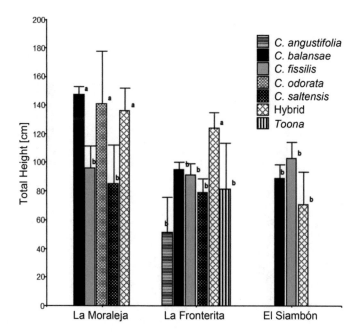

Fig. 14.7 Mean height at 3 years of age of the species assayed in the three study sites. Standard errors are presented. Means with the same letter are not significantly different ($p > 0.05$). (From Grignola 2014)

(AHGR) (San Antonio 71.6 ± 4 cm/y and Guaraní 60.0 ± 7 cm/y). These provenances did not show significant differences between them, and differed from all
those tested for *C. balansae* and the hybrid. In La Fronterita, on the contrary, three
of the four provenances of *C. fissilis* were the ones with the lowest AHGR (Guaraní
24.0 ± 4 cm/y, Las Marías 13 ± 5 cm/y and Eldorado 11 ± 4 cm/y). They did not
show significant differences between them and differed from all the provenances
tested for *C. balansae*, *C. saltensis*, the hybrid, and the control *Toona ciliata*
(Fig. 14.8). In Fig. 14.9, the species-by-site interaction becomes evident when
reducing the analysis to the main species analyzed in both trials.

BLUP analyses allowed identification of the families of the main evaluated species that were superior at both sites, despite the strong interaction between species
and sites. In Fig. 14.10, the performance of each essayed family is simultaneously
represented for both sites. Linear regression lines contribute to visualize performance trends for each species. In the upper-right quadrant, open-pollinated families
who had better performances in terms of height growth for the two sites are observed.
The families CM15, CM14, and CM17 of San Antonio provenance of the species
C. fissilis stand out. The Families BC05, BC02, BC04, BC01, BC10, and BC13 of
the Calilegua provenance of the species *C. balansae* were located in the same quadrant. The identification of these superior families might serve to go back to their
mothers, propagate them vegetatively, and infuse new genetics into existing clonal
seed orchards, or simply to create new ones.

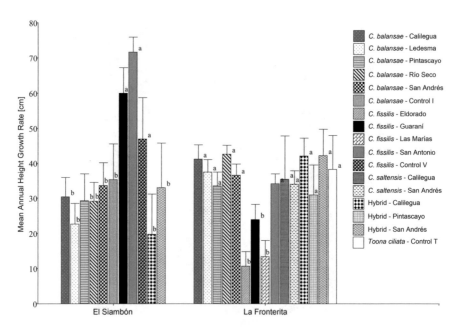

Fig. 14.8 Mean annual height growth rate (AHGR) at the eighth year for El Siambón and La
Fronterita trials for the different species and provenances studied. Standard errors are presented.
Means with different letters are significantly different (p > 0.05). (Grignola, data not published)

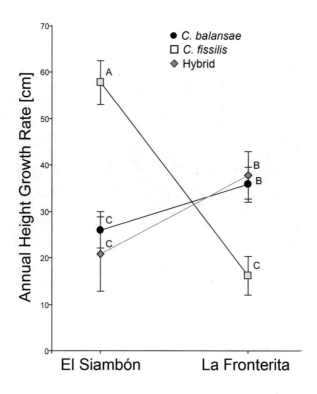

Fig. 14.9 Mean annual height growth rate (AHGR) at the eighth year for El Siambón and La Fronterita trials for the main species analyzed in both trials. Standard errors are presented. Means with different letters are significantly different ($p > 0.05$). (Grignola, data not published)

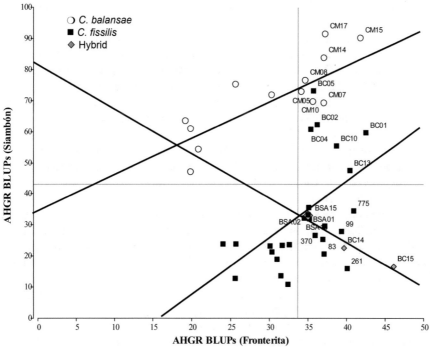

Fig. 14.10 Linear regression plot (oblique black lines) for the three species for AHGR BLUPs obtained from families growing in El Siambón vs. La Fronterita trials. All linear regressions were significant ($p < 0.05$). The dividing gray lines correspond to the averages AHGR BLUPs from each site. (Grignola, data not published)

Fig. 14.11 *Cedrela balansae* plantation in Yuto Experimental Station at 16 and 25 years of age, registered in INASE as seed production area. (Photos: Elvio Del Castillo and Ezequiel Balducci, respectively)

The first basic propagation material of *Cedrela* in Argentina for the supply of seeds to commercial nurseries is a seed production area of *C. balansae* planted in 1994 at the Yuto Experimental Station of INTA (23° 35′ 15.95″ S, 64° 30′ 27.69″ W, 350 m asl; Del Castillo, pers. comm.) (Fig. 14.11). It has a surface of 0.5 ha, began production in 2017, and has already been registered with the National Seed Institute (INASE).

The first clonal seed orchards (CSO) of *Cedrela* in Argentina were established in 2012 in Famaillá Experimental Station of INTA (Fornes et al. 2016): one for *C. balansae*, other for *C. angustifolia*, and a third one for *C. fissilis*. The trees were mass selected in the natural forest by height, stem rectitude, and sanity. Seeds, twigs of sexually mature branches, and leaves were collected from 107 trees belonging to eight natural populations of *C. balansae*, 52 trees from eight populations of *C. fissilis*, and 160 trees from 14 populations of *C. angustifolia*. Seeds were used for the abovementioned trials, twigs for the CSO, and leaves for genetic diversity studies with genetic markers (see Chap. 13). Each specimen was georeferenced and phenotypically described. This way, almost the entire range of the genus *Cedrela* in Argentina was covered.

The genus can be asexually propagated by several methods (Ortiz Morales and Herrera Tuz 2007). The method used to multiply the trees selected for these three CSOs was grafting by clefting. Scions 10 cm long and 1 cm in diameter were grafted onto rootstocks of the same species. In all three CSOs, a randomized complete block design with single-tree plots was used, and the number of blocks depended on the availability of ramets of each species. In the CSO of *C. balansae* (Fig. 14.12a), 48 clones were initially planted corresponding to four populations of Salta and two of Jujuy, with up to nine replicates per clone (N = 288). During the subsequent years, several genotypes were added, summing at present up to 66 clones in the seed orchard (N = 314). With a distance between plants of 5 m × 7 m, its total area is 2 ha, and its geographical coordinates are 27° 0′ 55.47″ S, 65° 22′ 16.84″ W (375 m asl).

Fig. 14.12 Individuals of *C. balansae* (**a**) and *C. fissilis* (**b**) from their respective clonal seed orchards at Famaillá Experimental Station of INTA. (Photos: Pablo Saravia)

In the CSO of *C. angustifolia*, 64 clones were initially planted, corresponding to seven populations of Salta, three of Tucumán, and two of Jujuy, with five replicates per clone (N = 320). Subsequent mortality reduced the number of clones to 44 (N = 85). The initial distance between plants was 4 m × 4 m, thus occupying an area of 0.5 ha (27° 1′ 13.89″ S, 65° 22′ 53.66″ W, 375 m asl).

C. fissilis CSO initially had 42 clones from five populations of Misiones and Corrientes, with nine replicates per clone (N = 378) (Fig. 14.12b). Currently, it consists in 77 clones since genotypes were added over time, although the number of ramets was reduced by mortality to 155. It occupies an area of 1.5 ha with a distance of 5 m × 7 m between plants and is located at 27° 1′ 50.22″ S, 65° 22′ 46.48″ W (365 m asl).

In 2016, seed production started in the orchards of *C. fissilis* and *C. balansae*. However, production still fluctuates from year to year, and several clones have not yet came to fruited. A significant portion of the seeds collected still has low germination capacity. It is estimated that production will stabilize in 2022.

References

Aquino DA, Tavares Texeira M, Balducci E, Baca V, Quintana de Quinteros S (2015) The microlepidopterous natural enemy *Brachymeria subrugosa* Blanchard, 1942 (Hymenoptera, Chalcididae): identity, hosts and geographic distribution. Zootaxa 4013:293–300

Aschero V (2006) Biología reproductiva e importancia de la polinización en *Cedrela lilloi*. In: Pacheco S, Brown A (eds) Ecología y Producción de Cedro (Género *Cedrela*) en Las Yungas Australes, Tucumán, Argentina, pp 41–50

Baca V, Lucia A, Balducci E, Sanchez E, Malizia L, Quintana de Quinteros S (2013) Dinámica poblacional del barrenador de las Meliaceas, *Hypsipyla grandella* (Zeller) y su asociación con los ataques ocasionados en plantaciones de Cedro en el norte de la provincia de Salta. 4to Congreso Forestal Argentino y Latinoamericano, 23–27 Sept. 2013, Iguazú.

Balducci E, Arturi M, Goya J, Brown A (2009) Potencial de plantaciones forestales en el pedemonte de las Yungas. Proyungas. Ediciones del Subtrópico, Yerba Buena, pp. 42. http://proyungas.org.ar/wp-content/uploads/2014/12/cartilla-Valle-Morado-1.pdf

Birchler T, Rose WR, Pardos M, Royo A (1998) La planta ideal: Revisión del concepto, parámentros definitorios e implementación práctica. Invest Agr Sist Recur For 7:109–121

Briceño Vergara AJ (1997) Aproximación hacia un manejo integrado del barrenador de las Meliaceas, *Hypsipyla grandella* (Zeller). Rev For Venez 41:23–28

Cozzo D (1995) Silvicultura de plantaciones maderables. Orientación Gráfica, Buenos Aires. Tomo II, pp. 899.

Del Castillo E, Tapia SN (2005) El barrenador de los brotes: *Hypsipyla grandella* Zéller, en plantaciones de importancia forestoindustrial en el NOA. Documentos INTA, Yuto. https://inta.gob.ar/sites/default/files/script-tmp-el_barrenador_de_los_brotes.pdf

Del Castillo ME, Zapater MA, Gil MN (2002) El Cedro rosado. Recolección de material genético. Viverización. Ensayos de implantación. SAGPyA, INTA EEA Yuto, pp. 23. https://inta.gob.ar/sites/default/files/script-tmp-cedro_rosado.pdf

Del Castillo EM, Zapater MA, Gil MN (2004) Resultados de plantaciones experimentales con *Cedrela Balansae* C. DC. (Cedro Orán) en INTA-Yuto: comparación con otras especies forestales nativas y exóticas. https://inta.gob.ar/sites/default/files/script-tmp-comparacin_cedrela_y_otras_especies.pdf

Dexter ST, Tottingham WE, Graber LF (1932) Investigations of the hardiness of plants by measurement of electrical conductivity. Plant Physiol 7:63–78

Engelbrecht BMJ, Kursar TA (2003) Comparative drought resistance of seedlings of 28 species of co-occurring tropical woody plants. Oecologia 136:383–393

Eskiviski E, Tapia S, Fornes L, Agostini J (2010) Evaluación de insecticidas en el control de *H. grandella* (Zeller) en condiciones de vivero. XIV Jornadas Técnias Forestales y Ambientales. Facultad de Ciencias Forestales. UNaM – INTA EEA Montecarlo, Eldorado, 10–12 junio 2010, pp. 10–14.

Fornes L (2012) Domesticación de especies de alto valor de las selvas subtropicales. Revista Producción Forestal MAGyP 4:28–52

Fornes L, Zelener N, Gauchat ME, Inza MV, Soldati MC, Ruíz V, et al. (2016) Subprograma Cedrela. In: Marcó M, Llavallol C (eds) Domesticación y Mejoramiento de Especies Forestales. Min. Agr. UCAR, pp 137–159

Garkoti S, Zobel D, Singh S (2003) Variation in drought response of sal (*Shorea robusta*) seedlings. Tree Physiol 23:1021–1030

Grau A, Zapater MA, Neumann RA (2006) Botánica y distribución del género *Cedrela* en el noroeste de Argentina. In: Pacheco S, Brown A (eds) Ecología y Producción de Cedro (Género *Cedrela*) en Las Yungas Australes. LIEY-ProYungas, Argentina, pp 19–30

Grignola J (2014) Plasticidad y tolerancia de diferentes especies y procedencias del género *Cedrela* a las bajas temperaturas. MSc thesis from Universidad Nacional de Córdoba, Argentina, pp. 116.

Horlent M, Monteverde D (2006) Crecimiento de *Cedrela balansae* en la plantación experimental de Valle Morado. In: Pacheco S, Brown A (eds) Ecología y Producción de Cedro (Género *Cedrela*) en Las Yungas Australes. LIEY-ProYungas, Argentina, pp 171–178

Lorenzi H (1992) Árvores brasileiras: manual de identificação e cultivo de plantas arbóreas nativas do Brasil. Instituto Plantarum de Estudos da Flora, Odessa, p 352

Meloni DA, Martínez CA (2011) Bases fisiológicas de la tolerancia al estrés térmico en especies del género *Cedrela*. XIV Reunión Latinoamericana de Fisiología Vegetal. Búzios, 19 y 22 de septiembre de 2011, p. 95.
Monteverde D (2006) Producción de plantines de cedro en vivero. In: Pacheco S, Brown AD (eds) Ecología y Producción de Cedro (Género *Cedrela*) en Las Yungas Australes. LIEY-ProYungas, Argentina, pp 155–160
Murchie EH, Lawson T (2013) Chlorophyll fluorescence analysis: a guide to good practice and understanding some new applications. J Exp Bot 64:3983–3998
Newton AC, Watt AD, Lopez F, Cornelius JP, Mesén JF, Corea EA (1999) Genetic variation in host susceptibility to attack by the mahogany shoot borer, *Hypsipyla grandella* (Zeller). Agric For Entomol 1:11–18
Orthen B, Popp M (2000) Cyclitols as cryoprotectants for spinach and chickpea thylakoids. Environ Exp Bot 44:125–132
Ortiz Morales ER, Herrera Tuz LG (2007) Cedro (*Cedrela odorata* L.). Protocolo para su colecta, beneficio y almacenaje. Comisión Nacional Forestal, Programa de Germoplasma Forestal, Estado de Yucatán, pp. 23. http://www.conafor.gob.mx:8080/documentos/docs/19/1299Cedro%20rojo%20Yucat%c3%a1n.pdf
Otieno DO, Schmidt MWT, Adiku S, Tenhunen J (2005) Physiological and morphological responses to water stress in two *Acacia* species from contrasting habitats. Tree Physiol 25:361–371
Pennington TD, Muellner AN (2010) A monograph of cedrela (*Meliaceae*). DH Books. The Manse. Chapel Lane, Milborne Port-England, pp. 112.
Rodríguez-Rey JA, Romero E, Gianfrancisco S, David S del C, Amado ME (2000) Evaluación de la capacidad de aclimatamiento a las bajas temperaturas de pimiento *Capsicum annuum* L. cultivado en invernadero sin calefacción. Rev Fac Agron (LUZ) 17:10–19
Ruiz VE, Meloni DA, Fornes LF, Ordano M, Prado FE, Hilal M (2013) Seedling growth and water relations of three *Cedrela* species sourced from five provenances: response to simulated rainfall reductions. Agrofor Syst 87:1005–1021
Tapia S (2012) El control del barrenador del brote de los cedros. Experiencias en el NOA. Revista Producción Forestal MAGyP 4:38–42
Tortorelli L (2009) Maderas y bosques argentinos, 2a edición. ed. Editora Orientación Gráfica, Buenos Aires, pp. 1111.
Wiersum KF (1981) Outline of the agroforestry concept. In: Wierzum KF (ed) Viewpoints on agroforestry. Agricultural University, Wageningen, pp 1–23

Chapter 15
Paraná Pine (*Araucaria angustifolia*): The Most Planted Native Forest Tree Species in Argentina

María Elena Gauchat, Natalia C. Aguirre, Fabiana Latorre, María Virginia Inza, Ector C. Belaber, Noga Zelener, Susana L. Torales, Luis F. Fornes, Martín A. Pinazo, Cristian A. Rotundo, Jorge C. Fahler, Norberto M. Pahr, and Hugo E. Fassola

15.1 Natural Distribution in Argentina: The Western Extreme of a Large Area

Araucaria angustifolia (Bertol.) Kuntze (Paraná pine) is one of the few native South American conifers. It belongs to the Araucariaceae family, which dominated both hemispheres during most part of the Mesozoic Era and the beginning of the Cenozoic

M. E. Gauchat (✉) · E. C. Belaber · M. A. Pinazo · C. A. Rotundo
N. M. Pahr · H. E. Fassola
INTA EEA Montecarlo, Montecarlo, Argentina
e-mail: gauchat.maria@inta.gob.ar

N. C. Aguirre
Instituto de Investigaciones Biotecnológicas (IB), Centro de Investigación en Ciencias Veterinarias y Agronómicas (CICVYA), INTA. Hurlingham, Buenos Aires, Argentina

F. Latorre
Instituto de Investigaciones Marinas y Costeras – CONICET-Universidad Nacional de Mar del Plata (FCEyN), Mar del Plata, Argentina

M. V. Inza · S. L. Torales
Instituto de Recursos Biológicos (IRB), Centro de Investigaciones de Recursos Naturales (CIRN), INTA. Hurlingham, Buenos Aires, Argentina

N. Zelener
Centro de Investigación de Recursos Naturales (CIRN), INTA. Hurlingham, Buenos Aires, Argentina

L. F. Fornes
INTA EEA Famaillá, Famaillá, Argentina

J. C. Fahler
Private Consultant, Puerto Esperanza, Argentina

© Springer Nature Switzerland AG 2021
M. J. Pastorino, P. Marchelli (eds.), *Low Intensity Breeding of Native Forest Trees in Argentina*, https://doi.org/10.1007/978-3-030-56462-9_15

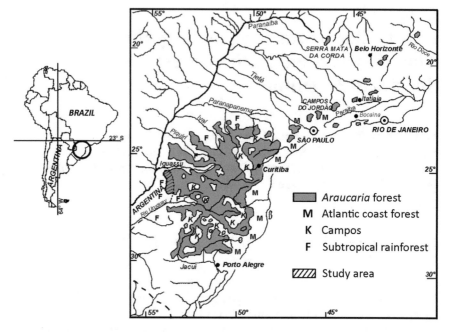

Fig. 15.1 Natural distribution area of *Araucaria angustifolia* in South Brazil and North Argentina (Rau 2005 according to Hueck 1966, modified)

until the appearance of angiosperms, which eventually prevailed. The genus *Araucaria* is currently present only in the southern hemisphere, but their fossils have been registered throughout the world (Kranitz et al. 2014). *A. araucana* and *A. angustifolia* are the only living species representative of the genus in South America.

Although several authors cited by Moura (1975) early established the natural range of *A. angustifolia*, Hueck (1952) was the one who made the most detailed description, providing accurate data of the presence of the species between 18° and 30° S and between 41° and 54° 30' W.

The geographical distribution of *A. angustifolia* (Fig. 15.1) includes a main continuous area in the eastern and central plateau of Southern Brazil (states of Rio Grande do Sul and Santa Catarina) and marginal forests that extend with a more dispersed pattern within the northeast extreme of Argentina (Fig. 15.2), eastern Paraguay, and the State of Paraná (Brazil). In addition, some sparse populations spread northward till the state of Rio de Janeiro (Di Bitetti et al. 2003; Sebbenn et al. 2003). The discontinued presence of *A. angustifolia* in its natural distribution area is usually associated with altitude. It commonly finds its lowest boundaries between 500 and 600 m asl, climbing up to 1200 m asl in the Serra da Mantiqueira and the Itatiaia, and even up to 1800 m asl in the Campos do Jordão region, at the north of its range. *A. angustifolia* forests are phytogeographycally classified as a district within the Paranaense biogeographic Province of the Amazonian Dominion (Cabrera and Willink 1980) but are more generally included within the Alto Paraná Rainforest.

Fig. 15.2 *Araucaria angustifolia* in remnant groups in the southwest of its natural distribution area, close to Bernardo de Irigoyen, Province of Misiones (Photo: Norberto Pahr)

According to different authors cited by Rau (2005), there would exist 11, six or five varieties of the species, most of which could be differentiated by the morphology of their mature seeds (Achten 1995). Other authors (Gurgel and Gurgel 1971, 1978; Fahler and Di Lucca 1980) have differentiated *A. angustifolia* into geographic races or ecotypes.

There are numerous local names for *A. angustifolia*, from which the most common ones are Pino de Misiones or Pino Paraná, in Argentina; Pinho Brasileiro or Pinheiro do Paraná, in Brazil; Kuri'y or Cury in both countries according to the Guarani indigenous language.

15.2   Uses and Conservation Status of the Natural Forests of *Araucaria angustifolia*

At the beginning of the twentieth century, *Araucaria angustifolia* covered about 200,000 km² mainly in Brazil. Nowadays, the species is "critically endangered" (Thomas 2013) due to overexploitation, which has severely fragmented and reduced its natural coverage (Auler et al. 2002), particularly in the last 70 years. According to the last forest inventory carried out in Brazil in 1980 (with data until 1977), in the south of that country remains only 3% of the original *A. angustifolia* forests surface (Machado and Siqueira 1980; Achten 1995).

In Argentina, a great deal of the original west marginal distribution of *A. angustifolia* forests appeared in small patches in the northeastern of the Province of Misiones, covering an area of approximately 210,000 ha (Cozzo 1960). In 1993, the remaining forest patches of *A. angustifolia* summed an estimated area of no more than 2000 ha, largely in reserves and provincial parks (Burkart 1993). Particularly, in this region, forest industry had almost exclusively used these natural stands, leading to a highly disturbed landscape. Despite the protection of the species by the Provincial Law N° 2380 of 1986, selective extraction has continued and has left only solitary, old, and often sick trees (Bertolini 2000; Chebez and Hilgert 2003; Rau 2005). This fact, together with the prohibition of silvicultural interventions in part of this area (Rau 2005), makes the development of techniques for the promotion of natural regeneration unfeasible.

Its high-quality wood and multiple applications have made *A. angustifolia* one of the most exploited species in the Province of Misiones. Together with other species of the region, such as *Cedrela fissilis*, *Pterogyne nitens*, and *Handroanthus heptaphyllus*, among others, its exploitation has facilitated the colonization of the region. During the last century, the forests of the Province of Misiones were deeply affected by drastic colonization policies. The increase in population since the 1930s, mainly due to a great influx of immigrants from Europe, had manifold impacts. With the new settlers, the necessity for more space, labor opportunities, and people's livelihood in general arose. Consequently, this immigration not only resulted in a rich exchange of cultures but also in the deforestation of ample areas of natural forests in a continuous process that lasts until today (Rau 2005).

Local aboriginal groups (i.e., Kaingang and Guarani) have used the seeds of *A. angustifolia* ancestrally as food (Cordenunsi et al. 2004). Pine nuts, known as kuri'î, are very appreciated in the region and highly consumed, both raw, roasted or boiled (Martínez Crovetto 1968), and this consumption might probably have an impact on the regeneration of the natural forests. The seeds have a high content of carbohydrates, mainly starch (near 73%), and a slight sweet taste (Ferreyra et al. 2006). However, little is known about its nutritional value and, in particular, about the antioxidant properties of its components. In this regard, there are studies that demonstrate that certain biflavonoids, present in the leaves of the species, would efficiently protect essential biological molecules from oxidative damage (Yamaguchi et al. 2005).

15.3 The Historical Use of *Araucaria angustifolia* in Productive Plantations

Together with a forestry industrial model developed in Argentina since the middle of the twentieth century, the need to increase the forest production led to extensive afforestation and reforestation plans, particularly in the Province of Misiones. Increasing the surface under cultivation is not synonymous with an increase in

Fig. 15.3 (**a**) Trial of intensity and opportunity of thinning at 21 years of age, Iguazú Department, Misiones (photo: Aldo Keller). (**b**) Pruned commercial plantation located in María Magdalena (Misiones), 8 years of age (photo: Rodolfo Martiarena)

productivity if certain basic rules of forestry practice are not taken into consideration, such as selection of the species, origin of the seeds, planting site, and the forest management plan of the plantation.

With the arrival of the first settlers toward 1930 to the north of the Province of Misiones, the cultivation of *A. angustifolia* began (Cozzo 1976). Its long fiber represented an opportunity for cellulose pulps production, of which Argentina is not self-sufficient yet. Toward 1945, a still working plant of cellulose was installed in Puerto Piray (at the side of Paraná River). It was initially feeded with *A. angustifolia* from the native forests but soon promoted large-scale afforestations with this species (Fig. 15.3).

At the same time, and to counteract the ongoing devastation suffered by this species, the National State started in 1946 a conservation program, in which an area of around 2000 ha with native *Araucaria* stands in northeastern Misiones was protected (Fernández et al. 2005). In 1948 this area was conferred the status of Forest Reserve, beginning activities of reforestation with *A. angustifolia*. Later, this reserve turned into the "Manuel Belgrano" Forest Station (26° 2′ 20″ S; 53° 46′ 48″ W; 540 m asl), founded with the purpose of promoting silvicultural research, and preserving *A. angustifolia* (mainly by means of plantations). In the framework of a FAO/UNDP project carried out by the National University of La Plata, in 1964, the Center of Studies for the Subtropical Forest was created in this forest station. Twelve years later, its administration was transferred to the Instituto Forestal Nacional (IFoNa), and finally to the Instituto Nacional de Tecnología Agropecuaria (INTA) in 1991. The Manuel Belgrano Forest Station harbors the current largest area of *A. angustifolia* old plantations (448 ha), which were installed at the end of the 1940s. These plantations are conformed by different provenances, most of them from Misiones (a minor part from Brazil), several unknown and probably already inexistant (Martín Pinazo and Hugo Fassola pers. comm.). This emblematic station is under sustainable management and conservation, according to the National Law 26,331 (Goya et al. 2012).

The apogee reached by the cultivation of *A. angustifolia* started to decline in the 1960s for several reasons. The main was related to the promotion of the "yellow pines" native to the south of USA: *Pinus elliottii* var. *elliottii* and *Pinus taeda*, which became the raw material for the pulp industry. These fast-growing species quickly replaced *A. angustifolia* not only for chemical transformation but also for sawn timber due to their lower prices, even in the large size pieces now achieved by means of multilaminated beams. The decorative value of the *A. angustifolia* lumber never got a reflection in its price. From a silvicultural point of view, some shortcomings should be mention, like its lower growth rate and its good-quality site requirements, the difficulty in obtaining seeds, and their recalcitrant character (they cannot be stocked for a long period).

The mentioned decline becomes evident through the evolution of the area covered by plantations of *A. angustifolia*, which went from 25,000 ha in 1975 (Cozzo 1975) to 20,000 ha in 1997 (SAGYP 1997), 18,000 ha in 2000 (Min. Ecol. 2001), and 16,000 ha in 2010 (SIFIP 2010) and has probably dropped even more nowadays. These data are not only evidence of the decline of the cultivation of the species but also reflect the fact that it is the most cultivated native species in Argentina.

In the 1970s, INTA initiated a series of investigations toward optimizing the cultivation of *A. angustifolia*. The first trials of geographical variation were installed (Fahler and Di Lucca 1980), and trials of planting density and thinning were also established (Crechi et al. 2001, 2009); studies on the determination of defective cylinder were performed (Fassi et al. 1993), and seed production was monitored as well (Fassola et al. 1999).

15.4 Reproductive Biology of *Araucaria angustifolia*

So far, natural trees of *A. angustifolia* are restricted to small forest fragments or to isolated individuals because of indiscriminate logging during the last century and the agricultural expansion (Luna and Fontana 2017). The drastic decline of its distribution area has reduced the connectivity between the population remnants (Pinazo et al. 2016) leading to a limited pollination and to a decrease in the genetic diversity. Consequently, the natural regeneration is scarce: seedlings are hardly found in the native forests, although the species is not strictly heliophilous and would be able to be established in the understory (Duarte et al. 2002). Reforestation, as an active restoration measure, could partially mitigate this critical situation, but this kind of practice has been severely restricted by the progressive decrease in the volume of seeds produced during the last years in Argentina (Fassola et al. 1999).

Thus, the fall in seed production may be due to the reproductive isolation after habitat fragmentation. A direct relationship between pollen and seed production often occurs in woody plants (Cour and van Campo 1980), then the smaller number of seeds detected could be attributed to a diminished pollination. According to Allison (1990), a deficient pollination negatively affects both fertilization of the ovules and seed production. Indeed, studies of controlled pollination carried out by

Anselmini and Zanette (2012) in this species showed that the maximum capacity to produce cones is given by the availability of pollen.

A. *angustifolia* is a dioecious and anemophilous species; the male gamete in the pollen grain produced by a male tree is windblown to the female cones of another tree. The reproductive cycle starts with strobilus formation in November and finishes when the main pollination occurs in October of the following year. Twenty months thereafter, the seeds are mature, so a total of about 30 months are needed to complete the entire process. During this long period, climatic conditions vary offering at each stage a new chance of failure of the reproductive process.

15.4.1   Limitations for the Pollination and Seed Formation

Important limitations in the pollen amounts for the pollination of A. *angustifolia* can be observed in two dimensions: (1) the decrease in pollen productivity that may be associated with previous temperature (Caccavari et al. 2000; Bittencourt and Sebbenn 2007) and (2) the scarce dispersion of pollen grains due to (2a) their substantial size and weight (Caccavari 2003), (2b) the low number of trees and the distance between them, and (2c) unfavorable weather conditions affecting the displacement of pollen grains (Latorre et al. 2013).

For anemophilous species with nonspecific pollination, one of the strategies that guarantees effective fertilization is the large amount of pollen (Faegri and van der Pijl 1979). Latorre et al. (2020) observed for A. *angustifolia* in Manuel Belgrano Forest Station using a 7-day Burkard aerobiologic trap, an alternation between 1 year of high pollen values (1685 pollen grains.day^{-1}.m^{-3}air, on average), followed by 2 years of a progressive reduction in intensity (456 pollen grains.day^{-1}.m^{-3}air, on average), a decline exceeding 50%. The relevance of detecting the years of increased production lies in the possibility of implementing management decisions, such as cleaning the understory during the year after the highest pollen values and collect the seeds for seedling production in the nursery or allow seedlings to germinate naturally and develop free of competition after the second year following the high pollen productivity.

Latorre et al. (2016) postulate the negative effect that high temperatures prior to pollination have on the number of pollen grains. The correlation between pollen production and the average minimum temperature of August is negative ($r = -0.75$, $p = 0.05$), supporting the key role that the environment plays on the species phenology (Fig. 15.4). The lower minimum temperature of August prior to pollination match with the higher pollen production recorded for that year, adjusting to a predictive equation in 80% of the cases, $p = 0.004$ (Latorre et al. 2020). Since seed production results from a successful fertilization that occurs 2 years before (Anselmini 2005), this statistical model could help estimate seed productivity 20 months ahead.

To evaluate the effect of climate on the pollen productivity of A. *angustifolia*, Simón et al. (2018) compared two old plantations growing under different climate

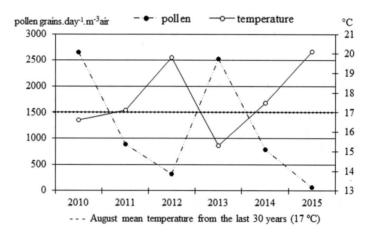

Fig. 15.4 Interannual variation of the pollen produced by *A. angustifolia* in Manuel Belgrano Forest Station and average of the minimum temperatures of August prior to pollination (horizontal dotted line: August mean temperature from the last 30 years, 17 °C) (from Latorre et al. 2016, with modifications)

regimes (Pérez et al. 2016). One is located in Manuel Belgrano Forest Station, within the natural range of the species, with a humid subtropical climate, and the other in 25 de Mayo Forest Station of INTA (35°28′ 45″ S; 60° 7′ 44″ W; 60 m asl: 25 M), Province of Buenos Aires (more than 1000 km south of the natural range) where the climate is temperate. The aerobiological indicators were measured during 3 years (period 2014–2016) with Tauber traps (Tauber 1965) placed under the tree canopy (Latorre et al. 2016). The annual mean pollen productivity in 25 M doubled that of Manuel Belgrano Forest Station (9440 and 5291 pollen grains m^{-2}.year^{-1}). In addition, low temperatures prior to pollination were reported to boost pollen productivity (as in temperate zones) and/or that high winter temperatures influenced negatively (as in subtropical zones), determining low pollen production. According to Caccavari et al. (2000), minimum temperatures should not exceed 10 °C during pollen grain formation and maturation, due to their negative impact.

Another phenological stage to consider is the development of the reproductive structures that in *A. angustifolia* starts on the summer before the pollination (Anselmini et al. 2006). Latorre et al. (2020) studied the species growing in Misiones during seven reproductive periods and noticed some climatic particularities of the year with the lowest amount of pollen (2012). The monthly rains from November 2011 to April 2012 (period of strobilus formation) had been 37 mm lower on average than the 1998–2010 statistical data. Likewise, the minimum temperatures of that summer were 2 °C lower, on average. As in Brazil, high temperatures and precipitations during cones formation period favored this phenological stage (Anselmini 2005).

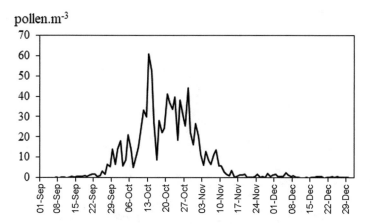

Fig. 15.5 Aerobiological curve of *A. angustifolia*. Mean daily concentrations, period 2010–2016

The aerobiological curves are important to represent the phenological reproductive dynamics and to analyze the interannual variations (Latorre 1999). The pollination curve of *A. angustifolia* growing in the southern extreme of its range begins in mid-September and ends in mid-November, with a peak in mid-October (Fig. 15.5).

In order to establish the relationship between the development of male cones and the phases of the aerobiological curve of *A. angustifolia*, the weekly pattern of atmospheric pollen was compared (Latorre et al. 2020) with phenological observations conducted directly on male cones with the help of binoculars (Latorre et al. 2014). A modified Anselmini scale (2005) was used, in which phenological state 2 corresponds to the initial phase of pollen release when the aments are yellow-green; phenological state 3 corresponds to pollen release phase in which the aments are more visible due to their size and have a darker color; phenological state 4 corresponds to the final stage of pollen release when the aments are brown and curved; and lastly, the phenological state 5 indicates the end of pollen release with the curved aments already fallen. The aerobiological curve corresponds to the development of male aments. The pollen release phenophase (2 to 4) occurred in October (weeks 42 and 43 of the year), in line with the month of pollen peak in the air (Fig. 15.6). These correspondence evidences that pollen records constitute a suitable indicator to infer the reproductive phases.

The aerial transport of anemophilous grains depends partially on its aerodynamic features. *A. angustifolia* has inaperturate, asaccate, and subspheroidal grains. Their nuclei and numerous starch granules make them heavy in relation to their volume, and the granulated exine has low resistance to the action of physical forces. Pollen grains collected in natural populations from Argentina and Brazil had an average size of 79 µm with an exine 2 µm thick (Del Fueyo et al. 2008). According to Simón and Latorre (PIA 14028 Report 2016), the grains from the plantations of Manuel

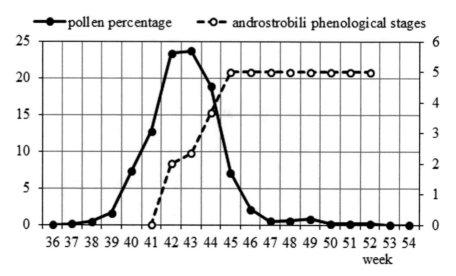

Fig. 15.6 Percentage of pollen of *A. angustifolia* in each week of the year, calculated from the weekly average of the years 2010–2016 (●). Phenological observations: phenophases 2, 3, and 4 correspond to different maturation stages of the aments during pollen release (○)

Belgrano Forest Station are relatively smaller (62 ± 7 μm) with a thinner wall (1.1 μm). In any case, Faegri and Van der Pijl (1979) established that the maximum grain size for optimum wind dispersal is 60 μm, with an average diameter of 20–30 μm. The larger size and heavier weight of *A. angustifolia* pollen grains play a negative role for this type of pollination and aid to explain why they travel a short distance.

On the other hand, Del Fueyo et al. (2008) observed the formation of broad depressions when the pollen grains were dehydrated. Niklas (1985) sustains that a pollen grain deformation by water loss could improve transport. Nonetheless, the high relative humidity of the subtropical environment in which *A. angustifolia* naturally grows decrease the probability of water loss. Even more, Latorre and Fassola (2014) did not observe deformed grains, though they did report broken pollen grains, with a torn "pac-man"-like exine and ejected cellular content in untreated samples collected directly from the air. The impact on the collecting surface could lead to its rupture in its weakest portion (Fig. 15.7c), similar to what happens when the pollen tube develops once the pollen grain reaches the female cone's ovule. In general, the grains found in the air are well hydrated and whole (Fig. 15.7a, b). The aerodynamic conditions of *A. angustifolia* pollen grains are not favorable for wind dispersal and would not be adapted to long-distance aerial travel (Caccavari et al. 2000).

In fragmented habitats, it is crucial to know the distance that could reach the pollen during its transport to infer its availability to pollination. The first study on the dispersal range of *A. angustifolia* pollen was conducted by Caccavari (2003) using the "fluxage" Cour method (1974) in the Province of Misiones, Argentina.

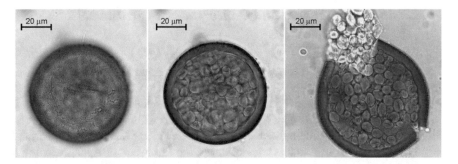

Fig. 15.7 Pollen grains of *Araucaria angustifolia* in their natural state in aerobiological samples from San Antonio (Province of Misiones, Argentina). Two views of the same grain in (a) and (b), where the granular ornamentation of the wall (**a**), and the size and shape (**b**) can be observed. Broken "pac-man"-type grain with ejected cellular content (**c**). (Photos: Fabiana Latorre)

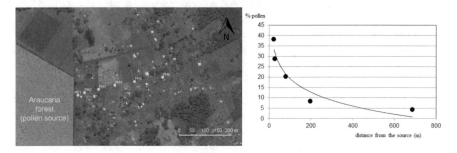

Fig. 15.8 Transect with sampling points from 20 to 700 m from the releasing pollen source (**a**). Pollen percentage of *A. angustifolia* at different distances from the releasing sources (**b**)

This study revealed that the transport capacity is scarce and lower than the 600 m that Niklas (1985) reports for other gymnosperms. Bittencourt and Sebbenn (2007) and Sant'Anna et al. (2013) established, by means of genetic techniques, that *A. angustifolia* pollen can travel as far as 2006 m between fragmented areas, or 318 m in a continuous forest. Nonetheless, these estimates of wind transport are indirect and do not account for the pollen abundance distribution from the emitting sources. In order to calculate the distance traveled by pollen and learn about pollen concentration during the journey, Latorre et al. (2016) performed mensurative experiments during 2 years, using portable Lanzoni aerobiological samplers at different distances from the sources (Fig. 15.8a). In line with the theoretical curve of pollen distribution, the pollen concentration of *A. angustifolia* was much higher near its sources and decreased logarithmically as it moved away from them (Fig. 15.8b). It is estimated that *A. angustifolia* pollen reaches a maximum transport distance of 455 m from its releasing sources; however, 60% of the pollen released reaches a distance of as much as 100 m, and 90% of the pollen falls before 200 m. Based on these results, it becomes clear that distance is a crucial factor for pollination, which although possible up to more than 2 km, is unlikely beyond 450 m.

15.4.2   Seed Production and Conservation of A. angustifolia's Recalcitrant Seeds

Seed production of the old plantation of Manuel Belgrano Forest Station was followed by Fassola et al. (1999) during 5 years on individual trees and mass plots. In 1993, the plots were between 35 and 40 years old; the basal area and the number of individuals ranged from 23 to 27 $m^2.ha^{-1}$ and 196 to 433 trees.ha^{-1}, respectively. The participation of female individuals was between 23.1% and 40.8% of the total stand trees. The fall of female cones began in the first week of April, reaching a maximum in late May and early June and ending in the first half of August. The largest seed production was in 1994, when 44.3 kg.ha^{-1} were collected, in the oldest plot with the highest proportion of female trees. Female cone production in individual trees was significantly associated with the size of the mother tree and its sociological position. The maximum recorded seed production was 1.5 kg for a tree with 52 cm of DBH. These results showed a consistent lower seed production than the natural forest located in Cruce Caballero Provincial Park, 50 km to the south, under the same environmental conditions.

A more recent study compared the seed production between plantations installed at contrasting sites (Simón et al. 2018). Between 10 and 15 female cones were collected in the old plantations of Manuel Belgrano Forest Station (subtropical climate) and in the already mentioned plantations of 25 de Mayo Forest Station (temperate climate), and the seeds per cone were counted. The average seed productivity was 10 times higher in the temperate climate site with 104 seeds.cone^{-1}, compared to 12 seeds.cone^{-1} obtained in the subtropical site. This contrasting seed productivity should be considered for conservation measures. The conditions prevailing during the postfertilization period also influence the seed productivity. High temperatures registered from December to April have a positive effect on seed maturation (Anselmini et al. 2006). The maximum temperatures during this period are lower in the subtropical climate of the Province of Misiones, where A. angustifolia is naturally distributed, and the number of seeds is lower if compared to what occurs in the temperate climate of 25 de Mayo (Simón et al. 2018).

When Simón et al. (2018) compared the relationship between pollination and seed yield, they found significant differences in the productivity of both events with respect to climatic conditions. In the subtropical climate of San Antonio, the amount of pollen produced was much lower, as well as the number of seeds, as compared to records from the temperate climate of 25 de Mayo. The reduced production of seeds was associated with the low number of pollen grains available to fertilize the ovules and form seeds; the difference between years for both events was smaller than the difference between environments (Fig. 15.9).

The recalcitrant character of A. angustifolia seeds is another important aspect for the conservation and use of this species (Tompsett 1984). Several methodological studies were performed during the 1990s to find an appropriate protocol to stock its seeds. The main results indicate the possibility of preserving a good germination capacity for a period of up to 24 months when the seeds are packaged in flexible plastic bags made with ethyl vinyl acetate and stocked at 0 °C (Chaves et al. 1999;

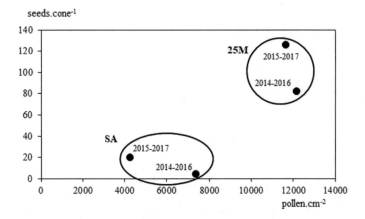

Fig. 15.9 Number of seeds per cone (2016 or 2017) based on the pollen produced (2014 or 2015) at each site. SA: San Antonio (Misiones) and 25 M: 25 de Mayo (Buenos Aires). (From Simón et al. 2018)

Piriz Carrillo et al. 2003, 2004). Recent results reinforced the convenience of storing at refrigerator temperatures (Garcia et al. 2014).

15.5   Genetic Variability of Natural Populations Estimated by AFLP Markers

Given the severe exploitation of the species in the last century and the conversion of its forests to agricultural lands, the genetic diversity of *A. angustifolia* is currently at risk. However, until recently, only populations of Brazil had been included in genetic studies of the species by means of markers (Sousa and Hattemer 2001; Auler et al. 2002; Sousa et al. 2004; Stefenon et al. 2007; Bittencourt and Sebbenn 2009; Ferreira de Souza et al. 2009). Inza et al. (2018) have conducted the first genetic variation study of the remnant *A. angustifolia* natural populations of Argentina using molecular markers. This work concludes that the geographical distribution and the demographic history were the main forces shaping the genetic variability of *A. angustifolia* populations in Argentina, and it remarks the populations with valuable gene pools for conservation and management purposes.

More in detail, Inza et al. (2018) studied the genetic diversity and differentiation of nine remnant natural populations with different logging history and the emblematic old plantation of the National Reserve of Manuel Belgrano Forest Station by means of AFLP markers (Table 15.1, Fig. 15.10). One pair of enzymes and three selective primer combinations generated 706 polymorphic markers over 312 *A. angustifolia* trees (Aguirre 2014; Inza et al. 2018). This number of AFLP loci is higher than similar studies of the species in Brazil (Stefenon et al. 2007; Ferreira de Souza et al. 2009).

Table 15.1 *Araucaria angustifolia* natural populations and the plantation of Manuel Belgrano Forest Station sampled in the Argentinean Alto Paraná rainforest and genetic diversity of nine natural populations by means of AFLP markers

Population	Code in map	N	Latitude S	Longitude W	Altitude (m asl)	Population status	He	PPL	EL
Manuel Belgrano plantation	MBp	77	26°02′20″	53°46′48″	530	PA, national reserve	–	–	–
Piñalito Norte	PiN	21	25°56′07″	53°56′55″	440	UA, private property	0.124	48.8	6
Manuel Belgrano nat. Forest	MBf	24	26°02′34″	53°47′06″	540	PA, national reserve	0.145	58.1	13
Gramado	Gr	28	26°12′12″	53°40′07″	700	UA, private property	0.143	62.4	30
Campiñas Américo	CA	28	26°16′33″	53°41′49″	790	UA, private property	0.121	52.8	11
PR El Piñalito	PPi	27	26°25′41″	53°50′04″	720	PA, provincial reserve	0.124	54.1	15
Santa Rosa	SR	20	26°26′14″	53°53′04″	640	UA, private property	0.139	53.7	4
PR Cruce caballero	PCC	31	26°31′01″	53°59′37″	610	PA, provincial reserve	0.119	49.7	11
PR Araucaria	PA	25	26°37′55″	54°06′00″	550	PA, provincial reserve	0.119	50.3	9
PR Caá Yarí	PCY	31	26°51′29″	54°13′20″	460	PA, provincial reserve	0.117	49.8	19

PR Provincial Reserve, *N* number of sampled trees, *PA* protected area, *UA* unprotected area, *He* mean expected heterozygosity, *PPL* percentage of polymorphic loci, *EL* number of exclusive loci

The genetic diversity of Argentinean natural populations of *A. angustifolia* was moderate to low ranging from He = 0.117–0.145 (Table 15.1), with a mean value of He = 0.128 (N = 235) (Inza et al. 2018). These values are lower than expected for long-lived perennial outcrossing species that are dispersed through gravity or by animals such as *A. angustifolia* (He = 0.16–0.27, Nybom 2004). Besides, they are even lower than those reported for Brazilian natural populations of the species by means of dominant markers (Ht = 0.26, Medri et al. 2003; Ht = 0.30, Stefenon et al. 2007; Ht = 0.27, Ferreira de Souza et al. 2009). Other *Araucaria* species, such as *A. cunninghamii* (Pye et al. 2009) and *A. bidwillii* (Pye and Gadek 2004) from Australia and *A. araucana* from Argentina (Bekessy et al. 2002), also presented

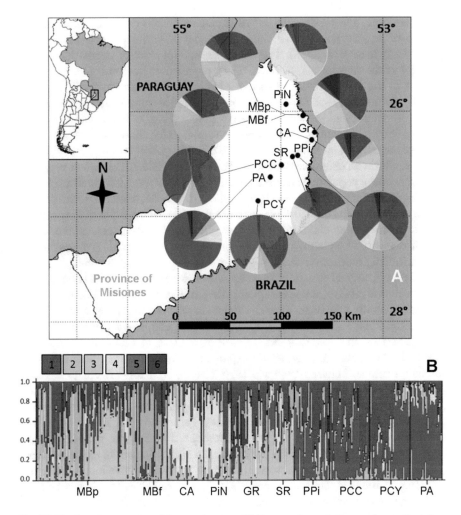

Fig. 15.10 Genetic structure of *Araucaria angustifolia* natural populations and one plantation, inferred from Bayesian analysis (K = 6). (**a**) Sampled populations and geographic distribution of genetic groups. Circles represent the percentages of assignment of populations to clusters. (**b**) Individual assignment of genetic clusters. Each vertical bar represents an individual

higher genetic diversity. However, all these works in *Araucaria* genus covered the whole species natural ranges, whereas Argentinean *A. angustifolia* comprised a marginal zone at the edges of its natural distribution. In agreement with this, Ferreira de Souza et al. (2009) reported low to moderate genetic diversity levels of isolated populations on the northernmost extreme of *A. angustifolia* distribution in Brazil (Ht = 0.16).

Furthermore, genetic diversity for each population was estimated with He, percentage of polymorphic loci (PPL), and number of exclusive loci (EL) indexes (Table 15.1; Inza et al. 2018). The results revealed a trend of decreasing genetic diversity of the Argentinean *A. angustifolia* populations from east to west, as the distance to the main area of the species distribution in southern Brazil increases. Thereby, the most diverse populations, Gramado and Manuel Belgrano natural forest, are those located closest to Brazil, whereas Piñalito Norte and provincial reserves Cruce Caballero, Caá Yarí, and Araucaria, which show less diversity, are at the edges of the species range (Inza et al. 2018). This result is consistent with haplotype studies with chloroplast markers (Ferrero Klabunde 2012), which suggested that Argentinean populations have migrated from glacial refuges in southern Brazil. This displacement would have led to a genetic drift process, associated with founder effects or population bottlenecks (Newton et al. 1999), which explains the observed genetic diversity gradient. Similarly, Marchelli et al. (2010) associated the loss of genetic diversity in western *A. araucana* populations of Patagonia (Argentina) to a recolonization process from the non-glaciated areas in the eastern vicinity.

Logging history of Argentinean *A. angustifolia* populations could also explain their current diversity (Mac Donagh and Rivero 2005), since the initial levels of genetic variability may be modified by fragmented landscapes (Degen et al. 2006). These effects were observed in *A. araucana* in Argentina and Chile by Bekessy et al. (2002) and in *A. angustifolia* in Brazil by Sousa et al. (2004) and Auler et al. (2002). In Argentina, *A. angustifolia* exploitation began and was more intense in southern Misiones between 1940 and 1970 (Rau 2005), likely affecting its genetic diversity. Despite some remnant populations from this area are protected, the reserves are small and/or recently created, showing different intensities of previous degradation (Bertolini 2000; Inza et al. 2018). Particularly, the fact that Cáa Yarí Provincial Reserve is within the Yabotí Biosphere Reserve that protects a larger continuous forest (SAyDS 2007) may explain its higher number of exclusive bands (EL = 19; Inza et al. 2018). However, lower values of exclusive loci in more disturbed populations of unprotected areas have been observed and could be explained by direct tree removal or a preliminary genetic drift process. In accordance to this, Bittencourt and Sebbenn (2009) observed loss of rare alleles in fragmented populations in contrast to continuous *A. angustifolia* populations, in Brazil. On the other hand, because of different logging histories, northern populations of the Argentinean range present more variable diversity levels than southern ones. Populations with long disturbance history, like Piñalito Norte and Campiñas de Américo, displayed much lower diversity than those less disturbed, like Manuel Belgrano natural forest, or historically more diverse due to their original high density of trees, like Gramado population (Ragonese and Castiglione 1946; Fernández et al. 2005). Nevertheless, if the valuable gene pool of the populations located in private properties remains unprotected, their singular genetic variability will drop, increasing the risk of local population extinction, according to different future scenarios that were predicted for the Province of Misiones by Izquierdo et al. (2011).

The genetic differentiation among Argentinean natural populations of *A. angustifolia* was examined by an analysis of molecular variance (AMOVA, GenAlEx 6.3 software, Peakall and Smouse 2006). Most of the genetic variation (91%) was distributed within populations, showing moderate and significant genetic differentiation among them ($\Phi_{PT} = 0.090$, $p \leq 0.001$; Inza et al. 2018). This is in accordance to the expected values for gymnosperms and long-lived perennials, woody, and outcrossing species (Hamrick et al. 1992; Nybom 2004). A moderate level of genetic differentiation might be associated with short distances of seed dispersal (Nybom 2004). On the other hand, while we have expected some restriction of pollen flow due to forest fragmentation in Argentina (Inza et al. 2018), this was not clearly evidenced because the estimated historical gene flow (Nm = 3.5, Crow and Aoki 1984) was above from minimum proposed (Nm = 1) to reduce genetic structuring by drift (Young et al. 1996). Besides, this value agreed with gene flow estimations from similar studies in Brazil (Auler et al. 2002; Bittencourt and Sebbenn 2009; Stefenon et al. 2009).

The differentiation among Argentinean populations was lower than that exhibited among Brazilian populations, both estimated with AFLP (10% by Stefenon et al. 2007; 19% and 12% by Ferreira de Souza et al. 2009) and with SSRs (13% by Stefenon et al. 2007). These differences could be explained by the geographical distances between the populations surveyed, which are quite larger in Brazil. Even though Bekessy et al. (2002) studied a similar range (~150 km) of *A. araucana* populations throughout Chile and Argentina with RAPDs, they observed a higher genetic differentiation (12.8%). However, clustering of the populations at both sides of the Andes may explain this result, acting the mountain range, instead of the distance, as a gene flow barrier.

Furthermore, an analysis of the genetic structure among the Argentinean populations was performed (Inza et al. 2018) by means of pairwise population differentiation (F_{ST} indexes), a dendrogram of Nei's genetic distances among populations (Rohlf 1998, data no shown, see Inza et al. 2018) and a Bayesian cluster analysis (Pritchard et al. 2000; Falush et al. 2007). The Bayesian analysis allowed us to recognize a structure of six genetic clusters (Fig. 15.10). As expected, the populations that registered more clusters in their composition were the nearest to Brazil and less disturbed, which have also shown the higher diversity (Manuel Belgrano natural forest and Gramado). On the other hand, the highly logged populations showed fewer clusters but with different composition. Although this could be associated to a genetic erosion process during migration events from Brazil, further studies including Brazilian populations should be performed to acquire comprehensive knowledge (Inza et al. 2018). All these results are evidence of a genetic structure according to geographical location and logging history. Besides, in agreement with studies of Brazilian *A. angustifolia* populations, genetic differentiation increased with the geographical distance between them (Stefenon et al. 2007; Ferreira de Souza et al. 2009).

15.5.1   The Outstanding Case of a Multi-provenance Plantation Over 70 Years Old

Given the loss of natural populations of *A. angustifolia* throughout its whole distribution area, plantations could be an important source of genetic diversity for the conservation and breeding of the species (Stefenon et al. 2009), in particular those older than 50 years, i.e., those planted when seed sources were much more diverse than nowadays. This is the reason why the old plantation of Manuel Belgrano Forest Station was included in the genetic diversity study performed by Inza et al. (2018).

The remaining stands of the old plantation of Manuel Belgrano Forest Station cover an area of 448 ha and were planted in the years 1947, 1949, and 1950, with seeds of multiple provenances. The still existing registers identify five different provenances (Pinazo and Fassola, pers. comm.):

1. Local: probably seeds from the natural forest of the Manuel Belgrano Forest Station
2. Regional: probably seeds from unidentified natural forests of the Province of Misiones, excluding Manuel Belgrano Forest Station
3. Brazil: seeds from unidentified stands from Brazil, probably commercialized for feeding
4. Barracón – Brazil: probably seeds from natural stands of the small village Barracão, some 30 km from Manuel Belgrano Forest Station
5. Buenos Aires, 25 de Mayo Forest Station: seeds from a plantation located in the old Forest Station of 25 de Mayo city, in the Province of Buenos Aires

This old plantation displayed higher genetic diversity indexes (He = 0.155; PPL = 80.2; EL = 32) than all the analyzed natural populations of Argentina (Table 15.1) and showed low genetic differentiation (3%, $p \leq 0.001$) with respect to all the natural populations taken together. This result is consistent with the multiple provenances comprised in this plantation, potentially including uncertain localities of the huge distribution of the species in Brazil. In contrast with these results, southern Brazilian plantations reported higher genetic diversity (Hj = 0.240, Stefenon et al. 2008), which could be due to seeds collected from the center of the species' distribution.

Furthermore, despite the fact that the plantation seems to contain a suitable representation of the genetic diversity of *A. angustifolia* from Misiones, its germplasm could principally come from few populations (Manuel Belgrano natural forest, Gramado, and Santa Rosa), as indicated by genetic structure and relationship evaluations (Fig. 15.10, Inza et al. 2018). However, since the admixture of seeds from several natural populations can lead to increase the variability (Sebbenn et al. 2003; Stefenon et al. 2008), this could explain the high genetic diversity found (Inza et al. 2018). The higher levels of genetic diversity of this plantation and its genetic similarity with the natural surrounded populations were also observed with SSRs focused only on the Manuel Belgrano Forest Station (Sarasola et al. 2011). In this work, the authors observed moderate to high levels of genetic diversity for this plantation (N = 99; He = 0.743) that were similar to natural stands (N = 31; He = 0.708), as well as very low genetic differentiation between them (F_{ST} = 0.013).

15.6 Breeding and Conservation Strategy for *Araucaria angustifolia*

15.6.1 First Steps Toward Improvement of the Species in Brazil

In the beginning of the 1970s, several provenance and progeny tests were planted in Brazil because of the productive potential of the species and the restriction on the exploitation of its native remnants (Aguiar and Sousa 2016). The main objective was conserving the genetic variability of the natural populations, disregarding the concern of establishing an experimental network for improvement programs in the short and long terms.

In 2010, taking advantage of these trials, EMBRAPA Florestas (Empresa Brasileira de Pesquisa Agropecuária) and its collaborators proposed an improvement program focused on the productivity and quality of wood and the production of pine nuts. The following activities were sequentially carried out: (1) genetic characterization of the tests at the molecular level with different types of markers (microsatellites, SNPs, etc.); (2) evaluation of quantitative characters (i.e., height, DBH, bark thickness, crown diameter, wood characteristics); (3) evaluation of physicochemical properties of wood using NIRS (near- infrared spectroscopy); (4) implementation of first- and second-generation progeny tests; and (5) cryopreservation of embryos, propagation, and early flowering. The results showed significant differences between provenances for quantitative characters, mainly between the provenances of the southern and southeastern region. In this way, two distinct groups were identified for improvement. According to Sousa and Aguiar (2012), the first group corresponds to the "northern" region with populations from the states of São Paulo and Minas Gerais, while the second group involves the "southern" region including the populations of the states of Paraná and Santa Catarina (Valgas 2008; Valgas et al. 2009; Sousa et al. 2009). These results have verified the tendencies observed in the provenance progeny tests implanted in 1973 and evaluated by Sebbenn et al. (2004). Thus, in Brazil, Sousa and Aguiar (2012) suggest using two different groups of populations (northern and southern region) in breeding programs, so that in the future, crosses between more distant individuals could be made aiming at the exploration of heterosis and incorporation of genes of adaptation for specific regions. In addition, these authors argue that taxonomic identification should be considered in the genetic improvement for this species. In this context, we can hypothesize that Argentinean populations could be considered as part of the southern region of Brazil, and exchange of genetic material could be an option to increase variability in the breeding program for both countries.

15.6.2   Provenance Tests in Argentina and Genetic Variation of Quantitative Traits

Observation of trees in even-aged natural stands or plantations reveals large differences among individual size, stem straightness, growth phenology, pest and disease tolerance, and other relevant traits. If only a small proportion of these differences are under genetic control, the extent of genetic variation for quantitative traits must be considered (Zobel and Talbert 1984). Knowledge of quantitative genetic variation plays a central role since breeding programs are conducted based on information provided by provenance trials (White et al. 2007).

In Argentina, during the 1970s, given the initial importance of *A. angustifolia* plantations, INTA initiated studies to evaluate the behavior of 32 provenances from the southern states of Brazil and 3 local provenances from Argentina. Fahler (1981) studied the variation between and within those 35 provenances of *A. angustifolia*, up to the age of 8 years by means of two trials. They were installed in 1972, one in Puerto Libertad (25° 55′ 48″ S, 54° 30′ 4″ W, 290 m asl), with all 35 provenances, and the other in San Vicente (26° 55′ 28″ S, 54° 25′ 50,6″ W, 534 m asl), with 31 provenances. Seed collection intended to sample most of the natural range, considering minimum distances between provenances (50 km) and trees within provenances (50 m). In both trials, provenances were represented by 10 open pollinated families, and, in all cases, seeds were collected in natural stands. The experimental design was compact family blocks with four individuals per family, 10 families per provenance, and four replicates. The plantation frame was 3.5 × 3.5 m, which allowed the study of genetic materials with low levels of competition between trees up to the 8 years of age and without needing thinning.

The main results of these trials were the following:

– Significant differences between provenances were shown for total height at the first year of planting and for diameter and volume from the fourth year in San Vicente and from the fifth year in Puerto Libertad. The heights of the provenances were markedly homogenous. In contrast, the diameters were different from each other, and the productivity measured as individual tree volume was much more sensitive to detect the differences between provenances.
– In both trials, the partition of the total variance in genetic and environmental variance showed that in all cases, the genetic effects, due to provenances, were those with the greatest participation in the observed phenotypic variation. For volume, between 59% and 67% of the variation was explained by the "provenance" factor.
– No genotype-environment interaction was detected. In both trials, the provenances located to the west of natural dispersal area of *A. angustifolia* were those with the best behavior, while those from the east were the ones with the lowest growth. Spearman correlation coefficients between sites for 31 provenances common to both trials reached values of 0.83, 0.88, and 0.87 for total height, diameter, and individual tree volume, respectively.

- Productivity was very influenced by the plantation site. Until the age of 8 years, the difference in individual volume production was 60% higher in San Vicente, when considering the general means for both trials. None of the provenances tested in San Vicente was surpassed in productivity by the same origin when it was tested in Puerto Libertad.
- The variation within provenances was lower than among provenances. For the volumetric production, 20% of the provenances tested in San Vicente showed highly significant differences between the progenies, while in Puerto Libertad, no origin revealed variation among the progenies for volume production.
- The phenotypic correlations between the different growth parameters at different ages showed that the height cannot be used to early evaluate the future behavior of the provenances. After 4 months, growth in height is still a function of the weight of the seed. After 1 year on the plantation, the height of the provenances was highly correlated with the height, diameter, and individual volume for any age up to 8 years.
- The correlations between the characters of growth in the progenies within the provenances at different ages were quite variable, and not always the height of 1 year was correlated with the characters at other ages, being necessary the individual study of families to establish that relationships.
- The associations between the provenance behavior in the trials and the home climatic and geographical characteristics of the provenances showed that the productivity was highly correlated with the annual average temperatures and the geographic longitude of the provenances. Thus, the higher the annual average temperature and the longitude of the provenances, the better growth was shown. In association with the longitude, the smaller distance between the provenance and the site of implantation, the better its behavior in both tested sites.

These results enabled invaluable guidance to farmers and companies about the best sources of seeds for use as propagation material in generating commercial plantations of the species, remarking that the best genetic material to generate them are the local provenance from Misiones or the nearest.

Later, in 1997, according to the management plan of the Provincial Reserve "Cruce Caballero," the Ministry of Ecology and Renewable Natural Resources (Province of Misiones), the Faculty of Forestry Sciences of Eldorado (FCF, National University of Misiones), and the Center for Forestry Research and Experiences (CIEF, association of private companies), have gathered around an in situ and an ex situ germplasm conservation project for *A. angustifolia* (Bertolini 1999). This agreement also aimed to guarantee high genetic quality of seed supply for the commercial plantations in the Province (a purpose of interest for the most important forest companies of the region, i.e., Alto Paraná SA (now Arauco SA), Celulosa Puerto Piray SA, Pecom Forestal, and others) by creating seed-producing areas in the properties of the Province, FCF, and private companies. In the frame of the mentioned agreement, three progeny trials with seedlings from 80 selected trees of the natural forest of Cruce Caballero Provincial Park were established in 1998 and 1999. Plantations were also carried out in degraded areas of that Provincial Reserve with seedlings of local provenance. The first progeny trial was installed in Eldorado (Beatriz Eibl, pers. comm.), the second one was established in Puerto Bossetti, and a third was located in the Province of Buenos Aires. So far, very few results about these trials have been published. In 1999, an internal report from Arauco SA Company showed that the trial located in Puerto Bossetti had 401 individuals alive corresponding to 33 progenies. The practically null seed production in the following years in that reserve made it impossible to continue with the installation of further trials as it was planned.

15.6.3   Base, Selected, Breeding, and Propagation Populations for an Incipient Improvement Program in Argentina

After the decline in the cultivation of the species (since 1970), a new impulse in the interest of cultivating native species as demanded by civil society started over the last 15 years. Then, institutions such as FCF and INTA have resumed conservation and breeding actions for *A. angustifolia* with varying intensity levels. Eibl et al. (2016, 2017) mentioned the installation of a trial in 2014 with progenies coming from Argentine natural stands of Bernardo de Irigoyen and San Pedro. Meanwhile, in 2011, INTA initiated a conservation and breeding program for *A. angustifolia* based on the old provenance and progeny tests, reproductive (palynological) studies, and genetic diversity analysis achieved with molecular markers. The new program intends to establish the bases for the management of the genetic resources of the species in Argentina.

The main objectives of the program are (1) the generation of a base population to initiate the genetic improvement of the species considering multiple cycles, (2) the conservation of its genetic resources through in situ and ex situ strategies, and (3) the creation of propagation populations for the production of seeds utilized in forestation and reforestation programs.

According to Zobel and Talbert (1984), the base population of a given cycle of improvement consists of all available trees that could be selected. Generally, the base population is very large, consisting of many thousands of genetically distinct individuals. For *A. angustifolia* in Argentina, the base population could be built by the old plantation and natural individuals of Manuel Belgrano Forest Station, the old commercial plantations that remain in private companies, the mentioned provenance tests installed in Argentina, and some dispersed, known individuals from the natural forests. So far, the first activity of this program was the massal selection of the best phenotypes within the Manuel Belgrano Forest Station, in the old plantations, since this source is shown to be adapted to the local environmental conditions. Also, some old commercial plantations belonging to private companies were also reviewed. The selected population was built based on independent selection criteria and with the help of data from preexisting inventories. The first selection criterion among the planted trees was the average DBH of the stand plus two standard deviations. Also, health and shape characters were considered, such as stem straightness and taper

Fig. 15.11 (**a**) Selected tree at Manuel Belgrano Forest Station (Misiones). (**b**) Seeds harvested to produce seedlings for progeny tests. (**c**) Scions from a selected tree to be grafted. (Photos: Cristian Rotundo)

and natural pruning. A total of 39 stands were screened, and 302 individuals were preselected. After checking their characteristics, 233 trees were selected (Fig. 15.11a) at Manuel Belgrano Forest Station and 17 in productive plantations of more than 50 years located in private properties.

According to Rotundo et al. (2014, 2016), the selected population surpasses the base population by almost 50% in DBH, obtaining an estimated genetic gain of 21.7%. This estimated value would imply 10 cm of superiority in average DBH of the offspring with respect to the base population. The selection differential was 22.5 cm (Fig. 15.12).

Following White et al. (2007), some or all of the individuals of the selected population will be included in the breeding population allowing them to intermate (crossbred) to regenerate genetic variability through recombination of alleles during sexual reproduction. Offspring from these intermatings will be planted in genetic tests to form the base population of the next breeding cycle.

In order to obtain breeding values of selected female trees, in 2012 and 2014, seeds were collected from all the selected trees with seeds (Fig. 15.11b). Since the seeds from *A. angustifolia* are recalcitrant, sowing must be done the same year of collection, and consequently progeny trials were installed in two different years. This sampling allowed installing three progeny trials representing 44 open pollinated families from the selected trees. The experimental design corresponds to randomized complete blocks with single-tree plots. The first trial was installed in 2013 in the Manuel Belgrano Forest Station with 11 progenies (Fig. 15.13), while the remaining two were implanted in 2015 with 34 and 28 progenies in the same Station and in a private property (PINDO SA) of Puerto Mado (26° 14′ 47″ S; 54° 30′ 27″ O), respectively. The main objectives of these tests are the evaluation of

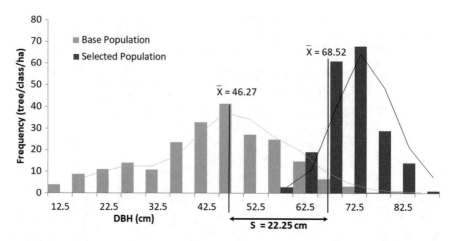

Fig. 15.12 Diametric distribution of base and selected population of *A. angustifolia*. S: selection differential

Fig. 15.13 Progeny test of *A. angustifolia* installed in Manuel Belgrano Forest Station at 4 years of age. (Photos: Mario Pastorino)

mothers and progenies for backward and forward selection in the future. In the long term, the expectation is to promote crossings for the generation of new variability, being the basis for the improvement of the species.

With the main objective of producing improved seeds, the first Clonal Seed Orchard (CSO) of *A. angustifolia* from Argentina is being installed since 2015 in Manuel Belgrano Forest Station. It also functions as a genetic bank of the species for conservation purposes. The genotypes that constitute this CSO correspond to the best female and male individuals of a phenotypic ranking, which includes all the selections made up to the present. The propagation of genotypes was carried out by grafting, using sprouts from primary branches with an orthotropic tendency as scions (Fig. 15.11c). The technique of full apical insertion was used on 2-year-old

Fig. 15.14 Apical grafting of selected *A. angustifolia* genotypes (photo: Cristian Rotundo) and clonal seed orchard at the age of 4 years installed in Manuel Belgrano Forest Station. (Photo: María Elena Gauchat)

rootstocks of the same species (Fig. 15.14), propagating 64 genotypes (male and female), achieving the production of 462 ramets. These grafts were established in the field in 2015 and 2016 (Fig. 15.14). Wendling (2011), using the same methodology, achieved male and female strobili flowering in 4 and 6.5 years respectively. The same trend was observed in Manuel Belgrano Forest Station.

15.7   Concerns and Actions for Conservation

Araucaria angustifolia natural stands are preserved in few and frequently small areas in Argentina (SAyDS 2007) and Brazil (Stefenon et al. 2009). These areas are not enough for conserving in situ the complete genetic pool of the species. Since the Argentinean *A. angustifolia* populations are at the edges of the species distribution, they likely harbor a great adaptive value, which is relevant to ensure the persistence of the species. From all remaining natural stands, conservation efforts should focus on the preservation of the natural populations with the highest variability, i.e., Manuel Belgrano Forest Station and Gramado, being the latter a private property.

We propose to increase the extension of the current Provincial Reserves, reorganizing the maintenance of their exclusive gene pools. These enlargements should enhance the connectivity of the fragments, and, in addition, restoration of degraded areas with the appropriate genetic materials should be considered. The old plantation of Manuel Belgrano Forest Station has been demonstrated to be highly valuable from a conservation genetics point of view and as a seed source for breeding programs. To improve its regeneration and future development, an enrichment with new material from natural populations from different genetic groups might be done.

According to the genetic structure analysis, more variable or differentiated fragments should be prioritized. Likewise, management of the plantation should ensure its natural regeneration. Understory cleanings could be scheduled for the productive pollen years in order to favor the development of seedlings after seed dispersal, thus contributing to the natural regeneration of the species.

On the other hand, in view of the scarce natural regeneration observed in the Argentinean stands (Fassola et al. 1999; Latorre et al. 2013), an ex situ conservation strategy must be considered. The ex situ production of seeds for the active restoration of native forests would be recommendable as well, based on the aerobiological pollen background presented in this chapter. Limitations of pollen transport for effective pollination in *A. angustifolia*, a direct relationship between pollination and seeds abundance, and the influence of temperature on the different reproductive stages of the species were confirmed. Simón et al. (2018) concluded that *A. angustifolia* stands growing in a temperate climate with cold winters produce a greater amount of pollen that leads to enhanced seed production. Also, they observed that the current climate conditions of the southern extreme of *A. angustifolia* natural range do not favor pollen production, thus affecting its reproduction. On the contrary, the temperate climate of the north center of the Province of Buenos Aires (like that of 25 de Mayo) could be a suitable location for the establishment of an ex situ seed production area. Ex situ conservation banks would be a good complementary strategy to the in situ reserves, becoming a "safeguard" of the genetic variability of the fragments in unprotected areas or close to the region of great demographic pressure (Bittencourt and Higa 2004). Thus, conservation banks and seed production areas can be combined in ex situ plantations located in appropriate sites. Consequently, a replicate of the CSO of *A. angustifolia* of Manuel Belgrano Forest Station was installed in 2020 in the 25 de Mayo Forest Station of INTA, 1450 km further south. It is constituted by 196 ramets of 30 of the 64 genotypes present in the CSO installed in Manuel Belgrano Forest Station, which were replicated by grafting.

In the context of climate change, progressive warming affecting the natural range of *A. angustifolia* (IPCC 2001–2007) has a deep impact into its regeneration. Therefore, the alternative of an assisted migration southward from its place of origin, to temperate climates where the environmental conditions favor its seed productivity, should be discussed. This strategy has been decided for the conservation of *Araucaria araucana* in Chile, where the natural populations of this sister species are being decimated by the occurrence of a foliar damage presumably related to the climate change (Ipinza 2017, Chapter 7A). In any case, it is advisable to maintain the existing plantations in temperate environments (i.e., 25 de Mayo, Buenos Aires) as genetic reservoirs and ex situ seed production areas for native forest enrichment or restoration programs, as well as for reforestation purposes in the Argentinean range of the species, where natural regeneration is minimal (Pinazo et al. 2016). In situ and ex situ conservation units are effective tools to prevent the extinction of the species natural populations.

References

Achten W (1995) Untersuchungen zur Ökologie und Schadwirkung der Kleinschmetterlinge an der Araukarie (*Araucaria angustifolia* (Bert.) O. Ktze.) in Südbrasiliens. Doctoral thesis, Albert-Ludwigs-Univesität, Freiburg i. Br

Aguiar AV, Sousa VA (2016) Propuesta de mejoramiento genético de *Araucaria angustifolia* en Brasil. In: XVII Jornadas Técnicas Forestales y Ambientales, Posadas, Misiones

Aguirre NC (2014) Diversidad genética de Pino Paraná (*Araucaria angustifolia*) en la Selva Paranaense: Análisis genómico mediante marcadores moleculares AFLPs. Graduate thesis, Fac. Cs. Exactas, Químicas y Naturales, Universidad de Morón, Buenos Aires

Allison TD (1990) Pollen production and plant density affect pollination and seed production in *Taxus canadensis*. Ecology 71:516–522

Anselmini J (2005) Fenología reproductiva da *Araucaria angustifolia* (BERT.) O. KTZE, NA Regiao de Curitiba-PR. MSc thesis, Universidad Federal de Paraná

Anselmini JI, Zanette F (2012) Controlled pollination in *Araucaria angustifolia*. Cerne 18:247–255

Anselmini JI, Zanette F, Bona C (2006) Fenologia reproductiva da *Araucaria angustifolia* (Bert.) O. Ktze, na região de Curitiba – PR. Floresta e Ambiente 13:44–52

Auler NMF, Reis MS, Guerra MP, Nodari RO (2002) The genetics and conservation of *Araucaria angustifolia*: I. Genetic structure and diversity of natural populations by means of non-adaptive variation in the State of Santa Catarina, Brazil. Genet Mol Biol 25:329–338

Bekessy SA, Allnutt TR, Premoli AC, Lara A, Ennos RA, Burgman MA et al (2002) Genetic variation in the vulnerable and endemic Monkey Puzzle tree, detected using RAPDs. Heredity 88:243–224

Bertolini MP (1999) Plan de Manejo del Parque Provincial Cruce Caballero. Ministerio de Ecología y R.N.R. de la Provincia de Misiones. pp 103

Bertolini MP (2000) Documento Base para la Discusión del Plan de Manejo del Parque Provincial de la Araucaria. Ministerio de Ecología y Recursos Naturales Renovables de la Provincia de Misiones. Administración de Parques Nacionales-Delegación Regional Nordeste Argentino

Bittencourt JVM, Higa A (2004) Final report of the IPGRI/BMZ-funded project, conservation, management and sustainable use of forest genetic resources with reference to Brazil an Argentina. International Plant Genetic Resource Institute (IPGRI), Rome

Bittencourt JVM, Sebbenn AM (2007) Patterns of pollen and seed dispersal in a small, fragmented population of the wind-pollinated tree *Araucaria angustifolia* in southern Brazil. Heredity 99:580–591

Bittencourt JVM, Sebbenn AM (2009) Genetic effects of forest fragmentation in high-density *Araucaria angustifolia* populations in Southern Brazil. Tree Genet Genomes 5:573–582

Burkart R (1993) Plan de Manejo para la producción sustentable de semilla de forestales nativas y la conservación de recursos genéticos. Informe final. Consejo Federal de Inversiones, Gobierno de la Provincia de Misiones

Cabrera A, Willink A (1980) Biogeografía de América Latina. 2a edición corregida. Monografía 13. Serie de Biología. Secretaría General de la Organización de los Estados Americanos. Washington DC. EEUU. 120 pp.

Caccavari MA (2003) Dispersión del polen de *Araucaria angustifolia* (Bert.) O'Kuntze. Rev. Mus. Argentino de Ciencias Naturales "B. Rivadavia" 5: 135–138

Caccavari M, Dome E, Del Fueyo G, Gauchat ME (2000) Biología reproductiva de *Araucaria angustifolia*. Estudios palinológicos: viabilidad del polen, fertilización; Fenología de la polinización. Relatorio PROYECTO IPGRI: Conservación, manejo y uso sustentable de forestas con *Araucaria angustifolia*, pp. 20

Chaves A, Mugridge A, Fassola H, Alegranza D, Fernandez R (1999) Conservación refrigerada de semillas de *Araucaria angustifolia* (Bert.) O. Kuntze. Bosque 20:117–124

Chebez JC, Hilgert N (2003) Brief history of conservation in the Parana forest. In: Galindo-Leal C, de Gusmão CI (eds) The Atlantic Forest of South America: biodiversity status, threats, and outlook. Island Press, Washington, pp 141–159

Cordenunsi BR, De Menezes WE, Genovese MIS, Colli C, De Souza GA, Lajolo FM (2004) Chemical composition and glycemic index of Brazilian pine (*Araucaria angustifolia*) seeds. J Agric Food Chem 52:3412–3416

Cour P (1974) Nouvelles techniques de détection des flux et des retombées polliniques: étude de la sédimentation des pollens et des spores a la surface du sol. Pollen Spores 16:103–142

Cour P, van Campo M (1980) Prévisions de récoltes á partir du contenu pollinique de l'atmosphere. C R Acad Sc Paris 290:1043–1046

Cozzo D (1960) Ubicación y riqueza de los bosques espontáneos de "pino" paraná (*Araucaria angustifolia*). Rev For Arg 4:46

Cozzo D (1975) Árboles forestales, maderas y silvicultura de la Argentina. In: Enciclopedia Argentina. Agricultura para Jardinería 2, (16-1): 1–156

Cozzo D (1976) Tecnología de la Forestación en Argentina y América Latina. Buenos Aires. Editorial Hemisferio Sur. 610 p

Crechi E, Friedl R, Fassola H, Fernández R, Dalprá L (2001) Efectos de la intensidad y oportunidad de raleo en *Araucaria angustifolia* (bert.) O. Ktze. sobre el crecimiento y la producción en el noroeste de misiones, argentina. Informe Técnico N° 34. INTA EEA Montecarlo. 30 p

Crechi E, Keller A, Fassola H, Fernández R, Friedl R (2009) Efectos del raleo sobre el crecimiento y la producción de madera de *Araucaria angustifolia* (Bert.) O. Ktze. en el noroeste de Misiones, Argentina. In: XIII World Forestry Congress

Crow JF, Aoki K (1984) Group selection for a polygenic behavioral trait: estimating the degree of population's subdivision. Proc Natl Acad Sci U S A 81:6073–6077

Degen B, Blanc L, Caron H, Maggia L, Kremer A, Gourlet-Fleury S (2006) Impact of selective logging on genetic composition and demographic structure of four tropical tree species. Biol Conserv 131:386–401

Del Fueyo GM, Caccavari MA, Dome EA (2008) Morphology and structure of the pollen cone and pollen grain of the *Araucaria* species from Argentina. Biocell 32:49–60

Di Bitetti MS, Placci G, Dietz LA (2003) Una Visión de Biodiversidad para la Ecorregión del Bosque Atlántico del Alto Paraná: Diseño de un Paisaje para la Conservación de la Biodiversidad y prioridades para las acciones de conservación. World Wildlife Fund, Washington, DC.

Duarte LS, Dillenburg LR, Rosa LMG (2002) Assessing the role of light availability in the regeneration of *Araucaria angustifolia* (Araucariaceae). Aust J Bot 50:741–751

Eibl BI, Lopez MA, Steffen C, Gonzalez C, Stadler NC (2016) Germinación y desarrollo de plantas de *Araucaria angustifolia* (Bertol.) Kuntze en tubetes, para progenies de dos orígenes de la Provincia de Misiones. In: Actas XVII Jornadas Técnicas Forestales y Ambientales, Posadas, Argentina. pp. 307–309

Eibl BI, Lopez MA, Stadler NC, Lopez MP, Gonzalez C (2017) Plantaciones productivas de *Araucaria angustifolia* (Bertol.) Kuntze utilizando progenies de dos orígenes de la Provincia de Misiones, Argentina. In: Libro de resúmenes IV Congreso Nacional de Ciencias Agrarias. San Lorenzo, Paraguay. pp 469–472

Faegri K, van der Pijl L (1979) The principles of pollination ecology. Pergamon Press, New York, p 24

Fahler JC (1981) Variación Geográfica entre y dentro de poblaciones de *Araucaria angustifolia* (Bert.) O. Ktze a los 8 años de edad en la provincia de Misiones, Argentina. MSc thesis, Universidad Federal de Paraná, Brasil

Fahler JC, Di Lucca CM (1980) Variación geografica de *Araucaria angustifolia* (Bert.) O. Ktze.: informe preliminar a los 5 años. In: IUFRO meeting of forestry problems of the genus *Araucaria*. FUPEF, Curitiba, pp 96–101

Falush D, Stephens M, Pritchard JK (2007) Inference of population structure using multilocus genotype data: dominant markers and null alleles. Mol Ecol Notes 7:574–578

Fassi G, Fassola HE, Marangoni G (1993) Determinación del cilindro con defectos en *Araucaria angustifolia* Bert. O. Ktze, mediante técnicas de aserrado. Actas: 1er Congreso Argentino y Latinoamericano. Comisión Bosque Nativo, Vol 5: 47-55. Paraná, Argentina

Fassola HE, Ferrere P, Muñoz D, Pahr N, Kuzdra H, Márquez S (1999) Observaciones sobre la producción de frutos y semillas en plantaciones de *Araucaria angustifolia* (Bert) O. K. (período 1993–1998). Informe Técnico N° 24 INTA EEA Montecarlo, Montecarlo, Misiones, Argentina. 12 p

Fernández R, Fassola H, Moscovich F, Pinazo M, Pahr N (2005) Campo Manuel Belgrano Compromiso Público con la Conservación Genética de la Araucaria. Idia XXI Forestales Año 5:272–275

Ferreira de Souza MI, Salgueiro F, Carnavale-Bottino M, Félix DB, Alves-Ferreira M, Bittencourt JVM, Margis R (2009) Patterns of genetic diversity in southern and southeastern *Araucaria angustifolia* (Bert.) O. Kuntze relict populations. Genet Mol Biol 32:546–556

Ferrero Klabunde GH (2012) Análise filogeográfica entre populações de *Araucaria angustifolia*. In: (Bert.) O. Kuntze em sua área de distribuição natural. MSc thesis, Universidade Federal de Santa Catarina, Brasil

Ferreyra RM, Viña SZ, Fassola H, Chaves A, Mugridge A (2006) Aportes nutricionales de semillas de *Araucaria angustifolia* (Bert.) O. Kuntze. In: 12as Jornadas Técnicas Forestales y Ambientales. Eldorado, Argentina

Garcia C, Coelho CMM, Maraschin M, Oliveira LM (2014) Conservação da viabilidade e vigor de sementes de *Araucaria angustifolia* (Bert.) O. Kuntze durante o armazenamento. Ciência Florestal 24:857–866

Goya J, Sandoval M, Arturi M, Burns S, Russo F, Santacá M, et al. (2012) Plan de Manejo Forestal del Campo Anexo Manuel Belgrano perteneciente a la EAA Montecarlo del INTA, Misiones. Laboratorio de Investigación de Sistemas Ecológicos y Ambientales (LISEA), Facultad de Ciencias Agrarias y Forestales. Universidad Nacional de La Plata, Buenos Aires

Gurgel JTA, Gurgel OA (1971) Raças geográficas em Pinheiro Brasilero, *Araucaria angustifolia* (Bert.) O. Ktze. In: FIEP (ed): Annais do Primeiro Congresso Florestal Brasileiro. Curitiba, pp 283–284

Gurgel JTA, Gurgel OA (1978) Ecotypes in Brazilian Pine, *Araucaria angustifolia* (Bert.) O. Ktze. In: Proceedings of the Seventh World Forestry Congress, Buenos Aires, Vol II (3): 2254–2255

Hamrick JL, Godt MJW, Sherman-Broyles SL (1992) Factors influencing levels of genetic diversity in woody plant species. New Forest 6:95–124

Hueck K (1952) Verbreitung und Standortsansprüche der brasilianischen Araukarie (*Araucaria angustifolia*). Forstwiss Centralbl 71:272–289

Hueck K (1966) Die Wälder Südamerikas. Fischer Vlg, Stuttgart

Inza MV, Aguirre NC, Torales SL, Pahr NM, Fassola HE, Fornes LF, Zelener N (2018) Genetic variability of *Araucaria angustifolia* in the Argentinean Parana Forest and implications for management and conservation. Trees 32:1135–1146

IPCC (2001, 2007) Climate change 2001, vol 1. Cambridge University Press. UK. http://www.ipcc.ch/ipccreports/assessments-reports.htmhttp://www.ipcc.ch/ipccreports/assessments-reports.htm

Ipinza R (2017) Migración asistida: una opción para la conservación de la Araucaria. Reunión Internacional Daño Foliar de *Araucaria araucana*. Villarrica, 7–9 Nov 2017. https://www.researchgate.net/publication/320991350_

Izquierdo AE, Grau HR, Mitchell Aide T (2011) Implications of rural-urban migration for conservation of the Atlantic forest and urban growth in Misiones, Argentina (1970–2030). Ambio 40:298–309

Kranitz ML, Biffin E, Clark A, Hollingsworth ML, Ruhsam M et al (2014) Evolutionary diversification of new Caledonian *Araucaria*. PLoS One 9(10):e110308. https://doi.org/10.1371/journal.pone.0110308

Latorre F (1999) El polen atmosférico como indicador de la vegetación y de su fenología floral. Doctoral thesis, Universidad de Buenos Aires. pp. 244

Latorre F, Fassola H (2014) Propuesta para estimar anticipadamente la cosecha de semillas de *Araucaria angustifolia* a partir del registro polínico. Relación con el clima. Impacto sobre las prácticas de manejo. INTA Edit. http://www.intaexpone.gob.ar/ MAGPyA. 1-8

Latorre F, Alarcón P, Fassola H (2013) Distribución temporal y espacial del polen de *Araucaria angustifolia* (Araucariaceae) en Misiones, Argentina. Bol Soc Argent Bot 48:453–464

Latorre F, Abud Sierra ML, Alarcón P, Fassola H (2014) Estudio aerobiológico de *Araucaria angustifolia* (Bertol.) Kuntze en San Antonio, Misiones. In: 16ª. Jornadas Técnicas Forestales y Ambientales. Eldorado, Argentina

Latorre F, Rotundo C, Abud Sierra ML, Carrizo S, Fassola H (2016) Estudio aerobiológico de la producción de polen de *Araucaria angustifolia* en Misiones (Argentina). XXVII Reunión Argentina de Ecología. 18–22 septiembre. Puerto Iguazú, Argentina. Pág. 252

Latorre F, Rotundo C, Abud Sierra ML, Fassola H (2020) Daily, seasonal, and interannual variability of airborne pollen of *Araucaria angustifolia* growing in the subtropical area of Argentina. Aerobiologia 36:277–290. https://doi.org/10.1007/s10453-020-09626-y

Luna CV, Fontana ML (2017) Estado de los bosques de *Araucaria angustifolia*: especie nativa en peligro crítico. Revista Estudios Ambientales (CINEA) 5:79–93

Mac Donagh P, Rivero L (2005) ¿Es posible el uso sustentable de los Bosques de la Selva Misionera? In: Brown A, Martínez Ortiz U, Acerbi M, Corcuera J (eds) La situación ambiental argentina 2005. Fundación Vida Silvestre Argentina, Buenos Aires, pp 210–217

Machado SA, Siqueira JDP (1980) Distribuição natural da *Araucaria angustifolia* (Bert.). In: Ktze O (ed) IUFRO meeting on forestry problems of genus *Araucaria*. FUPEF, Curitiba, pp 4–9

Marchelli P, Baier C, Mengel C, Ziegenhagen B, Gallo LA (2010) Biogeographic history of the threatened species *Araucaria araucana* (Molina) K. Koch and implications for conservation: a case study with organelle DNA markers. Conserv Genet 11:951–963

Martínez Crovetto R (1968) Introducción a la etnobotánica del Nordeste Argentino. Etnobiológica 11:1–10

Medri C, Ruas PM, Higa AR, Murakami M, de Fátima RC (2003) Effects of forest management on the genetic diversity in a population of *Araucaria angustifolia* (Bert.). O Kuntze. Silvae Genet 52:202–205

Ministerio de Ecología, Min. Ecol. (2001) Inventario Forestal de la Provincia de Misiones. Ministerio de Ecología, Recursos Naturales y Turismo. Provincia de Misiones

Moura VPG (1975) Capões remanescentes de *Araucaria angustifolia* (Bert.) O. Ktze, entre 19° e 20° de latitude, nas proximidades do Rio Doce, MG. Brasil Florestal 6:22–27

Newton AC, Allnutt TR, Gillies ACM, Lowe AJ, Ennos RA (1999) Molecular phylogeography, intraspecific variation and the conservation of tree species. Trees 14:140–145

Niklas KJ (1985) The aerodynamics of wind pollination. Bot Rev 51:328–386

Nybom H (2004) Comparison of different nuclear DNA markers for estimating intraspecific genetic diversity in plants. Mol Ecol 13:1143–1155

Peakall R, Smouse PE (2006) GENALEX6: genetic analysis in Excel. Population genetics software for teaching and research. Mol Ecol Notes 6:288–295

Pérez CF, Merino R, Gassmann MI, Latorre F (2016) Tendencias de anomalías de temperatura y precipitación en relación a la producción de semillas de pino Paraná. XVI Reunión Argentina de Agrometeororlogía. 20-23 setiembre. Puerto Madryn

Pinazo M, Inza MV, Latorre F, Rotundo C (2016) Antes había muchas. Vida Silvestre 136: 22–24. Edit. Buenos Aires, Fundación Vida Silvestre Argentina

Piriz Carrillo V, Chaves A, Fassola H, Mugridge A (2003) Refrigerated storage of seeds of *Araucaria angustifolia* (Bert.) O. Kuntze over a period of 24 months. Seed Sci Technol 31:411–421

Piriz Carrillo V, Fassola HE, Chaves AR (2004) Almacenamiento refrigerado de semillas de *Araucaria Angustifolia* (Bert.) O. Kuntze: conservación del poder germinativo. Buenos Aires: Red RIA. Revista de Investigaciones Agropecuarias 33:67–83

Pritchard JK, Stephens M, Donnelly P (2000) Inference of population structure using multilocus genotype data. Genetics 155:945–959

Pye MG, Gadek PA (2004) Genetic diversity, differentiation and conservation in *Araucaria bidwilli* (Araucariaceae), Australia's Bunya pine. Conserv Genet 5:619–629

Pye MG, Henwood MJ, Gadek PA (2009) Differential levels of genetic diversity and divergence among populations of an ancient Australian rainforest conifer, *Araucaria cunninghamii*. Plant Syst Evol 277:173–185

Ragonese A, Castiglione J (1946) Los pinares de *Araucaria angustifolia* en la República Argentina. Bol Soc Arg Bot 1:126–147

Rau MF (2005) Land use change and natural Araucaria forest degradation Northeastern Misiones, Argentina. Doctoral thesis, Faculty of Forestry, Albert-Ludwigs-University Freiburg in Breisgau, Germany

Rohlf FJ (1998) NTSYS-PC numerical taxonomy and multivariate analysis system version 2.0. Exeter Software, Setauket

Rotundo CA, Gauchat ME, Belaber E, Alarcón PC (2014) Avances en la selección de árboles plus de *Araucaria Angustifolia* (Bert.), en el NO de la provincia de Misiones. VI Reunión GEMFO/ Juan A. López (ed.); Silvia Cortizo (ed.). – Campana, Buenos Aires: Ediciones INTA

Rotundo C, Gauchat ME, Belaber E (2016) Mejoramiento Genético y Conservación de *Araucaria angustifolia* (Bert.) O. Kuntze en Argentina. Avances y perspectivas. In: XVII Jornadas Técnicas Forestales y Ambientales, Posadas, Misiones

SAGYP (Secretaría de Agricultura, Ganadería, Pesca y Alimentación de la Nación Argentina) (1997). Sector Forestal

Sant'Anna CS, Sebbenn AM, Klabunde GHF, Bittencourt R, Nodari RO, Mantovani A, dos Reis MS (2013) Realized pollen and seed dispersal within a continuous population of the dioecious coniferous Brazilian pine [*Araucaria angustifolia* (Bertol.) Kuntze]. Conserv Genet 14:601–613

Sarasola M, Zelener N, Fassola H, Pahr N, Fernandez R, Torales S (2011) Diversidad genética de *Araucaria angustifolia* (Bert.) o. Ktze en una Reserva Forestal Argentina. Análisis de semillas. Tomo 5, 18: 84–88

SAyDS (2007) Primer Inventario Nacional de Bosques Nativos. Informe Regional Selva Misionera. Secretaría de Ambiente y Desarrollo Sustentable de la Nación, Buenos Aires, Argentina

Sebbenn AM, Pontinha AAS, Giannotti E, Kageyama PY (2003) Genetic variation in provenance-progeny test of *Araucaria angustifolia* (Bert.) O. Ktze. In Saõ Paulo, Brazil. Silvae Genet 52:181–184

Sebbenn AM, Pontinha AAS, Freitas SA, Freitas JA (2004) Variação em cinco procedencias de *Araucaria angustifolia* (Bert.) O. Ktze. No sul do estado de Saõ Paulo. Rev Inst Flor 16(2):91–99

SIFIP (2010) Sistema de información foresto-industrial provincial. Min. del Agro y la producción, Misiones, Argentina. http://extension.facfor.unam.edu.ar/sifip/inventario.htm fecha de ingreso julio de 2011

Simón BE, Latorre F, Rotundo C (2018) Study of the reproductive phenology of *Araucaria angustifolia* in two environments of Argentina: its application to the management of a species at risk. Glob Ecol Conserv 16:e00483

Sousa VA, Aguiar AV (2012) Programa de melhoramento genético de araucária da Embrapa Florestas: situação atual e perspectivas. Embrapa Florestas Documentos, 237. 17 p

Sousa VA, Hattemer H (2001) Genetic variation in natural populations of *Araucaria angustifolia* (Bert.) O. Kuntze in Brazil. In: Müller-Starck G, Schubert R (eds) Genetic response of forest systems to changing environmental conditions, forestry sciences, vol 70. Springer, Dordrecht, pp 93–103

Sousa VA, Robinson IP, Hattemer HH (2004) Variation and population structure at enzyme gene loci in *Araucaria angustifolia* (Bert.) O. Ktze. Silvae Genet 53:12–19

Sousa VA, Valgas RA, Lavoranti OJ, Chaves Neto A, Shimizu JY (2009) Genetic differentiation among *Araucaria* populations in Brazil. In: XIII Congreso Forestal Mundial, FAO, 2009, Buenos Aires

Stefenon VM, Gailing O, Finkeldey R (2007) Genetic structure of *Araucaria angustifolia* (Araucariaceae) populations in Brazil: implications for the in situ conservation of genetic resources. Plant Biol 9:516–525

Stefenon VM, Gailing O, Finkeldey R (2008) Genetic structure of plantations and the conservation of genetic resources of Brazilian pine (*Araucaria angustifolia*). For Ecol Manage 255:2718–2725

Stefenon VM, Steiner N, Guerra M, Nodari R (2009) Integrating approaches towards the conservation of forest genetic resources: a case study of *Araucaria angustifolia*. Biodivers Conserv 18:2433–2448

Tauber H (1965) Differential pollen dispersion and the interpretation of pollen diagrams. Geological Survey of Denmark, 2nd. sec., 89: 34–41

Thomas P (2013) *Araucaria angustifolia*. The IUCN red list of threatened species 2013: e.T32975A2829141. https://doi.org/10.2305/IUCN.UK.2013-1.RLTS.T32975A2829141.en. Downloaded on 11 March 2019

Tompsett PB (1984) Desiccation studies in relation to the storage of Araucaria seed. Ann Appl Biol 105:587–589

Valgas RA (2008) Análise multivariada aplicada no mapeamento da divergência genética de subpopulações de *Araucaria angustifolia* por marcadores moleculares. MSc. thesis, Universidade Federal do Paraná, Curitiba. 139 p

Valgas RA, Chaves Neto A, Lavoranti OJ, Sousa VA (2009) Cluster analysis applied in mapping the genetic divergence of populations of *Araucaria angustifolia* (Bert.) O. Ktze by isoenzymatic markers. In: WSEAS international conference on mathematics and computers in biology and chemistry 10. WSEAS Press, Prague, pp 87–91

White TL, Adams WT, Neale DB (2007) Forest genetics. CABI, Wallingford/Cambridge, MA. 682 p

Yamaguchi LF, Giddings DV, Kato MJ, Di Mascio P (2005) Biflavonoids from Brazilian pine *Araucaria angustifolia* as potential protective agents against DNA damage and lipoperoxidation. Phytochemistry 66:2238–2247

Young AG, Boyle T, Brown T (1996) The population genetic consequences of habitat fragmentation for plants. Trends Ecol Evol 11:413–441

Zobel BJ, Talbert J (1984) Applied forest tree improvement. John Wiley, New York, p 505

Chapter 16
Peteribí (*Cordia trichotoma*), Lapacho Rosado (*Handroanthus impetiginosus*), and Cebil Colorado (*Anadenanthera colubrina* var. *cebil*): Three Valuable Species with Incipient Breeding Programs

Luis F. Fornes, Virginia Inza, María Victoria García, María Eugenia Barrandeguy, María Cristina Soldati, Sara Barth, Tilda Ledesma, Pablo Saravia, Christian Tarnowski, Gustavo Rodríguez, Ezequiel Balducci, Josefina Grignola, Patricia Schmid, Guadalupe Galíndez, and Adrián Trápani

16.1 General Considerations and Natural Ranges of the Species in Argentina

Like with other native species, the selective extraction, accessibility, agricultural frontier advance, and urbanization progress have drastically reduced the natural population existences of these species (Grignola et al. 2018). Due to a very intense and extensive exploitation during the twentieth century, most of their remnant forests currently present an open canopy, a disordered diametric structure, and a sanitary condition from regular to bad (Ortíz 2015).

Peteribí (*Cordia trichotoma* (Vell.) Arráb. ex Steud.), lapacho rosado (*Handroanthus impetiginosus* (Mart. ex DC.) Mattos), and cebil colorado (*Anadenanthera colubrina* (Vell.) Brenan var. *cebil* (Griseb.)Altschul are locally

L. F. Fornes (✉) · P. Saravia · J. Grignola
INTA EEA Famaillá, Tucumán, Argentina
e-mail: fornes.luis@inta.gob.ar

V. Inza · M. C. Soldati
Instituto de Recursos Biológicos (IRB), Centro de Investigaciones de Recursos Naturales (CIRN), INTA. Hurlingham, Buenos Aires, Argentina

M. V. García · M. E. Barrandeguy
Inst. Biología Subtropical, CONICET – Universidad Nac. de Misiones, Posadas, Misiones, Argentina

S. Barth · G. Rodríguez · P. Schmid
INTA EEA Montecarlo, Misiones, Argentina

© Springer Nature Switzerland AG 2021
M. J. Pastorino, P. Marchelli (eds.), *Low Intensity Breeding of Native Forest Trees in Argentina*, https://doi.org/10.1007/978-3-030-56462-9_16

known for their multipurpose applications, although their importance and uses vary according to the areas or regions from which they come from.

The genus *Cordia* (Boraginaceae) comprises 300 species of pantropical distribution, with centers of diversity in America and Africa. Four species of this genus are distributed in Argentina, out of which two are common to the NE and NW regions of the country. These species, *Cordia americana* and *C. trichotoma*, are of great value in forestry, as they provide quality timber, for which they are called noble species. *Cordia trichotoma* is a species with disjunct distribution (Fig. 16.1) that appears in both Argentinean rainforests. In the Alto Paraná Rainforest, it forms continuous forests occupying clearings produced mainly by anthropic disturbances, and it is known as "Peteribí" or "Loro negro." In the Yungas Rainforest, it is locally called "Afata" and forms monospecific forests in the first stages of ecological succession, but it is also found in the mixed mature forest stratum. *Cordia* species are pioneer trees but also show shade tolerance (Campanello et al. 2008, 2011; Montti et al. 2014), which would favor their consociation with other species in agroforestry systems (Fornes et al. 2016).

The related genera *Handroanthus* and *Tabebuia* (Bignoniaceae) are present in both Argentinean subtropical rainforests. Among the various species that exist of these genera in Argentina, the most important from a forestry point of view is *H. impetiginosus*, which is distributed in the North West of Argentina (NWA), across a narrow strip along the piedmont, slopes, and low mountains of the Yungas (Fig. 16.1), occupying an altitudinal range between 300 and 1300 m asl (Zapater et al. 2005; Lozano and Zapater 2008). It is well-known for its high valuable wood in the domestic market, together with that of *Cedrela* spp. and *Myroxylon peruiferum* (Eliano et al. 2009). Its wood is much appreciated for making floors, opening frames, and furniture, especially those coming from the northern region of the Yungas. The wood resistance makes it suitable for outdoor use as decks and garden furniture. This species is also highly appreciated for urban forestry and landscaping as it offers stunning flowering in different pinkish shades (Tortorelli 2009). In addition, the phenolic compound "lapachol" obtained from its bark has medicinal use, including a potential cancer treatment (Hussain and Green 2017).

Anadenanthera colubrina var. *cebil* is a pioneer leguminous species, widely distributed in subtropical forests and transition areas of South America. It grows on rocky hillsides in well-drained soils, often in the riverside. The geographic distribution comprises tropical and temperate countries from South America, including

T. Ledesma · E. Balducci
INTA EECT Yuto, Jujuy, Argentina

C. Tarnowski
INTA EEA La Consulta, San Carlos, Mendoza, Argentina

G. Galíndez
Universidad Nac. de Salta – CONICET, Salta, Argentina

A. Trápani
Fac. de Agronomía y Zootecnia, Universidad Nac. de Tucumán,
Tucumán, Argentina

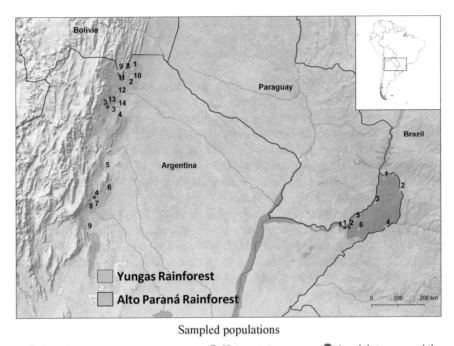

Sampled populations

○ *C. trichotoma*		○ *H. impetiginosus*	● *A. colubrina* var. *cebil*
1. Pto. Iguazú	8. Algarrobito	1. Río Seco	1. Candelaria
2. San Antonio	9. Candado Grande	2. Pocoy	2. Santa Ana
3. Montecarlo	10. Santa Barbara	3. Ledesma	3. Calilegua
4. El Soberbio	11. Rio Seco	4. El Talar	4. Horco Molle
5. Sto. Pipó	12. Cedro Solo	5. Rosario de la Frontera	
6. Oberá	13. Yuto	6. Burruyacu	
7. Posadas	14. Aguas Calientes	7. San Pablo	
		8. Lules	
		9. Huasapampa Sud	

Fig. 16.1 Sampled populations of *Cordia trichotoma*, *Handroanthus impetiginosus*, and *Anadenanthera colubrina* var. *cebil* in the two subtropical rainforests of Argentina for the genetic studies presented in this chapter.

Brazil, Paraguay, Bolivia, and Argentina (von Von Reis 1964). In the Yungas Rainforest it is distributed in the Selva Pedemontana community at the piedmont of the Andes and extends toward the ecotone with the western Chaco, where it also appears frequently. In the northernmost boundary of the Yungas, it can be found in an altitudinal range between 400 and 700 m asl (Fig. 16.1), where it is a dominant species of the community called "jungle of *Calycophyllum multiflorum* and *Phyllostylon rhamnoides*" (Brown et al. 2002). In the Selva Montana (the community of the Yungas between 700 and 1500 m asl), it is present up to 1200 m asl (Brown et al. 2002).

In Argentina the species is also present in the Alto Paraná Rainforest, at its southern boundary, being scanty in its north-northeast extreme (Fig. 16.1). Given its abundance and wood hardness (0.98 g/cm³), it is currently in high demand for poles, sleepers, rural carpentry, floors, opening frames, bodywork, naval constructions, and outdoor use. In addition, its bark contains tannins suitable for tannery (Tortorelli 2009). Given its greater existence in NWA, its exploitation is more accentuated in that region.

The three species considered in this chapter are currently under intense forest harvesting in the Yungas's piedmont, causing a deep impact on the natural forest. In the Alto Paraná Rainforest, a similar situation is found due to the productive transformation of the territory, which promoted the shift of many populations and generated fragmented landscapes. In addition, at a local level, populations of the three species would be facing specific threats that significantly affect the presence of seed trees. These factors could include selective logging, fire, and livestock pressure, among the most important (Ledesma 2014).

Due to the dynamics of land use change, the current fragmentation of the landscape, and the intense use of these species, domestication activities were initiated to carry out appropriate management and conservation strategies, both in situ and ex situ. Further details about the processes that gave rise to the current state of conservation of the Argentine subtropical forests in general can be found in the introduction to this section (Chap. 12).

16.2   Cordia trichotoma

16.2.1   Preliminary Molecular Analyses

Non-model species usually offer more difficulties for molecular characterization, mainly due to the lack of genomic information or to the presence in the sample material of unusual inhibitors that hinder, for example, obtaining a good DNA quality. Therefore, the first steps toward the analysis of the genetic variability of *C. trichotoma* were the adjustment of laboratory methodologies. Different protocols of DNA extraction, as well as different conservation methods for leaf material, were assessed. The selected DNA extraction protocol for the species was the traditional CTAB DNA extraction of Hoisington et al. (1994) with modifications according to Soldati et al. (2013), starting from fresh leaf material.

Cross-species amplification (transferability) among taxonomically related species is possible due to the homology of microsatellite (SSRs) flanking regions (Peakall et al. 1998; Oliveira et al. 2006). Transferability has been successful in several genus (Lemes et al. 2007; Corbo Guidugli et al. 2009; Tarnowski 2010; Soldati et al. 2014) and families (White and Powell 1997; Lemes et al. 2011; Soldati et al. 2014) of tropical trees. It is a valuable and cost-effective tool to assess genetic variability of important timber species and therefore to assist domestication, breeding, and conservation programs.

Then, the intra-genus transferability of 17 SSRs markers developed for *Cordia alliodora* (Marulanda et al. 2011) was evaluated (Table 16.1). The success of

Table 16.1 Seventeen primer pairs tested in *Cordia trichotoma* transferred from *Cordia alliodora*

SSRs	T°a	[MgCl₂]	DNA (ng)	Result
CaA103	57	1.5 mM	10 ng	P
CaA104	60	1 mM	5 ng	P
CaA105	54	3 mM	10 ng	NS
CaA3	57	2 mM	10 ng	P
CaA5	54	4 mM	10 ng	P
CaB101	54	2 mM	10 ng	NS
CaB102	54	3 mM	10 ng	P
CaB103	54	2 mM	10 ng	NS
CaB105	54	3 mM	10 ng	NS
CaB6	54	4 mM	10 ng	P
CaC103	57	1.5 mM	10 ng	P
CaC108	54	3 mM	10 ng	NS
CaC7	57	3 mM	10 ng	NS
CacC8	55	3 mM	10 ng	P
CaD1	54	4 mM	10 ng	P
CaD103	54	3 mM	10 ng	NS
CaD9	54	2 mM	10 ng	P

Best conditions are presented (annealing temperature and MgCl₂ and DNA concentrations), together with successful transference. *P* polymorphic, *NS* non-specific amplification

transferability as well as the clearness of resolution patterns and polymorphism levels was assessed on a sample of eight individuals randomly chosen, testing different polymerase chain reaction (PCR) conditions. Finally, ten polymorphic markers were successfully transferred (Table 16.1).

With the transferred markers, we begun to analyze material from natural populations. Since *C. trichotoma* is polyploid, with 2n = 104 and X = 13 (Las Peñas 2003), SSRs should be used as dominant markers. A preliminary study based on 21 individuals from two clonal seed orchards (see below) and 5 transferred SSRs markers showed moderate to low levels of polymorphism, with 61 polymorphic alleles per orchard. Nevertheless, further analyses are necessary to better understand *C. trichotoma*'s ploidy and to develop more suitable molecular markers to study this species.

16.2.2 Domestication Experiences

Several studies related to the domestication of *C. trichotoma* were initiated in different disciplines such as seed physiology, nursery, silviculture, breeding, and propagation techniques, among others, with variable degree of development.

Seeds of *C. trichotoma* show high viability (≥80%) and moisture content (MC: 9–14%), and they do not show either physical or physiological dormancy, at the time of dispersal (Galindez et al. 2019). Germination occurs in high percentages in light and darkness at both 25 °C and 30/20 °C regimes (≥ 90%), i.e., they are indifferent to light, indicating that seeds can germinate at both gaps and under forest canopy. Seeds tolerate desiccation to low moisture content ($\leq 5\%$), although it has been registered that some populations are more sensitive to seed desiccation than others. Seeds stored with low MC ($\leq 5\%$) at low temperatures (between -20 and $5\ ^{\circ}$C) do not show signs of aging after a year of storage, whereas seed deterioration occurs in seeds stored at $35\ ^{\circ}$C with 8–10% MC and even in seeds stored with 3–5% MC after a year of storage (Galindez et al. 2019). These results are consistent with previous observations (Ledesma 2014) and indicate that this species show orthodox storage behavior and can likely be preserved for medium- and long-term durations in seed banks. However, considering that both moisture content and temperature are key factors in seed deterioration, with some populations being more sensitive to storage conditions than others, it is essential to evaluate the behavior of each collected population before it is stored (Galindez et al. 2019) (Fig. 16.2).

There are several experiences of *C. trichotoma* plantations in both subtropical rainforests, some of them in agroforestry systems, for which the species has proved to adapt perfectly well. An interesting agroforestry trial was installed in 2010 in the town of Santo Pipó (Province of Misiones) (Munaretto et al. 2019). Yerba mate tree (*Ilex paraguariensis*) is one of the main industrial crops of the province, which is commonly conducted like a tea crop, that is, as bushes at full sun. The trial seeks to test the behavior of yerba mate in consociation with forest tree species of timber interest (nine native and exotic species were assayed). Until 7 years of age, the productivity of yerba mate does not decrease in consociation with the tree species tested. Thus, the agroforestry system not only maintained similar yields of yerba mate leaves but also improved the natural and work environments, increasing biodiversity and soil quality and providing more comfort to personnel working in the shade of the trees (Munaretto et al. 2019). Among the native tree species assayed, *Peltophorum dubium* and *C. trichotoma* stood out, the latter showing an average DBH of 16 cm and a mean total height of 8 m at the seventh year of implantation (Colcombet et al. 2019).

Another experience in the NEA was a silvicultural field trial with three native species assayed under pine canopy and at open sky. Crechi et al. (2010) reported that *Cordia trichotoma* and *Balfourodendron riedelianum* presented a good growth with monopodial architecture and straight stems (although less growth than *Enterolobium contortisiliquum*), resulting in promising species for the logging industry.

In NWA, there are few plantations at experimental scale with this species, being one of the most relevant located in a private property of Valle Morado (Province of Salta, 23°27′ S; 64°28′ W). In 2001 and 2002, two trials of *C. trichotoma* were installed, both purely and as mixed stands with other native species of the Selva Pedemontana of the Yungas, with a plantation frame of 4×3 m. In both conditions, the species presented a low mortality, close to 5% in the first year, proving to be a rustic species with good adaptation to plantation (Balducci et al. 2009). In mixed plantation with *Cedrela balansae*, the average diametric growth was 0.9 cm/year, whereas 1.4 cm/year in mixed plantation with *Handroanthus impetiginosus*. The density for *C. trichotoma* was 50 ind/ha, reaching the sixth year an average DBH of

Fig. 16.2 Morphologic traits of *Cordia trichotoma*. (**a**) Adult individual selected within the breeding program. (**b**) Adult individual in the flowering time. (**c, d**) Flower details. (**e**) Seed details. (**f**) Leaves and inflorescence. (**g**) Adult's bark. (**h**) Wood core sample

13.6 cm and a mean total height of 9.5 m. The species showed a very good behavior in full sun conditions at 6 years of age, being the straightness of the stem and good response to pruning the main characteristics that deserve to be highlighted for its use in forestry.

16.2.3 An Incipient Breeding Program

Given its promising characteristics, the Instituto Nacional de Tecnología Agropecuaria of Argentina (INTA) has recently begun a low-intensity breeding program for *Cordia trichotoma*. Planned strategy included the identification of the most representative populations, their phenotypic and molecular characterization, selection of plus trees, their vegetative propagation for the installation of seed orchards, and their open pollinated sexual propagation for testing the genetic merit of the mothers. Given its disjunct distribution, the strategy decided was to advance separately in each subtropical rainforest until having the molecular information to make further decisions related to the conformation of unique base populations and clonal gardens. The progeny tests are planned to be conducted in a network of trials including material from both rainforests, which will thus be tested together.

Since the natural range of the species is only coarsely known, a modeled niche distribution was performed by means of *Maxent* software (Ledesma 2014) to define the sites to search for plus trees (Fig. 16.3a). The predictive bioclimatic variables with the highest weight were the average temperature of the coldest quarter and temperature seasonality. This procedure allowed identifying numerous populations that were still unknown. The initial theoretical range was successively adjusted, first

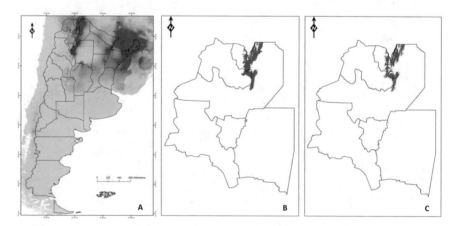

Fig. 16.3 (**a**) Potential distribution of *Cordia trichotoma* obtained by Maxent modeling (Ledesma 2014). The most probable areas are indicated with dark red. (**b**) Potential distribution in the Yungas adjusted with a cutoff threshold of 0.596. (**c**) The same potential distribution of B, but indicating the loss due to the change in land use

considering sites with higher probability (Fig. 16.3b) and then discarding the sites that are known to have already been transformed from forests to crops or urban developments, thus reaching a final area of 1,030,000 ha (Fig. 16.3c). It is interesting to note that this modeled distribution also serves to identify potential areas for cultivation of the species.

A given group of selected individuals represents each zone. So far, 133 individuals were selected in both Argentinean rainforests (Fig. 16.1). The locations belong to the Iguazú National Park, private reserves, and forest growers or companies. In the case of the NWA region, the populations were selected based on previous reports about presence and density (Ledesma 2014), and no evidence of the occurrence of the species was found south of 24° S.

The choice of phenotypically superior trees for the conformation of the base population is a key component of any genetic improvement program, as the set of selected trees will be the basis of future forest generations. This makes it necessary to reach a compromise vis-à-vis achieving outstanding genotypes in terms of desirable traits and preserving the population genetic variability, which will make it possible to accomplish improvement and conservation goals, respectively. Since the individuals were selected in the natural forest, which in this case is a complex ecosystem with difficult accessibility, this initial part of the program required the investment of high experimental efforts. The selection criteria were health, vigor, proportion of heartwood, and stem straightness (Rodríguez and Barth 2014). Vigor was estimated by a minimum DBH of 30 cm and a bole height over 6 m. The proportion of heartwood was measured by Pressler borer (only in the trees selected in NWA) (Fornes et al. 2016).

The selected trees were vegetatively propagated by grafting in the nurseries of the two experimental stations of INTA involved in this breeding program: Yuto and Monte Carlo. The development of vegetative propagation techniques is a highly useful tool for improvement programs. The use of indole butyric acid (IBA) hormone in doses lower than 1000 ppm was found to promote rooting rates of 96% in apical cuttings of *C. trichotoma* (Gonzalez et al. 2016), an extraordinary success rate that would make clonal forestry possible in the future. In addition, different types of grafts were evaluated, as well as the time between the scion harvest (from the canopy) and the moment of grafting. The best grafting performance was achieved using the apical full cleft technique, within 24 hours after scion harvesting, in order to prevent dehydration (Fornes et al. 2016).

With the clones of the trees selected in the Yungas, a Clonal Seed Orchard was installed in 2014 in Yuto Experimental Station (23° 35′ 22.15″ S; 64° 30′ 38.97″ W; 350 m asl), including 600 ramets corresponding to 69 genotypes of 7 provenances from the Yungas. Three years later, a new Clonal Seed Orchard was established in Laharrague Experimental Station of INTA (26° 33′ 35.61″ S; 54° 40′ 25.45″ W; 260 m asl) with 304 ramets corresponding to 64 genotypes of 7 provenances from the Alto Paraná Rainforest.

16.3   *Handroanthus impetiginosus*

16.3.1   *Preliminary Characterization of Its Genetic Variability*

The leaves of *H. impetiginosus* have high concentration of polyphenols (Jausoro et al. 2010) which leads to an extraction of DNA of poor quality and failures in the amplification of molecular markers. Three different dates of sampling (*start of sprouting*, October 2014; *90 days later*, January 2015; and *180 days from sprouting*, April 2015) were evaluated because it was suggested that the concentration of polyphenols in leaves decreases with increasing time since sprouting (Hernán Bach pers. comm.). Three protocols of DNA extraction (i, Hoisington et al. 1994; ii, Doyle and Doyle 1990 modified; iii, commercial Quiagen Kit) were assayed. Several features of the collected leaves were considered, including their conservation until laboratory processing. The DNA extraction was evaluated by the color/appearance of the pellet obtained and the quantity and integrity of DNA in agarose. In addition, DNA extraction from flowers was attempted, but although DNA could be extracted, amplification products could not be obtained.

Results indicated that 90 days from sprouting was the best date for sampling, and Doyle and Doyle (1990) modified with two steps of chloroform-isoamyl was the most appropriate protocol for DNA extraction. An additional proof of milling and conservation of sampled leaf material was conducted. Milling with liquid N of previously dried leaves in silica gel and immediate DNA extraction was finally selected. In fact, only for this protocol amplification products were obtained.

Transferability of ten SSRs (Tau 12, Tau 14, Tau 15, Tau 17, Tau 21, Tau 22, Tau 27, Tau 28, Tau 30, Tau 31) developed in *Tabebuia aurea* (Braga et al. 2007) to *H. impetiginosus* was examined (Inza et al. 2019). The success of transferability as well as the clearness of resolution patterns and the polymorphism levels was evaluated on a subset of individuals from different populations randomly chosen. Then, those SSRs transferred successfully that showed a polymorphic pattern were evaluated across a greater number of samples to better examine the level of polymorphism (number of alleles).

A set of five markers (Tau 12, Tau 14, Tau 17, Tau 28, and Tau 31) transferred from *T. aurea* (Braga et al. 2007) were finally selected to conduct population genetics studies. Conditions of polymerase chain reaction (PCR) were performed according to Braga et al. (2007). Molecular genetic variability of nine Argentine natural populations of *H. impetiginosus* was assessed using the five SSRs transferred (Inza et al. 2019). Young leaves from randomly chosen trees were collected considering a minimum distance of 80–100 m between them in order to reduce the probability of sampling-related individuals (Gillies et al. 1999). Collected leaves were dried in silica gel (Chase and Hills 1991) and total genomic DNA extracted and stored at −20 °C until its use.

Genetic diversity and differentiation parameters were estimated using the GenAlEx 6.3 software (Peakall and Smouse 2006). A total of 54 alleles were amplified with a range of 23 (Tau17) to 2 (Tau 28) alleles per microsatellite. Genetic diversity of the species over all populations was on average moderate to low (He = 0.390). The higher values were observed in San Pablo population (He = 0.473) in the south and in Río Seco population (He = 0.442) in the northern edge of the Yungas and the lower values in El Talar (He = 0.320) and Ledesma (He = 0.328), two nearby populations in the center of the Yungas (Table 16.2). The rest of the populations showed intermediate values of genetic diversity (in terms of He and Ho). The number of alleles per population was the highest in Río Seco (34), high in Pocoy (28) and Burruyacú (23), and the lowest in El Talar (13). In addition, Río Seco population showed the highest number of exclusive alleles (Ea = 11), thus resulting the most diverse population considering all the indexes as a whole. The high diversity of this population was firstly attributed to two trees that showed high number of rare alleles and greater genetic distance to the rest of the trees. However, after removing these outliers, Río Seco population still showed high genetic diversity values (He = 0.371, Ho = 0.386, A = 28). The high genetic diversity of Río Seco could be partly associated with the highest sample size of this population, but a biological meaning must be considered, since its location on the northern edge of Yungas has been associated with the highest levels of biological diversity in this ecosystem (Brown et al. 2001, 2006) and also with a high genetic diversity of other forest tree species (*Cedrela angustifolia*; Inza et al. 2012). Finally, it is important to note that trees from Río Seco and Pocoy populations (northernmost distribution) showed high values of exclusive alleles (Ea); therefore, these trees should be

Table 16.2 Genetic diversity and location of *Handroanthus impetiginosus* populations sampled in Argentine Yungas Rainforest

Population/province	N	Latitude (S)/longitude (W)	Altitude (m asl)	He	Ho	A	Ea	F_{IS}
Río Seco/Salta	29	22° 28′ 33.6″/63° 55′ 55.2″	666	0.442	0.401	34	11	0.08
Pocoy/Salta	22	22° 43′ 55.2″/63° 58′ 48.0″	440	0.376	0.341	28	6	0.07
Ledesma/Jujuy	18	23° 39′ 14.4″/64° 31′ 30.0″	385	0.328	0.350	18	1	−0.10
El Talar/Jujuy	9	23° 49′ 55.2″/64° 28′ 08.4″	483	0.320	0.344	13	0	−0.14
Rosario de la Frontera/Salta	9	25° 50′ 31.2″/64° 55′ 55.2″	984	0.383	0.451	17	2	−0.25
Burruyacú/Tucumán	18	26° 35′ 20.4″/64° 51′ 18.0″	591	0.409	0.440	23	3	−0.11
San Pablo/Tucumán	11	26° 52′ 03.7″/65° 20′ 47.4″	634	0.473	0.393	18	2	0.13
Lules/Tucumán	11	26° 53′ 42.0″/65° 22′ 48.0″	526	0.389	0.407	16	1	−0.10
Huasapampa Sud/Tucumán	14	27° 53′ 02.4″/65° 36′ 25.2″	604	0.416	0.386	15	0	0.04

N, sampled size; He, mean expected heterozygosity; Ho, observed heterozygosity; A, allelic richness; Ea, exclusive alleles; Fis, inbreeding index prioritized in domestication, breeding, and conservation programs of the species. In general, no inbreeding was detected, with almost all populations with F_{IS} near to 0.

In order to examine genetic differentiation among populations (Phi_{PT}), a hierarchical analysis of molecular variance (AMOVA, Excoffier et al. 1992) was performed by means of GenAlEx 6.2. A moderate to low genetic differentiation among populations was observed, and this was highly significant ($Phi_{PT} = 0.049$, $p \leq 0.001$). As it is generally expected for forest species, most of the genetic variation was observed within populations (95%), and this is consistent with the high level of observed historical gene flow, $Nm = 4.85$. To assess more in detail the population genetic structure, a Bayesian clustering approach by Structure 2.3.4 (Pritchard et al. 2000; Evanno et al. 2005; Falush et al. 2007) and Nei's genetic distances among populations and genetic similarities among individuals by NTSYS 2.0 (Nei 1978; Rohlf 1998) were performed (Inza et al. 2019).

Six genetic clusters were identified (K = 6), and they were generally distributed homogeneously between populations, meaning a weak genetic structure (Fig. 16.4). Clusters 1 (red), 2 (green), 3 (blue), 4 (yellow), and 6 (light blue) are present in proportions of 16–21% in all populations. However, for cluster 5 (pink), populations showed some differences with maximum values for Burruyacú and Río Seco (18 and 15%, respectively), which is principally due to the presence of this cluster in some trees. In Río Seco, trees 11 and 670 showed the highest proportion of this cluster which is consistent with the higher number of Ea and the greater genetic distance of these genotypes to the others. Cluster 5 showed a minimum proportion for El Talar (5%) and the three populations of southern Yungas (3%, San Pablo, Lules, and Huasapampa) that were very similar in their genetic composition. These results are consistent with the genetic distance among populations from the UPGMA dendrogram (r = 0.84, data not shown). The genetic similarity between some trees stresses the need to continue the study with a higher number of molecular markers that allows an unequivocal identification of all trees.

This is a preliminary study as a basis of knowledge of the genetic variability pattern of *H. impetiginosus* in the Argentine Yungas Rainforest to guide domestication, breeding, and conservation purposes. As general conclusions, Río Seco, Pocoy, and Burruyacú populations should be prioritized according to their higher genetic diversity and differentiation levels, and conservation measures (including ex situ strategies) should be implemented. San Pablo and Rosario de la Frontera populations have an important value due to their genetic diversity levels despite its low number of sampled trees. Some populations have an additional value because of their important geographical position. Rosario de la Frontera connects North-Central with Southern Yungas, and Huasapampa and Lules are on the southern edge of natural range of the species, and they could have an adaptive value. Future studies increasing both sample size and number of markers are needed for more conclusive results.

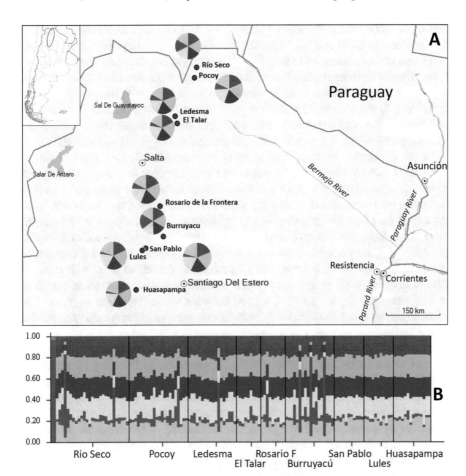

Fig. 16.4 Genetic structure of *Handroanthus impetiginosus* in the Argentine Yungas inferred by Bayesian model. (**a**) Geographic distribution of genetic clusters (K = 6). The circles represent the percentages of assignment of populations to clusters 1 (red), 2 (green), 3 (blue), 4 (yellow), 5 (pink), and 6 (light blue). Blue dots indicate the exact location of populations. (**b**) Population genetic structure with detailed individual composition. Populations are ordered from left to right in increasing order of latitude, and each vertical bar represents each individual

16.3.2 Domestication Experience and Breeding Strategy

According to nursery trials, *H. impetiginosus* seeds germinate 7–10 days after sowing, with a germination capacity of 90%. In addition to this excellent germination performance, the species had a great initial growth compared to other *Handroanthus* species, reaching the field transplant size at 150 days, with a height of 37–40 cm (Zapater et al. 2005).

Several experimental plots of the species have been installed in NWA; however many are not documented. A 6-year-old plantation installed in Yuto Experimental

Station of INTA with a plantation frame of 4 × 3.5 m showed an average growth of 1.31 cm/year in DBH and 1.07 m/year in total height (0.49 m/year of shaft height), yielding a timber volume of 1.096 m³/ha/year (Gil and Del Castillo 2004). Likewise, native forest enrichment trials (under forest canopy) were also carried out in single rows over 5 m wide strips open in the natural vegetation, with the seedlings separated by 4, 5, and 6 m in the row. At 8 years of age, the trees reached a mean DBH of 5.9 cm, 4 cm, and 5.9 cm for each plantation density, respectively, and a mean total height of 4.9 m, 4.4 m, and 5.4 m, respectively (Del Castillo et al. 2005).

In the abovementioned private property of Valle Morado (Province of Salta), three trials with *Handroanthus impetiginosus* were installed in 2001 and 2002, with a plantation frame of 4 × 3 m. One of them is a pure stand over an area of 1 ha. The two other trials are mixed, in combination with other native species of the Selva Pedemontana of the Yungas: one with *C. trichotoma* over an area of 2.7 ha and the other with *Pterogyne nitens* over an area of 0.9 ha. In all cases the mean diametric growth was quite promising, with 1.3 cm/year in the pure stand, 1.6 cm/year in combination with *C. trichotoma*, and 1.8 cm/year in the mixed stand with *P. nitens* (Balducci et al. 2009). In the first 5 years, mortality rate remained rather low, close to 5%, thus showing the species, a good behavior when implanted in block. The main concerns about the species growth from a genetic improvement perspective would be the stem straightness and the narrowness of the crown.

In view of the promising behavior of the species under a cultivation system, and parallel to the molecular diversity studies, a low-intensity breeding program was initiated for *Handroanthus impetiginosus* in Famaillá Experimental Station of INTA. The strategy decided was to select individuals in the natural forest, propagate them vegetatively, and install clonal seed orchards. Between 2011 and 2017, 189 plus trees were selected according to criteria defined by different private and public institutions of the forest-industrial sector, including stem straightness, self-pruning of the first portion of the trunk (up to 7 m height), stem taper, crown diameter, heartwood proportion, and health (especially wood health).

In practice, the selection procedure took advantage of the harvesting activities of industrial sawmills working in the natural forest, which kindly supported the breeding program. When outstanding exemplars of *H. impetiginosus* were harvested, upon notification from the sawmill company, we verified the quality of the candidates and eventually went back to the harvesting site to collected material for its vegetative propagation (Fig. 16.5). Given the harvesting activities are commonly performed in winter and early spring (the dry season of the year), we had a 2-month window to find branches still alive on the soil of the forest since the cutting down of the trees (Ignacio Sosa, pers. comm.). The selection carried out in the sawmill made it easier to consider wood features, like heartwood proportion and wood health.

Due to the selection procedure described, most of the genotypes of the base population of the breeding program are not anymore standing in the forest. This made necessary to preserve them in a clonal bank. Only the best individuals of the phenotypic ranking were finally included in a clonal seed orchard.

After locating the remains of the selected trees in the forest (by means of gps coordinates), at least 30 branches per tree were taken to the nursery in polyethylene

Fig. 16.5 (**a**) Stockyard at a sawmill; (**b**) Selection of the best trees in the field; (**c**) Ruling out individuals with wood diseases; (**d**) Search in the field of individuals selected in the sawmill; (**e**) Collection of sexually mature branches for grafts; (**f**) Selection of twigs for nursery grafts

bags, labeled and with appropriate moisture. Scions were prepared and grafting was finally done in the following days into 2-year-old rootstocks of the same species. The grafting is wedge type using the apical bud, cutting a small grafted branch 15–20 cm long with at least three buds.

Besides the grafting technique, other kinds of agamic propagation were reported for the genus *Handroanthus*. Oliveira et al. (2015 and 2016) analyzed the management of a multiclonal minigarden of *H. heptaphyllus* to optimize the production of cuttings and found that 5-cm-long mini-cuttings extracted in spring from the middle part of the shoot rooted quite well without application of auxin, with a successful rate of more than 80%. Later, Rodrigues (2019) also working with a clonal minigarden of this species observed that the best time to collect 5 cm mini-cuttings was when the shoots were no more than 40 cm long. Other authors were able to clonally propagate 15-year-old trees of this species by rooting 15-cm-long cuttings. They found that the best time of extraction was in spring, achieving a 52% survival of the cuttings with a rooting capacity of 30%, with application of indole butyric acid 3000 ppm (Ovando et al. 2013).

Using micropropagation techniques and from in vitro germinated *H. impetiginosus* seedlings, Larraburu (2014) adjusted a protocol which produced 83% of in vitro rooted plants, and the subsequent acclimatization allowed a survival of 50% of the obtained plants. Duarte et al. (2016) germinated in vitro *H. heptaphyllus* seeds and used different parts of the seedling to regenerate shoots. The rooting of the sprouts obtained in this way was achieved in a temporary immersion medium which showed a good response during acclimatization.

In vitro experiments on vegetative propagation of *H. impetiginosus* were conducted using nodal segments cultivated in both Murashige and Skoog salts with Gamborg vitamins (MSG) and Woody Plant Medium (WPM), with different concentrations of 6-benzylaminopurine (BA) and indole butyric acid (IBA). Morphogenic responses were differentially affected by salt compositions and their interactions with plant growth regulators in each micropropagation stage. According to response surface analysis, the optimum multiplication rate with 1 μM IBA ranged from 16.7 to 21.3 μM BA in WPM, and the inhibitors of endogenous auxins could increase multiplication rates. A pulse with 50 μM IBA in MSG produced 83% rooting with 3.2 roots per shoot and higher fresh and dry weights of shoots and roots. In the acclimatization stage, 50% of plants survived after 1 year (Larraburu et al. 2012).

Finally, in 2016 a Clonal Seed Orchard was installed in Famaillá Experimental Station of INTA (27° 0′ 40.11″ S; 65° 22′ 41.88″ W; 380 m asl), including 160 clones from nine provenances (those of Table 16.2). The design of the CSO was in randomize complete blocks with six repetitions of single tree plots with a distance between plants of 5x5 m. The CSO is currently starting to produce seeds. In the near future, those seeds will be used to establish a network of trials in order to evaluate each mother-tree in adaptive and productive traits.

16.4 *Anadenanthera colubrina* var. *cebil*

16.4.1 *Phenotypic and Genetic Studies: Development of Specific nuSSRs and Spatial Distribution of Phenotypic and Molecular Variability*

In Argentina, genetic diversity studies of *A. colubrina* var. *cebil* have considered the species as a model for population genetics studies of long-live species with disjunct distribution. Initially, these studies were performed using four universal microsatellites of the chloroplast genome due to the lack of nuclear specific primers (Barrandeguy et al. 2011). More recently, a set of eight polymorphic nuclear microsatellite primers were developed for the species using a genomic library enriched for tandemly repeated motifs (Barrandeguy et al. 2012; Barrandeguy and García 2016). At the same time, Feres et al. (2012) reported 14 polymorphic nuclear microsatellite markers for *A. colubrina*. Then, the genetic variability of Argentinean populations of *Anadenanthera colubrina* var. *cebil* from Selva Pedemontana of the Yungas and Alto Paraná Rainforest (Fig. 16.1) was analyzed using microsatellite markers from nuclear and chloroplast DNA, determining an important historical component in the distribution patterns of contemporary genetic variability in both genomes (Barrandeguy et al. 2014). Complementary, 13 quantitative traits (eight vegetative and five reproductive traits) were analyzed to characterize phenotypic variability (García et al. 2014).

Chloroplast genome variability studies by means of cpSSRs (Barrandeguy et al. 2011; Barrandeguy et al. 2014; Goncalves et al. 2014) and cpPCR-RFLPs (Goncalves et al. 2014) concluded that distribution patterns of chloroplast genetic diversity in the studied populations were defined by the action of genetic drift. The F_{ST} fixation index indicated a strong genetic structuring because it reached a value of 0.95, being statistically significant at a 95% confidence level (p < 0.05), while the Bayesian spatial model established that the most probable number of clusters was four ($P_{(k = 4)}$ = 0.99), with the individuals assigned to each cluster presenting the same geographical origin (Goncalves et al. 2014). In this way, the high level of population genetic structure detected in chloroplast genome would be explained by absence of seed-mediated gene flow between populations (Goncalves et al. 2014).

Combined analysis of genetic structure for both nuclear and chloroplast DNA showed that approximately 90% of the nuclear genetic variation was detected within populations, while the remaining variation was equally distributed among populations and phytogeographic provinces (~5%). Nearly 70% of the chloroplast genetic variation was found between the Yungas and Alto Paraná Rainforest, and 25% of variation was observed among populations. The estimates of global F_{ST} were statistically significant and indicated genetic differentiation in both genomes. According to Wright's qualitative guidelines for the interpretation of F_{ST} (Hartl and Clark 2007), these populations are moderately structured at the nuclear genome (F_{ST} = 0.11), and they are highly structured at the chloroplast genome (F_{ST} = 0.95). The highest values of pairwise F_{ST}-based on nuSSRs were observed between populations of the different phytogeographic provinces.

Also, two nuclear clusters were defined by the Bayesian model. Individuals were assigned to these clusters according to their phytogeographic province of origin (Fig. 16.6a). The intra-province-specific analyses found differentiation among the populations of the Alto Paraná Rainforest, but no differentiation among those of the Yungas (Fig. 16.6b). Also, six chloroplast clusters were defined by the Bayesian model [log(ml) = 0.957]. Individuals were assigned to each cluster according to their chloroplast haplotypes (Fig. 16.7a). The dendrogram showed two main groups of haplotypes corresponding to the phytogeographic origin of the individuals (Fig. 16.7b): one group integrated by the clusters 3, 4, and 6 containing all

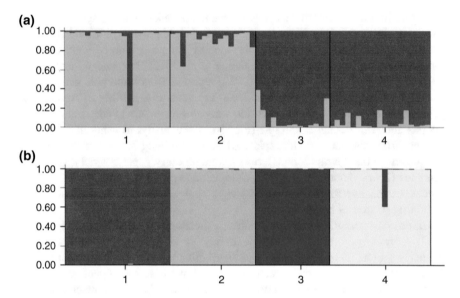

Fig. 16.6 Bayesian clustering of *Anadenanthera colubrina* var. *cebil* individuals using ncSSR genetic data in an admixture model: (**a**) global genetic structure (K = 2) and (**b**) substructure of each province

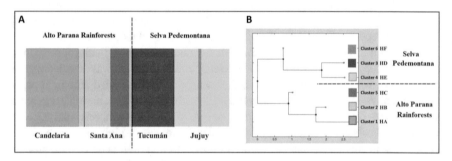

Fig. 16.7 (**a**) Bayesian clustering of individuals using cpSSR genetic data in a mixture model for linked loci. (**b**) Relationships among the cluster defined by the Bayesian analysis of genetic structure. HA to HF represent the different haplotypes, and the places of origin were included in the figure (From Barrandeguy et al. 2014)

individuals from the Yungas, and the second group integrated by the clusters 1, 2, and 5 containing all individuals from the Alto Paraná Rainforest. Furthermore, chloroplast genome keeps valuable information about historical population changes in this species emphasizing the role of ancient fragmentation in the species distribution (Barrandeguy et al. 2014).

Historical divergence in *A. colubrina* var. *cebil* was studied by means of an intronic region of chloroplast DNA allowing to detect genetic differences between the distribution fragments of this species and contextualize them in a time frame

(Calonga Solís et al. 2014). Five haplotypes were defined, while the divergence time between haplotypes corresponds to the Pleistocene epoch (Calonga Solís et al. 2014). Demographic analyses performed on the genetic variation in three cpSSRs support that the current distribution of chloroplastic haplotypes of *A. colubrina* var. *cebil* shows that the Yungas' populations maintain traces of a later expansion after its arrival in the region, whereas the Alto Paraná Rainforest shows historical stability (Barrandeguy et al. 2016).

Molecular and phenotypic variability studies (García et al. 2014) showed that genetic variation was higher in the Selva Pedemontana of the Yungas than in the Alto Paraná Rainforest, which translates into highly qualified populations for conservation. Seventy individuals from four populations were analyzed. Mean values of reproductive traits were higher in the Selva Pedemontana (6.66) than in the Alto Paraná Rainforest (5.45), whereas the mean values of vegetative traits were higher in the latter (57.80 and 62.90, respectively). Regarding phenotypic variability, populations from Alto Paraná Rainforest showed the highest mean value of number of seeds per fruit making them valuable as well for management strategies as a means to recover the impacted areas where these populations are located.

On the other hand, Mazo et al. (2014) analyzing adult individuals of *A. colubrina* var. *cebil* by means of phenotypic traits and molecular markers showed that polygenic traits did not evidence mean reduction in phenotypic values. This result indicates that there was no inbreeding depression, low levels of coancestry, and large effective sizes despite the small size of the analyzed populations, which are isolated as results of landscape fragmentation processes. Inbreeding and coancestry increase homozygosis and create autozygous genotypes, but they did not modify allelic frequencies. This allows to conclude that populations maintain genetic variability that could be considered as possible sources of adapted germplasm.

However, Goncalves et al. (2019) cautioned for using individuals from Alto Paraná Rainforest as a source of material for reforestation based on their study which included saplings and adult trees of *A. colubrina* var. *cebil* from Campo San Juan Nature Reserve (Southern Alto Paraná Rainforest). These authors reported high inbreeding and relatedness patterns of saplings suggesting a mixed mating system. Also, they found high genetic diversity, despite recent fragmentation, together with only moderate genetic structure and signatures of historical expansions, which suggest long-term population viability. In this way, high inbreeding and differentiated genetic clusters supported their warning to be cautious for using this stand as a source of material for reforestation.

16.4.2 Domestication Potential of Anadenanthera colubrina var. cebil in Argentina

The species has a great forestry potential because of its easy reproduction and highly satisfactory growth (Fig. 16.8) (Tortorelli 2009).

Fig. 16.8 Morphologic traits of *Anadenanthera colubrina* var. *cebil*. (**a**) Adult individual, (**b**) Fruits and composite leaf, (**c**) Long gland in the base of the petiole, (**d**) Mamelonated bark, (**e**) Globular inflorescence in bunch, (**f**) Discoidal flat seeds, and (**g**) Cross and longitudinal section of the trunk

In a review of promissory native forest tree species to be planted in Brazil, Carvalho (1998) reported for *A. colubrina* var. *cebil* a mean annual increment in solid volume with bark from 5 to 25 m³/ha/year⁻¹. There is evidence that native people from South America established small plantations of the species taking advantage of its simple reproduction by seeds (Viglione and Vallejo 2011). In the Province of Chaco, a 17-year-old experimental plantation located in Presidencia de la Plaza Forest Station of INTA (26° 55′ 30.0″ S; 59° 47′ 51.5″ W) showed 16 m height and 28 cm diameter in average. In addition, the species seems not to be attacked by ants and locusts, two common threats for forest plantations in the region (Tortorelli 2009). Another small plantation of 10 years of age located in a private property in the Province of Misiones showed 6 m of height and 34 cm diameter in average (María Eugenia Barrandeguy, pers. comm.).

In the already cited experience of Valle Morado (Province of Salta), also *A. colubrina* var. *cebil* was planted. Between years 2001 and 2002, a restoration trial was installed using 15 timber species from the Selva Pedemontana of the Yungas. The mixed plantation included the species with 32 individuals per ha, which reached at sixth year an average DBH of 11.6 cm and a total height of 7.1 m (Balducci et al. 2009).

Although *A. colubrina* var. *cebil* is not among the most valuable timber species, its potential is increasing as a substitute for hardwood species from Chaco and Selva Pedemontana (Eliano et al. 2009). Its wood has a specific weight of 0.980 g/cm^3 (Tortorelli 2009). Accordingly, it was classified as a promising wood species owing to its economic value (with valuable wood production), acceptable silviculture performance, and aptitude for artificial regeneration programs (Carvalho 1998).

In Argentina, the species shows great exploitation in the Northwest of the country. According to the official statistics, 12,567 tons were harvested there in 2017 (SGAyDS 2019). However, there are extraction data in the Northeast of Argentina for the year 1946 from both fiscal lands and private plantations (967 tons) (Tortorelli 2009), reflecting that it was also a lumber resource there in a recent past.

Nevertheless, so far there is not any domestication program in course for the species, which in addition is not specifically protected under any regulation (Richter and Dallwitz 2000) and is categorized as "least concern" with low risk of extinction by the IUCN (2012).

16.5   Breeding and Conservation Programs: Future Steps

The three species considered in this chapter show different degrees of progress in terms of breeding strategy and domestication studies. *Handroanthus impetiginosus* is perhaps the most advanced of the three species in studies on intra- and inter-population genetic diversity, selection of plus trees, agamic propagation, and creation of propagation materials. The genetic variability patterns of *H. impetiginosus* in the Argentinean Yungas Rainforest is an important input to guide domestication, breeding, and conservation efforts. However, it is very necessary in the near future to intensify sampling in populations currently represented by less than 20 trees in order to make populations more comparable, as well as increase the number of molecular markers for a better characterization. On the other hand, the CSO currently installed in Famaillá Experimental Station needs the backing of a network of progeny and provenances trials, which will be done at the time that it begins to produce seeds. A similar stage has been reached for *Cordia trichotoma*, with the two CSO installed. However, work on molecular characterization of populations is only just beginning in this species. On the other hand, in both species, it is also necessary to know the responses to abiotic stresses like low temperature and drought.

Anadenanthera colubrina var. *cebil* is the species with most incipient information, and there is a need for expanding the area of study, since its genetic resources status from Yungas Rainforest is, so far, barely known.

References

Balducci ED, Arturi FA, Goya JF, Brown AD (2009) Potential of Forest plantations in the foothills of the Yungas. Editions of the Subtropics. Proyungas Foundation

Barrandeguy ME, García MV, Argüelles CF, Cervigni GDL (2011) Genetic diversity of *Anadenanthera colubrina* Vell. (Brenan) var. *cebil*, a tree species from the South American subtropical forest as revealed by cpSSR markers. Silvae Genet 60:123–132

Barrandeguy ME, Prinz K, García MV, Finkeldey R (2012) Development of microsatellite markers for *Anadenanthera colubrina* var. *cebil* (Fabaceae), a native tree from South America. Am J Bot 99:e372–e374

Barrandeguy ME, García MV, Prinz K, Pomar RR, Finkeldey R (2014) Genetic structure of disjunct Argentinean populations of the subtropical tree *Anadenanthera colubrina* var. *cebil* (Fabaceae). Plant Syst Evol 300:1693–1705

Barrandeguy ME, García MV (2016) Chapter 3: Microsatellite as tool for the study of microevolutionary processes in native forest trees. In: Ibrokhim Abdurakhmonov (ed.) Microsatellite markers. Editorial: InTechOpen science

Barrandeguy ME, Prado D, Goncalves AL, García MV (2016) Demografía histórica de *Anadenanthera colubrina* var. *cebil* (Leguminosae) en Argentina. B Soc Argent Bot 51:689–703

Braga AC, Reis AMM, Leoi LT, Pereira RW, Collevatti RG (2007) Development and characterization of microsatellite markers for the tropical tree species *Tabebuia aurea* (Bignoniaceae). Mol Ecol Notes 7:53–56

Brown AD, Grau HR, Malizia RL, Grau A (2001) Argentina. In: Kappelle M, Brown AD (eds) Bosques nublados del neotrópico. Instituto Nacional de Biodiversidad, Santo Domingo de Heredia, Costa Rica, pp 623–659

Brown AD, Grau A, Lomáscolo T, Gasparri NI (2002) Una estrategia de conservación para las selvas subtropicales de montaña (yungas) de Argentina. Ecotropicos 15:147–159

Brown AD, Pacheco S, Lomáscolo T, Malizia L (2006) La situación ambiental en los bosques andinos yungueños. In: Brown A, Martínez OU, Acerbi M, Corchera J (eds) La situación ambiental argentina 2005. Fundación Vida Silvestre Argentina, Buenos Aires, pp 52–71

Calonga Solís V, Barrandeguy ME, García MV (2014) Divergencia histórica en *Anadenanthera colubrina* var. *cebil* (Leguminosae) analizando una región intrónica del ADN cloroplástico. B Soc Argent Bot 49:547–557

Campanello PI, Gatti MG, Goldstein G (2008) Coordination between water-transport efficiency and photosynthetic capacity in canopy tree species at different growth irradiances. Tree Physiol 28:85–94

Campanello PI, Gatti MG, Montti LF, Villagra M, Goldstein G (2011) Ser o no ser tolerante a la sombra: economía de agua y carbono en especies arbóreas del Bosque Atlántico (Misiones, Argentina). Ecol Austral 21:285–300

Carvalho PER (1998) Especies nativas para fins produtivos: recomendações silviculturais, potencialidades e uso da madeira. Seminário sobre espécies não tradicionais, Curitiba Brazil, Octover 06-08, 2011 https://core.ac.uk/download/pdf/45525675.pdf

Chase MW, Hills HH (1991) Silica gel: an ideal material for field preservation of leaf samples for DNA studies. Taxon 40:215–220

Colcombet L, Barth S, Gonzalez P, Loto M, Munaretto N, Rossner M, Ziegler A, Pachas N (2019) Aprendizajes de una parcela agroforestal para implementar sistemas silvopastoriles con especies latifoliadas en Misiones, Argentina. Actas X Congreso Internacional de Sistemas Silvopastoriles. Asunción, Paraguay

Corbo Guidugli M, de Campos T, Barbosa de Sousa AC, Feres JM, Sebbenn AM, Mestriner MA et al (2009) Development and characterization of 15 microsatellite loci for *Cariniana estrellensis* and transferability to *Cariniana legalis*, two endangered tropical trees species. Conserv Genet 10:1001–1004

Crechi E, Hennig A, Keller A, Hampel H, Domecq C, Eibl B (2010) Crecimiento de 3 especies latifoliadas nativas a cielo abierto y bajo dosel de pino hasta los 12 años de edad, en Misiones Argentina (*Cordia trichotoma* Vell. Arrab. ex Steudel, *Balfourodendron riedelianum* Engl., *Enterolobium contortisiliquum* Vell. Morong.). XIV Jornadas Técnicas Forestales y Ambientales. Facultad de Ciencias Forestales, UNaM. Eldorado, June 10-12 2010

Del Castillo E, Zapater MA, Gil MN, Tarnowski C (2005) Selva de Yungas del Noroeste Argentino. Recuperación Ambiental y Productiva. Lineamientos Silvícolas y Económicos para un Desarrollo Forestal Sustentable. Estación Experimental de Cultivos Tropicales INTA Yuto. Documento Técnico N°1, p 47

Doyle JJ, Doyle JL (1990) Isolation of plant DNA from fresh tissue. Focus 12:13–15

Duarte E, Sansberro P, Luna C (2016) In vitro propagation of *Handroanthus heptaphyllus* (Vell.) Mattos by shoot cultures. Afr J Biotechnol 15:1292–1292

Eliano PM, Badinier C, Malizia LR (2009) Sustainable forest management in Yungas. Protocol for the development of forest management and implementation in a pilot farm. Subtropical editions, Proyungas Foundation

Evanno G, Regnaut S, Goudet J (2005) Detecting the number of clusters of individuals using the software STRUCTURE: a simulation study. Mol Ecol 14:2611–2620

Excoffier L, Smouse P, Quattro J (1992) Analysis of molecular human mitochondrial DNA restriction data. Genetics 131:479–491

Falush D, Stephens M, Pritchard JK (2007) Inference of population structure using multilocus genotype data: dominant markers and null alleles. Mol Ecol Notes 7:574–578

Feres JM, Monteiro M, Zucchi MI, Pinheiro JB, Mestrine MA, Alzate-Marin AL (2012) Development of microsatellite markers for *Anadenanthera colubrina* (Leguminosae), a neotropical tree species. Am J Bot 99:e154–e156

Fornes L, Zelener N, Gauchat M, Inza MV, Soldati MC, Ruíz V et al (2016) Subprograma Cedrela. In: Domesticación y Mejoramiento de Especies Forestales. PROMEF, UCAR, pp 136–159

Galindez G, Ledesma T, Alvarez A, Pastrana-Ignes V, Bertuzzi T, Lindow-López L et al (2019) Intraspecific variation in seed germination and storage behaviour of *Cordia tree* species of subtropical montane forest of Argentina: implications for ex situ conservation. S Afr J Bot 123:393–399

García MV, Prinz K, Barrandeguy ME, Miretti M, Finkeldey R (2014) A unifying study of phenotypic and molecular genetic variability in natural populations of *Anadenanthera colubrina* var. *cebil* from Yungas and Paranaense biogeographic provinces in Argentina. J Genet 93:123–132

Gil NM, Del Castillo EM (2004) Forestación. Cartilla Teórico-Practica. Facultad de Cs Naturales, UNSA

Gillies ACM, Navarro C, Lowe AJ, Newton AC, Hernández M, Wilson J (1999) Genetic diversity in Mesoamerican populations of mahogany (*Swietenia macrophylla*), assessed using RAPDs. Heredity 83:722–732

Goncalves AL, Barrandeguy ME, García MV (2014) Estructura y representatividad genética cloroplástica en poblaciones naturales de *Anadenanthera colubrina* var. *cebil* (Leguminosae) de Argentina. B Soc Argent Bot 49:235–245

Goncalves AL, García MV, Heuertz M, González-Martínez SC (2019) Demographic history and spatial genetic structure in a remnant population of the subtropical tree *Anadenanthera colubrina* var. *cebil* (Griseb.) Altschul (Fabaceae). Ann For Sci 76:18

Gonzalez PA, Barth SR, Agostini JP, Krivenki M, Fassola H, Ferruchi R, Carvallo A (2016) Proyecto de investigación aplicada PIA 10047. Estudio ecológico y silvicultural de Cordia trichotoma, Cabraela canjerana y Picrasma crenata.: su potencialidad en la diversificación productiva. In: Investigación forestal 2011–2015. Los proyectos de investigación aplicada. UCAR-Proyecto Forestal BIRF 7520 AR-GEF09118. P. 202–205. https://www.magyp.gob.ar/sitio/areas/proyectos_forestales/pias/PIAS%20BAJA_con%20tapas.pdf

Grignola J, Fornes L, Saravia P, Trapani A, Ledesma T, Tarnowski C, et al. (2018) Rescate de *Handroanthus impetiginosus* de las Yungas argentinas y conservación *ex situ* en Huerto Semillero Clonal. 7° jornadas forestales del NOA, 6 y 7 Sept. 2018. San Salvador de Jujuy

Hartl DL, Clark AG (2007) Principles of population genetics, 3rd edn. Sinauer Associates, Inc. USA, 652 pág

Hoisington DA, Khairallah MM, Gonzales de Leon D (1994) Laboratory protocols. CIMMYT Applied Molecular Genetics Laboratory. CIMMYT, Hisfoa

Hussain H, Green IR (2017) Lapachol and lapachone analogs: a journey of two decades of patent research (1997–2016). Expert Opin Ther Pat 27:1111–1121

Inza MV, Zelener N, Fornes L, Gallo LA (2012) Effect of latitudinal gradient and impact of logging on genetic diversity of *Cedrela lilloi* along the argentine Yungas rainforest. Ecol Evol 2:2722–2736

Inza MV, Grignola J, Zelener N, Fornes L (2019) Variabilidad genética preliminar de *Handroanthus impetiginosus* (Mart. ex DC) Mattos (lapacho rosado) en las Yungas Argentinas. VIII Reunión Reunión Grupo de Genética y Mejoramiento Forestal INTA (GeMFo). López JA, Hernández MA, López JA (eds.), pp 73–76. Bella Vista, Corrientes, Argentina

IUCN (2012) Red list of threatened species. Retrieved from http://www.iucnredlist.org

Jausoro V, Llorente BE, Apóstolo NM (2010) Structural differences between hyperhydric and normal in vitro shoots of *Handroanthus impetiginosus* (Mart. Ex DC) Mattos (Bignoniaceae). Plant Cell Tissue Organ Cult 101:183–191

Larraburu EE (2014) Morfogénesis in vitro de *Handroanthus impetiginosus* (Mart. Ex DC.) Mattos (Bignoniaceae). Doctoral thesis. Universidad Nacional de Luján, Buenos Aires

Larraburu EE, Apóstolo NM, Llorente BE (2012) In vitro propagation of pink lapacho: response surface methodology and factorial analysis for optimisation of medium components. Int J For Res. Article ID 318258, 9 pp.

Las Peñas ML (2003) Los cromosomas somáticos de *Cordia ecalyculata* y *C. trichotoma* (Boraginaceae). Bol Soc Argent Bot 38:319–323

Ledesma T (2014) Las especies del género *Cordia* en el noroeste de Argentina: Conservación in situ y ex situ. MSc thesis, Facultad de Agronomía, Universidad de Buenos Aires, 84 p

Lemes MR, Martiniano T, Reis V, Faria C, Gribel R (2007) Cross amplification and characterization of microsatellite loci for three species of *Theobroma* (Sterculiaceae) from the Brazilian Amazon. Genet Resour Crop Evol 54:1653–1657

Lemes MR, Esashika T, Gaoue OG (2011) Microsatellites for mahoganies: twelve new loci for *Swietenia macrophylla* and its high transferability to *Khaya senegalensis*. Am J Bot 98:e207–e209

Lozano EC, Zapater MA (2008) Delimitación y estatus de *Handroanthus heptaphyllus* y *H. impetiginosus* (Bignoniaceae, Tecomeae). Darwin 46:304–317

Marulanda ML, Lopez AM, Uribe M, Ospina CM (2011) Genetic variability of *Cordia alliodora* (R. & P.) Oken progenies. Colombia Forestal 14:119–135

Mazo TM, Barrandeguy ME, García MV (2014) Endocría y coancestría en poblaciones naturales de *Anadenanthera colubrina* var. *cebil*. Revista de Ciencia y Tecnología 22:22–27

Montti L, Villagra M, Campanello PI, Gatti MG, Goldstein G (2014) Functional traits enhance invasiveness of bamboos over co-occurring tree saplings in the semideciduous Atlantic Forest. Acta Oecol (Montrouge) 54:36–44

Munaretto N, Barth S, Fassola H, Colcombet L, Gonzalez P, Comolli L, et al. (2019) Productividad de *Ilex paraguariensis* cultivada según disponibilidad de luz. XVIII Jornadas Técnicas Forestales y Ambientales 17–19 Oct. 2019, Eldorado. Misiones Argentina Páginas, pp 283–285

Nei M (1978) Estimation of average heterozygosity and genetic distance from a small number of individuals. Genetics 89:583–590

Oliveira EJ, Gomes Pádua J, Zucchi MI, Vencovsky R, Carneiro Vieira ML (2006) Origin, evolution and genome distribution of microsatellites. Genet Mol Biol 29:294–307

Oliveira TP, Barroso DG, Lamônica KR, Carneiro JG, Oliveira MA (2015) Productivity of poly-clonal minigarden and rooting of *Handroanthus heptaphyllus* Mattos minicuttings. Semina: Ciências Agrárias, Londrina 36:2423–2432

Oliveira TP, Barroso DG, Lamônica KR, Carvalho GC (2016) Aplicação de AIB e tipo de mini-estacas na produção de mudas de *Handroanthus heptaphyllus* Mattos. Ciência Florestal, Santa Maria 26:313–320

Ortíz G (2015) Characterization and perspective of the forest industry of the province of Jujuy. St ed. Autonomous City of Buenos Aires. Ministry of Agriculture, Livestock and Fishing of the Nation. Unit for Rural Change (UCAR)

Ovando GL, Enciso M, Ovelar G, Villalba N (2013) Propagación vegetativa de *Tabebuia hep-taphylla* (Vell.) Toledo (lapacho negro) mediante esquejes de raíz. Investigación Agraria 9:73–79

Peakall R, Smouse PE (2006) GENALEX6: genetic analysis in excel. Population genetics software for teaching and research. Mol Ecol Notes 6:288–295

Peakall R, Gilmore S, Keys W, Morgante M, Rafalski A (1998) Cross species amplification of soy-bean (*Glycine max*) simple-sequence repeats (SSRs) within the genus and other legume genera: implications for the transferability of SSRs in plants. Mol Biol Evol 15:1275–1287

Pritchard JK, Stephens M, Donnelly P (2000) Inference of population structure using multilocus genotype data. Genetics 155:945–959

Richter HG, Dallwitz MJ (2000) Commercial timbers: descriptions, illustrations, identification, and information retrieval In English, French, German, Portuguese, and Spanish Version: 17th Feb. 2019

Rodrigues MB (2019) Enraizamento de miniestacas e parâmetros de validação de clones de ipê-roxo para a propagação vegetativa. MSc Thesis. Centro de Ciências Rurais, Universidade Federal de Santa Maria, Rio Grande do Sul, Brasil

Rodríguez GH, Barth SR (2014) Selección fenotípica de "Peteribí" (*Cordia trichotoma* (Vellozo) Arrábida ex Steudel) en el noreste argentino. XVI Jornadas Técnicas Forestales y Ambientales. Mayo 2014. Eldorado. Misiones. Cd de Actas. https://www.jotefa.com/storage/actas/2014/Actas-Conferencias-JOTEFA-2014.pdf

Rohlf FJ (1998) NTSYS-PC numerical taxonomy and multivariate analysis system version 2.0. Exeter Software, Setauket

SGAyDS (2019) Anuario de Estadística Forestal 2017–2018. Buenos Aires, 175 pp. https://www.argentina.gob.ar/ambiente/tierra/bosques-suelos/manejo-sustentable-bosques/programa-nacional-estadistica-forestal

Soldati MC, Fornes L, van Zonneveld M, Thomas E, Zelener N (2013) An assessment of the genetic diversity of Cedrela balansae (Meliaceae) in Northwestern Argentina by means of com-bined use of SSR and AFLP molecular markers. BiochemSystEcol 47:45–55

Soldati MC, Inza MV, Fornes L, Zelener N (2014) Cross transferability of SSR markers to endan-gered Cedrela species that grow in Argentinean subtropical forests, as a valuable tool for popu-lation genetic studies. Biochem Syst Ecol 53:8–16

Tarnowski CG (2010) Desenvolvimento e caracterização de marcadores microssatélites (SSRs) para *Cedrela lilloi* C. de Candolle. MSc thesis, Centro de Ciências Agrarias, Universidade Federal de Santa Catarina, Brasil

Tortorelli LA (2009) Maderas y bosques argentinos. Orientación Gráfica Editora, Buenos Aires

Viglione A, Vallejo N (2011) CEBIL: an American hallucinogenic plant in botany books by differ-ent authors. 40th International Congress for the History of Pharmacy, Berlin, Germany, Sept. 14–17, 2011. http://www.histpharm.org/40ishpBerlin/L82F.pdf

Von Reis AS (1964) A taxonomic study of the genus *Anadenanthera*. Contributions from the Gray Herbarium of Harvard University 193:3–65

White G, Powell W (1997) Cross-species amplification of SSR loci in the Meliaceae family. Mol Ecol 6:1195–1197

Zapater MA, Califano LM, Del Castillo EM (2005) Avances en el estudio de las especies de *Tabebuia* (Bignoniaceae) y su cultivo en la selva subtropical del noroeste argentino. 3° Congreso Forestal Argentino y Latinoamericano, Corrientes

Part IV
New Tools and Final Considerations

Chapter 17
Application of High-Throughput Sequencing Technologies in Native Forest Tree Species in Argentina: Implications for Breeding

Susana L. Torales, Verónica El Mujtar, Susana Marcucci-Poltri, Florencia Pomponio, Carolina Soliani, Pamela Villalba, Maximiliano Estravis-Barcala, Lorena Klein, Martín García, Vivien Pentreath, María Virginia Inza, Natalia C. Aguirre, Máximo Rivarola, Cintia Acuña, Sergio González, Sabrina Amalfi, Micaela López, Pauline Garnier-Géré, Nicolás Bellora, and Verónica Arana

The high-throughput sequencing (HTS) or next-generation sequencing (NGS) technologies and associated bioinformatics tools have become a powerful approach for the development of entire genomes and transcriptomes, as well as molecular markers, on model and non-model organisms (Badenes et al. 2016). For non-model perennial trees, the HTS technologies provide a rapid way to access to genomic

S. L. Torales (✉) · F. Pomponio · M. V. Inza · M. López
Instituto de Recursos Biológicos (IRB), Centro de Investigaciones de Recursos Naturales (CIRN), INTA. Hurlingham, Buenos Aires, Argentina
e-mail: torales.susana@inta.gob.ar

V. El Mujtar · C. Soliani · V. Arana
Instituto de Investigaciones Forestales y Agropecuarias Bariloche (IFAB) INTA-CONICET, Bariloche, Argentina

S. Marcucci-Poltri · P. Villalba · M. García · N. C. Aguirre · M. Rivarola · C. Acuña
S. González · S. Amalfi
Instituto de Agrobiotecnología y Biología Molecular (IABiMO) INTA-CONICET, Buenos Aires, Argentina

M. Estravis-Barcala · N. Bellora
Instituto Andino Patagónico de Tecnologías Biológicas y Geoambientales (IPATEC)
CONICET–UNCo, San Carlos de Bariloche, Argentina

L. Klein
INTA EEA Saénz Peña, Saénz Peña, Argentina

V. Pentreath
Universidad Nacional de la Patagonia San Juan Bosco, Comodoro Rivadavia, Argentina

P. Garnier-Géré
INRAE, Paris, France

© Springer Nature Switzerland AG 2021   455
M. J. Pastorino, P. Marchelli (eds.), *Low Intensity Breeding of Native Forest Trees in Argentina*, https://doi.org/10.1007/978-3-030-56462-9_17

information crucial for breeding programmes considering their long generation times and long intervals of breeding cycle (Badenes et al. 2016). This chapter aims to examine the state-of-the-art scientific knowledge on the application of HTS technologies on forest trees, particularly on genomic resources and marker development for Argentinean native species.

17.1 High-Throughput Sequencing Technologies

The development of Sanger sequencing (Sanger et al. 1977) was the first step in recovering genomic information from different organisms. This is a method of DNA sequencing based on the selective incorporation of chain-terminating dideoxynucleotides by a DNA polymerase during in vitro DNA replication. The sequencing technology for the Sanger method has been progressively automated increasing the number of samples that could be sequenced at once (Adams 2008; Heather and Chain 2016). Therefore, Sanger technology has been widely adopted and employed in many large-scale sequencing projects such as the human genome as well as model and non-model species (e.g. Venter 2001; Chinwalla et al. 2002; Tuskan et al. 2006; Ueno et al. 2010).

After years of evolution, new high-throughput technologies emerged as the second generation of sequencing methods that exhibited better performance in aspects that include massively parallel sequencing, high throughput and reduced cost (Liu et al. 2012). The second-generation platforms vary with respect to throughput, cost, read lengths which are generally short (average read length: 75–700 bp) and the associated error rates varying according to platform and instrument (0,1-1) (Liu et al. 2012; Kchouk et al. 2017; Pfeiffer et al. 2018). In general, massive parallelization is facilitated by the creation of many millions of individual reaction centres, where DNA fragments are clonally amplified on a solid or liquid/emulsion medium to be able then to distinguish signal from background noise. A sequencing platform can collect information from all those reaction centres simultaneously and then sequencing many millions of DNA molecules in parallel (Goodwin et al. 2016). These HTS technologies are based on sequencing by ligation (SBL) and sequencing by synthesis (SBS). Fundamentally, in SBL approaches, a probe sequence with a fluorophore hybridizes to a DNA fragment and is ligated to an adjacent oligonucleotide for imaging. The emission spectrum of the different fluorophore probes gives the sequences (SOLiD technology). In SBS a polymerase is used, and a signal identifies the incorporation of a nucleotide into an elongating strand. The signal could be based on fluorescence (Illumina technology) which ranges from small to large ultra-high-throughput ones (Reuter et al. 2015), change in ion (proton, H^+) concentration (Ion Torrent technology, Reuter et al. 2015) or bioluminescence (Roche/454 technology, Goodwin et al. 2016). The third-generation sequencing platforms, based on generating longer DNA sequences, are single-molecule real-time (SMRT) developed by Pacific Biosciences (Roberts et al. 2013) and Oxford Nanopore Technologies (Wang et al. 2015). These long-read sequencing technologies deliver reads in excess of several kilobases (average 13–20 kb) allowing for the resolution of large structural features, such as complex or repetitive regions, and are also capable of spanning entire mRNA transcripts, discerning gene isoforms (Goodwin et al. 2016, Tyson et al. 2018).

The ideal situation for genomic research normally requires a reference genome and re-sequencing or genotyping data for hundreds to thousands of markers. For non-model species, however, reference genome and re-sequencing are more difficult to achieve due to their high investment or methodological issues (e.g. large and/or repetitive genome). The advances on these HTS technologies have also driven down the costs of DNA sequencing allowing the development of massive genotyping methods. These methods are particularly useful for species without any genomic data, for which previously the development and genotyping of molecular markers [e.g. simple sequence repeats (SSRs), single-nucleotide polymorphisms (SNPs)] was traditionally performed by using relatively expensive approaches (e.g. Eckert et al. 2013; Geraldes et al. 2013; Pavy et al. 2013).

One of the massive genotyping methods that emerged in recent years is called reduced representation, as it is focused only on a portion of the individuals' genome which reduces sequencing costs. Particularly, restriction enzyme-guided sequencing approaches have become the most popular reduced representation methods for non-model organisms (Parchman et al. 2018), being restriction site-associated DNA sequencing (RADseq, Baird et al. 2008; ddRADseq, Peterson et al. 2012), and genotyping by sequencing (GBS, Elshire et al. 2011) the most popular strategies. Briefly, in these methods, library preparation consists of genomic DNA digestion with one or two restriction enzymes, subsequent adapter ligation, selection, and high-throughput sequencing of DNA fragments. This approach simultaneously allows genotyping a large number of molecular markers (in particular SNPs) in a large number of individuals in the same assay. Massive genotyping methods prompted, therefore, genomic research in non-model species as they can be easily applied to any species for rapid marker discovery and genotyping without requiring previous genomic resources (Davey et al. 2011; Andrews et al. 2016; Parchman et al. 2018).

With the explosion of these enormous new datasets, the need for better and faster analyses increased dramatically. Bioinformatics is a discipline with a great expansion in the last decade as a way to deal with all these new approaches in genomic, transcriptomic and other -omics. The main objective in this new era is to use these large datasets, accounting for their biases and sequencing errors, to extract useful information and to gain better knowledge of the studied subject. In summary, to go from *data to information* is not an easy task to fulfill and is one of the major bottlenecks in -omics research (Greene et al. 2014).

The bioinformatics area is interdisciplinary from its beginning, going from database management, sequence analysis and/or statistics in genomics. In this new era, researchers can establish new experimental designs and tackle never-before-tried biological challenges, such as how do all genes change expression under a new condition, or genotype millions of positions in a genome for which no reference genome exist (Feuillet et al. 2011). These new technologies have not only scaled up datasets but have radically "modified/changed" the traditional paradigm of hypothesis-driven science, with a new paradigm of big data science (Stephens et al. 2015).

Data science is a specialized field that combines multiple areas such as statistics, mathematics, intelligent data capture techniques, data cleansing, mining, and programming to prepare and align big data for intelligent analysis to extract insights and information from them. Moreover, specifically in functional genomic analysis of species lacking a reference genome (de novo), this involves the integration of heterogeneous data sources including (but not limited to) transcriptome assembly and annotation, along with gene expression and genetic variants. In particular, it is important to establish that the volume and complexity of the data produced imply a technical, but also a scientific, challenge (Greene et al. 2014). Currently, many projects use big data HTS experiments, analyse the results and generate new hypotheses that are then further tested. Some strategies used in tree breeding which are of great impact, later covered in this chapter, are:

1. Novel SNP discovery and genome-wide genotyping of hundreds of individuals (De Pristo et al. 2011). With variant discovery, nowadays, researchers can sequence DNA or RNA samples from different individuals and without previous knowledge determine variant positions and genotype their samples with thousands of newly discovered SNPs. This opens a wide range of possibilities, never achieved before at such low cost.
2. De novo genomic and transcriptome assembly (Birol et al. 2013; Conesa et al. 2016). In de novo genomic or transcriptome assembly, researchers can identify new (near full length) transcripts or genes (loci) and provide functional annotations to better characterize their material, all with a lot less effort than ever thought few decades ago.
3. Reduced representation sequencing is a technique used to sequence an abridged representation of the genome in many individuals. Ideally, it allows sequencing the same reduced regions of the genome in all individuals and therefore identify variants in those regions.
4. Genome-wide association studies (GWAS) to identify associations between a genetic variant and a specific phenotype. It is basically a statistical correlation between a marker and a phenotype (Brachi et al. 2011; Beilsmith et al. 2019).
5. Genomic selection is a strategy which simultaneously estimates the effects of all available markers along the genome to predict individually breeding values, without specific knowledge of the genes that contribute to a particular character (Sewell and Neale 2000; Neale 2007).

17.2 Applications of HTS Technologies on Forest Tree Species

Genomics research in forest tree species has been prompted by the development of HTS technologies and bioinformatics tools. A systematic literature search from Scopus database combining two search word strategies[1] for the period 2016–2019

---

[1] Strategy A: (TITLE-ABS-KEY ("next-generation sequencing" OR "massive sequencing" OR "high-throughput sequencing") AND TITLE-ABS-KEY ("forest tree" OR "tree species" OR

provided 1343 articles (after filtering data), considering the relevance of each article to the focus of the literature search. This literature search revealed that scientific research includes a high number of tree species families, although families of model and worldwide cultivated species, such as Salicaceae, Pinaceae, Rosaceae, Fagaceae and Fabaceae, prevailed (Fig. 17.1a). Major research themes involving both HTS technologies and forest tree species included (1) the genetic bases of response to biotic and abiotic factors and quantitative trait variation, (2) marker development and characterization, (3) chloroplast characterization and phylogenetic analyses, (4) population genomics and evolutionary studies, (5) transcriptome characterization and (6) genome draft development (Fig. 17.1b). Illumina technology dominates high-throughput sequencing in forest tree species (Fig. 17.1c), with targeted sequencing being homogenously distributed among nuclear genome, organelle genome and transcriptome (Fig. 17.1d). Reduced genome sequencing has been less applied than whole-genome sequencing (Fig. 17.1e), with RADSeq and GBS representing around 50% of the applied reduced sequencing methods. The literature searches also revealed that both SSR and SNP markers have been developed from high-throughput sequencing in a similar percentage (around 50%) (Fig. 17.1f).

When considering genetic determinism studies of biotic and abiotic phenotypic responses that generate relevant information for breeding programmes, it can be noted that transcriptome sequencing (also called RNA-sequencing or RNA-Seq) prevailed (28 out of 39 articles). The transcriptome is an efficient tool for generating enormous sequence collections of expressed genes, in particular for forest tree species, which can have large or repetitive genome. It provides a valuable starting point for characterizing functional genetic variation and discovering functional molecular markers. Developments of gene database collections and bioinformatics tools have been crucial for the annotation of transcriptome sequences and the assignation of their putative biological functions.

Up to date, several transcriptomic studies on tree species have been carried out by means of Roche/454-technology strategies, mainly for the development of genomics resources of unexplored species and for secondary metabolic products discovery: in *Quercus* spp. (Ueno et al. 2010; Torre et al. 2014; Lesur et al. 2015), *Pinus contorta* (Parchman et al. 2010), *Pinus pinaster* (Canales et al. 2014), *Khaya senegalensis* (Karan et al. 2012), *Azadirachta indica* (Krishnan et al. 2012; Wang et al. 2016), *Nothofagus alpina* (Torales et al. 2012), *Prosopis alba* (Torales et al. 2013), *Carapa guianensis* (Brousseau et al. 2014), *Melia azedarach* (Wang et al. 2016), *Cedrela balansae* (Torales et al. 2018) and *Dalbergia odorifera* (Liu et al. 2019) between others. On the other hand, a review carried out by Parent et al. (2015) showed the species and approaches that were applied in order to find forest gene differential expression in wood formation, stress response among other conditions, through microarray and RNA-Seq analysis. They also highlighted the application of RNA-Seq analysis in many species without genome reference and strategies used to

"native tree")). Strategy B: (TITLE-ABS-KEY ("transcriptome" OR "genome" AND sequencing) AND TITLE-ABS-KEY ("forest tree" OR "tree species" OR "native tree")). Search date: 04-04-2019.

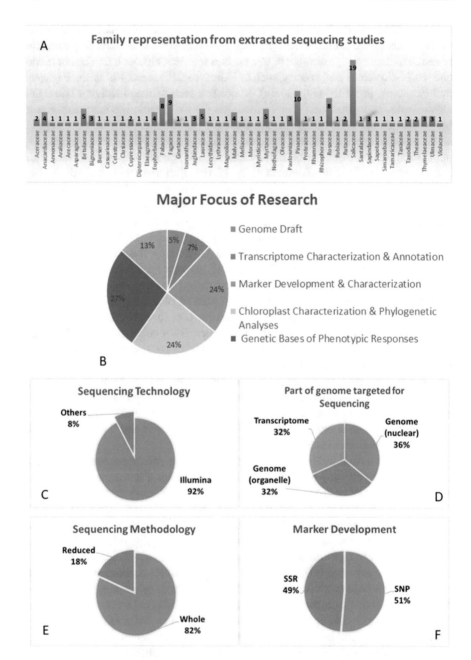

Fig. 17.1 (**a–f**) Number and proportions of literature searched genomics studies using NGS in tree species across families, major research focus, sequencing technologies and methodologies, part of the genome targeted for sequencing and type of markers developed

identify genome variation like SNP and the development of genotyping platforms. Neale and Wheeler (2019) completed and updated the review specifically for conifers. In López de Heredia and Vázquez-Poletti (2016), bioinformatics tools that are needed for the analysis of RNA-Seq data from forest trees are pointed out taking into account difficulties in de novo transcriptome assembly due to the high level of heterozygosity, large genome size particularly in conifers and annotation difficulties when based on intensively studied crops or herbs.

Recently, Estravis-Barcala et al. (2019) reviewed RNA-Seq works directed to study abiotic stress responses in trees, with focus on temperature and drought stresses. In the article, authors provide information about sequencing methods, bioinformatics tools employed, type of stress analysed, source of RNA and differentially Gene Ontology (GO) terms and Kyoto Encyclopedia of Genes and Genomes (KEGG) pathways overrepresented in each species/stress. Alternatively, high-throughput genotyping methods, such as RADSeq, ddRADSeq or GBS, offer powerful ways to generate genome-wide SNPs. They have been applied in several tree species for a diversity of research objectives such as linkage mapping, QTL analysis, markers development, phylogenetics, hybridization and introgression, phylogeography, genome scan, landscape genomics, genomic selection, association mapping, parentage analysis, dispersal distance estimation, species delimitation, heritability, population genetics and seed zone delineation (reviewed in Parchman et al. 2018).

In *Eucalyptus dunnii*, Aguirre et al. (2019) have published a detailed protocol of an optimized ddRADseq, which can be applied to other non-model plant species. Applying the technique GBS in two individuals of a breeding population of *E. dunnii* and a de novo analysis (Stacks 1.48 pipeline, Catchen et al. 2013), 25,778 SNPs in 14,423 genomic regions and 55 polymorphic SSRs were identified (Aguirre et al. 2019). Similarly, ddRADSeq is being applied to study phenotype-genotype association in *Nothofagus alpina* (Arias Rios, El Mujtar, Marchelli, in progress).

17.3   Genomic Strategies to Accelerate Tree Breeding

The development of forest tree breeding programmes faces multiple challenges, such as long breeding cycles, late flowering, emerging diseases and a constantly changing environment (Grattapaglia et al. 2018). Some of these challenges were simplified thanks to the development of DNA markers in the late 1980s and their subsequent application in marker-assisted selection (MAS) (Lande and Thompson 1990; Neale and Williams 1991; Strauss et al. 1992; Williams and Neale 1992).

The main benefit of the use of molecular markers in forestry programmes was reflected especially in the last decades, in the change from a traditional phenotype-based to a genotype-based selection. This allows delivering genetically improved material to the forestry sector (Grattapaglia et al. 2018). However, one crucial problem of MAS is that it is not suitable for complex traits that are controlled by a large number of genes (Bernardo 2008; Grattapaglia et al. 2018). High-throughput sequencing technologies (HTS) allowed a shift in the strategies of analyses to overcome the limitation of MAS (Minamikawa et al. 2018). One approach is the genome-wide association study (GWAS) which provides detection of quantitative trait loci (QTLs) for the target traits. However, if the objective is to select superior individuals for the breeding programmes based on their genomic estimated breeding values, the genomic selection (GS) is the best approach at present (Meuwissen et al. 2001; Cappa et al. 2019).

In wild forest tree species, where the gene pools have not been broadly changed for breeding purposes in comparison to crops and cultivated forest trees, the combination of these strategies (GWAS, GS and MAS) will not only accelerate the breeding cycles, but also improve the design of the breeding programmes (Grattapaglia et al. 2018; Minamikawa et al. 2018). For example, given that forest breeding programmes are conditional on the length of the breeding cycle, the application of genomic tools could maximize the genetic gain per unit of time by the early selection of juvenile individuals based on their genomic estimated breeding values. In addition, the genetic information obtained by the utilization of the HTS could help to study not only the genetic variation in the native populations but also to integrate the environmental data for evaluating the performance of the genetic material in different scenarios, which could improve the design of the breeding programmes.

17.3.1 Genome-Wide Association Studies

Finding the connection between the genotype information and complex phenotype has been one of the fundamental goals of plant genetics research, mainly to understand how this knowledge can be applied directly to the breeding programmes (Botstein and Risch 2003). This is so because in most of the complex traits of agricultural importance, the phenotypic variation and their interaction with the environment are controlled by multiple QTLs. Accordingly, in the last decade, there have been a lot of scientific publications where the search for molecular markers associated with QTLs was carried out to identify genes responsible for the quantitative variation of different complex traits (Thornsberry et al. 2001; Blott et al. 2003; Aranzana et al. 2005; Zhu et al. 2008).

Genome-wide association studies (GWAS) is a powerful tool to identify statistical associations between a genetic variant and a specific phenotype through the utilization of a large number of molecular markers distributed along the genome (Brachi et al. 2011; Beilsmith et al. 2019). This genomic approach consists in the use of individuals collected from natural populations or groups of individuals often of unknown ancestry, to detect the genetic variability by exploiting the historical recombination events and the natural genetic diversity (Risch and Merikangas 1996; Nordborg and Tavare 2002; Zhu et al. 2008). In other words, it relies on the linkage disequilibrium (LD) where nearby genetic polymorphisms are associated more frequently than by chance. This approach offers an interesting strategy for forest genetics since it allows studying simultaneously various regions of the genome without the construction of a specific mapping population and shows a high-resolution power, in particular, when the population under study has low LD (Alvarez et al. 2014). For example, 17 SNPs were potentially detected associated with serotiny in a local population of *Pinus pinaster* (Budde et al. 2014). This study showed not only the application of GWAS but also its promising utilization in non-model species in their natural environments. On the other hand, it was stressed the usefulness of combining common garden experiments of non-model species with genomewide association studies in several populations and especially for detecting genome markers for locally adapted traits that differentiate among populations (De Villemereuil et al. 2016).

GWAS analysis was reviewed in Naidoo et al. (2019) to explore tolerance to biotic stress in forest trees and in Du et al. (2018) to analyse wood properties traits. Some reports were carried out for identifying the loci that contribute to adaptive variation in several species like *Populus trichocarpa* (Chhetri et al. 2019), *Handroanthus impetiginosus* (Collevatti et al. 2019) and *Pinus contorta* (Mahony et al. 2019).

17.3.2 Genomic Selection

The perspective of accelerating and improving the accuracy of selection by using molecular markers in forest tree species has led to the application of different methodologies in order to locate the loci affecting the traits of interest (Sewell and Neale 2000; Neale 2007). By contrast, genomic selection (GS) is based on the simultaneous estimation of the effects of all available markers along the genome to predict individually breeding values, without specific knowledge of the genes that contribute to a particular character or location (Meuwissen et al. 2001).

The approach of GS uses a "training population", with genotypic and phenotypic data available, to fit a predictive model. This model is then used to predict the genomic breeding values of seedling progeny in future generations from its genotypic data. The prediction is used to generate a ranking to select superior individuals to be planted in the field. Compared with other methodologies, GS shows higher accuracy and allows shortening the improvement cycles by 50% (Grattapaglia and Resende 2011), which implies a gain associated with early elite clone selection. Also, it reduces the costs in the maintenance of the progeny tests and minimizes the measurements with the consequent reduction in logistics and personnel cost. Besides, GS is useful to improve multiple traits at one time (particularly for low heritability traits) by generating independent predictive models that allow the generation of selection indexes. This allows the breeder to select at an early age and optimize controlled crosses, by choosing mating pairs that provide their progenies with the best allelic composition for the traits to be improved (Toro and Varona 2010).

A "training population" for the breeding programme of a wild native species through GS implies the establishment of a seed orchard, as far as possible with clonal replicates, in view of the very probable absence of pedigree information. The presence of clones will allow measurements with less environmental noise, because a single plant may be subject to microsite effects. In its development, the registration of kinship relations is very important, for greater certainty of the identity by descent of the alleles of interest. The "application population" could be all trees available for selection in both natural stands and plantations.

There are still few studies on forest species in which the next generation of a breeding programme is predicted. Bartholomé et al. (2016) were the first to report genomic selection in *Pinus pinaster* with data from generations G0 and G1, to predict in G2 obtaining satisfactory results. In a more recent study, Thistlethwaite et al. (2019) performed cross-generational GS analysis on coastal Douglas-fir (*Pseudotsuga menziesii*) and obtained predictive accuracies almost as high as that of ABLUP (traditional BLUP based on the pedigree-based A matrix) and concluded that more markers or improved distribution of markers are required to capture LD in Douglas-fir (genome size ~16 Gpb). The latter supports the need for the development of mass genotyping methods in natives for better GS performance.

17.4 HTS Technologies Applied to Argentina's Native Tree Species

The native forest tree species of economic and ecological importance in Argentina such as *Nothofagus alpina (= N. nervosa)*, *N. obliqua*, *N. pumilio* and *Austrocedrus chilensis* of the Andean Patagonian forest; *Cedrela balansae*, *Cordia trichotoma* and *Handroanthus impetiginosus* from the subtropical rainforests of Argentina; and *Prosopis alba* of Chaco, among others, are being affected by global change (e.g. deforestation, forest exploitation, grazing, drought and thermal stresses (SAyDS 2007). Furthermore, predicted changes in temperature (Barros et al. 2015) may alter selective pressures on trees and, therefore, have the potential to influence evolutionary processes, which depend fundamentally on the genetic diversity existing in populations.

In this context, the conservation and management of the remaining forests, as well as the reforestation of degraded areas, requires knowledge of the genetic bases underlying adaptive traits. This knowledge is also crucial for breeding programmes. Therefore, the development of genomic tools, such as collections of genes and adaptive molecular markers, are required for monitoring the genetic diversity of natural populations, identifying individuals/populations with greater adaptation ability in the context of climate change and providing propagation of selected material for restoration and breeding programmes.

The lack of genomic information for native Argentinean forest species has been a limitation. Previous studies, like the candidate genes characterization of adaptive genetic variation in native species from Argentina [*Austrocedrus chilensis* (Pomponio et al. 2013) and *Prosopis alba* (Pomponio et al. 2014)], were performed based on genomic resources of model and non-model species (e.g. *Arabidopsis thaliana*, *Pinus taeda*, *Pinus pinaster*, *Cryptomeria japonica*, *Prosopis juliflora*, *Eucalyptus*) available in public databases [National Center for Biotechnological Information (NCBI) and The Institute for Genomic Research (TIGR)]. In the last decade, with the advent of HTS, scientific research has taken advantage of the new, low-cost and fast-sequencing technologies to acquire genomic information of native forest species from Argentina.

17.4.1 Transcriptome and Genome Sequencing of Nothofagus *Species*

In Argentina, the genus *Nothofagus* is represented by six endemic species (*N. alpina*, *N. obliqua*, *N. antarctica*, *N. dombeyi*, *N. betuloides* and *N. pumilio*) distributed on the Patagonian Andes Range. *Nothofagus alpina* (Poepp. and Endl.) Oerst. (= *N. nervosa* (Phil.) Dim. et Mil.) and *N. obliqua* (Mirb.) Oerst. are two closely related species occurring in sympatry over approximately 18,000 ha (Sabatier et al. 2011), but presenting ecological and physiological differences (Varela et al. 2010; Azpilicueta et al. 2013) that suggest adaptive variation. *Nothofagus pumilio* is the most abundant native tree of the Patagonian temperate forest. Its distribution spans from 35° S to 56° S, occupying the high altitudes and building the tree line. Photoperiod and temperature change considerably towards southern latitudes, marking *N. pumilio* as a suitable model to test local adaptation and clinical variation in genetic and adaptive traits. This species' preference for colder environments indicates a particularly high sensitivity towards future changes in temperature as predicted in a Global Climate Change context (IPCC 2014). In spite of their ecological and economical importance, little genomic resources were available for this species (Azpilicueta et al. 2004; Marchelli et al. 2008; Soliani et al. 2010).

In the last decade, a first transcriptome sequencing of *N. alpina* was performed on young leaf tissue using Roche/454 technology (Torales et al. 2012). The methodology included de novo assembly, functional annotation, and in silico discovery of potential molecular markers (Fig. 17.2 and Table 17.1). The 82% of the identified unigenes showed high similarity to unigenes from the Fagaceae Genomic database (http://www.fagaceae.org/), while the others could be novel genes. Unigenes with putative function (15,497) represented 50% of the *N. alpina* genes catalogue considering that the average number of genes encoded in a plant nuclear genome is about 30,000 (Logacheva et al. 2011). RNA-Seq on this species also allowed the identification of SSRs, with 55% of them located within gene coding regions (functional SSR markers). Seventy-three (73) SSRs located in sequences with annotation function related to response to abiotic stress were selected, validated in *N. alpina* and transferred to other South American *Nothofagus* species (*N. obliqua*, *N. pumilio*, *N. antarctica* and *N. dombeyi*). A set of 27 SSRs was then validated for these species providing a useful panel of markers for conservation and evolutionary studies and

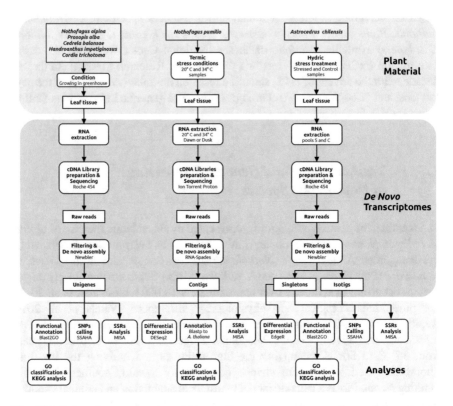

Fig. 17.2 Schematic representation of the overall sequencing and annotation workflow of native forest tree species' transcriptome (*Nothofagus alpina*, *Prosopis alba*, *Cedrela balansae*, *Handroanthus impetiginosus*, *Cordia trichotoma*, *Nothofagus pumilio* and *Austrocedrus chilensis*)

demonstrating that discriminant markers can be developed efficiently by mining functional annotation of transcriptome sequences (El Mujtar et al. 2017a). In this sense, the comparison between markers from unigenes of the abiotic stress category and anonymous markers revealed a greater discriminant capacity in the functional markers (Torales et al. 2012; El Mujtar et al. 2017a).

As stated previously in this Section, *N. pumilio* thrives in relatively cold environments, marking it as a potentially sensitive species to temperature increases such as those predicted by Global Climate Change. In this context, in order to study the overall effect of high temperatures in the physiology and gene expression of this species, Estravis-Barcala et al. (in prep.) performed RNA-Seq in contrasting experimental conditions, consisting of two temperatures (20 and 34 °C) and two moments of the day (dawn and dusk). Two biological replicates from independent experiments were sequenced for each combination of temperature and time of the day. The total RNA was extracted from whole leaves following Chang et al. (1993), and poly-A-selected mRNA libraries were prepared for sequencing in an Ion Torrent Proton device (Table 17.1). Bioinformatic workflow is summarized in Fig. 17.2.

Table 17.1 Summary of HTS results obtained for the Argentinean native tree species

Species	System	# Raw reads	# Unigenes	# SSRs	# SNPs	# BLAST matches	# GO terms
Nothofagus alpina	Transcriptome (454 GS-FLX)	146,267 (408)[a]	24,886	3,048 (2,517)[b]	ND	15,497	11,834
N. alpina-N. obliqua	Genome (454 GS-FLX)	361,438 (313)[a]	32,595	2,274 (769)[b]	8,669	5,134	4,750
N. pumilio	Transcriptome (IonTorrent Proton dev.)	222,828,783 (133)[a]	81,761	8,456	ND	36,371	6,153
Prosopis alba	Transcriptome (454 GS-FLX)	1,103,231 (421)[a]	54,814	5,992 (4,593)[b]	6,236	37,563	22,095
Cedrela balansae	Transcriptome (454 GS-FLX)	212,589 (434)[a]	27,111	2,663 (1,932)[b]	1,243	20,953	19,029
Austrocedrus chilensis	Transcriptome (454 GS-FLX)	C: 395,835 S: 421,884	32,255 (C+S)	1,450(C+S)[b]	C:253 S:186	ND[c]	12,382 (C+S)
Austrocedrus chilensis	Genome (454 GS-FLX)	358,803 (378)[a]	ND	3,563 (2,056)[b]	ND	11,032	ND
Handroanthus impetiginosus	Transcriptome (454 GS-FLX)	318,156 (408)[a]	44,263	9,745 (7,324)[b]	ND[c]	31,958	8,792
Cordia trichotoma	Transcriptome (454 GS-FLX)	413,968 (465)[a]	14,700	1,500 (903)[b]	ND[c]	ND[c]	ND[c]

C control, *S* stress, *ND* no data
[a]Mean length (bp)
[b]SSRs with designed primers
[c]Not yet analysed

Annotation against the *Arabidopsis thaliana* proteome resulted in 36,371 contigs (44.48% of the total) with a name and function assigned. On the other hand, preliminary differential expression (DE) analysis revealed more than 1000 contigs differentially expressed between temperatures, regardless of the moment of the day (FDR-corrected p-value <0.05). This first step in the -omic study of *N. pumilio* will be of great help to current and future studies related to functional genomics, local genetic variation and adaptation and gene discovery and characterization in this important native species.

In angiosperm native species, genome sequencing has also only been performed on *N. alpina* and *N. obliqua* using Roche/454 technology (El Mujtar et al. 2014; Table 17.1 see also SeqQual pipeline, available at https://github.com/garniergere/SeqQual). Several local and online bioinformatics tools were combined for the assembly, annotation and classification of produced genomic resources (Fig. 17.3). Although it was a partial genome sequencing of low coverage (13.4 Mb of

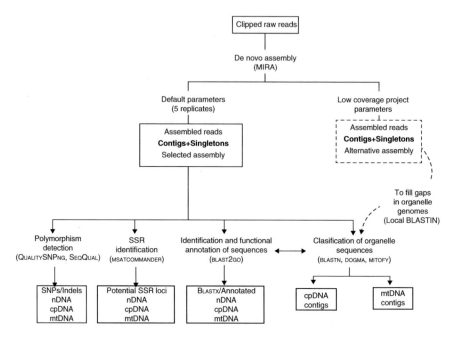

Fig. 17.3 Summary of the strategy followed for assembly, annotation and classification of genomic resources derived from partial genome sequencing of *Nothofagus alpina/N. obliqua* and *Austrocedrus chilensis*

assembled sequences with an average length of 412 bp), genomic resources developed allowed (1) to generate information about genic and non-genic nuclear and organelle (chloroplast and mitochondrial) regions, (2) to identify and characterize SSR and SNP markers, (3) to determine using developed SSRs a higher genetic diversity for *Nothofagus obliqua* (average number of alleles 4.89) than for *N. alpina* (2.89) and (4) to reveal a strong conservation of the organelle genome in contrast to numerous exclusive nuclear polymorphisms in both species.

Unigenes identified were mostly (around 70%) not characterized in the *N. alpina* transcriptome sequencing (Torales et al. 2012), contributing therefore additional genic information of non-expressed genes on leaf tissue. The sequence data produced allowed the construction of the first detailed map of the chloroplast genome in both *Nothofagus* species, which could be used as a reference for the assembly of chloroplast genomes of related species and for filtering organelle data from future sequencing genome or transcriptome projects. Moreover, 29 SSRs were validated in 42 individuals per species, among which 8 SSRs were efficient to discriminate among *N. obliqua* and *N. alpina* species. These markers further allowed to identify and classify hybrid individuals in natural mixed forests of both species (El Mujtar et al. 2017b).

SSR markers were also transferred to other *Nothofagus* species contributing to the generation of a panel of SSRs for these species. This panel includes neutral and functional markers, located in coding and non-coding regions of nuclear and extra-nuclear genomes (chloroplast and mitochondrion), presenting polymorphism and species-specific alleles (high divergence between species). This genomic information is useful to investigate demographic, adaptive and hybridizing events in *Nothofagus* species, taking advantage of their remarkable eco-physiological differences, and also to assist conservation, management and breeding programmes.

17.4.2   Transcriptome Sequencing of Prosopis alba

Prosopis alba is a multipurpose native tree of great economic value adapted to arid and semi-arid regions of Northwestern Argentina. However, this species had few genomic resources: 6 microsatellites from *Prosopis chilensis* (Mottura et al. 2005) and 12 from a bulk sample of several species of *Prosopis* (Bessega et al. 2013). Therefore, RNA-Seq was performed to identify genes controlling important traits and to develop functional markers (Torales et al. 2013). Assembled and annotated unigenes (Table 17.1 and Fig. 17.2) indicated a good proportion of the *P. alba* gene catalogue, considering the average number of genes encoded in a plant nuclear genome (Logacheva et al. 2011). The remaining sequences (32%), without significant identity to any other protein sequences in the existing databases, could be novel genes and may still be informative for identifying putative biological functions, which may be considered as *P. alba*-specific ones. Furthermore, the limited set of molecular markers currently available in this species was significantly increased with the thousands of new markers identified (Table 17.1).

The SSRs developed in *P. alba* were easily transferred to other *Prosopis* species (*P. denudans*, *P. hassleri*, *P. flexuosa*, *P. chilensis*) and the interspecific hybrids (*P. flexuosa x P. chilensis*), which did not have functional markers (Pomponio et al. 2015). In addition, high-confidence nuclear SNPs were identified, some of them related to the production of cellulose, together with the lignin biosynthetic pathway and with abiotic stress, among others. These markers will be useful for breeding programmes. On the other hand, contigs/isotigs of chloroplast reads composed of 59,040 bp were identified through alignment analyses to related chloroplast sequences. The partial *Prosopis* cp genome represents ~40% of the total cp genome of *Vigna radiata* (151,271 bp) (Tangphatsornruang et al. 2010). In addition, there were 55 intra scaffold gaps in *P. alba* cp genome.

The development of genomics resources represents a starting point for functional genetic studies in natural and cultivated populations of *Prosopis denudans* and *P. alba* (see below). It also represents useful tools to conduct comparative genomics studies with other *Prosopis* species, taking advantage of their remarkable eco-physiological differences. This work highlights the utility of transcriptome high-performance sequencing as a fast and cost-effective way of obtaining rapid information on the coding genetic variation in *Prosopis* genus.

17.4.3   Transcriptome Sequencing of Cedrela balansae

Cedrela balansae C. DC. is an endangered and economically important native species that inhabit the subtropical rainforests in Northwestern Argentina. Its wood has
a great value in the market, and consequently it has been intensively exploited. This
context has generated severe fragmentation processes with an impact on its levels of
genetic diversity (Kageyama and Gandara 1998). Molecular research was scarce
until recently because of the limited genomic information available. In order to
complement strategies of conservation and breeding in this species with great
potential for forest plantations, *C. balansae* transcriptome was developed through
RNA-Seq using 454 GS FLX (Torales et al. 2018). The assembling and annotation
process (Fig. 17.2) generated 92.3 Mb of sequence data and an important number of
unigenes, SSRs and SNPs (Table 17.1).

A subset of 70 SSRs allocated in sequences of important "stress tolerance" genes
[transcription factors (Zinc fingers, Myb), peroxidases, among other categories]
were selected for PCR-validation in eight genotypes from eight populations of the
species. The validated and polymorphic SSRs showed putative biological functions
corresponding to response to cold (64%) and other stress stimulus (36%).
Furthermore, the transferability of functional markers to other *Cedrela* species
(*C. fissilis*, *C. saltensis* and *C. angustifolia*) inhabiting in northwest and northeast of
Argentina was performed. As expected in congeneric species, high levels of cross-
amplification were obtained, reaching values between 77% (*C. angustifolia*) and
93% (*C. fissilis*). Finally, the 70 SSRs analysed resulted in about 46% (32) polymorphic markers for all the studied species.

In general, RNA-Seq has speeded up the investigation of the complexity of gene
transcription patterns, functional analyses and gene regulation networks. *Cedrela
balansae* is a cold-sensitive species; therefore, low temperatures and freezing conditions are factors that limit their cultivation. The knowledge of the molecular
mechanisms of response to abiotic stress is relevant to identify genes linked to cold
in the seedling development stages. The *C. balansae* transcriptome database
obtained will be useful for discovering genes of interest related to abiotic and biotic
stresses, to assist breeding programmes and develop genetic markers, to investigate
functional diversity in remnant natural populations and to identify key populations
for its preservation.

17.5   Ongoing HTS Projects

17.5.1   Transcriptome Sequencing of Austrocedrus chilensis, Handroanthus impetiginosus *and* Cordia trichotoma

Austrocedrus chilensis (D. Don) Pic. Serm. et Bizzarri (Patagonian cypress) is a
native conifer of the Andean-Patagonian Forests, with high adaptive potential to
water stress conditions. Its range covers a wide gradient of precipitation in the

west-east direction, with mean annual values ranging from 2500 mm in the west, on the border with Chile, to 400 mm in the east, in the Patagonian steppe (Pastorino and Gallo 2002). This species has a great capacity to adapt to the marginal, arid and semi-arid zones of North Patagonia, and therefore it represents an ideal species for studies of adaptive variability, constituting also a productive alternative for semi-arid temperate regions. An exploratory study using candidate genes from model species allowed identifying genes potentially involved in the adaptive response to drought stress. Nucleotide diversity and neutrality tests were applied to search for evidence of adaptive selection in natural populations of *Austrocedrus chilensis* (Pomponio et al. 2013). However, the main limitations of this study included the lack of genomic resources in the species and the low transferability of genes from model species.

Currently, there are no studies examining the transcriptome-wide gene and allelic expression patterns under water stress conditions of *A. chilensis*. Therefore, RNA-Seq was performed using Roche/454 technology to achieve a global view of differential expression genes that contribute to the tolerance to water-restrictive environments (Torales et al. unpublished data). Assembly and annotation is presented in Table 17.1 and Fig. 17.2. Ninety percent of the isotigs and 50% of the singletons had at least one blast hit, of which approximately 60–80% had at least one assigned ontology (GO) gene term (Table 17.1). An analysis of potential molecular markers was performed, obtaining SNP and SSR markers (Table 17.1). The differential expression was estimated using the 1731 isotigs (transcripts), taking as an expression measure the reading counts for each isotig and each condition, normalizing by the total amount of readings of each library. After this analysis, 354 transcripts showed differential expression between control and stress conditions with values of $-2 > \log2 (FC) > 2$ (stress/control comparison): 11 stress-related gene categories with upregulation and 126 metabolic and cell wall organization gene categories with downregulation. Validation of 100 SSRs in 6 individuals from 3 *A. chilensis* populations is currently under analysis, and until now 10 (ten) microsatellite markers are polymorphic.

Cordia trichotoma (Vell.) Steud. is a species of high timber value with the potential of cultivation in its distribution area. In Argentina, it is present in the Yungas and in the Alto Paraná rainforests. It is also cultivated as an ornamental plant and is widely used by traditional communities for medicinal purposes for the treatment of respiratory and urogenital diseases, such as anti-inflammatory, anti-microbial and anti-helminthic (Menezes et al. 2004; Matias et al. 2015; Oza and Kulkarni 2017). Another native forest species of interest is *Handroanthus impetiginosus* (Vell.) Mattos also valued for its high-quality wood. Its bark has pharmacological applications, and it is used as ornamental in gardens and public areas (Ettori et al. 1996; Neto and de Morais 2003; Martins 2012; Silva et al. 2012). Transcriptome sequencing of these two subtropical tree species has been performed on leaf tissue using Roche/454 technology. The de novo assembly allowed to generate isotigs (around 400 pb) and to identify SSR markers (Table 17.1). The total of SSR loci from *H. impetiginosus* for which primers were designed are 7324 and the number of unique markers was 6178. Of those markers, a total of 50 SSRs were selected and are currently being transferred to other species of the genus (Pomponio et al. 2019). Functional annotation of both transcriptomes is reported in Table 17.1.

17.5.2   Genome Sequencing of Austrocedrus chilensis

In gymnosperm native species, partial genome sequencing has only been performed on *A. chilensis* using 454 GS FLX Titanium technology (El Mujtar, unpublished data; Table 17.1). Genomic resources were assembled, classified and annotated using a procedure that combined several local and online bioinformatics tools (Fig. 17.3). A total of 89,562 contigs (including 34,581 singletons) was generated, representing nuclear and organelle (1418 contigs) genomes. Annotation was only possible for 14.5% of the contigs, with a high proportion corresponding to repetitive sequences (e.g. transposon, gag-protease poly, gag-pol poly). This is not surprising however for partial sequencing of species with large and complex genomes (high proportion of repetitive sequences) such as gymnosperms (Pellicer et al. 2018). Contigs assigned to cpDNA genome were realigned using *Cunninghamia lanceolata* chloroplast genome as reference, allowing the reconstruction of seven consensus contigs coding for 116 annotated chloroplast genes. Genomic resources were also used to identify SSRs markers located in 33.4%, 16.2% and 1.4% of nuclear, chloroplast and mitochondrial contigs, respectively. These genome resources expand the -omic resources for *Austrocedrus chilensis.*

17.5.3   Transcriptome Sequencing of Nothofagus alpina and N. obliqua *Under Water Stress Conditions*

Nothofagus alpina and *N. obliqua*, as stated previously in this chapter, are two closely related species which occur in sympatry in part of their Argentinian natural range. Particularly, in the Lake Lácar basin (Lanín National Park), both species are present along the pluviometrical gradient from west to east. Given their physiological differences in drought tolerance (see, e.g. Varela et al. 2010) and Global Climate Change predictions of higher temperature and lower precipitation for the region (Barros et al. 2015), a drought stress assay was performed in order to gain insight into the global, transcriptomic responses of these species to water stress (Marchelli, Arana, Varela, et al. unpublished data). The experiment was performed in a greenhouse with two contrasting provenances of each species, one humid and one drier, and was repeated with different plants in a different season. The experiment setting was periodically measured for soil humidity, air temperature, leaf stomatal conductance and pre-dawn water potential in order to ensure proper water stress at the time of harvesting the stressed plants.

For the first step of this sequencing project, leaves from plants of the drier provenances of both species, which are expected to be more severely affected by climate change in the near future, were harvested for RNA-Seq. In total, ten samples were sequenced on an Illumina HiSeq4000 after RNA extraction and library preparation in Novogene (Davis, California, USA). These samples were control group and stressed group at the beginning and end of the experiment, two biological replicates (eight samples) for *N. alpina* and stressed group at the beginning and end of the experiment, one biological replicate (two samples) for *N. obliqua*. The sequencing emphasized *N. alpina* because it was shown to be less tolerant to drought stress (Varela et al. 2010) and potentially more affected by climate change in its natural range (Marchelli et al. 2017). Around 22 million paired-end 150 bp reads were obtained for each sample, which are currently being analysed in a similar fashion to the *N. pumilio* transcriptomes related to high-temperature stress (López Laphitz, Bellora et al. unpublished data). Apart from further enlarge the -omic resources for *Nothofagus*, this study will give insight into these species' response to drought stress, which was never until now studied in a genome-wide approach.

17.6 Current and Future Genetic Studies on Native Forest Tree Species and Implications for Breeding

Breeding programmes have to deal with complex problems, such as diversity evaluation and selection of individuals adapted to stressing conditions or expressing better phenotypes, identification of clones, kinship information, genetic diversity and/ or genetic contamination in seed orchards, among others. Genetic diversity information of natural populations of the focus species is crucial for any breeding programme. Moreover, genetic breeding programmes require versatile and low-cost tools that allow a better selection of the material to be reproduced in less time. For South American native tree species, breeding programmes are, with some variations, in their initial stages. This section highlights how genomic resources and molecular markers recently developed by HTS technologies in these species can contribute to deal with inherent problems and, therefore, to assist and accelerate breeding programmes.

Selection assisted by molecular markers in combination with progeny tests could allow faster and more accurate results in the selection of complex characters such as growth rate and quality of the wood. In *Prosopis alba*, Klein et al. (2019) identified progenies that showed better behaviour and genetic stability in 19 families from the Province of Chaco. They performed morphological (stem straightness, stem height, stem bifurcation height and diameter) and genetic (17 SSRs) characterization on progenies of a 5-year-old plantation from selected parent trees with desired phenotypic characteristics. Genetic characterization revealed (1) low probability of inbreeding, suggesting a low level of kinship between progenies, (2) higher genetic differentiation within families than between families and (3) good differentiation among individuals. These results allowed the selection of trees, showing high stability at molecular and morphological level but conserving a percentage of variability, based on high genetic distance individual criteria for selection. These trees will be used in the *Prosopis alba* breeding.

In the case of *Prosopis denudans*, studies of genetic diversity and genetic structure of natural populations have been started in order to assist ecological restoration and breeding programme (for biomass product) (Vivien Pentreath unpublished data). It is a shrub (or small tree) of Patagonia adapted to the cold and semi-desert steppe, thus showing tolerance to drought, salinity, alkalinity and cold. Neutral and putative functional SSRs transferred from *P. alba* were used, revealing most of them moderate to high degree of genetic diversity (e.g. average observed and expected heterozygosities ranging from 0.324 to 0.732) and low genetic differentiation between populations ($F_{ST} = 0.064$). This work will be continued with more analyses using molecular, morphometric and environmental data, in order to study the influence of the different factors on the genetic diversity of the species and to identify/select better genotypes.

Among *Nothofagus* species, several studies have been carried out or started in the last years based on the genomic resources and molecular markers developed by high-throughput sequencing. Azpilicueta et al. (2016) using SSRs and Sanger sequencing of nuclear Adh and ITS genome regions evaluated the taxonomic status of Lagunas de Epulauquen population, the northernmost population of *N. obliqua* in Argentina, which has been questioned due to its morphological, architectural and genetic distinctiveness. The performed analysis revealed that individuals from this population were closely related to other *N. obliqua* individuals, while the presumed hybridization with *N. alpina* was not detected. This result supports the inclusion of Lagunas de Epulauquen within Argentinean *N. obliqua* domestication and breeding programmes, increasing the genetic diversity available for individual selection.

El Mujtar et al. (2017b) used six species-specific nuclear SSRs for the identification and classification of hybrids, in natural mixed forests of *N. obliqua* and *N. alpina*, based on estimates of ancestry and interclass heterozygosity. Results indicated that (1) introgressive hybridization occurs at a global rate of 7.8%; (2) F1 hybrids occur at a global rate of 3.8% and are fertile, as the detection of first- and late-generation hybrids indicates; (3) patterns of natural introgression are correlated with altitude; and (4) differences on the patterns were not detected between adults and regeneration, suggesting early-acting reproductive isolating barriers. Knowledge about natural occurrence of introgressive hybridization is relevant for seed collection, and it could be a possible source of contamination but also a source of genetic variation for breeding programmes. For example, five hybrid individuals have been detected using species-specific nuclear SSRs in a selected seed stand (SSS) of *N. alpina* (251 individuals) installed in 1999 at Trevelin Forest Station of INTA and registered with INASE (National Seed Institute) in 2017 (Pastorino et al. 2016; see Chap. 3). Besides, the use of SSRs allowed determining that genetic diversity of the SSS is similar to the average value of the natural distribution of the species in Argentina. Moreover, genotype data of individuals from this SSS were also used to construct a relationship matrix that is being used for the genetic evaluation of the trees combining phenotypic, pedigree and marker information and for the evaluation of the accuracy of the predicted breeding values (Cappa et al. unpublished data).

The use of SSR markers in *Nothofagus* also allowed to determine (1) restricted gene flow distance (<45 m) and significant seed and pollen immigration for continuous forests of *N. obliqua* and *N. alpina*, (2) the influence of tree size (DBH and height) on female and/or male fertility and (3) the occurrence of significant fine-scale spatial genetic structure (FSGS), which is consistent with the restricted seed and pollen dispersal (Sola et al. 2020). All this knowledge has several implications both for understanding the evolution of the species and for breeding actions. For example, significant FSGS suggests that, for future ex situ conservation, restoration and breeding strategies, seeds need to be collected from selected mothers belonging to different genetic clusters at a distance >45 m to avoid collecting seeds from genetically related individuals.

Collections of annotated genes have been used to select 25 candidate genes (CG) related to water stress responses in order to characterize their genetic variation along a precipitation gradient and to identify putative adaptive molecular markers for *Nothofagus* species (Soliani et al. 2020). The study revealed (1) moderate SNP diversity within and between *N. alpina* and *N. obliqua*; (2) higher nucleotide and haplotype diversity in *N. obliqua* than in *N. alpina*, consistently with the results obtained by El Mujtar et al. (2014) using SSRs; (3) different patterns of variation along a precipitation gradient, with higher diversity at the east (lower precipitation) for *N. alpina* and at the west (higher precipitation) for *N. obliqua*; (4) signal of balancing selection in two populations of *N. obliqua* at one CG; and (5) a similar pattern for *N. pumilio* with moderate nucleotide and haplotype diversity and higher diversity in western population. The screening of functional SSR markers for the same species (and gradients) revealed (1) genetic diversity trends differing between species and showing an opposite pattern of variation with respect to the CGs, being correlated with mean annual precipitation only in *N. alpina* and (2) signal of balancing or positive selection for some SSRs. These results provide preliminary evidence of gene regions functionally related to key responses of trees inhabiting environmental gradients.

On the other hand, although the capacity to apply massive genotyping to native forest tree species exists in Argentina, its application is yet scarce mostly due to the high costs involved. *Nothofagus* species are perhaps the exception, as massive genotyping has been used or is being included on research projects on species of the genus. Hasbún et al. (2016) using GBS investigated more than 10K SNP loci in Chilean *N. dombeyi* populations, revealing that genome-wide patterns of genetic diversity and differentiation varied widely across the genome. The study allowed the identification of numerous genomic regions exhibiting signatures of divergent selection and provided strong evidence of substantial genetic differentiation associated with both temperature and precipitation gradients, suggesting local adaptation. This strategy is also being used in *N. dombeyi* populations of Argentina to detect genomic markers associated to drought response (Fasanella et al. 2019). Three other projects focused on phenotype-genotype association studies will perform massive genotyping (RADSeq, GBS and exome capture) for common garden, populations of the natural distribution of *Nothofagus* species in Argentina and/or natural forests assays along environmental gradients (Arias Rios *(N. alpina)*, Cagnacci (*N. pumilio* and *N. obliqua*) and Sekely (*N. pumilio*) PhD theses).

Furthermore, the use of massive genotyping for population genomics studies in *Nothofagus* species has been also validated, combining Ion Torrent PGM sequencing with the Fluidigm platform (48 individuals × 48 amplicons) as a cost-effective strategy to produce re-sequencing data (El Mujtar et al. 2015). In total, ~50 barcoded libraries were sequenced on an Ion Torrent PGM across 4 runs, each being either a single individual or different sizes (6 or 12 individuals) pools. Variation between runs, techniques (either tagged individuals after mixing single PCR amplicons versus Fluidigm multiple PCRs) and single or pooled libraries for the same technique and the impact of primer purification (desalting/PAGE) and of concentration of amplicons (equimolar/non-equimolar) were tested.

All these projects suggest that high-throughput sequencing and genotyping techniques could be more frequently used in Argentina for genome-wide association studies (GWAS) and genomic selection (GS) in the near future accelerating forest tree breeding of native Argentinean species.

References

Adams J (2008) DNA sequencing technologies. Nature Education 1:1931

Aguirre N, Filippi C, Zaina G, Rivas J, Acuña C, Villalba P et al (2019) Optimizing ddRADseq in non-model species: a case study in *Eucalyptus dunnii*. Agronomy 9:484

Alvarez M, Mosquera T, Blair M (2014) The use of association genetics approaches in plant breeding. In: Janick J (ed) Plant breeding reviews, vol 38. Wiley, Hoboken, pp 17–68

Andrews K, Good J, Miller M, Luikart G, Hohenlohe P (2016) Harnessing the power of RADseq for ecological and evolutionary genomics. Nat Rev Genet 17:81–92

Aranzana M, Kim S, Zhao K, Bakker E, Horton M et al (2005) Genome-wide association mapping in *Arabidopsis* identifies previously known flowering time and pathogen resistance genes. PLoS Genet 1:e60

Azpilicueta M, Caron H, Bodénès C, Gallo L (2004) SSR markers for analyzing South American *Nothofagus* species. Silvae Genet 53:240–243

Azpilicueta M, Soliani C, Gallo L, van Zonneveld M, Thomas E, Moreno C, Marchelli P (2013) Definición de zonas genéticas en cuatro especies de *Nothofagus* de los bosques andino-patagónicos IV Congreso Forestal Argentino y Latinoamericano. Iguazú, 23–27 September

Azpilicueta M, El Mujtar V, Gallo L (2016) Searching for molecular insight on hybridization in *Nothofagus* spp. forests at Lagunas de Epulauquen, Argentina. Bosque 37:591–601

Badenes M, Fernández I, Ríos G, Rubio-Cabetas M (2016) Application of genomic technologies to the breeding of trees. Front Genet 7:198

Baird N, Etter P, Atwood T, Currey M, Shiver A, Lewis Z et al (2008) Rapid SNP discovery and genetic mapping using sequenced RAD markers. PLoS One 3:e3376

Barros V, Boninsegna J, Camilloni I, Chidiak M, Magrín G, Rusticucci M (2015) Climate change in Argentina: trends, projections, impacts and adaptation. Wiley Interdiscip Rev Clim Change 6:151–169

Bartholomé J, Van Heerwaarden J, Isik F, Boury C, Vidal M, Plomion C, Bouffier L (2016) Performance of genomic prediction within and across generations in maritime pine. BMC genomics 17:604

Beilsmith K, Thoen M, Brachi B, Gloss A, Khan H, Bergelson J (2019) Genome-wide associations studies on the phyllosphere microbiome: embracing complexity in host-microbe interactions. Plant J. 97:164–181

Bernardo R (2008) Molecular markers and selection for complex traits in plants: Learning from the last 20 years. Crop Sci 48:1649–1664

Bessega C, Pometti C, Miller J, Watts R, Saidman B, Vilardi J (2013) New microsatellite loci for *Prosopis alba* and *P. chilensis* (Fabaceae). Appl Plant Sci 1:1200324

Birol I, Raymond A, Jackman S, Pleasance S, Coope R, Taylor G et al (2013) Assembling the 20 Gb white spruce (*Picea glauca*) genome from whole-genome shotgun sequencing data. Bioinformatics. https://doi.org/10.1093/bioinformatics/btt178

Blott S, Kim J, Moisio S, Schmidt-Kuntzel A, Cornet A et al (2003) Molecular dissection of a quantitative trait locus: a phenylalanine-to-tyrosine substitution in the transmembrane domain of the bovine growth hormone receptor is associated with a major effect on milk yield and composition. Genetics 163:253–266

Botstein D, Risch N (2003) Discovering genotypes underlying human phenotypes: past successes for mendelian disease, future approaches for complex disease. Nat Genet 33(Suppl):228–237

Brachi B, Morris G, Borevitz J (2011) Genome-wide association studies in plants: the missing heritability is in the field. Genome Biol 12:232

Brousseau L, Tinaut A, Duret C, Lang T, Garnier-Gere P, Scotti I (2014) High-throughput transcriptome sequencing and preliminary functional analysis in four Neotropical tree species. BMC Genomics 15:238

Budde K, Heuertz M, Hernández-Serrano A, Pausas J, Vendramin G, Verdú M, González-Martínez S (2014) In situ genetic association for serotiny, a fire-related trait, in Mediterranean maritime pine (*Pinus pinaster* Aiton). New Phytol 201:230–241

Canales J, Bautista R, Label P, Gómez Maldonado J, Lesur I, Fernández Pozo N et al (2014) De novo assembly of maritime pine transcriptome: Implications for forest breeding and biotechnology. Plant Biotechnol J 12:286–299

Cappa E, Marco de Lima B, da Silva-Junior O, Garcia C, Mansfield S, Grattapaglia D (2019) Improving genomic prediction of growth and wood traits in *Eucalyptus* using phenotypes from non-genotyped trees by single-step GBLUP. Plant Sci 84:9–15

Catchen J, Hohenlohe P, Bassham S, Amores A, Cresko WA (2013) Stacks: an analysis tool set for population genomics. Mol Ecol 22:3124–3140

Chang S, Puryear J, Cairney J (1993) A simple and efficient method for isolating RNA from pine trees. Plant Mol Biol Rep 11:113–116

Chhetri H, Macaya-Sanz D, Kainer D, Biswal A, Evans L, Chen J et al (2019) Multi-trait genome-wide association analysis of Populus trichocarpa identifies key polymorphisms controlling morphological and physiological traits. New Phytol 223:293–309

Chinwalla A, Cook L, Delehaunty K, Fewell G, Fulton L, Fulton R et al (2002) Initial sequencing and comparative analysis of the mouse genome. Nature 420:520–562

Collevatti R, Novaes E, Silva-Junior O, Vieira L, Lima-Ribeiro M, Grattapaglia D (2019) A genome-wide scan shows evidence for local adaptation in a widespread keystone Neotropical forest tree. Heredity 123:117–137

Conesa A, Madrigal P, Tarazona S, Gomez-Cabrero D, Cervera A, McPherson A et al (2016) A survey of best practices for RNA-seq data analysis. Genome Biol 17:13

Davey J, Hohenlohe P, Etter P, Boone J, Catchen J, Blaxter L (2011) Genome-wide genetic marker discovery and genotyping using next-generation sequencing. Nat Rev Genet. 12:499–510

De Pristo M, Banks E, Poplin R, Garimella K, Maguire J, Hartl C et al (2011) A framework for variation discovery and genotyping using next-generation DNA sequencing data. Nat Genet 43:491–498

De Villemereuil P, Gaggiotti O, Mouterde M, Till-Bottraud I (2016) Common garden experiments in the genomic era: new perspectives and opportunities. Heredity 116:249–254

Du Q, Lu W, Quan M, Xiao L, Song F, Li P, Zhou D, Xie J, Wang L, Zhang D (2018) Genome-Wide Association Studies to Improve Wood Properties: Challenges and Prospects. Front Plant Sci 9:1912

Eckert A, Wegrzyn J, Liechty J, Lee J, Cumbie W, Davis J et al (2013) The Evolutionary Genetics of the Genes Underlying Phenotypic Associations for Loblolly Pine (*Pinus taeda*, Pinaceae). Genetics 195:1353–1372

El Mujtar V, Gallo L, Lang T, Garnier-Géré P (2014) Development of genomic resources for *Nothofagus* species using next-generation sequencing data. Mol Ecol Resour 14:1281–1295

El Mujtar V, Guichoux E, Boury C, Pilliet M, Delcamp A, Salin F, Garnier Géré P (2015) Validating the use of a genotyping by sequencing approach for population genomic studies in non-model species. 3ème Colloque de Génomique Environnementale, "Le vivant à l'ère des nouvelles technologies de séquençage des génomes". 26–28 Octubre, Montpellier.

El Mujtar V, López M, Amalfi S, Pomponio F, Marcucci Poltri S, Torales S (2017b) Characterization and transferability of transcriptomic microsatellite markers for *Nothofagus* species. N Z J Bot 55:347–356

El Mujtar V, Aparicio A, Sola G, Gallo L (2017a) Pattern of natural introgression in a *Nothofagus* hybrid zone from South American temperate forests. Tree Genet Genomes 13:49

Elshire R, Glaubitz J, Sun Q, Poland J, Kawamoto K, Buckler E, Mitchell S (2011) A Robust, Simple Genotyping-by-Sequencing (GBS) Approach for High Diversity Species. PLoS One 6:e19379

Estravis-Barcala M, Mattera M, Soliani C, Bellora N, Opgenoorth L, Heer K, Arana M (2019) Molecular bases of responses to abiotic stress in tres. J Exp Bot. https://doi.org/10.1093/jxb/erz532

Ettori L, Siqueira A, Sato A, Campos O (1996) Variabilidade genética em populações de Ipê-roxo – *Tabebuia heptaphylla* (Vell.) Tol. – para conservação ex situ. Revista do Instituto Florestal 8:61–70

Fasanella M, Suarez M, Hasbún R, Premoli A (2019) Genomic markers as indicators of potential drought adaptation in *Nothofagus dombeyi*. Topwood Conference, Bariloche, 12–15 March

Feuillet C, Leach J, Rogers J, Schnable P, Eversole K (2011) Crop genome sequencing: lessons and rationales. Trends Plant Sci 16:77–88

Geraldes A, Difazio S, Slavov G, Ranjan P, Muchero W, Hannemann J et al (2013) A 34K SNP genotyping array for *Populus trichocarpa*: Design application to the study of natural populations and transferability to other *Populus* species. Mol Ecol Res 13:306–323

Grattapaglia D, Resende M (2011) Genomic selection in forest tree breeding. Tree Genet Genomes 7:241–255

Grattapaglia D, Silva-Junior O, Resende R, Cappa E, Müller B, Tan B et al (2018) Quantitative genetics and genomics converge to accelerate forest tree breeding. Front Plant Sci 9:1693

Greene C, Tan J, Ung M, Moore J, Cheng C (2014) Big data bioinformatics. J Cell Physiol 229:1896–1900

Goodwin S, McPherson J, McCombie WR (2016) Coming of age: ten years of next-generation sequencing technologies. Nat Rev Genet 17:333–351

Hasbún R, González J, Iturra C, Fuentes G, Alarcón D, Ruiz E (2016) Using genome-wide SNP discovery and genotyping to reveal the main source of population differentiation in *Nothofagus dombeyi* (Mirb.) Oerst. in Chile. Int J Genomics. Article ID 3654093. https://doi.org/10.1155/2016/3654093

Heather J, Chain B (2016) The sequence of sequencers: The history of sequencing DNA. Genomics 107:1–8

IPCC (2014) Cambio climático. Quinto Informe de Evaluación del Grupo Intergubernamental de Expertos sobre el Cambio Climático [Equipo principal de redacción, RK Pachauri y LA Meyer (eds.)]. Ginebra, Suiza, 157 págs

Kageyama P, Gandara F (1998) Consecuencias de la fragmentación sobre poblaciones de especies arbóreas, Camará – Centro de Apoio às Sociedades Sustentáveis. Serie técnica IPEF 12:65–70

Karan M, Evans D, Reilly D, Schulte K, Wright C, Innes D et al (2012) Rapid microsatellite marker development for African mahogany (*Khaya senegalensis*, Meliaceae) using next-generation sequencing and assessment of its intra-specific genetic diversity. Mol Ecol Res 12:344–353

Kchouk M, Gibrat J, Elloumi M (2017) Generations of sequencing technologies: from first to next generation. Biol Med (Aligarh) 9:3

Klein L, Spoljaric M, Torales S (2019) Identificación de genotipos estables en 19 familias de *Prosopis alba* usando marcadores de microsatélites y parámetros de productividad. Quebracho 27:26–36

Krishnan N, Pattnaik S, Jain P, Gaur P, Choudhary R, Vaidyanathan S et al (2012) A draft of the genome and four transcriptomes of a medicinal and pesticidal angiosperm *Azadirachta indica*. BMC Genomics 13:464

Lande R, Thompson R (1990) Efficiency of marker-assisted selection in the improvement of quantitative traits. Genetics 124:743–756

Lesur I, Le Provost G, Bento P, Da Silva C, Leplé J-C, Murat F et al (2015) The oak gene expression atlas: insights into Fagaceae genome evolution and the discovery of genes regulated during bud dormancy release. BMC Genomics 16:112

Liu L, Li Y, Li S, Hu N, He Y, Pong R, Lin D, Lu L, Law M (2012) Comparison of next-generation sequencing systems. J Biomed Biotechnol. Article ID 251364. https://doi.org/10.1155/2012/251364

Liu F, Hong Z, Yang Z, Zhang N, Liu X, Xu D (2019) De novo transcriptome analysis of *Dalbergia odorifera* T. Chen (Fabaceae) and transferability of SSR markers developed from the transcriptome. Forests 10:98

Logacheva M, Kasianov A, Vinogradov D, Samigullin T, Gelfand M, Makeev V, Penin A (2011) De novo sequencing and characterization of floral transcriptome in two species of buckwheat (*Fagopyrum*). BMC Genomics 12:30

López de Heredia U, Vázquez-Poletti J (2016) RNA-seq analysis in forest tree species: bioinformatic problems and solutions. Tree Genet Genomes 12:30

Mahony CR, MacLachlan IR, Lind BM, Yoder JB, Wang T, Aitken SN (2019) Evaluating genomic data for management of local adaptation in a changing climate: a lodgepole pine case study. https://doi.org/10.1101/568725

Marchelli P, Caron H, Azpilicueta M, Gallo L (2008) A new set of highly polymorphic nuclear microsatellite markers for *Nothofagus nervosa* and related South American species. Silvae Genet 57:82–85

Marchelli P, Thomas E, Azpilicueta M, van Zonneveld M, Gallo L (2017) Integrating genetics and suitability modelling to bolster climate change adaptation planning in Patagonian *Nothofagus* forests. Tree Genet Genom 13:119

Matias E, Ferreira A, Nascimento Silva M, Alencar Carvalho V, Melo Coutinho H, Martins da Costa J (2015) The genus *Cordia*: botanists, ethno, chemical and pharmacological aspects. Revista Brasileira de Farmacognosia 25:542–552

Martins S (2012) Restauração ecológica de ecossistemas degradados. Visçosa, Brasil. UFV 293 p

Menezes J, Machado F, Lemos T, Silveira E, Filho R, Pessoa O (2004) Sesquiterpenes and a Phenylpropanoid from *Cordia trichotoma*. Z Naturforsch 59:19–22

Meuwissen T, Hayes B, Goddard M (2001) Prediction of total genetic value using genome-wide dense marker maps. Genetics 157:1819–1829

Minamikawa M, Takada N, Terakami S, Saito T, Onogi A, Kajiya-Kanegae H et al (2018) Genome-wide association study and genomic prediction using parental and breeding populations of Japanese pear (*Pyrus pyrifolia* Nakai). Sci Rep 8:2045–2322

Mottura M, Finkeldey R, Verga A, Gailing O (2005) Development and characterization of microsatellite markers for *Prosopis chilensis* and *Prosopis flexuosa* and cross-species amplification. Mol Ecol Not 5:487–489

Naidoo S, Slippers B, Plett J, Coles D, Oates C (2019) The road to resistance in forest trees. Front Plant Sci 10:273

Neale D, Williams C (1991) Restriction-Fragment-Length-Polymorphism mapping in conifers and applications to forest genetics and tree improvement. Can J Forest Res Revue Can Rech Forest 21:545–554

Neale D (2007) Genomics to tree breeding and forest health. Curr Opin Genet Dev 17:539–544

Neale D, Wheeler N (2019) The conifers: genomes, variation and evolution. Springer; Edición: 1st ed
Neto G, de Morais R (2003) Recursos medicinais de espécies do cerrado de Mato Grosso: um estudo bibliográfico. Acta Botânica Brasileira 17:561–584
Nordborg M, Tavare S (2002) Linkage disequilibrium: What history has to tell us. Trends Genet 18:83–90
Oza M, Kulkarni Y (2017) Traditional uses, phytochemistry and pharmacology of the medicinal species of the genus *Cordia* (Boraginaceae). J Pharm Pharmacol 69:755–789
Parchman T, Geist K, Grahnen J, Benkman C, Buerkle C (2010) Transcriptome sequencing in an ecologically important tree species: assembly, annotation, and marker discovery. BMC Genomics 11:180
Parchman T, Jahner J, Uckele K, Galland L, Eckert A (2018) RADseq approaches and applications for forest tree genetics. Tree Genet Genomes 14:39
Parent G, Raherison E, Sena J, MacKay J (2015) Forest Tree Genomics: Review of Progress. Adv Bot Res 74:39–92
Pastorino M, Gallo L (2002) Quaternary evolutionary history of *Austrocedrus chilensis*, a cypress native to the Andean-Patagonian Forest. J Biogeogr 29:1167–1178
Pastorino M, El Mujtar V, Azpilicueta M, Aparicio A, Marchelli P, Mondino V et al (2016) Subprograma *Nothofagus*. Domesticación y Mejoramiento de Especies Forestales. INTA-UCAR 6:161–188
Pavy N, Gagnon F, Rigault P, Blais S, Deschenes A, Boyle B et al (2013) Development of high-density SNP genotyping arrays for white spruce (*Picea glauca*) and transferability to subtropical and nordic congeners. Mol Ecol Res 13:324–336
Pellicer J, Hidalgo O, Dodsworth S, Leitch I (2018) Genome size diversity and its impact on the evolution of land plants. Genes 9:88
Peterson B, Weber J, Kay E, Fisher H, Hoekstra H (2012) Double digest RADseq: an inexpensive method for de novo SNP discovery and genotyping in model and non-model species. PLoS One 7:e37135
Pfeiffer F, Gröber C, Blank M, Händler K, Beyer M, Schultze J, Mayer G (2018) Systematic evaluation of error rates and causes in short samples in next-generation sequencing. Sci Rep 8:10950
Pomponio M, Torales S, Gallo L, Pastorino M, Marchelli P, Cervera M, Marcucci Poltri S (2013) DNA Sequence Variation of Drought-Response candidate genes in *Austrocedrus chilensis*. Electron J Biotechnol 16(2)
Pomponio M, Marcucci Poltri S, López Lauenstein D, Torales S (2014) Identification of single nucleotide polymorphisms (SNPs) at candidate genes involved in abiotic stress in two *Prosopis* species and hybrids. For Syst 23:490–493
Pomponio M, Acuña C, Pentreath V, López Lauenstein D, Marcucci Poltri S, Torales S (2015) Characterization of functional SSR markers in *Prosopis alba* and their transferability across *Prosopis* species. For Syst 24:2
Pomponio M, Fornés L, Marcucci S, Torales S (2019) Desarrollo del transcriptoma foliar y marcadores microsatélites del Lapacho Rosado (*Handroanthus impetiginosus*). XVIII Reunión GEMFO 2019 (Bella Vista, Corrientes, Argentina) Trabajos Técnicos.
Reuter J, Spacek D, Snyder M (2015) High-Throughput Sequencing Technologies. Mol Cell 58:586–597
Risch N, Merikangas K (1996) The future of genetic studies of complex human diseases. Science 273:1516–1517
Roberts RJ, Carneiro MO, Schatz MC (2013) The advantages of SMRT sequencing. Genome Biol 14:405
Sabatier Y, Azpilicueta M, Marchelli P, González Peñalba M, Lozano L, García L et al (2011) Distribución natural de *Nothofagus alpina* y *Nothofagus obliqua* (nothofagaceae) en Argentina, dos especies de primera importancia forestal de los bosques templados norpatagónicos. Bol Soc Argent Bot 46:131–138

Sanger F, Nicklen S, Coulson A (1977) DNA sequencing with chain-terminating inhibitors. Proc Natl Acad Sci USA 74:5463–5467

Secretaría de Ambiente y Desarrollo Sustentable – Dirección de Bosques (2007) Informe sobre desforestación en Argentina

Sewell M, Neale D (2000) Mapping quantitative traits in forest trees. Molecular biology of woody plants. For Sci 64:407–423

Stephens Z, Lee S, Faghri F, Campbell R, Zhai C, Efron M et al (2015) Big data: astronomical or genomical. PLoS Biol 13:e1002195

Silva A, Ribeiro de Paiva S, Figueiredo M, Coelho Kaplan M (2012) Biological activity of naphthoquinones from Bignoniaceae species. Revista Fitos 7:4

Sola G, El Mujtar V, Gallo L, Vendramin G, Marchelli P (2020) Staying close: short local dispersal distances on a managed forest of two Patagonian *Nothofagus* spp. Forestry 93(5):652–661

Soliani C, Sebastiani F, Marchelli P, Gallo L, Vendramin G (2010) Development of novel genomic microsatellite markers in the southern beech *Nothofagus pumilio* (Poepp. Et Endl.) Krasser. Mol Ecol Res 10:404–408

Soliani C, Azpilicueta M, Arana M, Marchelli P (2020) Clinal variation along precipitation gradients in Patagonian temperate forests: unravelling demographic and selection signatures in three *Nothofagus* spp. Ann For Sci 77:4

Strauss S, Lande R, Namkoong G (1992) Limitations of molecular-marker-aided selection in forest tree breeding. Can J For Res 22:1050–1061

Tangphatsornruang S, Sangsrakru D, Chanprasert J, Uthaipaisanwong P, Yoocha T, Jomchai N, Tragoonrung S (2010) The chloroplast genome sequence of mungbean (*Vigna radiata*) determined by high-throughput pyrosequencing: structural organization and phylogenetic relationships. DNA Res 17:11–22

Thistlethwaite F, Ratcliffe B, Klápště J, Porth I, Chen C, Stoehr M, El-Kassaby Y (2019) Genomic selection of juvenile height across a single-generational gap in Douglas-fir. Heredity 122:848

Thornsberry J, Goodman M, Doebley J, Kresovich S, Nielsen D, Buckler E (2001) Dwarf polymorphisms associate with variation in flowering time. Nat Genet 28:286–289

Torales S, Rivarola M, Pomponio M, Fernández P, Acuña C, Marchelli P et al (2012) Transcriptome survey of Patagonian southern beech *Nothofagus nervosa* (= *N. alpina*): assembly, annotation and molecular marker discovery. BMC Genomics 13:291

Torales S, Rivarola M, Pomponio M, Gonzalez S, Acuña C, Fernández P et al (2013) De novo assembly and characterization of leaf transcriptome for the development of functional molecular markers of the extremophile multipurpose tree species *Prosopis alba*. BMC Genomics 14:705

Torales S, Rivarola M, Gonzalez S, Inza M, Pomponio M, Fernández P et al (2018) De novo transcriptome sequencing and SSR markers development for *Cedrela balansae* C.DC., a native tree species of northwest Argentina. PLoS One 13:e0203768

Toro M, Varona L (2010) A note on mate allocation for dominance handling in genomic selection. Genet Sel Evol 42:33

Torre S, Tattini M, Brunetti C, Fineschi S, Fini A, Ferrini F, Sebastiani F (2014) RNA-Seq analysis of *Quercus pubescens* leaves: de novo transcriptome assembly, annotation and functional markers development. PLoS One 9:e112487

Tuskan G, DiFazio S, Jansson S, Bohlmann J, Grigoriev I, Hellsten U et al (2006) The genome of black cottonwood, *Populus trichocarpa* (Torr. and Gray). Science 313:1596–1604

Tyson J, O'Neil N, Jain M, Olsen H, Hieter P, Snutch T (2018) MinION-based long-read sequencing and assembly extends the *Caenorhabditis elegans* reference genome. Genome Res 28:266–274

Ueno S, Le Provost G, Léger V, Klopp NC, Frigerio J et al (2010) Bioinformatic analysis of ESTs collected by Sanger and pyrosequencing methods for a keystone forest tree species: oak. BMC Genomics 11:650

Varela S, Gyenge J, Fernández ME, Schlichter T (2010) Seedling drought stress susceptibility in two deciduous *Nothofagus species* of NW Patagonia. Trees 24:443–453

Venter J (2001) The sequence of the human genome. Science 291:1304–1351

Wang Y, Yang Q, Wang Z (2015) The evolution of nanopore sequencing. Front Genetics 5:449

Wang Y, Chen X, Wang J, Xun H, Sun J, Tang F (2016) Comparative analysis of the terpenoid biosynthesis pathway in *Azadirachta indica* and *Melia azedarach* by RNA-seq. SpringerPlus 5:819

Williams C, Neale D (1992) Conifer wood quality and marker-aided selection – a case-study. Can J Forest Res Revue Can Rech Forest 22:1009–1017

Zhu C, Gore M, Buckler E, Yu J (2008) Status and prospects of association maping in plants. Plant Genome 1:5–20

Chapter 18
Questions, Perspectives and Final Considerations of Planting Native Species Under the Climate Change Conditioning

Mario J. Pastorino, Paula Marchelli, Verónica Arana, and Alejandro G. Aparicio

18.1 Genetic Considerations of Planting Native Species: Maladaptation and Genetic Contamination

There is a growing interest in Argentina (and in the world) for planting native tree species. This has triggered the development of domestication programs for those species with greater productive potential or a key ecological value. Domestication involves the development of technologies to bring wild species to plantation on an industrial scale, for commercial purposes or for the restoration of ecosystems, involving all stages of the biological/productive cycle. Knowledge is needed for industrial production of seedlings, methods and designs of plantation, intermediate silvicultural treatments, and the definition of rotation periods and adjustment of harvesting practices. As well, it also includes a perhaps inconspicuous but essential stage that predates all of these: the choice of genetic material to be reproduced.

The sympathy that society currently expresses for native species may lead to believe that their plantation will always be positive. However, some genetic risks are specifically caused by planting native species. If seedlings used in the plantation are produced with seeds from a different forest than the local one, then processes may occur that attempt against the planted seedlings themselves or, what is even worse, against forests of the same species surrounding the plantation.

A wrong choice can lead to maladaptation processes that depress the vigor of the implanted trees and even compromise their survival (McKay et al. 2005). In the case of a commercial plantation, it can decrease its productivity to the point of represent-

M. J. Pastorino (✉) · P. Marchelli · V. Arana · A. G. Aparicio
Instituto de Investigaciones Forestales y Agropecuarias Bariloche (IFAB) INTA-CONICET, Bariloche, Argentina
e-mail: pastorino.mario@inta.gob.ar

© Springer Nature Switzerland AG 2021
M. J. Pastorino, P. Marchelli (eds.), *Low Intensity Breeding of Native Forest Trees in Argentina*, https://doi.org/10.1007/978-3-030-56462-9_18

ing an economic failure. In an enrichment plan, trees may express poor competitive aptitude and be suppressed under pre-existing vegetation, whether they be trees, shrubs, or grasses.

Maladaptation can occur very early, even when the seedlings face the first environmental stress, but it can also be verified belatedly, when the plantation faces extreme climatic events of low recurrence. Thus, an atypical drought or an exceptional frost can kill all the planted seedlings, although their initial survival and growth has been adequate so far. Possible biological attacks can also reveal maladaptation processes. An eventual outbreak of a pest or the proliferation of a disease due to predisposing environmental conditions may put at risk the survival or alter the tree shape of the planted individuals, even though they have been vigorous and healthy up to that point.

On the other hand, the wrong choice of the genetic pool to propagate can have a detrimental effect not only on the implanted mass but also on pre-existing forests. If afforestation is carried out in regions where the same species grows naturally, the use of a genetic pool from another region entails a "genetic contamination" risk (Bischoff et al. 2010; Vander Mijnsbrugge et al. 2010), that is, the introgression of "exotic" (non-local) genes or genotypes in natural forest stands. This process can occur through both free pollination and natural seed dispersal, and its effect becomes irreversible, leading to possible irrecoverable loss of the local genetic pool.

The mixing of exotic and local genes can produce the effect of "outbreeding depression" (Fenster and Galloway 2000), by which advanced hybrid generations express a lower fitness than that of the progenitors, even when hybrid vigor has been verified in the first generation of hybridization. This effect appears to be due to the loss of heterozygosity after F1 and the disruption of co-adapted gene complexes (Fenster and Dudash 1994; Frankham et al. 2011). The long life cycle of trees may delay the verification of this effect, being perhaps too belatedly. This risk is particularly important in restoration programs that carry out enrichment afforestation in degraded forests in order to recompose the forest continuity between the surviving remnants after fires or other catastrophic events.

In the particular case of enrichment afforestation, extensive planting in degraded forests, in which the remaining individuals of the species to be planted are relatively few, can cause what is called "genetic swamping." In this process, the local gene pool is diluted in the gene pool brought by the plantation, and if these are different, loss of allelic variants or multigene complexes may occur, even those with adaptive value (Lesica and Allendorf 1999). There may also be a swamping process when plants of different provenances are mixed in the plantation, and some of them have a significantly better performance than the others, including the local one, leading in the long term to the imposition of this genetic pool (Falk et al. 2001).

In this sense, the most conservative decision (the one that involves the least risks) when choosing the genetic material to propagate, both for a productive project and for a restoration program, is the use of local genetic pools (Vander Mijnsbrugge et al. 2010). Collecting seeds from the forests that will be restored (or in their immediate vicinity) is a basic rule of ecosystem restoration.

However, in some cases forests could have been affected to the point of local extinction, thus forcing foresters to use exotic seed sources to carry out the restoration. Even without going to that extreme, the degradation is often so strong that only few trees remain from the original forest, which makes it inadvisable to obtain seeds from these remnants. Using seeds exclusively from those few trees would cause a significant reduction in the genetic diversity of the future afforestation due to genetic drift, with a resulting reduction in their adaptability. In addition, inbreeding could be also a consequence, leading to the appearance of deleterious phenotypes (Carr and Dudash 2003).

Genetic drift is the aleatory variation of allelic and genotypic frequencies in the population from one generation to the next, which occurs by random crossing between parents. This common phenomenon in any sexually reproducing population has drastic consequences for small ones, such as the population composed of the few remaining trees in a strongly degraded forest. It is highly probable that some (at random) low frequent alleles are not represented in the next generation simply because the few parents that carry them randomly fail to leave offspring. The consequence is the irreversible loss of those allelic variants in the population. Low frequent alleles commonly have low relevance for the adaptation to the current conditions, but might be quite important for the adaptation to the future environment. If the seedlings used in the plantation come exclusively from the seeds of a population composed by few trees, this drift effect would occur in the plantation itself.

Inbreeding refers to self-pollination and to crossbreeding between related individuals and often leads to depressed phenotypes due to the expression of recessive alleles. In this case, this process would occur in the plantation, because selfing is accentuated due to the low availability of pollinators, and also in the next generations of the established forest, because a large portion of the planted seedlings would have a close kinship.

All these risks must be considered in a forestation program with native species, to weigh the probability of occurrence of any of these undesirable effects based on the particular conditions of each case, thus assisting decision-making (Frankham et al. 2011; Byrne et al. 2011).

18.2   Should We Think About Productivity or the Preservation of Genetic Identity?

In a program of active restoration of degraded ecosystems, two basic objectives can be identified from a genetic point of view: that of "authenticity" and that of "functionality" (Clewell 2000; Falk et al. 2001). Authenticity refers to replicating the original genetic pool and involves ethical principles. Although this cannot be fully achieved, the degree of authenticity can be measured, and therefore a certain goal can be arbitrarily set.

Functionality aims to attain the persistence, resilience, and stability of the restored ecosystem, once the intervention is finished (Falk et al. 2001). Therefore, it focuses on the current adaptation of the gene pool used and on the adaptability to the future environmental conditions (unknown) of the implanted trees and of the next generations that will succeed them naturally. Consequently, a high level of genetic variation is necessary to ensure functionality, which can be detrimental to authenticity. Some authors recommend the inclusion of non-local gene pools in the restoration program, thus adding genetic diversity, particularly when local populations are genetically impoverished (Vergeer et al. 2004; Broadhurst et al. 2008; Vander Mijnsbrugge et al. 2010). Others, instead of promoting the increase of genetic diversity, focus on avoiding its loss (Basey et al. 2015), recommending management measures for all the stages of seedling production (collection, cleaning, storage and germination of seeds, and the maintenance of seedlings in the nursery). In any case, the solution to the authenticity vs. functionality dilemma will be a balance that the restoration program must decide in a more or less arbitrary way.

In turn, the objectives of authenticity and functionality should also be considered in afforestation for productive purposes, provided it is carried out with a native species, that is, in a place where the same species is naturally present. In this case, the functionality must be understood in a productive rather than ecological sense. Adaptation to current conditions will continue to be of primary interest, but instead, the adaptability of the next generations will lose relevance. This is not the case with the risk of genetic contamination, which could be large if the plantation size is significant compared to the surrounding natural populations.

Several real cases of productive use of native species in the current Argentine scenario require caution regarding genetic risks. Some of them refer to subtropical rainforest species with disjoint distribution ranges. For example, *Cordia trichotoma* is present in both the Yungas and the Alto Paraná Rainforest, and we still do not know the level of genetic differentiation between these two regions. Until knowledge is gained, the ongoing breeding program for *C. trichotoma* maintains both origins separate, in fact considering them to belong to different breeding zones, and consequently a particular program for each of them is being developed.

Likewise, there are also cases of possible interspecific hybridization of species of the same genus with disjoint distribution, for example, in the genera *Cedrela* and *Handroanthus*. In fact, *Cedrela fissilis* from the Alto Paraná Rainforest has been planted successfully in the Yungas, for the enrichment of degraded forests. We do not know yet if this may have any impact on the natural populations of *C. balansae*, *C. angustifolia*, or *C. saltensis*. The case of *Handroanthus*, with *H. impetiginosus* in the Yungas and *H. heptaphyllus* in the humid Chaco, is of less urgent resolution since until now, these species have only been planted as urban ornamental trees and in experimental trials.

Another case that deserves special attention has the Chaco as its setting. The ongoing *Prosopis alba* breeding program has been able to identify a provenance with ostensibly superior productive qualities: Campo Durán. This provenance belongs to the "Salteño" morphotype of the species, which differs from the "Chaqueño" and "Santiagueño" morphotypes (the former has a much smaller distribution range in Argentina) (Verga et al. 2009). Its excellent productive performance in the multiple environments in which it has been tested (it even exhibits better growth than the local morphotypes; López Lauenstein et al. 2019; see Chap. 10) encourages its cultivation throughout the region. However, we still do not know if this outstanding behavior during the first years of implantation (the oldest trials with this provenance are 7 years old) will continue until the final cutting. Likewise, we have no knowledge about the tolerance of this genetic pool to withstand possible extreme climatic events of low recurrence that might characterize the implantation sites but do not occur at the home environment.

Returning to the dilemma between authenticity and functionality, the balance between these two objectives can be found by resorting to the definition and use of Operational Genetic Management Units (OGMU: Pastorino and Gallo 2009; Pastorino et al. 2015). These are groups of geographically contiguous natural populations with a certain genetic similarity, which are expected to respond similarly to a given management decision. Among those management decisions is the use of their genetic pools for planting the species in a specific site. In other words, it is expected that the genetic pools of the different provenances of an OGMU will show the same adaptation and adaptability to a plantation site considered.

A first approach to the OGMU definition can be based on genetic information provided by neutral genetic markers. The OGMU obtained in this way are what we call Genetic Zones (GZ). Although neutral markers allow us to deduce demostochastic processes and thus reconstruct the demographic history of the populations that comprise the species, they do not provide information about adaptation processes they have undergone or about their adaptability to future conditions. For this, information based on adaptive markers or variation in quantitative traits, that is, with expression in the phenotype, should be added. We called Provenance Regions (PR) to the OGMU so defined. Alternatively, we can infer PR through the environmental conditions of each natural population, assuming that adaptation of the last generation to those conditions by itself alone modulates the genetic pools of each population (e.g., Vergara 2000).

According to its definition, each of the OGMU comprises multiple natural populations. In this way, applying the concept of authenticity at the OGMU level, instead of restricting it to the population level, would allow mixing the genetic pools of different populations and thus gain genetic diversity even while maintaining a genetic pool with its own identity (Pastorino et al. 2017). We have started to walk the path of OGMU definition in Argentina with species such as *Austrocedrus chilensis* (Pastorino and Gallo 2009; Pastorino et al. 2015; see Chap. 6), *Nothofagus alpina* and *Nothofagus obliqua* (Azpilicueta et al. 2013, 2017; see Chaps. 3 and 4), and *Nothofagus pumilio* and *Nothofagus antarctica* (Soliani et al. 2017; see Chap. 5).

The definition of OGMU can also be useful to guide the selection of suitable provenances for a plantation site when the local provenance no longer exists, or as

explained, when the remaining trees are extremely scarce. Thus, the local provenance can be replaced (or enriched) by other provenances from the same OGMU.

18.3   Climate Change in Argentina

Global climate change, resulting from the greenhouse effect, has become an
unavoidable determining factor in tree cultivation. In Argentina, the most recent
information in this regard corresponds to the Third National Communication to the
United Nations Framework Convention on Climate Change, prepared by the Center
for Research on the Sea and the Atmosphere (CIMA 2015). Between 1960 and
2010, in most of non-Patagonian continental Argentina, there was an increment in
average temperature of up to 0.5 °C, which represents a lower increase than the
world average in continental regions, probably due to the oceanic conditions of the
South American extreme. Only in some places in the center of the country did a
mean temperature decrement occur, also up to 0.5 °C. In the Patagonian region the
increase was greater, surpassing in some places 1 °C. In this region, the maximum
temperature increased more than the minimum, contrary to what happened in the
rest of the country.

For the same period, precipitation increased in almost the entire Argentina, with
the largest absolute increments exceeding 200 mm per year in the east of the country, but with a greater relative increase in some semi-arid areas. Conversely, the
level of precipitation has decreased in the Patagonian Andes. Beyond average volumes, in general the intensity and frequency of precipitation have also increased in
almost the whole country, although in the northern Andean Patagonia and in sectors
of the Puna periods without precipitation have become longer.

For the climate forecast, the period 1981–2005 was taken as a reference, with a
horizon for the near future (2015–2039) and another for the distant future
(2075–2099). For the country as a whole, 42 models of the CMIP5 base of climate
scenarios were averaged. The average temperature would increase throughout
Argentina for both futures, with values between 0.5 °C and 1 °C in almost the entire
country (Fig. 18.1). The maximum increase would occur in the northwest, reaching
over 3.5 °C in the most severe scenario.

Regarding precipitation, the forecasted changes are not very large, with a tendency to increase in the north and east of Argentina between 10% and 20% depending on the scenario and the future and a trend toward decrease in Patagonia and
Andean Patagonia (with the exception of Tierra del Fuego), also between 10% and
20% (Fig. 18.2).

In addition to this general projection for the whole of Argentina, there are
regional projections for which a little more detail and reliability are achieved, since
they are based on models that are better suited to locally observed climates.

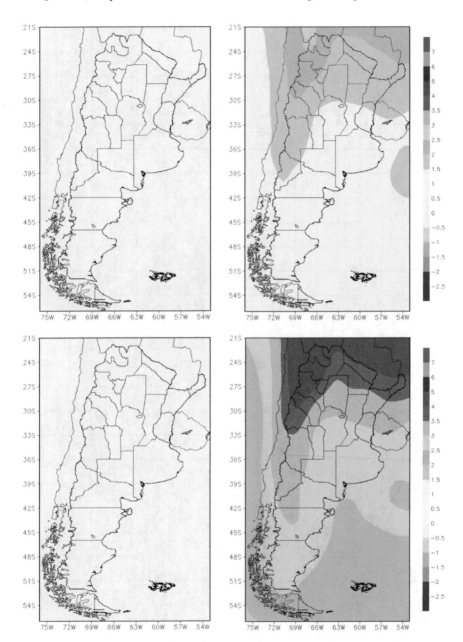

Fig. 18.1 Change in mean annual temperature with respect to the period 1981–2005. Averages of 42 models of the CMIP5 base. Top panel: scenario RCP4.5; bottom panel: scenario RCP8.5. Left: near future (2015–2039); right: distant future (2075–2099) (CIMA 2015)

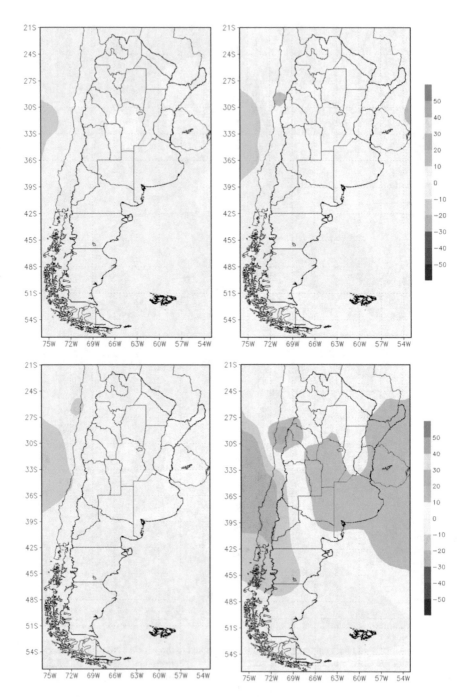

Fig. 18.2 Relative change in annual precipitation with respect to the period 1981–2005. Averages of 42 models of the CMIP5 base. Top panel: scenario RCP4.5; bottom panel: scenario RCP8.5. Left: near future (2015–2039); right: distant future (2075–2099) (CIMA 2015)

18.4 Shifting of the Natural Range of Forest Tree Species Associated to Climate Change Forecast

Distributional shifts in tree species populations are already taking place (Lloyd et al. 2011; Fisichelli et al. 2013; Boisvert-Marsh et al. 2014). However, uncertainty remains about the ability of trees to effectively colonize new habitats in pace with a rapidly changing climate, particularly in light of widespread habitat fragmentation due to agriculture expansion and urbanization (Hampe et al. 2013; Alfaro et al. 2014). In fact, failure of trees to track climate warming, especially at the sapling stage, warns about the ability of forests to surpass global change (Sittaro et al. 2017). Although there are controversies about the reasons for these distributional changes (Máliš et al. 2016), the concern is general.

We present here two studies carried out in Argentina as examples of the possible effect of the climate change on the natural distribution of forest tree species.

Example 1: Range Prediction Through Ecological Niche Modeling
In Argentina *Nothofagus* forests are constrained to a narrow longitudinal area determined by the rain shadow effect of the Andes Mountain range. To the east, there is a transition to steppe vegetation, and particularly this transitional zone between Patagonian forest and steppe is expected to undergo rapid and pronounced shifts in distribution and species composition as a consequence of projected climate change (Allen and Breshears 1998; Suarez and Kitzberger 2010). Climatic models predicted that the precipitation gradient would become sharper due to expected drops in rainfall (Castañeda and González 2008; Barros et al. 2015) and increases in extreme climatic events, like severe droughts (Rusticucci and Barrucand 2004). Therefore, current distribution ranges might be compromised.

Ecological niche modeling (ENM) is a useful tool to investigate about potential distribution patterns of species. This approach is based on relating known occurrences of species to landscape level variation in environmental parameters of importance to species' distribution ecology, providing models of inferred environmental requirements (Peterson 2006; Waltari et al. 2007). By reconstructing future (2050s) habitat suitability scenarios of *N. alpina* and *N. obliqua*, the possible shifts on their current distribution ranges was investigated (Marchelli et al. 2017). As both species have sympatric distributions, but occupy differential niches along the precipitation gradient, owing to their inherent temperature and hydric tolerances, they can serve as a model for gauging the expected impact of climate change on the eastern Patagonian forests. Suitability mapping was based on ensembles of modeling of 11 algorithms (for details see Marchelli et al. 2017). For characterizing future climate conditions, 19 and 31 downscaled climate models for the period 2040–2069 based on the SRES-A2 scenario of greenhouse gas emissions, obtained from CMIP5, were used (Ramirez-Villegas and Jarvis 2010).

The modeling approach showed that *N. alpina* might be impacted more severely by climate change than *N. obliqua*. No single area in Argentina exists where all future climate model projections unanimously predict habitat suitability by the 2050s for *N. alpina*. Only a very tiny fraction of its current distribution is expected to remain suitable according to at least half of all future climate model projections (Fig. 18.3).

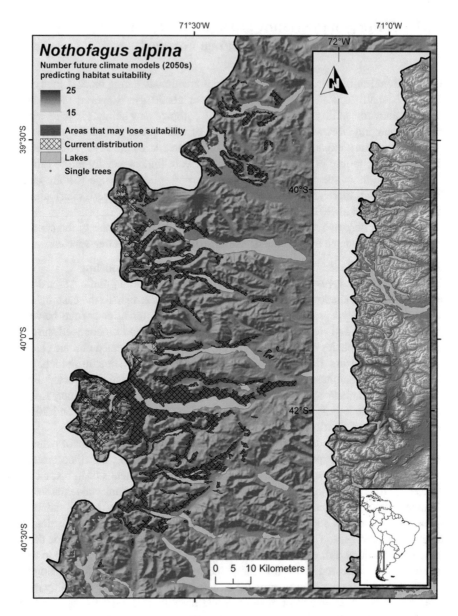

Fig. 18.3 Future (2050s) habitat suitability of *Nothofagus alpina*. Areas identified as suitable by at least 15 of 31 future CMIP5 climate model projections (modified from Marchelli et al. 2017)

For *N. obliqua* the predicted impact is less dramatic with potential suitability losses for the northern, eastern, and southern-most populations (Fig. 18.4). On the other hand, according to the model, an increase in suitable areas is expected for this species, mainly in most of the areas that may become unsuitable for *N. alpina*. In addition, new potential areas are identified for *N. obliqua* toward southern latitudes

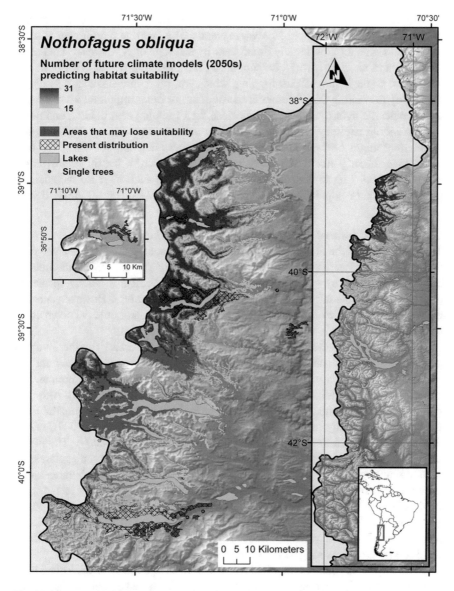

Fig. 18.4 Future (2050s) habitat suitability of *Nothofagus obliqua*. Areas identified as suitable by at least 15 of 31 future CMIP5 climate model projections. (Modified from Marchelli et al. 2017)

(around 42° S) where the species does not exist currently (Fig. 18.4). Provenance trials for both species have demonstrated good survival and growth at these latitudes (Gallo et al. 2000, Chapters 3 and 4), a basic measure of adaptation (Thomas et al. 2014). Therefore, these areas might be considered for translocation from northern areas that currently hold similar climatic conditions than those predicted for the future in the south.

The trends indicated by the models imply that the current mixed forests may gradually be replaced by pure *N. obliqua* stands. The fact that recruitment at some localities is already now largely dominated by saplings of *N. obliqua* (Sola et al. 2015; Sola et al. 2016) would be supporting the modeling results. However, this might also be the legacy of past logging, which was more intense toward *N. alpina* thus likely promoting this differential recruitment. Another important issue is natural interspecific hybridization. Hybrids close their growing season later, and under the current climate they are more susceptible to early frosts (Crego 1999). In this sense, they might be favored by a warming climate, if it implies a less frequent occurrence of early frosts. A successional change toward a hybrid swarm might already be occurring in the mixed forest and could as such increase the adaptive potential of these populations (Rieseberg et al. 2003; Lexer et al. 2004; Becker et al. 2013).

Example 2: Altitudinal Shifting of Neighboring *Nothofagus* Species

The elucidation of the physiological and molecular responses of trees to changes in the environment becomes a priority in the context of the aforementioned global climate change. Many works report physiological responses of woody species to different environmental factors such as water and temperature. However, most of these studies were performed under controlled conditions, and little is known about the responses of trees to changes in the environment in the wild.

Environmental gradients can work as "natural laboratories" for the study of growth and development of tree species in relation to the conditions they endure. Particularly, altitudinal gradients constitute a useful tool to study the influence of natural climatic conditions on the distribution of species (Körner 2000). Although numerous environmental factors vary with altitude (Körner 2007), temperature is the cue that is usually more correlated with elevation (Loarie et al. 2009), with a drop of air temperature of 5 to 6.5 °C per 1000 m (Jump et al. 2009). The use of these natural laboratories constitutes a promising approach to predict fitness of woody plant species in contrasting sites and the adjustment of their ranges to a changing environment. In this regard, we present here a case study in Northern Patagonia with three *Nothofagus* species that grow separately along an altitudinal gradient: *N. obliqua* at 650–850 m asl, *N. alpina* between 850 and 1000 m asl, and *N. pumilio* above 1000 m asl. In these ecosystems, temperature is the environmental factor that shows the strongest association with altitude.

The sharp separation of altitudinal habitats of these species suggests a clear preference for particular thermal environments. This can be tested combining laboratory experiments, in which the particular effect of temperature can be evaluated under controlled conditions, with experiments across different altitudes aiming to test species' fitness inside and outside their natural distribution range. For this purpose, it is necessary to identify the particular processes across the whole plant life cycle that are susceptible to changes in the natural habitat and study their environmental regulation. It is known that germination and seedling establishment usually show a strong environmental vulnerability (Grubb 1977; Bykova et al. 2012; Green et al. 2014); therefore they can be included among the potential candidate traits. In this context, the combination of threshold models with ecological studies along an altitudinal gradient provided a useful approach for the prediction of seed behavior of *N. obliqua*, *N. alpina*, and *N. pumilio* in the forest, from dispersion to germination (Arana et al. 2016). This conceptual framework provided insights into the plasticity of species to changes in the natural thermal environment.

Threshold models allow to predict changes in phenology in response to changes in the environment, and by incorporating population variance, they can investigate mechanisms of intraspecific variation or population synchronization (Donohue et al. 2015). For the particular example of temperature, thermal time models are threshold models that describe changes in phenology in response to the thermal environment (Bradford 1996). A brief description of thermal time models can be found in Chap. 3.

For the particular case of *N. obliqua*, *N. alpina*, and *N. pumilio*, experiments under controlled conditions established that the thermal time model that showed the best fit with the germination behavior in response to different temperatures was based on a constant base temperature (Tb) and variable lower limit temperature for germination (Tl). In this case, Tl of the population is characterized by the population parameters $Tl_{(50)}$ that is the mean lower limit temperature for germination of the seed population and its standard deviation θ_{Tl}. Tl means the minimum temperature at which seeds are able to accumulate developmental units (thermal time), and therefore germination is possible. The rate of accumulation of developmental units will be defined by the differences between the soil temperature and Tb (Arana et al. 2016). By using thermal time models, it was possible to determine that the germination of the three *Nothofagus* species showed specific adaptations to the thermal environment of their preferred ecological niches. This was evidenced by the fact that once dormancy was lost after a period of 110 days of cold stratification, the species that were more abundant at higher (i.e., colder) positions of the gradient showed lower Tb (Tb *N. obliqua* = 4.5 °C > *N. alpina* = 0.5 °C > *N. pumilio* = 0 °C). Similarly, non-dormant seeds of species inhabiting the colder positions of the gradient showed lower values of the mean lower temperature limit for germination ($Tl_{(50)}$). This means that species of the higher altitudes are able to germinate at very low temperatures (around 0 °C), and this behavior is not present in the species of the warmer (i.e., lower) zones. In this particular case, the estimation of the population thermal time parameters Tb and $Tl_{(50)}$ allowed to quantify the germination sensitivity to the temperature and suggest the existence of specific adaptation of the different species to temperature. Both $Tl_{(50)}$ and $\sigma_{(Tl)}$ are high at the moment of seed dispersal and decreased with the cold stratification in the range between 0.5 and 6 °C (Fig. 18.5).

These laboratory observations allowed to set up a model that could be tested in nature, in ecological experiments along an altitudinal gradient. In this model, it was proposed that seeds of the three *Nothofagus* species are dormant at dispersal time, and their level of dormancy is broken during overwintering, when soil temperatures are around or below 5 °C. During spring, interspecific variation in $Tl_{(50)}$, in combination with natural gradients of soil temperature across altitudes, would strongly regu-

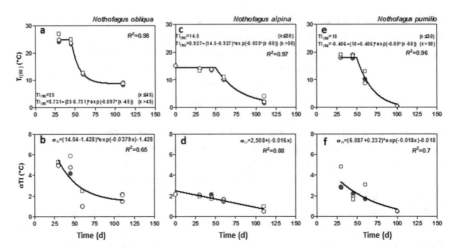

Fig. 18.5 Estimated values of the mean lower limit temperature ($Tl_{(50)}$), standard deviation of the lower limit temperature (θ_{Tl}) in *Nothofagus obliqua* (**a–b**), *N. alpina* (**c–d**), and *N. pumilio* (**e–f**) seeds plotted as function of stratification time. Seeds were stratified at 0.5 °C (open black circles), 1.5 °C (closed blue circles), or 4 °C (open red squares) for the indicated time in days (**d**). Mathematical equations describing the relationship between the estimated thermal time parameters with time of seed stratification and R^2 are indicated inside each figure panel. (Adapted from Arana et al. 2016)

late the phenology of germination across the altitudinal gradient, and this phenomenon would be dependent on the species and the altitude (Fig. 18.6).

Experiments in field plots along an altitudinal gradient, conducted to test this hypothesis, were able to validate the thermal models created in the lab and showed that, in their natural distribution range, germination of all three species temporally overlapped and was restricted to early and mid-spring. By contrast, there was large variation in the timing of germination of the different species for seeds buried outside of their natural distribution zone. In particular, the species inhabiting the lower altitudes showed a delay in germination at higher altitudes, and the opposite pattern was observed for the species of the higher altitudes, confirming the proposed hypothesis (Fig. 18.6).

These data suggest that the species-specific responses to temperature affect the timing of germination, probably contributing to fitness and therefore maintaining patterns of species distribution along the altitudinal gradient. Given that *Nothofagus* spp. do not form persistent seed banks in natural conditions (Cuevas and Arroyo 1999), the described model provides a mechanistic framework for the prediction of seed behavior in the forest from dispersion to germination.

However, when predicting the dynamics of forest regeneration in a context of global climate change, it is necessary to note that other factors, in addition to seed responsiveness to temperature, are also likely to contribute. For example, interactions of *Nothofagus* with different organisms (Caccia et al. 2006; Kitzberger et al. 2007; Garibaldi et al. 2011), heterogeneity of forest understory (Giordano et al. 2006), and natural gradients in UV radiation (Caldwell 1968; Körner 2007) may

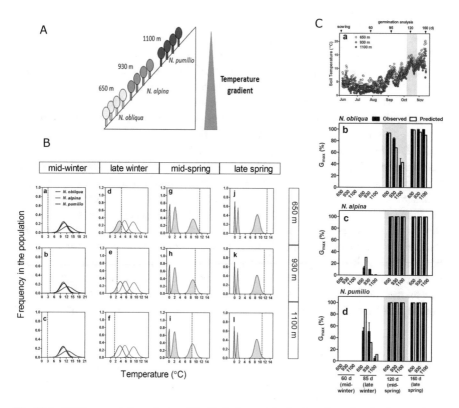

Fig. 18.6 (**A**) Representative scheme of the *Nothofagus* altitudinal gradient. Species that are dominant at each elevational range are indicated into the scheme, with their respective altitudinal values of highest abundance. (**B**) Effects of the lower limit temperature of germination (Tl) and soil temperature in the seasonal timing of germination of *Nothofagus obliqua*, *N. alpina*, and *N. pumilio* across the altitudinal range. (a–l) Shaded areas indicate the population fraction predicted to germinate based on average soil temperature (interception of x axes by the vertical dotted line). (**C**) Evaluation of germination behavior of *Nothofagus* seeds across altitudes. (a) Daily mean soil temperature at different altitudes during the period of seed overwintering and germination in the forest (autumn-spring). Observed (closed bars) and predicted (open bars) germination of *N. obliqua* (b), *N. alpina* (c), and *N. pumilio* (d) seeds sown for 60 (midwinter), 85 (late-winter), 120 (mid-spring), or 160 days (late spring) across the altitudinal gradient is also indicated. Altitude (m asl), days of seed incubation in the field, and the season of seed exhumation are indicated below (d). Predicted thermal time parameters for theoretical data calculation are listed in Arana et al. (2016). Vertical bars indicate the SEM. (b–d) Gmax (%) is the final percentage of germination. (Adapted from Arana et al. 2016)

exert strong influences, and their relevance might vary according to the climate characteristics of future scenarios. In addition, future climatic scenarios may influence dormancy through maternal effects. Another relevant characteristic of *Nothofagus* spp. is their seed-masting habit (Richardson et al. 2005), which represents a concentrated sporadic reproductive effort. Masting/dormancy dynamics may change in future climatic scenarios.

Despite all these additional factors, the wide range of germination temperatures shown by *Nothofagus* seeds during dormancy alleviation (Fig. 18.6) suggests that germination behavior could adapt quickly to climate change, what would have strong consequences on plant distribution shift. These observations have important implications in the context of future climate scenarios because they indicate that the predicted shifts in temperature for the Patagonian Andes (Barros et al. 2015) will affect the timing of germination in their current distribution areas. Since the phenology of germination influences plant survival and the selection of post-germination traits (Donohue et al. 2010), it is possible that the observed changes in the timing of germination will affect seedling survival, with consequences in plant recruitment, which may affect the current pattern of abundance and distribution of *Nothofagus* spp. across environmental gradients. In fact, the transition from germinated seed to seedling constitutes one of the most susceptible developmental stages during the early steps of the regeneration process of these species. The most critical temporal window for survival corresponds to the first 60 days after germination. In this period, the survival is highly influenced by the environment, particularly by temperature (Cagnacci et al. 2020). In this context, shift in the timing of germination due to changes in the climate can modify the environment perceived by the seedlings in this first critical period, with strong consequences on their survival ability.

18.5 Assisted Migration: Breeding for the Present or for the Future Climate?

Due to their longevity, the trees we plant today will have to adapt to the environmental conditions of the future. On the other hand, we can predict in which direction the climate will change. Therefore, it becomes evident to wonder if we should select the trees according to the current conditions or to the conditions of the future. In other words, should we plant the species and the provenances adapted to the current or to the future climate?

Since the beginning of the century and as the evidence of climate change became more obvious, there has been a debate in the scientific community about the risks and potential benefits of an assisted migration (AM) strategy for the persistence of threatened species (Lawler and Olden 2011). AM involves the intentional translocation of specimens of a certain species into an area outside its natural distribution in order to protect the species from anthropogenic threats that can lead to its extinction (Seddon 2010). While some authors argue that AM is an essential tool for species conservation in the face of a changing climate, others highlight the unpredictable risk of biological invasions that could have a devastating impact of the translocated species on the recipient ecosystems (Ricciardi and Simberloff 2009).

There are undoubtedly examples for both extremes: species on the way to extinction due to the absence of actions to mitigate the effects of climate change and also ecosystems that are strongly impacted, and even lost, by the invasion of exotic species. As examples of invasions of tree species in Argentinean forests, we can mention *Ligustrum lucidum* in the Yungas (Aragón and Groom 2003; Ayup et al. 2014) and in the Chaco (Gavier-Pizarro et al. 2012), *Morus alba* in the Yungas (Zamora-Rivera 2002), *Melia azedarach* in the Espinal (Micou 2003), *Hovenia dulcis* in the Alto Paraná Rainforest (Lazzarin et al. 2015), and *Pseudotsuga menziesii* in Patagonian Andean forests (Orellana and Raffaele 2010; Sarasola et al. 2006). For a regional example of mitigation, we can mention the case of *Araucaria araucana* assisted migration program recently initiated in Chile (Ipinza et al. 2019) and already described in Chap. 7. The hindered natural propagation of *Araucaria angustifolia* in the northeast of Argentina has also been addressed, in Chap. 15. It is presumably related to climate change, and the movement of the species toward temperate climates, such as that of the Province of Buenos Aires, has been suggested to overcome that restriction.

McLachlan et al. (2007) proposed a useful framework for debating the opportunities and risks involved in AM. They schematically recognize three possible positions in the decision-making debate on AM, which are graphically presented in Fig. 18.7.

Position 1 involves a perception of high risk of extinction if AM is not carried out and is supported by great confidence in the predicted path that the ecosystem will follow after inclusion of the migrated species. On the opposite side is Position 2, which recognizes the enormous uncertainty in ecological understanding of which factors control species distribution and abundance and therefore emphasizes awareness of the unintended consequences of well-intentioned human interference.

Fig. 18.7 Schematic diagram for decision-making on assisted migration (from McLachlan et al. 2007)

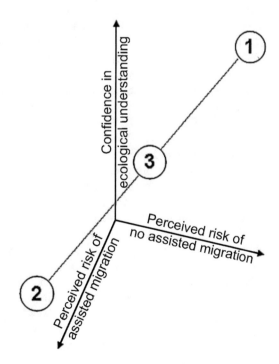

Facilitating natural population spread would be its preferred option to try to save the threatened species. Position 3 represents an intermediate alternative that may range broadly between positions 1 and 2, stating that AM is necessary to preserve biodiversity despite recognized risks. Those risks can and must be minimized through careful restrictions on actions, planning, monitoring, and adaptive management. These constraints will require scientific knowledge and a board of experts to conduct the AM process. Obviously, this will come at a significant cost and could therefore only be implemented for a few species of highest concern.

As a complementary and even alternative strategy to AM, to increase connectivity between threatened ecosystems and threat-free sites has been proposed (Lawler and Olden 2011; Thomas et al. 2014). Greater connectivity would be achieved through the elimination or at least the permeation of artificial barriers that impede the free movement of species, ultimately achieving an effect similar to that of AM.

Although AM is a strategy that focuses on species, it can also be thought in an intraspecific dimension (Fady et al. 2015). It has been called within-species assisted migration (Richardson et al. 2009) or assisted gene flow (Aitken and Whitlock 2013). Thus, the provenance to be planted could be chosen in reference to the environmental conditions of the future, in such a way as to deliberately use an exotic provenance, that is, different from the local one, considering that the local one is on the way to extinction. In this sense, the method would require making an analogy between the predicted future environment and a current environment for which the adaptation of the genetic pool (provenance) to be used has been proven.

Foresters have utilized exotic provenances since at least the nineteenth century in Europe for productive purposes, and maladaptation experiences are abundant (e.g., Johnson et al. 2004; Timbal et al. 2005). Therefore, it is now well recognized that assisted gene flow is risky both for the target species themselves and for the ecosystems they will compose (Fady et al. 2015). However, it could be a reasonable solution when local genetic pools are not able to adapt to the ongoing changes and show undoubted signs of decline, such as adult tree mortality above an acceptable threshold and severe drop or lack of natural regeneration, even failing assisted regeneration with local resources. In this case, assisted gene flow would be a sort of genetic rescue through the hybridization of local and exotic genetic pools. Outbreeding depression could still be a result, and ecological risks cannot be discarded either (Fady et al. 2015). In this sense, the guidance of Operational Genetic Management Units, such as Seed Zones, Genetic Zones, or Provenance Regions, could help to minimize the genetic risks inherent in the use of exotic germplasm.

Although assisted gene flow is the strategy implemented, since adaptation to a predicted but unknown environment is the key issue, two main considerations must guide the election of the propagation material: (i) the genetic pool to be used must be highly variable so as to give the chance for the future generations to count with different alternatives to adapt, and (ii) phenotypic plasticity of the genetic material must be particularly valorized, since it will be key for the current generation to survive and produce offspring. Phenotypic plasticity of fundamental adaptive traits should perhaps be the most relevant selection criterion in a low intensity breeding program ruled by the pressure of climate change.

References

Aitken SN, Whitlock MC (2013) Assisted gene flow to facilitate local adaptation to climate change. Annu Rev Ecol Evol S 44:367–388

Alfaro RI, Fady B, Vendramin GG, Dawson IK, Fleming RA, Saénz-Romero C et al. (2014) The role of forest genetic resources in responding to biotic and abiotic factors in the context of anthropogenic climate change. For Ecol Manag 333:76–87

Allen CD, Breshears DD (1998) Drought-induced shift of a forest–woodland ecotone: rapid landscape response to climate variation. Proc Natl Acad Sci U S A 95:14839–14842

Aragón R, Groom M (2003) Invasion by *Ligustrum lucidum* in NW Argentina: plant characteristics in different habitat types. Rev Biol Trop 51:59–70

Arana MV, Gonzalez-Polo M, Martinez-Meier A, Gallo LA, Benech-Arnold R, Sanchez RA, Batlla D (2016) Seed dormancy responses to temperature relate to *Nothofagus* species distribution and determine temporal patterns of germination across altitudes in Patagonia. New Phytol 209:507–520

Ayup MM, Montti LF, Aragón R, Grau HR (2014) Invasion of *Ligustrum lucidum* (Oleaceae) in the Southern Yungas. Changes in habitat properties and decline in bird diversity. Acta Oecol 54:72–81

Azpilicueta MM, Gallo LA, van Zonneveld M, Thomas E, Moreno C, Marchelli P (2013) Management of *Nothofagus* genetic resources: definition of genetic zones based on a combination of nuclear and chloroplast marker data. For Ecol Manag 302:414–424

Azpilicueta MM, Marchelli P, Gallo LA, Umaña F, Thomas E, van Zonneveld M et al. (2017) Zonas genéticas de raulí y roble pellín en Argentina: herramientas para la conservación y el manejo de la diversidad genética. Ediciones INTA

Barros VR, Boninsegna JA, Camilloni IA, Chidiak M, Magrín GO, Rusticucci M (2015) Climate change in Argentina: trends, projections, impacts and adaptation. Wiley Interdisciplinary Rev Clim Change 6:151–169

Basey AC, Fant JB, Kramer AT (2015) Producing native plant materials for restoration: 10 rules to collect and maintain genetic diversity. Native Plants J 16:37–52

Becker M, Gruenheit N, Steel M, Voelckel C, Deusch O, Heenan PB et al. (2013) Hybridization may facilitate in situ survival of endemic species through periods of climate change. Nat Clim Chang 3:1039–1043

Bischoff A, Steinger T, Müller-Schärer H (2010) The importance of plant provenance and genotypic diversity of seed material used for ecological restoration. Restor Ecol 18:338–348

Boisvert-Marsh L, Périé C, de Blois S (2014) Shifting with climate? Evidence for recent changes in tree species distribution at high latitudes. Ecosphere 5:art83

Bradford KJ (1996) Population-based models describing seed dormancy behaviour: implications for experimental design and interpretation. In: Lang GA (ed) Plant dormancy: physiology, biochemistry, and molecular biology. CAB International, Wallingford, pp 313–339

Broadhurst LM, Lowe A, Coates DJ, Cunningham SA, McDonald M, Vesk PA, Yates C (2008) Seed supply for broad scale restoration: maximizing evolutionary potential. Evol Appl 1:587–597

Bykova O, Chuine I, Morin J, Higgins SI (2012) Temperature dependence of the reproduction niche and its relevance for plant species distributions. J Biogeogr 39:2191–2200

Byrne M, Stone L, Millar MA (2011) Assessing genetic risk in revegetation. J Appl Ecol 48:1365–1373

Caccia FD, Chaneton EJ, Kitzberger T (2006) Trophic and non-trophic pathways mediate apparent competition through post-dispersal seed predation in a Patagonian mixed forest. Oikos 113:469–480

Cagnacci J, Estravis-Barcala M, Lia MV, Martínez-Meier A, Gonzalez PM, Arana MV (2020) The impact of different natural environments on the regeneration dynamics of two *Nothofagus* species across elevation in the southern Andes. For Ecol Manag 15 May 2020 Article 118034. https://doi.org/10.1016/j.foreco.2020.118034

Caldwell MM (1968) Solar ultraviolet radiation as an ecological factor for alpine plants. Ecol Monogr 38:243–268

Carr DE, Dudash MR (2003) Recent approaches into the genetic basis of inbreeding depression in plants. Phil Trans R Soc Lond B 358:1071–1084

Castañeda M, González M (2008) Statistical analysis of the precipitation trends in the Patagonian region in southern South America. Atmósfera 21:303–317

CIMA (2015) Cambio climático en Argentina; tendencias y proyecciones. Tercera Comunicación Nacional de la República Argentina a la Convención Marco de las Naciones Unidas sobre Cambio Climático. http://3cn.cima.fcen.uba.ar/informe/ModClim_Indice.pdf

Clewell AF (2000) Restoring for natural authenticity. Ecol Restor 18:216–217

Crego P (1999) Variación genética en el comportamiento fenológico y el crecimiento juvenil de progenies puras e híbridas de raulí, *Nothofagus nervosa* (Phil.) Dim. et Mil. Graduate thesis. Fac. Cs. Biológicas, CRUB, Universidad Nacional del Comahue

Cuevas JG, Arroyo MTK (1999) Absence of a persistent seed bank in *Nothofagus pumilio* (Fagaceae) in Tierra del Fuego, Chile. Rev Chil Hist Nat 1:73–82

Donohue K, Rubio de Casas R, Burghardt L, Kovach K, Willis CG (2010) Germination, post-germination adaptation and species ecological ranges. Annu Rev Ecol Evol Syst 41:293–219. https://doi.org/10.1146/annurev-ecolsys-102209-144715

Donohue K, Burghardt LT, Runcie D, Bradford KJ, Schmitt J (2015) Applying developmental threshold models to evolutionary ecology. Trends Ecol Evol 30:66–77

Fady B, Cottrell J, Ackzell L, Alía R, Muys B, Prada A, González-Martínez SC (2015) Forests and global change: what can genetics contribute to the major forest management and policy challenges of the twenty-first century? Reg Environ Chang 16:927–939

Falk DA, Knapp EE, Guerrant EO (2001) An introduction to restoration genetics. Society for Ecological Restoration. U.S. Environmental Protection Agency, p 33

Fenster CB, Dudash MR (1994) Genetic considerations for plant population restoration and conservation. In: Bowles ML, Whelan CJ (eds) Restoration of endangered species. Cambridge University Press

Fenster CB, Galloway LF (2000) Inbreeding and outbreeding depression in natural populations of *Chamaecrista fasciculata* (Fabaceae). Conserv Biol 14:1406–1412

Fisichelli NA, Frelich LE, Reich PB (2013) Temperate tree expansion into adjacent boreal forest patches facilitated by warmer temperatures. Ecography (Cop) 37:152–161

Frankham R, Ballou JD, Eldridge MDB, Lacy RC, Ralls K, Dudash MR, Fenster CB (2011) Predicting the probability of outbreeding depression. Conserv Biol 25:465–475

Gallo LA, Marchelli P, Crego P, Oudkerk L, Izquierdo F, Breitembücher A et al. (2000) Distribución y variación genética en características seminales y adaptativas de poblaciones y progenies de raulí en Argentina. In: Domesticación y Mejora Genética de raulí y roble. Universidad Austral de Chile-Instituto Forestal, Valdivia, pp 133–156

Garibaldi LA, Kitzberger T, Chaneton EJ (2011) Environmental and genetic control of insect abundance and herbivory along a forest elevational gradient. Oeocologia 167:117–129

Gavier-Pizarro GI, Kuemmerle T, Hoyos LE, Stewart SI, Huebner CD, Keuler NS, Radeloff VC (2012) Monitoring the invasion of an exotic tree (*Ligustrum lucidum*) from 1983 to 2006 with Landsat satellite data and a support vector machine in Córdoba, Argentina. Remote Sens Environ 122:134–145

Giordano CV, Sánchez RA, Austin AT (2006) Gregarious flowering of bamboo opens a red: far red window of opportunity for forest regeneration in a temperate forest of Patagonia. New Phytol 181:880–889

Green PT, Harms KE, Connell JH (2014) Nonrandom, diversifying processes are disproportionately strong in the smallest size classes of a tropical forest. Proc Natl Acad Sci U S A 111:18649–18654

Grubb PJ (1977) The maintenance of species-richness in plant communities: the importance of the regeneration niche. Biol Rev 52:107–145

Hampe A, Pemonge MH, Petit RJ (2013) Efficient mitigation of founder effects during the establishment of a leading edge oak population. Proc R Soc London Ser B Biol Sci 280:20131070

Ipinza R, Gutiérrez B, Muller-Using S, Molina MP, Gonzalez J (2019) La migración asistida de la *Araucaria araucana*, plan operacional. Ciencia e Investigación Forestal INFOR Chile 25:75–88

Johnson GR, Sorensen FC, St Clair JB, Cronn RC (2004) Pacific Northwest forest tree seed zones: a template for native plants? Nat Plants J 5:131–140

Jump AS, Mátyás C, Peñuelas J (2009) The altitude-for-latitude disparity in the range retractions of woody species. Trends Ecol Evol 24:694–701

Kitzberger T, Chaneton EJ, Caccia FD (2007) Indirect effects of prey swamping: differential seed predation during bamboo masting event. Ecology 88:2541–2554

Körner C (2000) Why are there global gradients in species richness? Mountains might hold the answer. Trends Ecol Evol 15:513–514

Körner C (2007) The use of 'altitude' in ecological research. Trends Ecol Evol 22:569–574

Lawler JJ, Olden JD (2011) Reframing the debate over assisted colonization. Front Ecol Environ 9:569–574

Lazzarin LC, da Silva AC, Higuchi P, Souza K, Perin JE, Pereira Cruz A (2015) Biological invasión by *Hovenia dulcis* Thunb. in forest fragments in upper-Uruguay region, Brazil. Rev Árvore 39:1007–1017

Lesica P, Allendorf FW (1999) Ecological genetics and the restoration of plant communities: mix or match? Restor Ecol 7:42–50

Lexer C, Heinze B, Alia R, Rieseberg LH (2004) Hybrid zones as a tool for identifying adaptive genetic variation in outbreeding forest trees: lessons from wild annual sunflowers (*Helianthus* spp.). For Ecol Manag 197:49–64

Lloyd AH, Bunn AG, Berner L (2011) A latitudinal gradient in tree growth response to climate warming in the Siberian taiga. Glob Chang Biol 17:1935–1945

Loarie SR, Duffy PB, Hamilton H, Asner GP, Field CB, Ackerly DD (2009) The velocity of climate change. Nature 462:1052–1055

López Lauenstein D, Vega C, Verga A, Fornes L, Saravia P, Feyling M, et al. (2019) Evaluación de diez orígenes de Algarrobo para establecer sistemas silvopastoriles en el Chaco semiárido argentino. In: Proceedings of X Congreso Internacional sobre Sistemas Silvopastoriles. Asunción, Sept. 24–26. Editorial CIPAV, Cali

Máliš F, Kopecký M, Petřík P, Vladovič J, Merganič J, Vida T (2016) Life stage, not climate change, explains observed tree range shifts. Glob Chang Biol 22:1904–1914

Marchelli P, Thomas E, Azpilicueta MM, Van Zonneveld M, Gallo LA (2017) Integrating genetics and suitability modelling to bolster climate change adaptation planning in Patagonian *Nothofagus* forests. Tree Genet Genomes 13:119

McKay JK, Christian CE, Harrison S, Rice KJ (2005) "How local is local?" A review of practical and conceptual issues in the genetics of restoration. Restor Ecol 13:432–440

McLachlan JS, Hellmann JJ, Schwartz MW (2007) A framework for debate of assisted migration in an era of climate change. Conserv Biol 21:297–302

Micou AP (2003) Riesgo ambiental por invasiones biológicas en una zona con alto valor de conservación. Graduate thesis, Facultad de Fiolosofía y Letras, Universidad de Buenos Aires, p 157

Orellana IA, Raffaele E (2010) The spread of the exotic conifer *Pseudotsuga menziesii* in *Austrocedrus chilensis* forests and shrublands in northwestern Patagonia, Argentina. NZ J Forestry Sci 40:199–209

Pastorino MJ, Gallo LA (2009) Preliminary operational genetic management units of a highly fragmented forest tree species of southern South America. For Ecol Manag 257:2350–2358

Pastorino MJ, Aparicio AG, Azpilicueta MM (2015) Regiones de Procedencia del Ciprés de la Cordillera y bases conceptuales para el manejo de sus recursos genéticos en Argentina. Ediciones INTA, Buenos Aires, 107 pp

Pastorino MJ, Aparicio AG, Azpilicueta MM, Soliani C, Marchelli P (2017) Genética de la restauración: tendiendo puentes entre la investigación y la gestión. In: Zuleta G, Rovere A, Mollard F (eds) SIACRE-2015, Aportes y Conclusiones: Tomando Decisiones para Revertir la Degradación Ambiental. Vázquez Mazzini Editores, Buenos Aires, pp 147–152, 240 pp

Peterson AT (2006) Uses and requirements of ecological niche models and related distributional models. Biodivers Informatics 3:59–72

Ramirez-Villegas J, Jarvis A (2010) Downscaling global circulation model outputs: the delta method. Decision and policy analysis working paper no. 1

Ricciardi A, Simberloff D (2009) Assisted colonization is not a viable conservation strategy. Trends Ecol Evol 24:248–253

Richardson SJ, Allen RB, Whitehead D, Carswell FE, Ruscoe WA, Platt KH (2005) Climate and net carbon availability determine temporal patterns of seed production by *Nothofagus*. Ecology 86:972–981

Richardson DM, Hellmann JJ, McLachlan JS, Sax DF, Schwartz MW, Gonzalez P, Brennan E, Camacho A et al. (2009) Multidimensional evaluation of managed relocation. Proc Natl Acad Sci U S A 106:9721–9724

Rieseberg LH, Raymond O, Rosenthal DM, Lai Z, Livingstone K, Nakazato T et al. (2003) Major ecological transitions in wild sunflowers facilitated by hybridization. Science 301:1211–1216

Rusticucci M, Barrucand M (2004) Observed trends and changes in temperature extremes over Argentina. Am Meteorol Soc 17:4099–4107

Sarasola MM, Rusch VE, Schlichter TM, Ghersa CM (2006) Invasión de coníferas forestales en áreas de estepa y bosque de ciprés de la cordillera en la región Patagónica. Ecol Austral 16:143–156

Seddon PJ (2010) From reintroduction to assisted colonization: moving along the conservation translocation spectrum. Restor Ecol 18:796–802

Sittaro F, Paquette A, Messier C, Nock CA (2017) Tree range expansion in eastern North America fails to keep pace with climate warming at northern range limits. Glob Chang Biol 23:3292–3301

Sola G, Attis Beltran H, Chauchard L, Gallo LA (2015) Efecto del manejo silvicultural sobre la regeneración de un bosque de *Nothofagus dombeyi*, *N. alpina* y *N. obliqua* en la Reserva Nacional Lanín (Argentina). Bosque 36:113–120

Sola G, El Mujtar V, Tsuda Y, Vendramin GG, Gallo LA (2016) The effect of silvicultural management on the genetic diversity of a mixed *Nothofagus* forest in Lanín natural reserve, Argentina. For Ecol Manag 363:11–20

Soliani C, Umaña F, Mondino VA, Thomas E, Pastorino M, Gallo LA, Marchelli P (2017) Zonas genéticas de lenga y ñire en Argentina : y su aplicación en la conservación y manejo de los recursos forestales. Ediciones INTA, Bariloche

Suarez ML, Kitzberger T (2010) Differential effects of climate variability on forest dynamics along a precipitation gradient in northern Patagonia. J Ecol 98:1023–1034

Thomas E, Jalonen R, Loo J, Boshier D, Gallo L, Cavers S et al. (2014) Genetic considerations in ecosystem restoration using native tree species. For Ecol Manag 333:66–75

Timbal J, Bonneau M, Landmann G, Trouvilliez J, Bouhot-Delduc L (2005) European non boreal conifer forests. In: Andersson FA (ed) Ecosystems of the world (6): coniferous forests. Elsevier, Amsterdam, pp 131–162

Vander Mijnsbrugge K, Bischoff A, Smith B (2010) A question of origin: where and how to collect seed for ecological restoration. Basic Appl Ecol 11:300–311

Verga A, López-Lauenstein D, López C, Navall M, Joseau J, Gómez C et al. (2009) Caracterización morfológica de los algarrobos (*Prosopis* sp.) en las regiones fitogeográficas Chaqueña y Espinal norte de Argentina. Quebracho 17:31–40

Vergara R (2000) Regiones de procedencia de *N. alpina* y *N. obliqua*. In: Ipinza R, Gutierrez B, Emhart V (eds) Domesticación y Mejora Genética de raulí y roble. Universidad Austral de Chile-Instituto Forestal, Valdivia

Vergeer P, Sonderen E, Ouborg NJ (2004) Introduction strategies put to the test: local adaptation versus heterosis. Conserv Biol 18:812–821

Waltari E, Hijmans RJ, Peterson AT, Nyári ÁS, Perkins SL, Guralnick RP (2007) Locating Pleistocene refugia: comparing phylogeographic and ecological niche models predictions. PLoS One 2:e563

Zamora-Rivera SV (2002) Efecto de la dominancia de las especies exóticas invasoras sobre la sucesión de bosques secundarios de las yungas argentinas. MSc thesis. Facultad de Agronomía, Universidad de Buenos Aires, p121

Index

© Springer Nature Switzerland AG 2021
M. J. Pastorino, P. Marchelli (eds.), *Low Intensity Breeding of Native Forest
Trees in Argentina*, https://doi.org/10.1007/978-3-030-56462-9

Printed in the United States
by Baker & Taylor Publisher Services